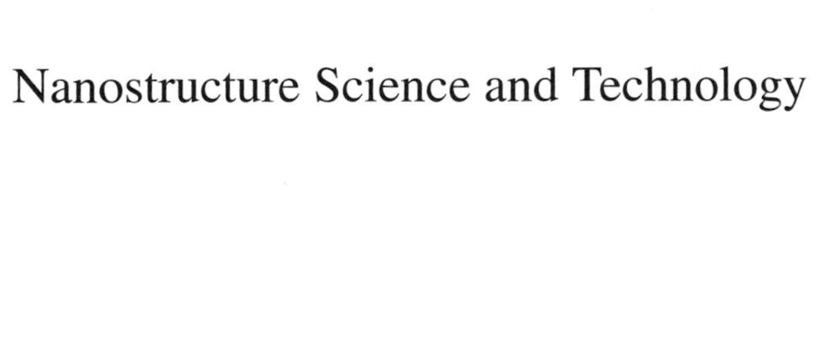

Nanostructure Science and Technology

For other titles published in this series go to,
http://www.springer.com/series/6331

Patrik Schmuki · Sannakaisa Virtanen
Editors

Electrochemistry at the Nanoscale

Editors
Patrik Schmuki
University of Erlangen-Nuernberg
Erlangen, Germany
schmuki@ww.uni-erlangen.de

Sannakaisa Virtanen
University of Erlangen-Nürnberg
Erlangen, Germany
virtanen@ww.uni-erlangen.de

ISSN 1571-5744
ISBN 978-0-387-73581-8 e-ISBN 978-0-387-73582-5
DOI 10.1007/978-0-387-73582-5

Library of Congress Control Number: 2008942146

Printed on acid-free paper

springer.com

Preface

For centuries, electrochemistry has played a key role in technologically important areas such as electroplating or corrosion. Electrochemical methods are receiving increasing attention in rapidly growing fields of science and technology, such as nanosciences (nanoelectrochemistry) and life sciences (organic and biological electrochemistry). Characterization, modification, and understanding of various electrochemical interfaces or electrochemical processes at the nanoscale have led to a huge increase of scientific interest in electrochemical mechanisms as well as in application of electrochemical methods to novel technologies. Electrochemical methods carried out at the nanoscale lead to exciting new science and technology; these approaches are described in 12 chapters.

From the fundamental point of view, nanoscale characterization or theoretical approaches can lead to an understanding of electrochemical interfaces at the molecular level. Not only is this insight of high scientific interest, but also it can be a prerequisite for controlled technological applications of electrochemistry. Therefore, the book includes fundamental aspects of nanoelectrochemistry.

Then, most important techniques available for electrochemistry on the nanoscale are presented; this involves both characterization and modification of electrochemical interfaces. Approaches considered include scanning probe techniques, lithography-based approaches, focused-ion and electron beams, and procedures based on self-assembly.

In classical fields of electrochemistry, such as corrosion, characterizing surfaces with a high lateral resolution can lead to an in-depth mechanistic understanding of the stability and degradation of materials. The nanoscale description of corrosion processes is especially important in understanding the initiation steps of local dissolution phenomena or in detecting dissolution of highly corrosion-resistant materials. The latter can be of crucial importance in applications, where even the smallest amount of dissolution can lead to a failure of the system (e.g., release of toxic elements and corrosion of microelectronics).

Since many electrochemical processes involve room-temperature treatments in aqueous electrolytes, electrochemical approaches can become extremely important whenever living (bio-organic) matter is involved. Hence, a strong demand for electrochemical expertise is emerging from biology, where charged interfaces play an important role. Interfacial electrochemistry is crucially important for understanding the interaction of inorganic substrates with the organic material of biosystems. Thus, recent developments in the field of bioelectrochemistry are described, with a focus on nanoscale phenomena in this field.

Electrochemical methods are of paramount importance for fabrication of nanomaterials or nanostructured surfaces. Therefore, several chapters are dedicated to the electrochemical creation of nanostructured surfaces or nanomaterials. This includes recent developments in the fields of semiconductor porosification, deposition into templates, electrodeposition of multilayers and superlattices, as well as self-organized growth of transition metal oxide nanotubes. In all these cases, the nanodimension of the electrochemically prepared materials can lead to novel properties, and hence to novel applications of conventional materials.

We hope that this book will be helpful for all readers interested in electrochemistry and its applications in various fields of science and technology. Our aim is to present to the reader a comprehensive and contemporary description of electrochemical nanotechnology.

Contents

Contributors

Rabah Boukherroub Biointerfaces Group, Interdisciplinary Research Institute (IRI), FRE 2963, IRI-IEMN, Avenue Poincaré-BP 60069, 59652 Villeneuve d'Ascq, France, rabah.boukherroub@iemn.univ-lille1.fr

Thierry Djenizian Laboratoire MADIREL (UMR 6121), Université de Provence-CNRS, Centre Saint Jérôme, F-13397 Marseille Cedex 20, France, thierry.djenizian@univ-provence.fr

A. Floris van Driel Condensed Matter and Interfaces, Debye Institute, Utrecht University, P.O. Box 80000, 3508 TA, Utrecht, The Netherlands, a.f.vandriel@hotmail.com

Steven H. Goods Dept. 8758/MS 9402, Sandia National Laboratories, Livermore, CA 94550, USA, shgoods@sandia.gov

James J. Kelly IBM, Electrochemical Processes, 255 Fuller Rd., Albany, NY 12203, USA, mjklly@us.ibm.com

John J. Kelly Condensed Matter and Interfaces, Debye Institute, Utrecht University, P.O. Box 80000, 3508 TA, Utrecht, The Netherlands, J.J.Kelly@phys.uu.nl

Christoph Lehrer Lehrstuhl für Elektronische Bauelemente, Universität Erlangen-Nürnberg, Cauerstr. 6, 91058 Erlangen, Germany, Christoph.Lehrer@brose.com

Ezequiel P.M. Leiwa Universidad Nacional de Cordoba, Unidad de Matematica y Fisica, Facultad de Ciencias Quimicas, INFIQC, 5000 Cordoba, Argentina, eleiva@mail.fcq.unc.edu.ar

Olaf M. Magnussen Institut für Experimentelle und Angewandte Physik, Christian-Albrechts-Universität zu Kiel, 24098 Kiel, Germany, magnussen@physik.uni-kiel.de

Philippe Marcus Laboratoire de Physico-Chimie des Surfaces, CNRS-ENSCP (UMR 7045) Ecole Nationale Supérieure de Chimie de Paris, Université Pierre et Marie Curie, 11, rue Pierre et Marie Curie, 75005 Paris, France, philippe-marcus@enscp.fr

Charles R. Martin Department of Chemistry, University of Florida, PO Box 117200, Gainesville, FL 32611, USA, crmartin@chem.ufl.edu

Vincent Maurice Laboratoire de Physico-Chimie des Surfaces, CNRS-ENSCP (UMR 7045), Ecole Nationale Supérieure de Chimie de Paris, Université Pierre et Marie Curie, 11, rue Pierre et Marie Curie, 75231 Paris Cedex 05, France, vincent-maurice@enscp.fr

Ana Maria Oliveira Brett Departamento de Quimica, Universidade de Coimbra, 3004-535 Coimbra, Portugal, brett@ci.uc.pt

Tetsuya Osaka Department of Applied Chemistry, School of Science and Engineering, Waseda University, 3-4-1 Okubo, Shinjuku-ku, Tokyo, 169-8555, Japan, osakatets@waseda.jp

Sudipta Roy School of Chemical Engineering and Advanced Materials, Newcastle University, Newcastle Upon Tyne NE1 7RU, UK, s.roy@newcastle.ac.uk

Wolfgang Schmickler Department of Theoretical Chemistry, University of Ulm, D-89069 Ulm, Germany, Wolfgang.Schmickler@uni-ulm.de

Patrik Schmuki University of Erlangen-Nuremberg, Department for Materials Science, LKO, Martensstrasse 7, D-91058 Erlangen, Germany, schmuki@ww.uni-erlangen.de

Charles R. Sides Gamry Instruments, Inc., 734 Louis Drive, Warminster, PA 18974-2829, USA, Rsides@gamry.com

Sabine Szunerits INPG, LEMI-ENSEEG, 1130, rue de la piscine, 38402 Saint Martin d'Herès, France, sabine.szunerits@gmail.com

Theories and Simulations for Electrochemical Nanostructures

E.P.M. Leiva and Wolfgang Schmickler

1 Introduction

Electrochemical nanostructures are special because they can be charged or, equivalently, be controlled by the electrode potential. In cases where an auxiliary electrode, such as the tip of a scanning tunneling microscope, is employed, there are even two potential drops that can be controlled individually: the bias potential between the two electrodes and the potential of one electrode with respect to the reference electrode. Thus, electrochemistry offers more possibilities for the generation or modification of nanostructures than systems in air or in vacuum do. However, this advantage carries a price: electrochemical interfaces are more complex, because they include the solvent and ions. This poses a great problem for the modeling of these interfaces, since it is generally impossible to treat all particles at an equal level. For example, simulations for the generation of metal clusters typically neglect the solvent, while theories for electron transfer through nanostructures treat the solvent in a highly abstract way as a phonon bath. Therefore, a theorist investigating a particular system must decide, in advance, which parts of the system to treat explicitly and which parts to neglect. Of course, to some extent this is true for all theoretical research, but the more complex the investigated system, the more difficult, and debatable, this choice becomes.

There is a wide range of nanostructures in electrochemistry, including metal clusters, wires, and functionalized layers. Not all of them have been considered by theorists, and those that have been considered have been treated by various methods. Generally, the generation of nanostructures is too complicated for proper theory, so this has been the domain of computer simulations. In contrast, electron transfer through nanostructures is amenable to theories based on the Marcus [1] and Hush [2] type of model. There is little overlap between the simulations and the theories, so we cover them in separate sections.

W. Schmickler (✉)
Department of Theoretical Chemistry, University of Ulm, D-89069 Ulm, Germany
e-mail: Wolfgang.Schmickler@uni-ulm.de

P. Schmuki, S. Virtanen (eds.), *Electrochemistry at the Nanoscale,* Nanostructure Science and Technology, DOI 10.1007/978-0-387-73582-5_1,

Nanostructures in general are a highly popular area of research, and there is a wealth of literature on this topic. Of course, electrochemical nanostructures share many properties with nonelectrochemical structures; in this review we will focus on those aspects that are special to electrochemical interfaces, and that therefore depend on the control of potential or charge density. We shall start by considering simulations of electrochemical nanostructures, and shall then turn toward electron transfer through such structures, in particular through functionalized adsorbates and films.

2 Simulations of Electrochemical Nanostructures

As mentioned in the introduction, electrochemistry offers an ample variety of possibilities for the generation of nanostructures, some of which are illustrated in Fig. 1. In this figure, we concentrate on the type of nanostructures that can be obtained by the application of a scanning probe microscope (SPM), which is the general denomination of a device used to image materials at the nanometer scale. In the case of surface imaging, the sensitivity of the SPM device to some property of the surface (local electronic density of states, chemical nature, electronic dispersive forces) is used to get some topological information. However, our current interest is related to the possibility of using the SPM arrangement to change the atomic ordering and eventually the chemical nature of the surface.

Figure 1a shows the so-called tip-induced local metal deposition, where a scanning tunneling microscope (STM) tip is first loaded via electrochemical deposition with a foreign metal; then the tip is moved to the surface of a substrate, and, upon retraction, a cluster is left behind on the surface. This method works for a number of adsorbate/substrate combinations, but for others it does not, and computer

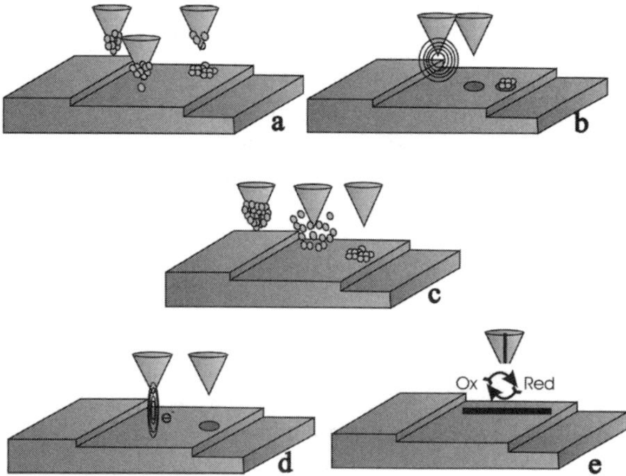

Fig. 1 Schematic illustration of several electrochemical techniques employed for surface nanostructuring: (**a**) Tip-induced local metal deposition; (**b**) defect nanostructuring; (**c**) localized electrochemical nucleation and growth; (**d**) electronic contact nanostructuring; and (**e**) scanning electrochemical microscope

simulations have contributed to understanding the formation of the clusters. Figure 1b illustrates a different procedure, where a hole is generated by applying a potential pulse to the STM tip. Then the potential of the substrate is changed to allow controlled metal electrodeposition into the cavity. Computer simulations of this type of processes are in a developing stage, with promising results. The method depicted in Fig. 1c also starts with metal deposition on the tip, but the metal is then dissolved by a positive potential pulse that produces local oversaturation of the metal ions, which subsequently nucleate on the surface. Some aspects of this problem have been discussed by solving the diffusion equations for the ions generated at the tip. Combinations of techniques (b) and (c) exist, where a foreign metal is deposited on the tip, and a double pulse is applied to it. The first pulse generates a defect on the surface, and the second pulse desorbs adatoms from the tip that nucleate into the defect. This method has been simulated using Brownian dynamics. The method depicted in Fig. 1d) is employed to generate defects on the surface by means of some type of electronic contact between the tip and the surface and has not been modeled so far. Finally, the method depicted in Fig. 1e involves the generation of some species on the tip that further may react with the surface. This is the basis of the technique denominated scanning electrochemical microscopy. While it can be used to image regions of the surface with different electrochemical properties, it can also be applied to modify, at will, the surface if the latter reacts with the species generated at the tip. Due to technical limitations, this technique has generally been applied in the micrometric rather than in the nanometric scale, but it is a matter of improvement to shift its application into the nanometer range.

2.1 Computer Simulation Techniques for Nanostructures

The term computer simulation is so widely used that here a short comment is required to clarify the matter. The term simulation is often applied to various numerical methods for studying the time dependence of processes. In fact, computers may be used to solve numerically numerous problems that range from phenomenological equations to sophisticated quantum mechanical calculations. An electrochemical example for the former is the simulation of voltammetric profiles by solving the appropriate diffusion equations coupled to an electron-transfer reaction. On the other end of the simulation methods, the time-dependent Schrödinger equation may be solved to study dynamic processes in an ensemble of particles. Here we refer only to simulations on an atomic scale.

In the simulations of electrochemical nanostructures, we have to deal with a few hundreds or even thousands of atoms; purely quantum-mechanical calculations of the *ab initio* type are still prohibitive for current computational capabilities. However, since it is expected that at the nanometric scale the atomic nature of atoms may play a role, it is desirable that the methodology should reflect atomic nature of matter, at least in a simplified way. We shall return to this point below.

A typical atomistic computer simulation consists in the generation of a number of configurations of the system of interest, from which the properties in which we are interested are calculated. From the viewpoint of the way in which the configurations are generated, the simulations methods are classified into deterministic or stochastic. In the first case, a set of coupled Newton's equations of the type:

$$m_i \frac{d^2 r_i}{dt^2} = f_i(t) \tag{1}$$

is solved numerically, where the atomic masses m_i and the positions r_i are related to the forces $f_i(t)$ experimented by the particle i at the time t. These forces are calculated from the potential energy of the system $U(\mathbf{r}_n)$:

$$f_i = -\nabla U(\mathbf{r}_n) \tag{2}$$

where we denote with $\mathbf{r}_n$ the set of coordinates of the particles constituting the system. Since Eq. (1) contains the time, it is clear that this type of method allows obtaining dynamic properties of the system. A typical algorithm that performs the numerical task is, for example, the so-called velocity Verlet algorithm:

$$r_i(t+\Delta t) = r_i(t) + v_i(t)\Delta t + \frac{1}{2}\frac{f_i(t)}{m_i}\Delta t^2 \tag{3}$$

where $v_i(t)$ is the velocity of the particle at time t and Δt is an integration step. The recurrent application of this equation to all the particles of the system leads to a trajectory in the phase space of the system, from which the desired information may be obtained. We can get a feeling for the Δt required for the numerical integration by inserting into Eq. (3) some typical atomic values. If we want to get a small atomic displacement $\Delta r_i = r_i(t+\Delta t) - r_i(t)$, say of the order of 10^{-3} Å for a mass of the order of 10^{-23} g, subject to forces of the order of 10^{-1} eV/Å , we get:

$$\Delta t = O\left(\sqrt{\frac{10^{-23} g \times 10^{-11} cm}{10^{7}\,\frac{\mathrm{eV}}{\mathrm{cm}} \times 1.6 \times 10^{-12}\,\frac{\mathrm{erg}}{\mathrm{eV}}}}\right) = O(10^{-15} s) \tag{4}$$

where 1.6×10^{-12} erg/eV is a conversion factor. The result obtained in Eq. 4 means that the integration time required is of the order of a femtosecond, so that a few million of integration steps will lead us into the nanosecond scale, which are the typical simulation times we can reach with current computational resources in nanosystems.

The simulation method discussed so far is the so-called *atom* or *molecular dynamics* (MD) procedure and can be employed to calculate both equilibrium and nonequilibrium properties of a system.

In molecular dynamics the generated configurations of the system follow a deterministic sequence. However, the trajectory of the system in configuration space may be chosen differently, following some rules that allow one to obtain sets of coordinates of the particles that may later be employed to calculate equilibrium properties. Thus, the main idea underlying stochastic simulation methods is, as in MD, to generate a sequence of configurations. However, the transition probabilities between them are chosen in such a way that the probability of finding a given configuration is given by the equilibrium probability density of the corresponding statistical thermodynamic ensemble. Making a short summary of equilibrium statistical thermodynamics, we recall that the first postulate states that an average over a temporal behavior of given a physical system can be replaced by the average over a collection of systems or *ensemble* that exhibits the same thermodynamic but different dynamic properties. The second postulate refers to isolated systems, where the volume V, the energy E, and the number of particles N are fixed, and states that the systems in the ensemble are uniformly distributed over all the quantum states compatible with the NVE conditions. Besides the NVE or microcanonical ensemble, some other popular ensembles are the canonical ensemble (constant number of particles, volume, and temperature), denoted as NVT, the isothermal isobaric ensemble or NPT; and the grand canonical ensemble, where the variables fixed are the chemical potential, volume, and the temperature (μVT ensemble). The molecular dynamics procedure described above is usually run in the (NVE) or *NVT* ensemble, but other conditions are also possible.

Returning to the stochastic simulation methods, and taking the NVT ensemble as an example, the transitions between the different configurations are chosen in such a way that the equilibrium probability density of a certain configuration $\mathbf{r}_n$ will be given by:

$$\rho(\mathbf{r}_n) = \frac{\exp(-U(\mathbf{r}_n)/kT)}{\int \exp(U(\mathbf{r})/kT)\, d\mathbf{r}}$$

where the integral in the denominator runs over all the possible configurations of the system. A way to generate such probability transitions between configurations is that proposed by Metropolis:

$$\begin{aligned} \prod_{mn} = \alpha_{mn}\frac{\rho(\mathbf{r}_n)}{\rho(\mathbf{r}_m)} = \alpha_{mn}\exp\left[\left(U(\mathbf{r}_m) - U(\mathbf{r}_n)\right)/kT\right] \quad &\text{if } \rho(\mathbf{r}_n) < \rho(\mathbf{r}_n) \\ \prod_{mn} = \alpha_{mn} \quad &\text{if } \rho(\mathbf{r}_n) < \rho(\mathbf{r}_n) \end{aligned} \tag{5}$$

where Π_{mn} is the transition probability from state m to state n, and α_{mn} are constants which are restricted to be elements of a symmetric matrix. Thus, a given transition probability Π_{mn} is obtained by calculating the energy of the system in state m, say $U(\mathbf{r}_n)$ and the energy in the state n, say $U(\mathbf{r}_m)$. The configuration in the n state, given by the vector $\mathbf{r}_n$, is obtained from $\mathbf{r}_m$ by allowing some random motion of the system. For example, a particle i may be selected at random, and its coordinates are displaced from their position r_n^i with equal probability to any point r_m^i inside a small cube surrounding the particle. To accept the move with a probability of $\exp\left[-\left(U(\mathbf{r}_n) - U(\mathbf{r}_m)\right)/kT\right]$, a random number ξ is generated with uniform probability density between 0 and 1. If ξ is lower than $\exp\left[-\left(U(\mathbf{r}_n) - U(\mathbf{r}_m)\right)/kT\right]$ it is accepted, otherwise it is rejected. Alternatively, the simultaneous move of all the particles may be attempted. Similarly to the procedure described above, the energy values before and after the move attempt to determine the transition probability. The intensive use this technique makes of random (or better stated pseudo-random) numbers has earned it the name of Monte Carlo(MC) method.

In the case of grand canonical or μVT simulations, in addition to the motion of the particles, attempts are taken to insert into or remove particles from the system. In this case, all simulation moves must be done in such a way that the probability density of obtaining a given configuration $\mathbf{r}_n$ of a system of N particles is given by:

$$\rho(\mathbf{r}_n, N) = \frac{\exp\left[-\left(U(\mathbf{r}_n) + N\mu\right)/kT\right]}{\sum_N \int \exp\left[-\left(U(\mathbf{r}) + N\mu\right)/kT\right] d\mathbf{r}}$$

Grand canonical simulations are very useful for simulating electrochemical systems, because a constant electrode potential is equivalent to a constant chemical (or electrochemical) potential.

2.2 Interaction Potentials

The main difficulty of computer simulations is that a knowledge of the function potential energy functions $U(\mathbf{r}_n)$ is required to calculate the forces in atom dynamics or to calculate the transition probabilities in MC. Strictly speaking, $U(\mathbf{r}_n)$ stems from the quantum-mechanical interactions between the particles of the system, so that it should be obtained from first-principles calculations. However, for large ensembles this is not possible, so approximations, or model potentials, are needed. Such potentials are available for many systems: for example, for ionic oxides, closed-shell molecular systems, and fortunately for the problem at hand, for transition metals and their alloys. To the latter class of models belong the so-called glue model or the embedded atom method (EAM) [3].

Within the pair-functional scheme, the EAM proposes that the total energy $U(\mathbf{r}_n)$ of any arrangement of N metal particles may be calculated as the sum of individual particle energies E_i

$$U(\mathbf{r}_n) = \sum_{i=1}^{N} U_i$$

where the U_i are

$$U_i = F_i(\rho_{h,i}) + \frac{1}{2}\sum_{j\neq i} V_{ij}(r_{ij}) \tag{6}$$

F_i is the embedding function and represents the energy necessary to embed atom i into the electronic density $\rho_{h,i}$. This latter quantity is calculated at the position of atom i as the superposition of the individual atomic electronic densities $\rho_i(r_{ij})$ of the other particles in the arrangement as:

$$\rho_{h,i} = \sum_{j \neq i} \rho_i(r_{ij})$$

The attractive contribution to the energy is given by the embedding function F_i, which contains the many-body effects in EAM. The repulsive interaction between ion cores is represented as a pair potential, $V_{ij}(r_{ij})$ which depends exclusively on the distance between each pair of interacting atoms and has the form of a pseudo-coulombic repulsive energy:

$$V_{ij} = \frac{Z_i(r_{ij})Z_j(r_{ij})}{r_{ij}}$$

where $Z_i(r_{ij})$ may be considered as an effective charge, which depends upon the nature of particle i. This potential has been adequately parametrized from experimental data so as to reproduce a number of parameters as equilibrium lattice constants, sublimation energy, bulk modulus, elastic constants and vacancy formation energy.

2.3 The Creation of Atomic Clusters with the Aid of an STM Tip

As shown in Fig. 1a, an efficient electrochemical method to generate metal clusters on a foreign metal surface consist in moving a metal-loaded tip toward a surface of different nature. The onset of the interaction between the tip and the surface produces an elongation of the tip at the atomic scale – the so-called *jump to contact* – that generates a connecting neck between the tip and the surface. After this, the tip may approach further, and can be retracted at different penetration stages, with various consequences for the generated nanostructure. Experiments with several systems show that this procedure works with several tip/sample combinations, but with others it fails to produce well-defined surface features. For example, a model system for this type of procedure is Cu, deposited on the STM tip, and squashed against a Au(111) surface; for this combination of metals, an array of 10,000 clusters may be produced in a few minutes [4]. Similarly, with the system Pd/Au(111) well-defined nanostructures have been generated [5]. On the other hand, attempts to generate Cu clusters on Ag(111) failed; only dispersed monoatomic-high islands have been observed [6]. The systems Ag/Au(111) and Pb/Au(111) exhibit of an intermediate behavior, presenting a wide scatter of cluster sizes and poor reproducibility [6].

Since the creation of the clusters is a dynamic process, the natural computational tool to study this process appears to be molecular (or more properly stated) atom dynamics. In this respect, it is worth mentioning the pioneering work of Landman et al. [7], who employed this method to investigate the various atomistic mechanisms that occur when a Ni tip interacts with a Au surface. This investigation established a number of features that were relevant to understand the operating mechanism: on the one hand, during the onset of the tip–surface interaction, the occurrence of a mechanical instability leading to the formation of the connecting neck between tip and surface mentioned above. On the other hand, and in the subsequent stages of the nanostructuring process, the existence of a sequence of elastic and plastic deformation phases that determine the final status of the system. Typically, due to the time-step limitations mentioned above, the simulations are restricted to a few nanoseconds. However, the electrochemical generation of nanostructures involves times of the order of milliseconds, so that some long-time features of the experimental problem will be missing in the simulations. This must be taken into account for a careful interpretation of the experimental results in terms of simulations, in the sense that some slow processes, not observed in the simulations, may occur in experiments. However, the information obtained in the computational studies is important in the sense that if some processes are observed, they will certainly occur in the experiments.

A typical atomic arrangement employed for a MD simulation of the generation of clusters is shown in Fig. 2. The simulated tip consists of a rigid core, in the present case of Pt atoms, from which mobile atoms of type *M* to be deposited are suspended. These amount typically to a couple of thousand atoms, the exact figures for the different systems can be found in reference [8]. Different structures were assumed for the underlying rigid core. In some simulations a fcc crystalline structure was assumed, oriented with the (111), or alternatively the (100), lattice planes facing the substrate. We shall call them [111] and [100] tips, respectively.

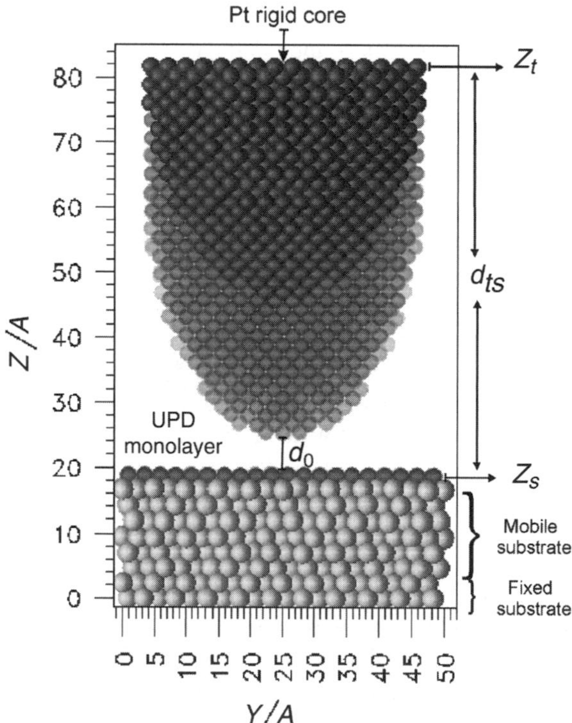

Fig. 2 $x - y$ section of a typical atomic arrangement employed in the simulation of tip-induced local metal deposition. *Dark circles* represent atoms belonging to the STM tip. *Transparent circles* denote the atoms of the material *M* being deposited, and *light gray circles* indicate the substrate atoms. The presence of a monolayer adsorbed on the electrode is depicted in this case. d_{ts} indicates the distance between the surface and the upper part of the tip. d_0 is the initial distance between the surface and the lower part of the tip. Mobile and fixed substrate layers are also marked in the figure. Taken from reference [8].

In other simulations, an amorphous structure was assumed [9]. The substrate S was represented by six mobile atom layers on top of two static layers with the fcc(111) orientation. An adsorbed monolayer of the same material as that deposited on the tip was occasionally introduced, so simulate nanostructuring under underpotential deposition conditions. Periodic boundary conditions were applied in the $x-y$ plane, parallel to the substrate surface. The tip was moved in z direction in 0.006 Å steps, performing first a forward motion toward the surface followed by a stage of backward motion. The distances given below will be referred to the jump-to-contact point described above, and the turning point of the motion of the tip, which determines the magnitude of the indentation, will be denoted with d_{ca}. The metallic systems so far investigated were Cu/Au(111), Pd/Au(111), Cu/Ag(111), Pb/Au(111), Ag/Au(111), and Cu/Cu(111).

Figure 3 depicts several snapshots of a typical simulation run where cluster creation is successful, in this case for the system Pd[110]/Au(111). The simulation starts at a point where the tip–surface interaction is negligible (Fig. 3a). The

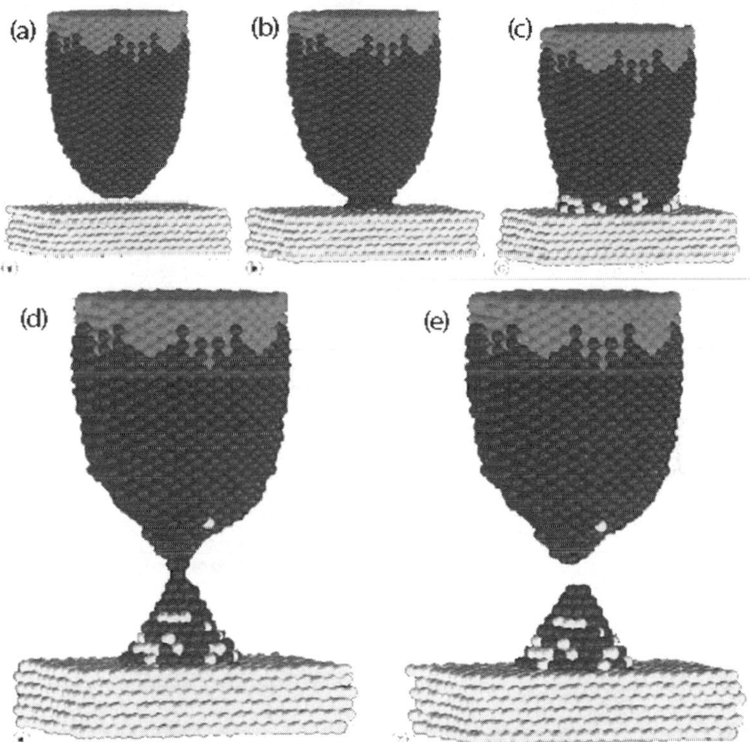

Fig. 3 Snapshots of an atom dynamics simulation taken during the generation of a Pd cluster on Au(111): (**a**) initial state; (**b**) jump-to-contact from the tip to the surface, (**c**) closest approach distance; (**d**) connecting neck elongation; and (**e**) final configuration. The resulting cluster has 256 atoms, is eight layers high and its composition is 16% in Au. Taken from reference [10]

approach of the tip to the surface leads to the jump-to-contact process, where mechanical contact between the tip and the surface sets in (Fig. 3b). A further approach of the tip results in an indentation stage (Fig. 3c), where the interaction between the tip and the surface is strong enough to produce mixing between *M* and *S* atoms. Figure 3d and e present the stages preceding and following the breaking of the connecting neck, leaving a cluster on the surface. A systematic study using different d_{ca} allows a more quantitative assessment of the nature of the nanostructures generated. Figure 4 shows the cluster size and height as a function of the distance of closest approach. It is clear that a deeper indentation of the tip generates larger and higher clusters. However, this is accompanied by an enrichment of the cluster in Au content, which in the simulations reported here reaches up to 16%. Extensive simulations showed that the nature of the clusters formed depends on the fact whether or not the surface is covered by a monolayer of Pd. In the former case, rather pure Pd clusters are formed with the [110]-type tip. However, the structure of the tip also plays a role in the mixing between substrate and tip atoms. In fact, the protective effect of the adlayer disappears when a [111] Pd tip is employed, and alloyed clusters are obtained. Simulation studies performed with a tip and a substrate of the same nature allowed to study pure tip structure effects [9]. In the case of the Cu/Cu(111) homoatomic system adatom exchange is found to take place in the case of the [111] tip almost exclusively. On the other hand, when nanostructuring is achieved with [110] type tips, the easy gliding along (111) facets allows the transfer of matter to the surface without major perturbations on the substrate.

Figure 5 shows snapshots of a simulation where cluster creation fails, in this case Cu[111]/Ag(111). As in the previous case, Fig. 5a shows the appearance of

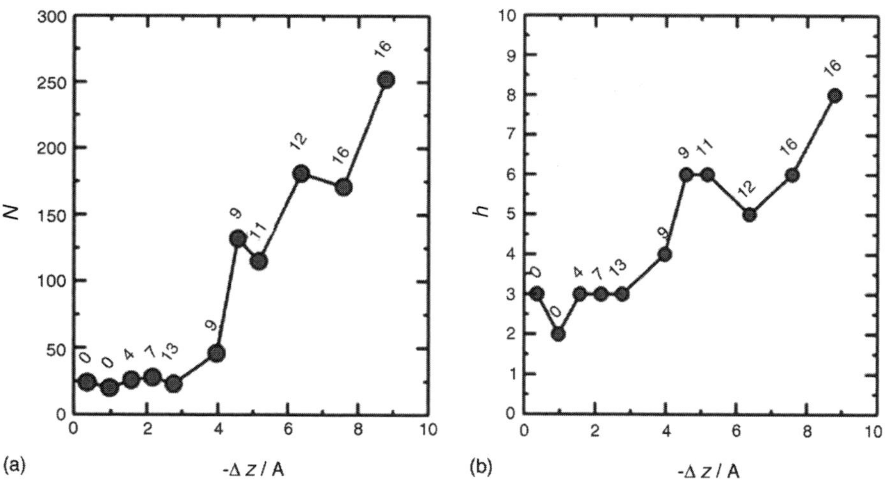

Fig. 4 Cluster size and composition for Pd nanostructuring on Au(111), as a function of deepest tip penetration *z* for a typical set of runs: (**a**) number of particles in the cluster and (**b**) cluster height in layers. The numbers close to the circles denote the Au atomic percentage in the clusters. Taken from reference [11]

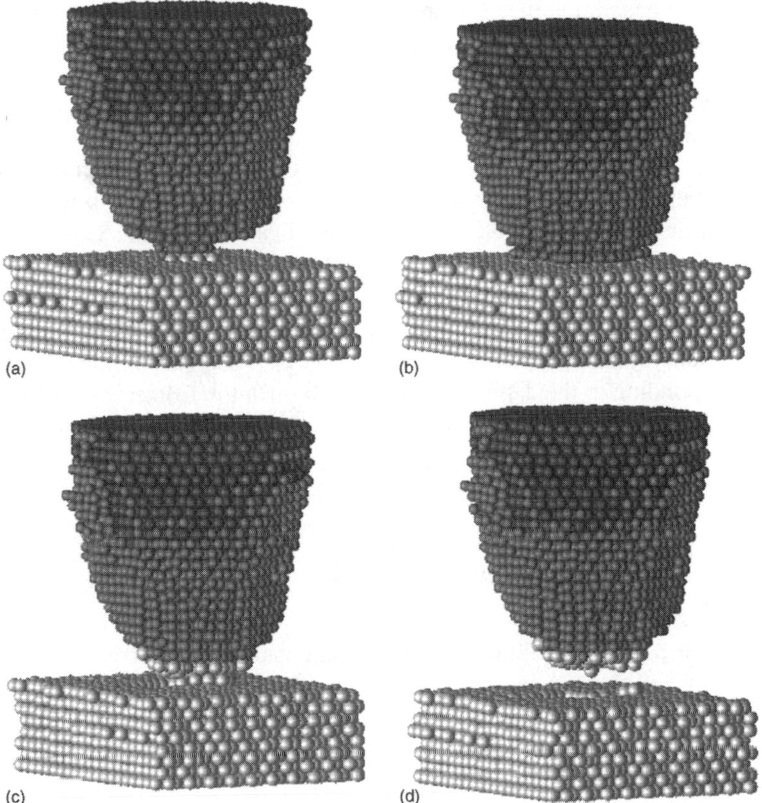

Fig. 5 Snapshots of an atom dynamics simulation taken during a failed attempt to generate a Cu cluster on Ag(111): (**a**) jump-to-contact from the surface to the tip. Note that the jump-to-contact goes in the opposite direction to that of Fig. 4; (**b**) closest approach distance; (**c**) retraction of the tip to 4.8 Å above the closest approach distance; and (**d**) final state after the rupture of the connecting neck. Thirty-eight Ag atoms were removed from the surface. Taken from reference [10]

a mechanical instability that generates the tip–surface contact. However, it must be noted that in the present case the surface atoms of the substrate participate actively in the processes, being lifted from their equilibrium position. In the present simulation, although the motion of the tip is restricted to a gentle approach to the surface ($d_{ca} = -3.6$ Å, Fig. 5b), the interaction is strong enough to dig a hole on the surface. Other simulations where d_{ca} is more negative lead to larger holes, and some Cu atoms are dispersed on the surface. On the other hand, when a Cu[110] tip is used for this system, pure small Cu clusters result, which are, however, unstable.

Simulations for the Cu/Au(111) system also lead to successful nanostructuring of the surface, with results qualitatively similar to those of Pd/Au(111). On the other hand, the Ag/Au(111) system yields poorly defined nanostructures, and something similar happens with the Pb/Au(111) system, where only pure Pb, tiny, unstable clusters are obtained [8, 12].

Besides the generation mechanism of the clusters, the question of their stability is highly relevant for technological applications. Experiments show that they are surprisingly resistant against anodic dissolution [6]. For example, a Cu cluster on Au(111) at a potential of 9–10 mV Versus Cu/Cu^{+2} was found to be stable for at least 1 h. This fact is rather surprising, since due to surface effects, clusters should be less stable than the bulk material. In fact, atoms at the surface of a cluster are less coordinated than those inside the nanostructure, so that the average binding energy of the atoms in the cluster should be smaller than that of bulk Cu.

A meaningful concept of the theory of the electrochemical stability of monolayers is the so-called *underpotential shift* $\Delta\phi_{up}$, which was originally defined by Gerischer and co-workers [13] as the potential difference between the desorption peak of a monolayer of a metal M adsorbed on a foreign substrate S and the current peak corresponding to the dissolution of the bulk metal M. A more general definition of $\Delta\phi_{up}$ can be stated in terms of the chemical potentials of the atoms adsorbed on a foreign substrate at a coverage degree Θ, say $\mu_{M_\Theta(S)}$, and the chemical potential of the same species in the bulk μ_M according to:

$$\Delta\phi_{upd}(\Theta) = \frac{1}{ze_0}(\mu_M - \mu_{M_\Theta(S)}) \tag{7}$$

Following the same line, the electrochemical stability of a given nanostructure may be analyzed through the chemical potential of its constituting atoms. Thus, since the stability of the cluster on the surface is given by the chemical potential of the atoms μ, a grand-canonical simulation with μ as control parameter appears as the proper tool to study cluster stability. Further parameters in the experimental electrochemical systems are the temperature T and the pressure P, so that the proper simulation tool appears to be a μPT simulation, where the ensemble partition function is given by:

$$\Upsilon = \sum_{N,V,E_i} e^{-U_i/kT} e^{-pV/kT} e^{N\mu/kT} \tag{8}$$

However, for the usual experimental conditions, where solid species are involved and $p = 1$ atm, the sum in (8) can be replaced by:

$$\Upsilon = \sum_{N,V,E_i} e^{-U_i/kT} e^{N\mu/kT} \tag{9}$$

where we have set $p = 0$. This equation contains volume-dependent terms through U_i, since the energy of the nanostructure depends on its volume. However, as shown above in Eq. (5), MC simulations do not involve the sum (9), but rather the ratio of probability densities between two states, that in the present case will be given by:

$$\frac{\rho_n}{\rho_m} = \beta_{mn} \exp\left[\left(U_m(V) - U_n(V)\right)/kT\right] \exp\left[N\mu/kT\right]$$

where we have written $U_m(V) - U_n(V)$ to emphasize the fact that the energy is a function of the volume of the nanostructure. Although the partition function involved is a different one, this equation is formally identical to that obtained for a grand-canonical (GC, $\mu V_{box} T$) simulation, where particles are created in and removed from a simulation box with volume V_{box} containing the nanostructure. Note that although the volume V_{box} is held constant when calculating transition probabilities in the GC simulations, these allow for the the fluctuation of the volume V , with the concomitant changes in the energy U_m.

As we have seen above, depending on the penetration of the tip into the surface, different degrees of mixing between the material of the tip and that of the substrate may be obtained, which can affect the stability of the clusters. The behavior of pure and alloyed clusters could be in principle studied by comparing the behavior of the different clusters formed in the MD simulations. However, since the energy of the clusters (and thus their chemical potential) is a function not only of their composition but also of their geometry (i.e., surface-to-volume ratio), and the clusters formed under different conditions present different geometries, direct comparison appears somewhat complicated. A way to circumvent this problem consists in considering an alloyed cluster, and replacing the atoms coming to the substrate by atoms of the nanostructuring material. In this way, the geometry at the beginning of the simulation should be approximately the same, apart from a slight stress that may be relaxed at the early stages of the simulation.

Figure 6 shows in snapshots of a GCMC simulation, the comparative behavior of pure Pd, and alloyed Pd/Au clusters upon dissolution. In the case of the pure Pd clusters, it can be seen that, as the simulation proceeds, dissolution takes place from all layers, while the cluster tries to keep its roughly pyramidal shape. Hexagonally shaped structures, constituted by seven atoms, are found to be particularly stable. In the case of alloyed clusters, they become enriched in Au atoms as the simulation proceeds; simultaneously, the dissolution process is slowed down. This can be seen more quantitatively in Fig. 7, where the dissolution of a pure Pd and of an alloyed Pd/Au cluster is studied at a constant chemical potential. It can be noticed that the cluster remains relatively stable at chemical potentials close to 3.80 eV. This value is higher than the cohesive energy of Pd, −3.91 eV, showing that, as expected, the cluster is less stable than the bulk material. At higher values of μ the cluster starts to grow, at lower values it dissolves. Concerning the chemical potentials that allow cluster growth, both the pure and the alloyed clusters behave similarly. However, marked differences occur for values of μ at which the clusters are dissolved. While the pure Pd clusters dissolve rapidly, the alloyed clusters become more stable as the simulation proceeds as a consequence of their enrichment in Au atoms.

The results compiled from simulations for different systems are summarized in Table 1, together with experimental information. We can draw two main conclusions from experiments and simulations. On the one side, the best conditions for cluster generation exist in those systems where the cohesive energies of the metal being deposited and the substrate are similar. On the other side, in systems where the alloy heats of solution ΔH_d for single substitutional impurities is negative, the mixing of the atoms will be favored on thermodynamic grounds, and thus cluster stability will be improved.

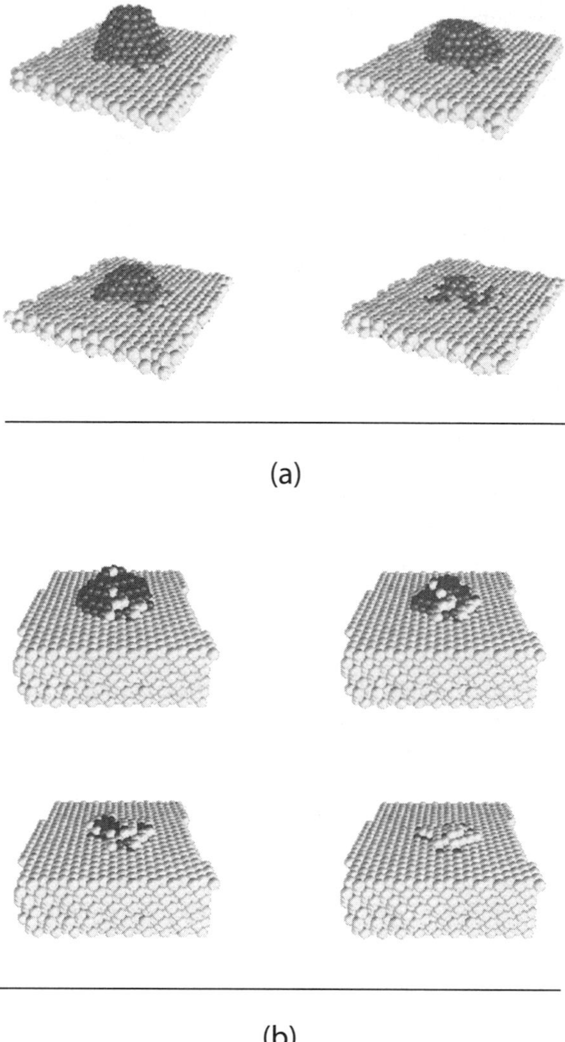

Fig. 6 Different stages of a dissolution of a pure (**a**) and alloyed (**b**) Pd cluster on a Au(111) surface. The alloyed cluster was 16% in content of Au atoms and contained an initial number of 252 atoms

2.4 The Filling of Nanoholes

In the previous section, we have seen how grand-canonical Monte Carlo simulations can be employed to understand experimental results for the localized metal deposition on surfaces. The same tool can be employed to consider the filling of surface defects like nanocavities, as explained above in Fig. 1b. Schuster et al. [20] have shown that the application of short voltage pulses to a scanning tunneling

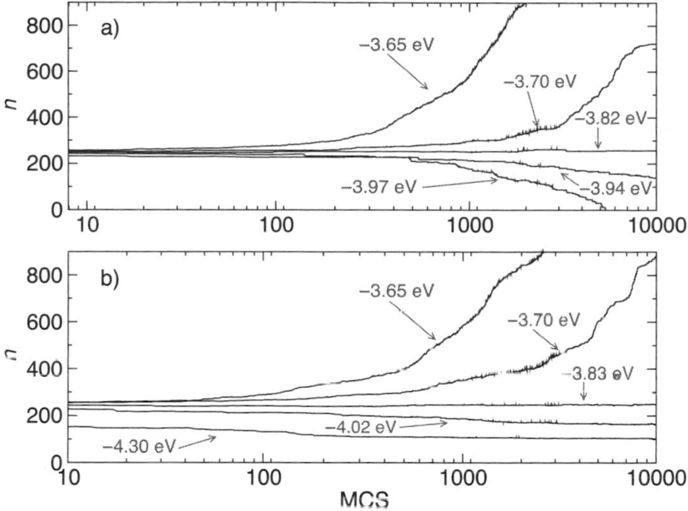

Fig. 7 Evolution of cluster size (expressed in terms of number of atoms) as a function of the number of Monte Carlo steps at different chemical potentials. (**a**) Pure Pd cluster initially containing 252 atoms. (**b**) Alloyed Pd/Au cluster, initially containing 211 Pd and 41 Au atoms. Taken from reference [8]

Table 1.1 Compilation of experimental and simulated systems according to their difference binding energy ΔE_{coh} and alloy heats of solution ΔH_d for single substitutional impurities. Experimental results were taken from ref. [4, 14, 15, 16, 6, 17], the information corresponding to Ni–Au(111) simulations was taken from [7, 18]. ΔE_{coh} and ΔH_d were taken from [19]. *3D-stable* means that the clusters generated in the experiments endure dissolution. *3D-unstable* means that experimental clusters dissolve easily. *2D* means that only 2D islands were obtained. *Hole* refers to the fact that only holes were generated on the surface after the approach of the tip. Concerning the column of simulation, *alloying* means that the cluster contained not only atoms of the metal being deposited but also some coming from the substrate. Taken from reference [8]

System	Experiment	Simulation	$\Delta E_{coh}/eV$	$\Delta H_d/eV$
Pd/Au(111)	3D-Stable	3D-alloying	0.02	−0.36, −0.20
Cu/Au(111)	3D-Stable	3D-alloying	−0.38	−0.13, −0.19
Ag/Au(111)	3D-Stable	3D-alloying	−1.08	−0.16, −0.19
Pb/Au(111)	3D-Unstable	3D-pure	−1.84	0.03, 0.01
Cu/Ag(111)	2D hole	hole, 3D small	0.69	+0.25, +0.39
Ni/Au(111)		hole	0.52	+0.22, +0.28

microscopic tip close to a Au(111) surface may be employed to generate nanometer-sized holes. While the depth of the nanoholes only amounts 1–3 monolayers, their lateral extension is about 5 nm. The subsequent controlled variation of the potential applied to the surface immersed in a Cu^{+2}-containing solution leads to a progressive filling of the nanoholes in a stepwise fashion. This is illustrated in Fig. 8, where the time evolution of the height of the nanostructure (hole + cluster inside it) is given. It

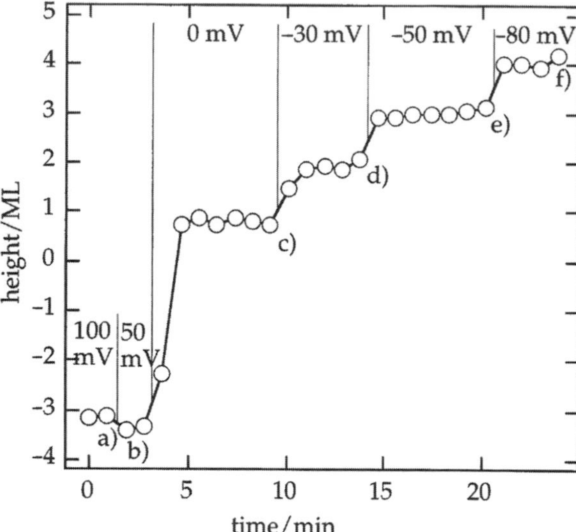

Fig. 8 Time evolution of the height of the nanostructure obtained by polarization at different potentials of a Au(111) surface with nanometer-sized holes. The nanostructure considered consists of a ca. 8-nm-wide hole in which Cu atoms are deposited by potential control. The potentials are referred to the reversible Cu/Cu^{+2} electrode in the same solution. Taken from reference [22]

is clear that the equilibrium height depends on the applied overpotential, indicating that the surface energy of the growing cluster is balanced by the electrochemical energy. The idea put forward by the authors was that the overpotential was related to the Gibbs energy change associated with the cluster growth. From this hypothesis, the authors estimated that boundary energy amounts 0.5 eV/atom for the present system. More recently, Solomun and Kautek [21] have studied the decoration of nanoholes on Au(111) by Bi and Ag atoms. While the Au nanoholes were completely filled by Bi at underpotentials, the Ag nanoholes remained unfilled, even at low overpotentials. In the latter case, only at the equilibrium potential and at overpotentials did the nanoholes start to be filled flat to the surface, showing no protrusion over it.

As stated in the section on interaction potentials, a simulation containing all the electrochemical ingredients should involve in principle interfacial charging effects, solvent, and importantly the contribution of the long-ranged interaction of adsorbed anions, which in some systems such as Cu/Au play a key role in the stabilization of the underpotential deposits. However, simulations with the EAM potentials, which properly take into account interactions between metallic atoms, will give at least a qualitative idea of the leading contribution to the defect nanostructuring process. Figure 9 shows results of such a simulation, where the chemical potential was stepped to simulate the negative polarization of the surface, together with snapshots of the simulation.

It can be seen that at chemical potentials μ close to the binding energy of bulk Cu (-3.61 eV), the Au nanohole, initially decorated by Cu atoms, suddenly fills up to a level close to substrate surface. Then, a further step in the height is found, and later the growth proceeds with the cluster aquiring a pyramid-like shape, slowly expanding beyond the borders of the nanohole. The calculation of boundary energy

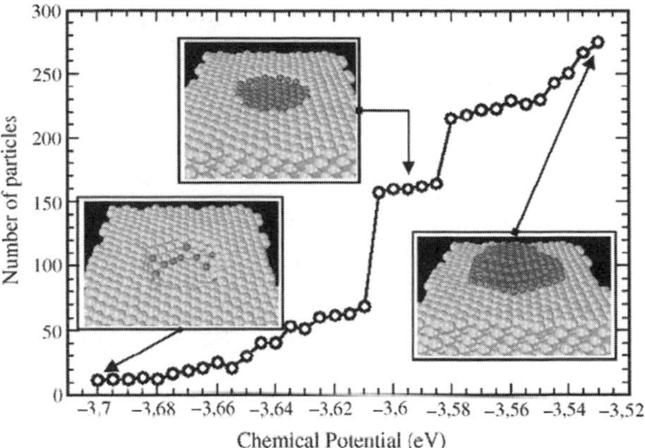

Fig. 9 Growth of a Cu cluster inside a defect on a Au(111) surface. The number of Cu atoms is given as a function of the chemical potential of the Cu atoms μ. The snapshots were taken at the points marked by the arrows. Taken from reference [10]

using EAM yields values of the order of 0.1 eV atom, considerably lower than the experimental estimation. This low value is partially due to the inaccuracy of the potentials employed for the simulations – the higher experimental values are reasonable. If the experimental values were as low as the calculated ones, the stepwise growth of the nanostructure would not be observed clearly. In fact, the stepwise growth is a consequence of the fact that the border atoms make a sizable contribution to the free energy of the growing cluster, and this relative contribution becomes smaller as cluster size increases. In the limit of infinitely large clusters, no steps in the growth of the cluster would be observed. Furthermore, the experimental system presents a Cu layer deposited at underpotentials. The simulation do not show this layer, since no upd is predicted for this system. Rather than to an excess of the binding energy of the Cu/Au(111) system with respect to bulk Cu, the existence of the Cu upd overlayer has been attributed to anion effects [23].

3 Electron Transfer Through Functionalized Adsorbates and Films

3.1 Electron Exchange with a Monolayer

Monolayers that self-assemble on electrode surfaces (SAMs) are a fascinating area of research. Of particular interest are films whose ends have been functionalized with electron donors or acceptors that can exchange electrons with the electrode surface. In the simplest case, in which no other electrodes are present, electron exchange between the active species and the electrode is well described by the Marcus–Hush type of theories and their modern extensions [24]. The main difference compared to normal electron exchange with a solvated species is the presence of the intervening layer, which impedes electron transfer. The transfer of an electron along a chain molecule belongs to the more general subject of molecular conductance, on which there is a rich literature (see, e.g., [25 – 27]), and which lies outside of this review. In particular, the dependence of the electron-transfer rate on the chain length has been investigated by several groups (e.g., [28]); often, an exponential decay is observed, which is usually expressed in the form:

$$k \propto \exp -\beta l \tag{10}$$

where l is the chain length; the decay constant β is often found to be of the order of 1 Å^{-1}; this value reflects both the electronic properties of the wire and a possible length dependence of the reorganization energy associated with the electron transfer.

Since the presence of the chain decreases the reaction rate, it is possible to measure the rate constant over a large range of potentials. This is usually not possible on bare electrodes since the rate becomes too fast at high overpotentials. In fact, the best proofs for the sigmoidal dependence of the rate on the potential, predicted by electron-transfer theories [29] have been obtained on film-covered electrodes.

3.2 Metal–Adsorbate-Metal Systems

Adding a second electrode, such as the tip of an STM, to the system greatly increases the possibilities for investigations. A possible configuration, consisting of an electrode and a tip, is shown in Fig. 10. The electro-active centers now serve as an intermediate state for the electron exchange between the two metals. Various mechanisms for the indirect electron exchange have been proposed, and we shall discuss them in turn.

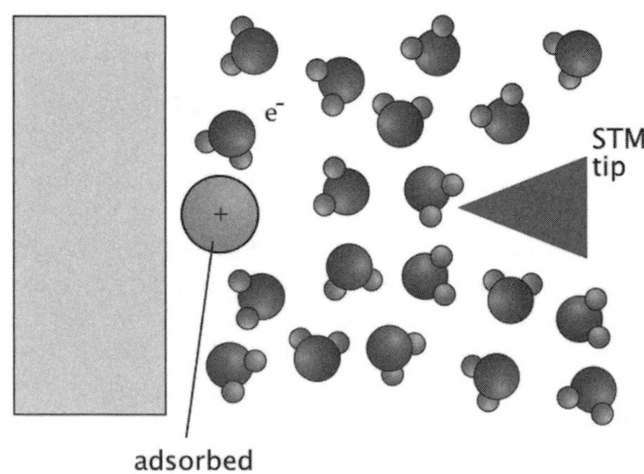

Fig. 10 Basic setup for investigating an adsorbed redox center with the tip of an STM

3.2.1 Two-Step Mechanism

Conceptually, the simplest mechanism for electron exchange between two metals and an intermediate consists of two chemical reactions. The overall rate can then be obtained from formal chemical kinetics. In Fig. 10, we denote the rate constant for electron transfer from the electrode to the intermediate by k_1, and for the back reaction by k_{-1} , and by k_2, k_{-2} the rates from the intermediate to the tip and back. Under stationary conditions, rate of electron transfer from the metal to the tip is:

$$v = \frac{k_1 k_2 - k_{-1} k_{-2}}{\sum_i k_i} \tag{11}$$

As with all electrochemical reactions, the rate constants depend on the potential. We choose, as our reference potential for the reactant the value, when the electron exchange between the left metal electrode and the reactant is in equilibrium, and denote the deviation from this value by the overpotential η. Under normal experimental conditions, η will be of the order 100 mV, so that we may use the usual Butler–Volmer equation for the potential dependence of the rate constants. For simplicity, we take a value of 1/2 for the transfer coefficient, so that there is a symmetry between the forward and the backward reaction; both theory and experiment give values close to 1/2 for simple electron exchange. Thus we have:

$$k_1 = k_{10} \exp \frac{e_0 \eta}{2k_B T}, \qquad k_{-1} = k_{10} \exp -\frac{e_0 \eta}{2k_B T} \tag{12}$$

where e_0 is the unit of charge, k_B is Boltzmann's constant, and T the temperature.

Electron exchange between the reactant and the tip is in equilibrium, when the bias V between tip and electrode is canceled by the overpotential. Therefore, we may write:

$$k_2 = k_{20} \exp \frac{e_0(V - \eta)}{2k_B T}, \qquad k_{-2} = k_{20} \exp -\frac{e_0(V - \eta)}{2k_B T} \tag{13}$$

Experiments are usually either performed at constant bias V and variable potential η, or at fixed η and variable V. Figure 11 shows calculated current–potential curves corresponding to the former type of experiments. The current shows a maximum when η lies between zero and V; its position is determined by the condition that the average occupancy of the intermediate state is one half. Therefore, the maximum is at the potential where the faster of the two electron exchanges is in equilibrium. Calculated curves for variable bias are shown in Fig. 12. As may be expected, the rate increases strongly with the bias. Here the two standard rates were set equal, $k_{10} = k_{20} = 1$, so that for a given bias, the optimum value for the overpotential lies at about $\eta = V/2$.

Electron exchange in solutions generally require thermal activation, usually connected to the reorganization of the solvent. When the electronic coupling is weak, the system will subsequently relax to a new thermal equilibrium, and a real

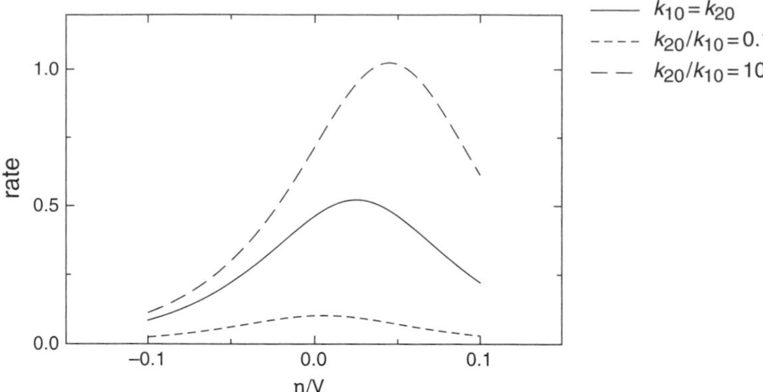

Fig. 11 Calculated current–potential curves at a constant bias; the standard rate for exchange between the electrode and the intermediate was set to $k_{10} = 1$

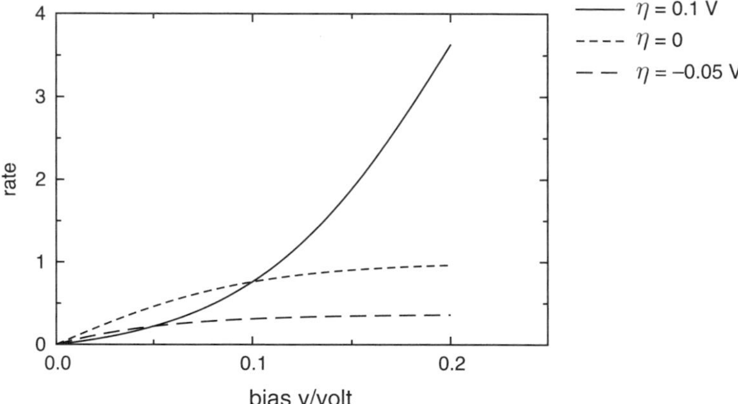

Fig. 12 Calculated current-bias curves at a constant overpotential; the standard rates were set to $k_{10} = k_{20} = 1$

intermediate state is formed. Then the mechanism discussed above will prevail. However, if the electronic coupling is strong another electron transfer may occur before relaxation; the corresponding mechanisms will be considered below.

3.2.2 Resonant Transition

When thermal relaxation is slower than electron transfer, the reactant can serve as a virtual intermediate state. In this section we discuss the basic mechanism.

To be specific, we consider the current for electron transfer from the electrode to the tip. In contrast to ordinary electron-transfer reactions, the electronic energy need not be conserved in the process. If the transferring electron couples to vibrations localized in the reactant, it can transfer energy to them; this constitutes an inelastic exchange, in contrast to elastic processes where the electronic energy is conserved. In principle, the electron can also pick up energy from an excited vibration, but this process, which is similar to anti-Stokes transitions in Raman scattering, is so unlikely that it can be disregarded. In general, the current associated with the transfer can be written in the form (see Fig. 13):

$$j = \int f(\epsilon)[1 - f(\epsilon' - e_0 V)]T(\epsilon, \epsilon') \approx \int_{-e_0 V}^{0} T(\epsilon, \epsilon') \tag{14}$$

where $f(\epsilon)$ denotes the Fermi–Dirac distribution, V is the bias between the two electrodes, and the Fermi level of the left electrode has been taken as the energy zero [30]. In the approximate version the Fermi–Dirac distributions have been replaced by step functions.

The form of the transmission function depends on the mechanism of electron transfer; the process is similar to resonant transitions in solid-state systems [31]. In the simplest case the intermediate state just interacts electronically with the two bulk metals. Due to these interactions, the intermediate state, with energy ϵ_r, attains a resonance width $\Delta = \Delta_1 + \Delta_2$, with contributions from both electrodes. In the so-called wide-band approximation the resonance width is independent of the electronic energy ϵ; as the name indicates, this approximation should be valid if the widths of the bands in the two bulk metals, to which the intermediate state couples, is much larger than Δ. Usually this is a safe approximation, and the transmission function then takes the form of a Lorentz distribution:

$$T(\epsilon, \epsilon') = \frac{\Delta^2}{(\epsilon - \epsilon_r)^2 + \Delta^2}\, \delta(\epsilon - \epsilon') \tag{15}$$

This equation is easily interpreted: The intermediate state serves as a resonance of total width Δ, and only elastic scattering is allowed, as is indicated by the delta-function. The same equation has been derived for the conductance of a mono-atomic

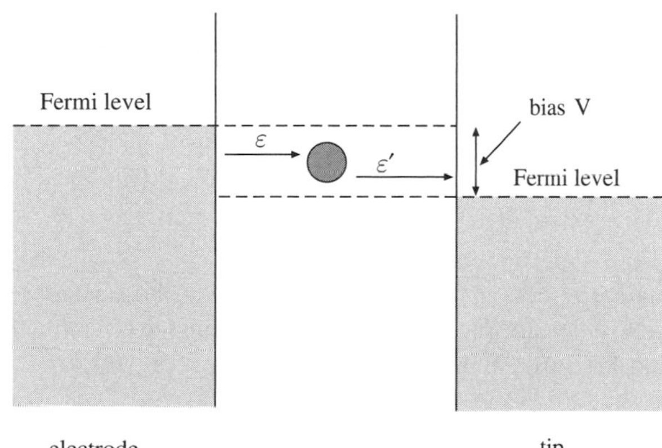

Fig. 13 Elementary process of electron transfer from an electrode to a tip

molecular wire [32]. In an electrochemical environment the electron always couples to solvent modes, so that the resonance energy becomes a function of the solvent coordinates q_ν, where ν labels the solvent modes. These are usually classical, and if they are the only modes with which the electron interacts is sufficient to introduce an explicit dependence $\epsilon_r(q_\nu)$ into eq. (15) and perform a thermal average over all solvent configurations. The details of this averaging depend on the situation. There are two possibilities: (1) The electron transfer to the acceptor is so fast that there is no electronic equilibrium between the electrode and the reactive center. In this case electron transfer occurs through the empty orbital, and the average is performed over the corresponding solvent configurations. (2) Electron exchange between the electrode and the center is much faster than the exchange with the tip, so that the redox species is in electronic equilibrium with the electrode, and the transfer can either occur from a filled state or through an empty one. We consider the former situation first and return to the other below.

The thermal average over the solvent configurations corresponding to an empty center is readily accomplished. The result can be written in the form [30]:

$$T_{\rm th}(\epsilon,\epsilon') = \Gamma_1\Gamma_2\delta(\epsilon-\epsilon')\sqrt{\frac{\pi}{\lambda k_B T}}\int \exp\frac{(x-(\epsilon-\epsilon_r))^2}{4\lambda k_B T}\frac{1}{x^2+\Delta^2}dx \qquad (16)$$

and is thus a convolution of the Lorentz distribution with the Marcus-type thermal broadening [1, 33]:

$$T_{\rm reorg}(\epsilon) = \sqrt{\frac{\pi}{\lambda k_B T}}\exp -\frac{(\epsilon-\epsilon_r)^2}{4\lambda k_B T} \qquad (17)$$

Thus, the electronic level of the resonant state is broadened by two effects: The electronic coupling induces the width Δ, and the interaction with the solvent fluctuations results in a thermal broadening of width $(4\lambda kT)^{1/2}$. The electrons move much faster than the solvent; therefore, the electronic energy is conserved, and the transition is elastic.

The integral in Eq. (16) can be expressed through the complex error function $\mathrm{W}(z) = \exp(-z^2)\,\mathrm{erfc}(-iz)$ to yield:

$$D(\epsilon) = \left(\frac{\pi}{kT\lambda}\right)^{1/2}\Re\left[\mathrm{W}\left(\frac{\epsilon_r-\epsilon+i\Delta}{2\sqrt{kT\lambda}}\right)\right] \qquad (18)$$

where $\Re$ denotes the real part. $D(\epsilon)$ can be considered as a generalization of the redox density of states for electron transfer as defined by Gerischer [34]. With its aid Eq. (16) can be written in the compact form:

$$T_{\rm th}(\epsilon,\epsilon') = \Gamma_1\Gamma_2\delta(\epsilon-\epsilon')D(\epsilon) \qquad (19)$$

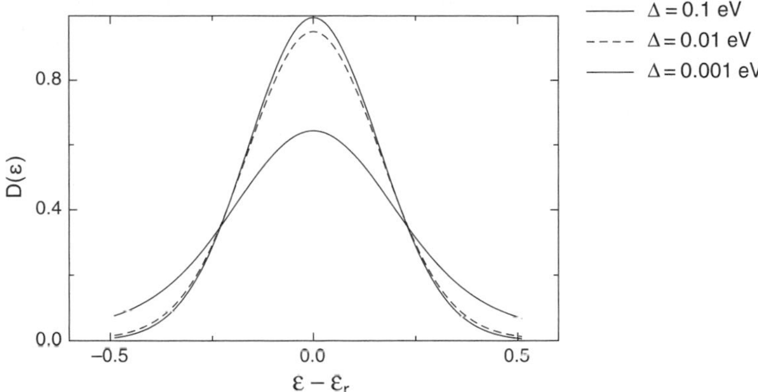

Fig. 14 Some examples for the generalized redox density of states $D(\epsilon)$; the reorganization energy was set to $\lambda = 0.5$ eV, and the temperature to $T = 293$ K

A few typical examples for $D(\epsilon)$ are shown in Fig. 14. The width of the Gaussian is $(k_B T)^{1/2}$, that of the Lorentzian is Δ, and the shape of $D(\epsilon)$ is governed by the broader of the two distributions.

Quantum modes move on the same timescale as electrons. Therefore, when the transferring electrons couple to vibrations on the reactant, inelastic transitions can occur. As mentioned above, the most likely inelastic process is the excitation of a vibration. The transition function can then be written as a sum over the number of phonons that are excited:

$$T(\epsilon, \epsilon') = \sum_{n=0}^{\infty} A_n(\epsilon)\delta(\epsilon - \epsilon' - n\hbar\omega) \tag{20}$$

where ω is the frequency of the excited mode. Explicit expressions for the corresponding amplitudes have been given in the literature [30]. Of course, the electron may couple to more than one quantum mode; the corresponding generalization of Eq. (20) is cumbersome, though not difficult.

Since the transferring electron must come from an occupied state and go into an empty state, the number of phonons n emitted must obey $n\hbar\omega \leq e_0 V$ (see Fig. 13). Therefore, the number of inelastic channels that are open depends on the bias potential. In principle, this offers the possibility to perform spectroscopy on the quantum modes that couple to the electron transfer by measuring the current as a function of the bias. Every time the bias passes a multiple of $\hbar\omega$, a new elastic channel opens, with a corresponding increase in the current. In the second derivative $\mathrm{d}^2 I/\mathrm{d}V^2$ the opening of a channel will appear as a peak. Figure 15 shows a few calculations exhibiting this effect. The difference between two peaks gives the energy of the emitted photon, and the amplitude indicates the coupling strength. Depending on the magnitude of the latter, a sizable fraction of the current can pass through inelastic channels at high bias voltages.

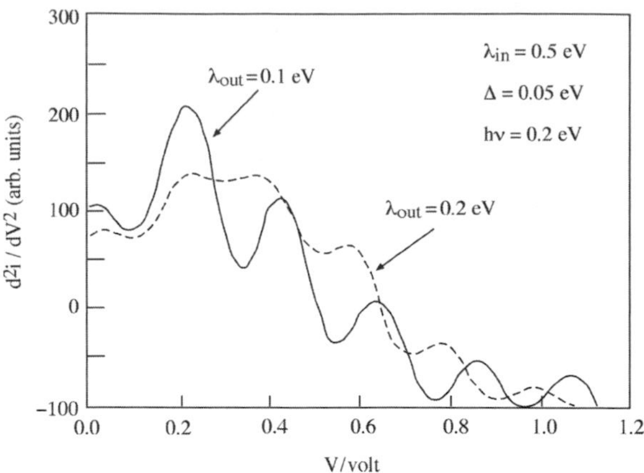

Fig. 15 Calculated second derivatives of the current as a function of the bias voltage

3.2.3 Adiabatic Limit

Kuznetsov and Schmickler [34] have considered the case in which the electron exchange of the reactant with the two metals is adiabatic, so that the electronic system, for a given value of the solvent coordinates, is in equilibrium. If only classical solvent modes are coupled to the transfer, they can be represented by a single effective solvent coordinate q [35]. It is convenient to normalize q in such a way that $q = 0$ corresponds to the oxidized and $q = 1$ to the reduced state. In the adiabatic limit, for each value of q, there is a constant flux of electrons from one metal to the other, resulting in an average occupation $\langle n_r \rangle$ of the intermediate orbital, which is given by:

$$\langle n_r \rangle = \frac{1}{\pi} \int_{-\infty}^{E_F^{\mathrm{el}}} \frac{\Delta_1}{(\omega - \tilde{\epsilon}(q))^2 + \Delta^2} \, d\omega + \frac{1}{\pi} \int_{-\infty}^{E_F^{\mathrm{tip}}} \frac{\Delta!_2}{(\omega - \tilde{\epsilon}(q))^2 + \Delta^2} \, d\omega \tag{21}$$

where $\tilde{\epsilon}(q) = \epsilon_r - 2\lambda q$ is the electronic energy of the adsorbate level as a function of the solvent coordinate. E_F^{el} is the Fermi level of the electrode and E_F^{tip} that of the tip.

The rate of electrons passing from the electrode to the tip is then given by [30]:

$$k(q) = \frac{1}{\pi \hbar} \int_{E_F^{\mathrm{tip}}}^{E_F^{\mathrm{el}}} \frac{\Delta_1 \Delta_2}{(\omega - \tilde{\epsilon}(q))^2 + \Delta^2} \, d\omega \tag{22}$$

The effective potential-energy surfaces for the system are markedly affected by the presence of the tip. Figure 16 shows some examples for various values of the couplings Δ_1 and Δ_2 for the case where the intermediate is in equilibrium with the left electrode. In contrast to the Marcus theory for simple electron transfer, in which the adiabatic potential energy curves are composed of two parabolic sections, the activation region is markedly broadened by the interaction with the tip. The corresponding

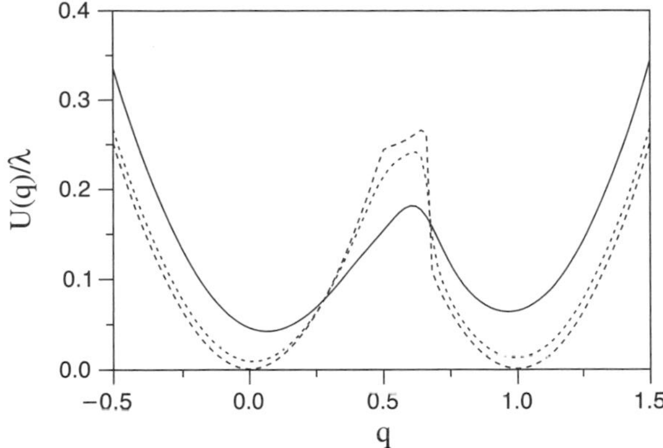

Fig. 16 Potential–energy curves for the case where the intermediate is in equilibrium with the left electrode. System parameters: $\lambda = 0.6\,\text{eV}$; $V_b = 0.2\,\text{V}$. Full curve: $\Delta_1 = \Delta_2 = 0.05\,\text{eV}$; short dashes: $\Delta_1 = \Delta_2 = 0.01\,\text{eV}$; long dashes: $\Delta_1 = \Delta_2 = 0.001\,\text{eV}$

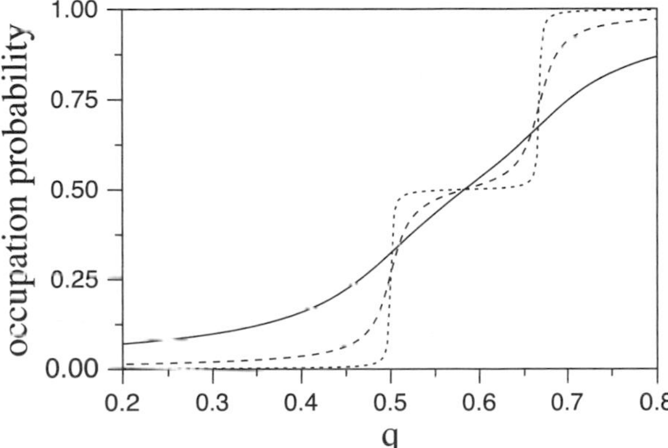

Fig. 17 Occupation probability of the intermediate for the case depicted in Fig. 16. System parameters: Full curve: $\Delta_1 = \Delta_2 = 0.05\,\text{eV}$; long dashes: $\Delta_1 = \Delta_2 = 0.01\,\text{eV}$; short dashes: $\Delta_1 = \Delta_2 = 0.001\,\text{eV}$

occupation probabilities $\langle n_r \rangle$ are shown in Fig. 17. For the chosen parameters, there is an extended saddle-point region in which the occupation probability is close to 1/2.

The rate of electron exchange can either be determined by performing a thermal average over Eq. (22) or by stochastic molecular dynamics; the latter was the route chosen by Kuznetsov and Schmickler [34]. Unfortunately, the relationships between the current and the bias, or the current and the potential, are qualitatively similar to those obtained from the simple two-step mechanism discussed above. As an example we show the relation between the current and the overpotential with respect to the left electrode, calculated at constant bias (see Fig. 18). However, the exact shape of such current-overpotential curves depends on the energy of reorganization λ for electron transfer to the intermediate state. This can be used to extract λ from experimental data. This is shown in Fig. 19, where experimental data from Tao [37], which were obtained from doped protoporphyrins films adsorbed on highlyoriented graphite electrodes at a constant bias of 50 mV, are compared with calculated

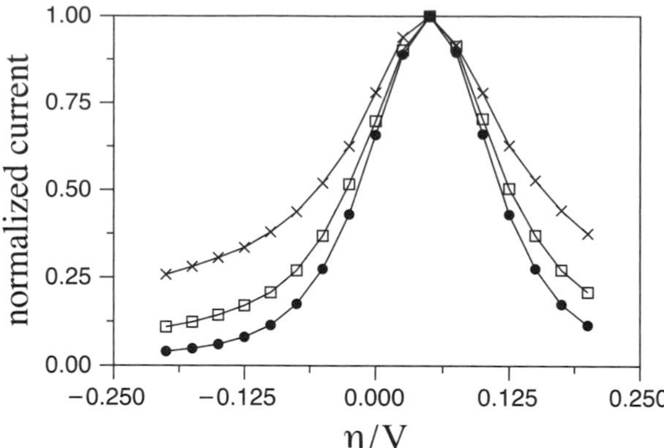

Fig. 18 Current as a function of the overpotential η with respect to the left electrode. System parameters: $\Delta_1 = \Delta_2 = 0.01$ eV; $V_b = 0.1$ V. (•) $\lambda = 0.2$ eV; (□) $\lambda = 0.4$ eV; (×) $\lambda = 0.6$ eV. The currents have been normalized such that the maximum current is unity

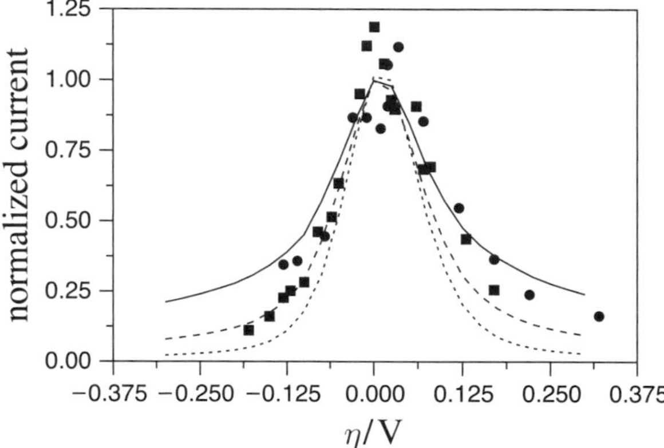

Fig. 19 Current as a function of the overpotential. System parameters: $\Delta_1 = 0.01$ V, $\Delta_2 = 0.001$ V, $V_b = 0.05$ V. *Full line*: $\lambda = 0.6$ eV; *dashed line*: $\lambda = 0.4$ eV; *dotted line*: $\lambda = 0.2$ eV. The points are the experimental values from Tao [36]

curves. As can be seen, the experimental data can be fitted to a reorganization energy of the order of $\lambda = 0.4 - 0.6$ eV. However, the same data can also be explained by assuming a nonadiabatic process [37].

3.3 Two Intermediate States

Progress in nanofabrication has made it possible to study electron exchange between two electrodes covered by electroactive adsorbate layers (see Fig. 20). Again, the simplest mechanism consists of normal chemical reactions, in this case consisting of two electrochemical and one chemical step. If the active sites are not fixed, diffusion may also play a role. For experimental details and interpretation in terms of a series of reactions we refer to [38, 39, 40]. We focus here on the exchange through virtual intermediates, or superexchange. First we discuss the elementary act, and then interpret a specific example.

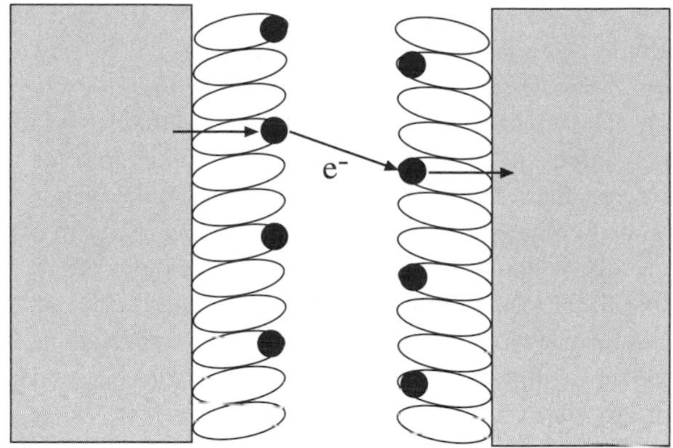

Fig. 20 Electron exchange between two electrodes mediated by two adsorbate films with active sites

3.3.1 Electron Exchange via Two Bridge States

The electronic interactions are now governed by three terms: the resonance widths Δ_1 and Δ_2 for electron exchange of the left film with the left electrode and of the right film with the right electrode, and by the matrix element V_{12} for exchange between the two films. Similarly, there are two energies of reorganization λ_1 and λ_2 for the two electroactive species; in addition, there can be an overlap λ_{12} between the two solvation spheres. We focus on the basic case, in which the overlap between the solvation spheres is negligible, and in which the coupling V_{12} between the two sites is smaller than the coupling Δ_1 and Δ_2. A fuller treatment has been given by Schmickler [41].

Under the conditions given above, the current from the left electrode to the right can be written in the form:

$$i_f = \frac{2\pi}{\hbar} e_0 |V_{12}|^2 \int_{-e_0 V_b}^{0} d\epsilon \, D_1(\epsilon) D_2(\epsilon) \tag{23}$$

where the redox densities of states have been given in eq. (18). Equation (23) is easily interpreted: Only electronic levels between the two Fermi levels contribute; the contribution of each energy level is proportional to the overlap in the density of states of the two participating states and to their coupling. With two redox systems participating, there are a fair number of system parameters, and various shapes of current-potential curves are possible; some particular cases have been discussed in the literature [41]. We restrict our discussion here to the case where the two redox systems are equal, because this is a case which has been investigated both experimentally and theoretically.

3.3.2 Electron Exchange via Two Equal Reactants

The groups of Rampi and Whitesides have conducted numerous experimental investigations into indirect electron exchange between two metals in close contact. Here we consider in greater detail their work with two chemically equal intermediate states, and interpret this in the light of the previously presented model.

The basic system is thus as shown in Fig. 20, where the adsorbates consisted of monolayers of $[Ru(NH_3)_5(NC_5H_4CH_2NHCO(CH_2)_{10}SH][PF_6]$ Compared with the general results of the previous section, there are some modifications. First, since the intermediate states are well separated from the metal, the level widths Δ_1 and Δ_2 are small compared to the thermal broadening, and can hence be neglected in the redox density of states. Further, the experimental data are consistent with the assumption, that the Rucenters are in electronic equilibrium with the metals on which they are attached. If we assume that the current flows from the left to the right electrode, this means that the electrons are transferred from occupied states on the left electrode to empty states on the right electrode. Therefore, $D_1(\epsilon)$ in Eq. (23) must be replaced by the density of occupied, or reduced, levels, which is [29, 33]:

$$D^1_{\text{red}}(\epsilon) = \left(\frac{4\pi}{\lambda k_B T}\right)^{1/2} \exp -\frac{(\lambda + \epsilon - e_0\eta_1)^2}{4\lambda k_B T} \tag{24}$$

where η_1 is the deviation of the potential of the right electrode from the standard equilibrium potential for electron exchange with the adsorbed species. Similarly, $D_1(\epsilon)$ must be replaced by the density of empty, or oxidized, levels, which is:

$$D^2_{\text{ox}}(\epsilon) = \left(\frac{4\pi}{\lambda k_B T}\right)^{1/2} \exp -\frac{(\lambda - \epsilon - e_0\eta_2)^2}{4\lambda k_B T} \tag{25}$$

The current exchanged between the two metals is then proportional over that part of the product that lies between the two Fermi levels:

$$i \propto \int_{E^1_F}^{E^2_F} D_p(\epsilon)\mathrm{d}\epsilon = \int_{E^1_F}^{E^2_F} D^1_{\text{red}}(\epsilon) \times D^2_{\text{ox}}(\epsilon)\, \mathrm{d}\epsilon \tag{26}$$

The situation is depicted in Fig. 21. Two kinds of experiments have been performed: In one type the potential of the left electrode was kept constant, and the bias was increased. In Fig. 21 this corresponds to lowering the Fermi level of the right electrode. Obviously, the current will increase with the bias and tend to a constant value, once the maximum of the product $D_p(\epsilon)$ lies between the two Fermi levels. This has indeed been observed (see Fig. 22).

The second type is more interesting: here the bias is kept constant at a small value, while the electrode potential is varied. This amount to shifting $D_p(\epsilon)$ with respect to the two Fermi levels, so that the current essentially maps out $D_p(\epsilon)$. This makes it possible to obtain the energy of reorganization λ from the current–potential curves. A corresponding plot is shown in Fig. 23, which shows that the experimental data can be fitted with a value of $\lambda = 0.4$ eV, which is of a reasonable order of magnitude for a redox species attached to an adsorbate.

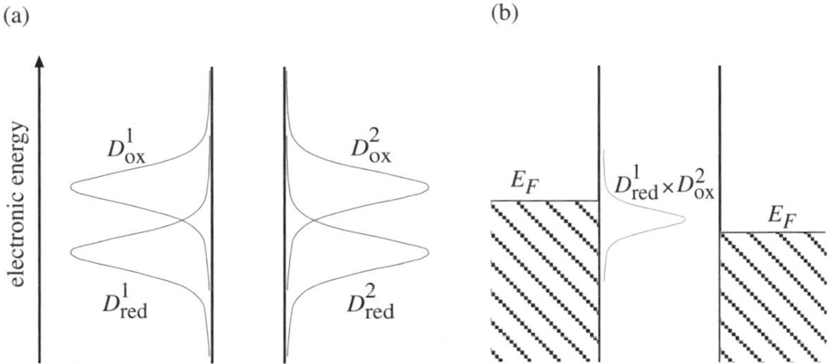

Fig. 21 Indirect electron exchange via two identical species

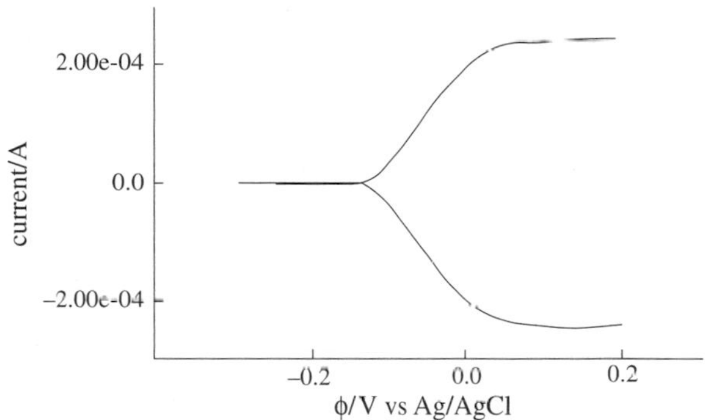

Fig. 22 Currents at both electrodes as a function of the potential of electrode 2; the potential of the first electrode was kept constant at -0.2 V

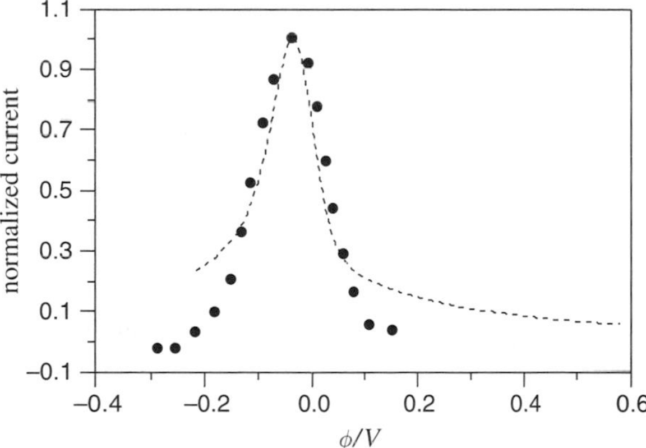

Fig. 23 Comparison between experimental and theoretical data. In the calculations, $\lambda = 0.4$ eV

However, even though this explanation in terms of the theory is consistent, and the experimental data fit well, an explanation in terms of a two-step mechanism, similar to the one discussed in Section 3.2.1, cannot be ruled out at this stage.

4 Conclusion

As mentioned in the introduction, electrochemistry, at least in principle, offers excellent opportunities for the creation of nanostructures. As is evidenced by two recent issues of surface science [42, 43] devoted to this topic, or by the recent work of the Wandlowski group [44], there is ample experimental activity in this area. Of the related theoretical work we have focused here on the two aspects on which most work has been done: the simulation of metal nanostructures and theories of electron transfer through adsorbates. The former work seems quite mature, and has been successful in explaining the properties of several types of structures. In particular, it has explained the unusual stability of some clusters, which had erroneously been attributed to quantum confinement [45], by alloy formation. This is especially important, since electrochemical scanning probe techniques do not give information on the chemical nature of atom that is observed and do not allow one to tell if an observed metal structure is pure or alloyed.

Electron transfer through electrochemical structures is a more speculative topic. The theories are quite advanced, and there are good experiments, but it is difficult to decide, what the electron-transfer mechanism is specific situations is. Often the data can be explained by more than one model.

Besides the topics that we have treated here, there are theories and models for other electrochemical nanostructures such as steps [46, 47] and islands [48] or nanowires [49, 50]. However, we have not included them here because of restrictions of space, and because it would have made this review too inhomogeneous.

Acknowledgments W.S. acknowledges financial support by the Deutsche Forschungsgemeinschaft. E.P.M.L. acknowledges financial support from CONICET, Agencia Córdoba Ciencia, Secyt U.N.C., Program BID 1201/OC-AR PICT No. 06-12485.

References

1. R. A. Marcus, J. Chem. Phys. **24** (1956) 966.
2. N. S. Hush, J. Chem. Phys. **28** (1958) 962.
3. S. M. Foiles, M. I. Baskes, M. S. Daw, Phys. Rev. B **33** (1986) 7983.
4. G.E. Engelmann, J. Ziegler, D.M. Kolb, Surf. Sci. Lett **401** (1998) L420.
5. D.M. Kolb, F.C. Simeone, Electrochim. Acta **50** (2005) 2989.
6. D.M. Kolb, R. Ullmann and J.C. Ziegler, Electrochim. Acta **43** (1998) 2751.
7. U. Landman, W.D. Luedtke, N.A. Burnham, R.J. Colton, Science **248** (1990) 454.
8. M.G. Del Pópolo, E.P.M. Leiva, M.M. Mariscal, W. Schmickler, Surf. Sci. **597** (2005) 133.
9. M.M. Mariscal, C.F. Narambuena , M.G. Del Popolo, E. P. M. Leiva, Nanotechnology **16** (2005) 974.
10. N.B. Luque, E.P.M. Leiva, Electrochimica Acta **50** (2005) 3161.

11. M.G. Del Pópolo, E.P.M. Leiva, H. Kleine, J. Meier, U. Stimming, M. Mariscal, W. Schmickler, Appl. Phys. Lett. **81** (2002) 2635.
12. M. Mariscal, W. Schmickler, J. Electroanal. Chem. **582** (2005) 64.
13. D.M. Kolb, M. Przasnyski, H. Gerischer, J. Electroanal. Chem. **54,** (1974) 25 .
14. G.E. Engelmann, J.C. Ziegler, D.M. Kolb, J. Electrochem. Soc. **45** (1998) L33.
15. R. Ullman, PhD. Thesis, Fakultät für Naturwissenchaften der Universität Ulm (1997).
16. J.C. Ziegler, PhD Thesis, Fakultät für Naturwissenschaften der Universität Ulm (2000).
17. D.M. Kolb, G.E. Engelmann, J.C. Ziegler, Solid State Ionics, **131** (2000) 69.
18. U. Landman, W.D. Luedtke, *Scanning tunnelling microscopy*, Chapter 3, Springer Verlag, (1993).
19. S.M. Foiles, M.I. Baskes, M.S. Daw, Phys. Rev. B 33 (1986) 7983.
20. R. Schuster, V. Kirchner, X.H. Xia, A.M. Bittner, G. Ertl, Phys. Rev. Lett. **80** (1998) 5599.
21. T. Solomun, and W. Kautek, Electrochimica Acta **47** (2001) 679.
22. X.H. Xia, R. Schuster, V. Kirchner and G. Ertl, J. Electroanal. Chem. **461** (1999) 102.
23. C. Sánchez, E.P.M. Leiva, J. Electroanal. Chem **458** (1998) 183.
24. C.E.D. Chidsey, Science **251** (1991) 919.
25. *Molecular Electronics – Science and Technology*, edited by A. Aviram and M. Ratner, Ann. N.Y. Acad. Sci. **852** (1998).
26. S. Roth, C. Joachim, *Atomic and Molecular Wires*, Kluwer, Dordrecht (1997).
27. *Molecular Devices*, edited by F. L. Carter, Marcel Decker, N.Y. (1982).
28. J.F. Smalley, S.W. Feldberg, C.E.D. Chidsey, M.R. Lindford, M.D. Newton, Yi-Ping Liu, J. Phys. Chem **99** (1995) 13149.
29. W. Schmickler, *Interfacial Electrochemistry*, Oxford University Press, New York, 1996.
30. W. Schmickler, Surf. Sci. **295** (1993) 43.
31. N.S. Wingreen, K.J. Jacobsen, J.W. Wilkins, Phys. Rev. Lett. **61** (1988) 1396; Phys. Rev. B **40** (1989) 11834.
32. L.E. Hall, J.R. Reimers, N.S. Hush, K. Silverbrook, J. Chem. Phys. **112** (2000) 1510.
33. H. Gerischer, Z. Phys. Chem. NF **6** (1960) 223.
34. A.N. Kuznetsov, W. Schmickler, Chem. Phys. **282** (2002) 371.
35. M.T.M. Koper, W. Schmickler, *A Unified Model for Electron and Ion Transfer Reactions on Metal Electrodes*, in: Frontiers of Electrochemistry, ed. by J. Lipkowski and P.N. Ross, VCH Publishers (1998).
36. N. Tao, Phys. Rev. Lett. **76** (1996) 4066.
37. W. Schmickler, N. Tao, Electrochim. Acta **42** (1997) 2809.
38. E, Tran, M.A. Rampi, G.M. Whitesides, Angew. Chem. Int. Ed. **43** (2004) 3835.
39. M.L. Chabinyc, X. Chen, R.E. Holmlin, H.O. Jacobs, H. Skulason, C. D. Frisbie, V. Mujica, M.A. Ratner, M. A., Rampi, G.M. Whitesides, J. Am. Chem. Soc. **124** (2002) 11730.
40. C. Grave, E. Tran, P. Samor, G. M. Whitesides, and M. A. Rampi, Synthetic Metals **147** (2004) 11.
41. W. Schmickler, Chem. Phys. **289** (2003) 349.
42. Surface Science **573** (2004).
43. Surface Science **597** (2005).
44. Z. Li, B. Han, L.J. Wan, T. Wandlowski, Langmuir **21** (2005) 6915.
45. D. M. Kolb, G. E. Engelmann, J. C. Ziegler, Angewandte Chemie, Int. Ed. 39 (2000) 1123.
46. H. Ibach, W. Schmickler, Phys. Rev. Lett., **91** (2003) 016106.
47. M. Giesen, H. Ibach, W. Schmickler, Surf. Sci. **573** (2004) 24.
48. M. Giesen Progr. Surf. Sci. **68** 1 (2001).
49. E. Leiva, P. Vélez, C. Sanchez, W. Schmickler, subm. to Phys. Rev. **B74** (2006) 035422.
50. M.G. Del Pópolo, PhD. Thesis, Facultad de Ciencias Químicas de la Universidad Nacional de Córdoba (2002).

SPM Techniques

O.M. Magnussen

1 Introduction

Scanning probe microscopy (SPM) techniques are of central importance to nanoscience and nanotechnology. They allow characterization of the local structure and properties of surfaces and interfaces as well as surface modification on the micro- to subnanometer scale, opening up the possibility to perform physical, chemical, and biological experiments on individual nanoscale objects. Contrary to many surface science and electron microscopy methods, SPM techniques are not restricted to vacuum conditions, but can be performed in gaseous and liquid environment, including electrolyte solutions. They have therefore become a major, indispensable tool for the *in situ* study of electrochemical interfaces and interface processes and play an important role in the development of electrochemical surface- and nanoscience.

All SPMs are based on a common, simple principle (Fig. 1a): the sample surface is approached by a probe, typically a cone-shaped or pyramidal tip with an apex of (sub) micrometer dimensions, which interacts with the sample by a distance-dependent, local interaction. Depending on the SPM technique this interaction is different in nature, for example, electrical, magnetic, mechanical, chemical, and optical. It provides a signal (e.g., an electric signal or a force), which is measured by a sensor, compared with a setpoint value, and used to control the probe position in z direction via a mechanical actuator. Sensor, control electronics, and actuator thus form a negative feedback loop that keeps the probe-sample distance constant. By scanning a rectangular area of the sample line by line and simultaneously recording the mechanical motion of the probe, a three-dimensional (3D) "image" of the surface can be generated. Technically, this 3D motion is usually performed by applying appropriate voltages to piezoelectric actuators that move the probe and sample

O.M. Magnussen (✉)
Institut für Experimentelle und Angewandte Physik, Christian-Albrechts-Universität zu Kiel, 24098 Kiel, Germany
e-mail: magnussen@physik.uni-kiel.de

P. Schmuki, S. Virtanen (eds.), *Electrochemistry at the Nanoscale,* Nanostructure Science and Technology, DOI 10.1007/978-0-387-73582-5_2,

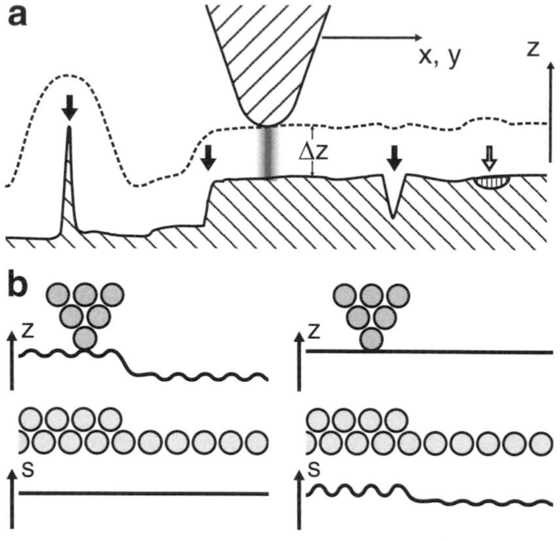

Fig. 1 Schematic illustration of (**a**) the basic principle of SPM measurements and (**b**) the main operating modes, that is, operation at constant current (STM) or constant force (AFM), respectively, (*left*) and operation at constant height (*right*)

relative to each other. To suppress mechanical vibrations and thermally induced drift motion, a major problem in any high-resolution microscopy, compact setups with high mechanical stiffness are used. A detailed treatment of the theoretical underpinnings and experimental realization of the most common scanning probe methods can be found in several textbooks on SPM [1–4].

The microscopic resolution of an SPM technique is determined by the effective size of the interacting areas on tip and sample, which, in turn, depends strongly on the variation of the interaction with probe-sample distance. In the case of very pronounced distance dependence, the measured signal can be dominated by the interaction with the outmost atom of the probe, resulting in true atomic resolution (see below). In general, the generated image is a convolution of probe and sample shape and can substantially deviate from the real sample topography if the surface exhibits features that are sharper than the tip apex (Fig. 1a, black arrows). This can cause characteristic artifacts, such as broadening of steep steps and hillocks, problems in imaging pores and crevices, and multiple imaging of prominent features with probes that end in several tips. For this reason, preparation of sharp probes with well-defined shape and the use of sufficiently smooth sample surfaces are important prerequisites for SPM experiments.

As shown above, an SPM image is a map of local properties rather than merely an accurate reproduction of the surface topography, that is, the positions of the ion cores. This is a major strength of SPM techniques and has been extensively used for the characterization of nanoscale structures. It allows, for example, to distinguish areas of different chemical composition or electronic properties (see Fig. 1a, white arrow), such as in alloys (see Section 2.2), coblock polymers, or standing electron waves as in the famous quantum corral structures [5]. An even more comprehensive sample characterization can be obtained by recording several signals in parallel in an SPM experiment. For example, in scanning near-field optical microscopy (SNOM;

see also Section 4), a force signal is typically used to keep the probe-sample distance constant and to record the sample topography, while simultaneously measuring the local optical properties [2].

SPM is a scanning technique, where the image is generated by sequential recording of data points. Its temporal resolution is therefore limited by the maximum scanning speed, which in turn is determined by the bandwidth of the signal acquisition and the requirements of the control electronics for stable negative feedback operation. In the most common mode of operation (called "constant current" and "constant force" mode in scanning tunneling microscopy (STM) and atomic force microscopy (AFM), respectively) the signal s is kept constant during the scan and the height z of the probe is recorded (Fig. 1b, left panel). This requires the feedback to be sufficiently fast to follow the contours of the surface. Usually, only scanning speeds up to several hundred nanometers (nm) per second are possible in this mode, since the limiting frequency of the feedback loop has to be lower than the lowest mechanical resonance frequency of the SPM (typically several kHz) for stable feedback. Higher image-acquisition rates can be obtained in the so-called "constant height" mode (Fig. 1b, right panel), where the time constant of the feedback loop is much lower than the fluctuations in the signal, caused by the spatial variations in probe-sample distance during scanning, and only the average height of the probe with respect to the surface is kept constant. In this mode the image is generated from the fluctuations in the signal s, which often provides a good lateral contrast, but is more difficult to interpret quantitatively. The main application of "constant height" mode SPM is the study of atomic-scale surface dynamics (see Section 2.3). Image-acquisition rates in the video range and above have been obtained by this method [6–10].

Many SPM techniques can be readily applied *in situ* in electrochemical environment. Obviously, the presence of the electrolyte medium between sample and probe alters their mutual interactions and in turn the SPM signals, as will be discussed below for electrochemical STM and AFM. If electrical signals are measured, the probe has to be treated as an additional electrode (see Section 2.1). For operation in solution usually a vertical setup with the sample placed in a small electrochemical cell (often formed by a Teflon ring pressed on the sample surface) and the probe entering from the top is used. Special attention has to be paid to the employed probes, which have to be stable in this environment against modification or corrosion by chemical reactions (see below). Furthermore, various other technical problems caused by the presence of the electrolyte have to be addressed, for example, the protection of piezo elements and other electronic components against condensation of moisture, which can cause unwanted voltage fluctuations.

The following chapter focuses on STM and AFM, the most important varieties of scanning probe microscopy. In addition, some other SPM techniques relevant to electrochemistry will be briefly treated, followed by a discussion of nanostructuring methods based on electrochemical SPM. More detailed overviews on the application of SPM techniques for *in situ* electrochemical studies are given in several previous reviews [11–19].

2 Electrochemical STM

2.1 Principle of Operation and Experimental Considerations

STM is the oldest SPM technique and still offers the highest spatial resolution of all SPMs. After its invention by Binnig and Rohrer [20], it was first operated under ultrahigh vacuum (UHV) conditions, but applied to studies in liquid environment only a few years later [21]. In STM, a conductive sample is scanned with a very sharp metallic tip (usually an etched W or PtIr wire) kept at nm distances. Due to quantum mechanical tunneling of electrons a tunnel current I_t flows between the tip and sample, if a voltage U_{bias} is applied. The high resolution of STM stems from the steep exponential decay of I_t with increasing tip-sample distance Δz, which in the most simple 1D model is (for small U_{bias}) approximately:

$$I_t \propto U_{bias} \cdot \exp(-A \cdot \sqrt{\bar{\Phi}} \cdot \Delta z)$$

where $A \equiv 2 \cdot \sqrt{2m_e}/\hbar \approx 1\,eV^{-1/2}\text{Å}^{-1}$ and $\bar{\Phi}$ is the effective height of the energy barrier for the tunneling electrons. Typical distances Δz are 5–20 Å, resulting in tunnel currents in the pA to nA range. Since the effective barrier height in the tunnel regime is typically several eV, a change of Δz by 1 Å changes I_t by an order of magnitude. Hence, the tunnel current can be effectively restricted to the apex of atomically sharp protrusions ("microtips") at the very end of the STM tip, even if the overall radius of tip curvature is much larger (typically more than several hundred nm). This is the basis of the capability of STM to resolve individual atoms at the surfaces of metallic or semiconducting substrates. Meso-scale STM images recorded in constant current mode reflect the substrate topography, whereas atomic-resolution imaging requires a more precise treatment (see Section 2.2).

The tunnel barrier can be formed by a vacuum, gas, or (nonmetallic) liquid phase, that is, also an electrolyte solution. However, in the latter case, the STM tip has to be considered as an additional electrode, as illustrated schematically in Fig. 2. This results (i) in faradayic currents $I_{f,tip}$ due to electrochemical reactions at the tip surface and (ii) a contribution to the tip capacity C_t caused by the electrochemical

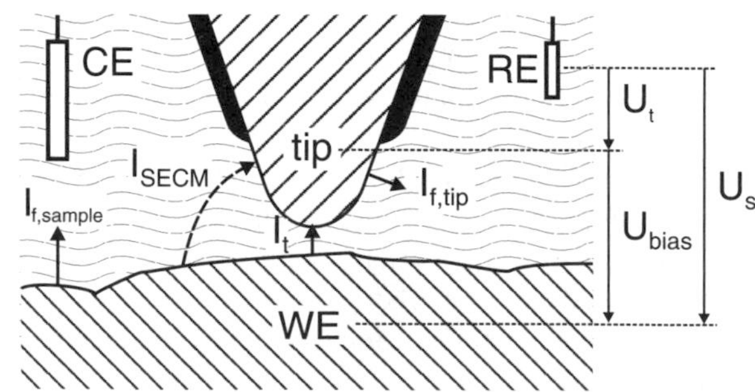

Fig. 2 Schematic illustration of the electric configuration of *in situ* STM, showing the tunnel tip and working (WE), reference (RE), and counter (CE) electrode, as well as relevant voltages and currents (see text). In addition, the tip coating is indicated in black

double layer of the tip, which both interfere with the STM measurement: First, electrochemical currents at the tip have to be largely suppressed, since only the sum of tunnel and electrochemical current $I_t + I_{f,tip}$ can be measured. Stable STM operation is only possible if the tunnel current is the dominant contribution, that is, $I_t > I_{f,tip}$. Second, the noise in the high-gain current–voltage converter used to measure the tip current increases with increasing input impedance. For an optimized signal-to-noise ratio, C_t should therefore be minimized, in particular for studies at low tunnel current or high recording speeds. Both $I_{f,tip}$ and C_t can be drastically reduced by coating the tip with an electric insulator, leaving only a μm-sized area at the tip apex uncovered. The coating material should exhibit a high chemical stability in the electrolyte solution, to avoid contamination of the sample surface. In practice, Apiezon wax, polymer coatings, or electrophoretic paints are the most commonly employed, allowing routine preparation of tips, where $I_{f,tip}$ is reduce below 10 pA and C_t to a few pF [22, 23].

In addition, $I_{f,tip}$ can be lowered further by keeping the tip potential in the double-layer regime. This also prohibits adverse structural changes in the tip during the measurements due to electrochemical reactions, such as oxidation, dissolution, or deposition. It requires that not only the potential U_s of the sample (working electrode), but also the tip potential U_t is controlled with respect to a reference electrode, which is accomplished by a bipotentiostatic electronic setup with a design adapted to the rather different current ranges of tip (pA to nA) and sample (typically $\geq$μA). During changes of the sample potential U_s, which is common in *in situ* electrochemical studies, either the tip bias U_{bias} or the tip potential U_t can be kept constant. The first mode ensures constant tunnel conditions but may require limiting the accessible range of U_s to maintain U_t in the double-layer regime of the tip. In the second mode, the electrochemical conditions at the tip, and consequently $I_{f,tip}$. are independent of U_s. The variation of U_{bias} in this mode usually does not have a strong effect on the STM imaging of metal substrates, where the dependence on U_{bias} is weak. However, for *in situ* STM studies of semiconductors, where both tunneling and electrochemical conditions have to be considered, the choice of U_t may be difficult and potentials that allow stable imaging cannot always be found [24, 25].

The electrochemical reactions at tip and sample can also be coupled, if a species generated at one of the electrodes partakes in a reaction at the other electrode (Fig. 2, dashed arrow). This is the basis of scanning electrochemical microscopy (SECM, see Section 4) as well as of some SPM-based electrochemical nanostructuring methods. Due to the strong exponential distance dependence of I_t, this current (I_{SECM}) becomes dominant if the tip is withdrawn by a few nm only. This allows combined experiments where the electrode structure is characterized microscopically by STM imaging, and the same tip is used subsequently in an SECM mode to measure the local electrochemical reactivity [26, 27]. Inversely, the presence of the tip also restricts mass transport to the surface, resulting in reduced reaction rates (e.g., considerably lower deposition rates, see below) in the surface area imaged by the scanning tip. Furthermore, due to the small tip-sample distance, the local electrochemical behavior of the sample may be distorted by the presence of the tip.

While such tip effects are unwanted in *in situ* STM studies of the electrode surface structure, they are of great importance for *in situ* nanostructuring and will be discussed in more detail in Section 5.

2.2 High-Resolution In Situ Studies of Electrode Surface Structure

STM owns its popularity in no small part to the spectacular atomic-scale images of a wide variety of surface and adsorbate structures obtained by it in the last 25 years. This capability has been of tremendous importance for studies of electrochemical interfaces, where most traditional surface-science methods (e.g., electron diffraction) cannot be employed. The demonstration of *in situ* atomic-resolution imaging of graphite [21, 28] and metal electrodes [29, 30] and the following extensive applications of this technique [14, 15] promoted significantly the development of electrochemical surface science. Up to now, STM is the most widely used *in situ* method for the atomic-resolution characterization of the electrode surface structure.

The examples presented in Fig. 3 illustrate the capability of *in situ* STM for true atomic resolution. As visible in Fig. 3a, where a reconstructed Au(111) electrode surface is shown, the atomic lattice at the surface is fully resolved, including the small lateral and vertical modulation in the position of the Au surface atoms caused by the surface reconstruction. The variation of the positions within the unit cell of the surface reconstruction (indicated by rectangular box) is only $\approx$0.2 Å in vertical direction and $\approx$0.5 Å within the surface [31]. Similar images have been obtained on various single-crystal surfaces of the coinage metals and of many transition metals (e.g., Pt, Ir, Rh, Pd, and Ni) as well as on some semiconductor surfaces [14, 15]. As an example of the latter, an image of hydrogen-terminated Si(111) in NaOH solution is shown in Fig. 3b [25]. In this image, not only the (1×1) Si–H lattice, but also individual Si–OH are discernible (manifesting as protrusions, surrounded by an apparent depression), demonstrating clearly that individual surface atoms can be resolved by *in situ* STM.

To understand and correctly interpret such images the imaging mechanism of STM has to be considered in more detail. As illustrated in the energy diagram in Fig. 4, in the case of a negatively biased sample, electrons tunnel elastically through the barrier from an occupied electronic state in the sample to an empty state in the tip. Inversely, electrons tunnel from occupied tip states into empty sample states for positive sample bias. Although all electrons in the energy window defined by U_{bias} participate, the strongest contribution comes from electrons near the Fermi level ε_{F}, for which the tunnel barrier is lowest. According to the standard model of the STM by Tersoff and Hamann [32], where the tunnel current is calculated by perturbation theory for a spherical tip with s-type wave functions, the tip follows in constant current mode lines of constant local density of states (LDOS; $D_s(\mathbf{r}, \varepsilon_F)$) of the sample at the Fermi level. On metal surfaces, the contours of $D_s(\mathbf{r}, \varepsilon_F)$ resemble those of the total electron density, so that images in this simplified interpretation depict the metal ion cores (i.e., the topography). This model seems to provide a reliable description of STM images, if the feature size of the imaged surface structures is

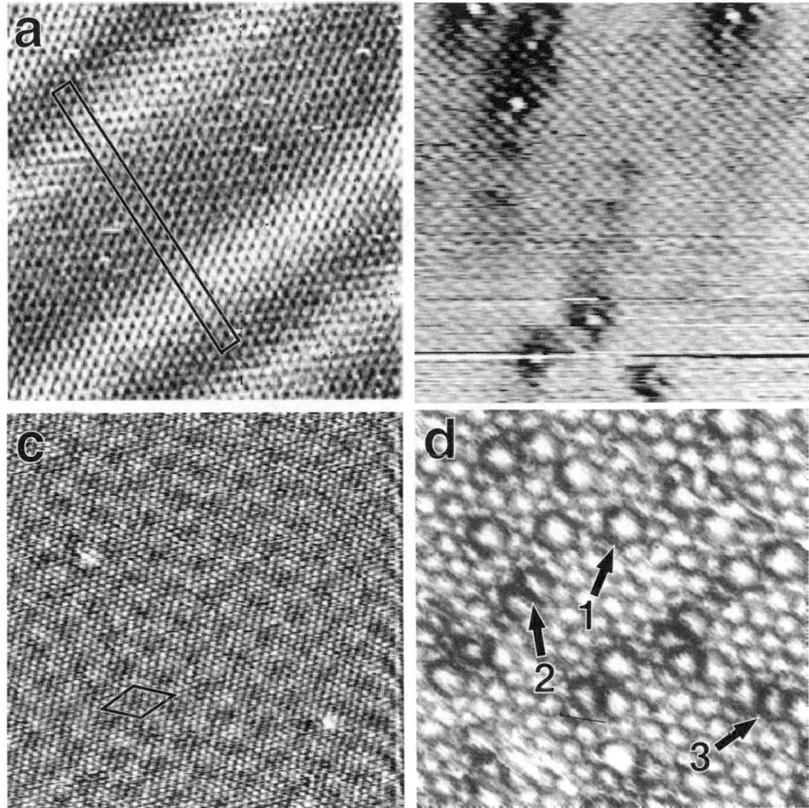

Fig. 3 Atomic-resolution *in situ* STM images of electrode surfaces. (**a**) reconstructed Au(111) surface in 0.1 M H_2SO_4 solution, exhibiting two pairs of surface-dislocation lines ($10 \times 10\,nm^2$, from ref. [192]). (**b**) Hydrogen-terminated Si(111) in 2 M NaOH ($9 \times 9\,nm^2$. Reprinted from ref. [25] with permission by the author. Copyright (1996) by American Physical Society). (**c**) Pb monolayer on Ag(111) in Pb-containing 0.01 M $HClO_4$ solution, exhibiting a *moiré* pattern ($18 \times 18\,nm^2$. Reprinted from ref. [38] with permission by the author). (**d**) $Au_{93}Pd_{07}(111)$ alloy surface in 0.1 M H_2SO_4 solution, showing Pd monomers, dimers, and trimers (marked by arrows 1, 2, and 3, respectively) which appear more prominent in the image than the Au surface atoms ($5.5 \times 5.5\,nm^2$) [F. Maroun, F. Ozanam, O.M. Magnussen, R.J. Behm, unpublished]. In (**a**) and (**d**) the superstructure unit cells are indicated

well above the spatial extension of the electron states of the tip (e.g., for surface reconstructions) and if no substantial chemical interactions between tip and sample occur [4, 33]. Later this approach was generalized by considering p- and d-state tip orbitals [3]. More advanced models consider tip and sample as an interacting system and describing the tunnel event as a scattering process [4, 33, 34]. These latter, numerically very demanding approaches, which include multiple pathways of the tunneling electrons and also may consider inelastic effects, provide the most accurate description of the tunneling junction to date [4, 33].

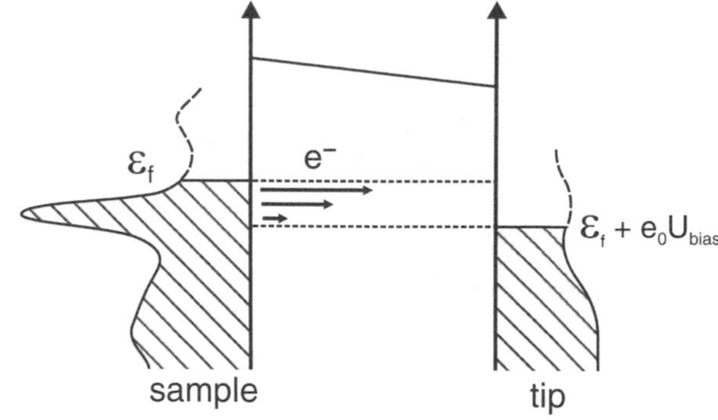

Fig. 4 Energy diagram of the tunnel contact. The local densities of states at tip and sample surfaces are indicated perpendicular to the axis indicating the energy

Resolution of the atomic lattice of close-packed metal surfaces, such as in Fig. 3a, usually requires gap resistances $\leq 10^7\,\Omega$, that is, very small tip-sample distances. The Tersoff–Hamann model is not expected to be applicable under these conditions and indeed cannot explain the observed high atomic corrugations. Modern calculations indicate that at distances <5 Å, the mutual chemical interactions result in significant deformations of the atoms in tip and sample surface, which contribute significantly to the corrugations [4, 33, 35]. Upon approaching the tip only 0.5 Å closer to the surface, the tunnel junction mechanically collapses ("jump to contact"), resulting in the formation of an electric point contact. The small range of distances (and consequently of I_t) where such dynamic enhancement of the corrugation occurs may explain why stable atomic resolution of bare metal surfaces is difficult to achieve. These considerations should also hold for high-resolution imaging of the metal-surface lattice in electrochemical environment, since the electrolyte is largely excluded from the tunnel gap at such small tip-sample distances and at most can contain one water layer (see Section 2.5). This is supported by the experimentally observed corrugations in *in situ* studies, which do not differ noticeable from those under UHV conditions. However, atomic resolution is generally less reproducibly achieved *in situ* than in UHV–STM experiments, where various standard methods for preparing a defined tip apex exist (e.g., field evaporation, sputtering, and high field treatment; see ref. [3]). Although electrochemical methods (e.g., electrodeposition on the tip) or controlled mechanical collisions of tip and sample can improve the *in situ* imaging, the development of well-defined procedures for tip sharpening and cleaning in electrolyte solutions which allow the preparation of well-defined tip states capable of atomic resolution still remains a challenge.

The possibility to *in situ* resolve close-packed metal lattices has been particular important for elucidating the structure of electrodes surface, especially of surface reconstructions [36], and ultrathin metal layers, electrodeposited at over- or underpotentials (see below) [37]. In both cases the interatomic distances or even the in-plane symmetry of the metal surface layer(s) often deviate from that of the underlying substrate. This results in an additional long-range periodic structure, where surface atoms occupy different sites of the bulk lattice. The corresponding vertical and lateral buckling of the surface lattice are directly visible in the STM images. Examples are uniaxial or 2D dislocation network structures, such as the Au(111) surface reconstruction (Fig. 3a), where these deviations from the bulk lattice parameters are restricted to domain walls in the surface layer (brighter stripes in Fig. 3a), or *moiré* pattern that exhibit a more uniform lattice mismatch, as illustrated in Fig. 3c for a Pb monolayer on Ag(111) [38, 39]. By analyzing the surface atom positions with respect to the superstructure lattice, the interatomic distances and angles can be obtained with significantly increased precision (up to ≈0.01 Å) [39–41].

A special case is high-resolution imaging of alloy surfaces, where STM has been shown to be capable of distinguishing chemically different surface atoms ("chemical contrast") [42]. As demonstrated by Maroun et al. for AuPd(111) electrode surfaces, this is also possible in *in situ* STM studies, which allows to characterize the structure and distribution of different sites on these heterogeneous surfaces (Fig. 3d) [43]. Imaging with chemical contrast may be a valuable tool to clarify the local reactivity of heterogeneous surfaces in electrocatalysis and corrosion science, where alloy electrodes are of great importance. In the study by Maroun et al., the chemical-contrast STM images of the bimetallic AuPd model catalyst revealed at low Pd content a significantly enhanced surface concentration of Pd monomers (arrow 1) as compared to random distributions, whereas the concentrations of Pd dimers (arrow 2) and trimers (arrow 3) were lowered. By correlation of these structural data with electrochemical and spectroscopic results, critical ensemble sizes for the adsorption of hydrogen and CO, important species in fuel cell reactions, could be identified [43].

One of the earliest and most important applications of high-resolution *in situ* STM has been the investigation of adsorbates on electrode surfaces. In fact, observations of ordered iodine adlayers on Pt(111) [29] and underpotential deposited (UPD) Cu on Au(111) [30] were the first examples of *in situ* atomic-resolution imaging on metal electrodes. Since then a large number of systems have been studied, including metal adlayers [14, 15, 37], anion adlayers [14, 15, 44], and various molecular adlayers, ranging from simple inorganic molecules (e.g., CO and CN) to large supramolecular assemblies [14, 15]. In particular, STM has significantly helped to elucidate the lateral ordering within these adlayers, revealing an unexpectedly complex 2D-phase behavior. Simple or high-order commensurate structures as well as uniaxial and 2D incommensurate structures with potential-dependent lattice parameters were observed in these studies. A detailed overview on the structural data obtained by STM as well as other techniques can be found elsewhere [14, 15, 37, 44, 45].

STM images of pure metal adlayer such as Pb on Ag(111) (Fig. 3c), which are strongly electronically coupled to the substrate, can be interpreted in a similar way as images of the metal substrate (see above). However, the interpretation of atomic-scale images of nonmetallic adsorbates is not straightforward. As shown by UHV–STM studies and corresponding detailed calculations, maxima in the STM images, especially in images of molecules, not necessarily are located at the positions of atoms and the observed heights of the maxima are not at all related to the adsorbate geometry (and may substantially depend on the tunneling conditions) [34, 46].

Low-temperature STM studies of isolated adsorbates on metal surfaces revealed that even adsorbates that would seem to have no electronic states in the vicinity of the Fermi level, such as noble gases or water, can provide pronounced contrast. In fact, contributions to the STM images come from the adsorbate as well as the underlying substrate, both perturbed by their mutual interaction [34, 46]. While on substrates that interact weakly with the adsorbate (e.g., graphite) the contribution of the substrate can be neglected, on metal substrates the shift of surface levels by the presence of the adsorbate decreases the contribution of the substrate to I_t, corresponding to a depression in the STM image. The total shape of the adsorbate is given by the interference of the substrate contribution with the direct contribution of the adsorbate, which would create a bump. In the most common case, the adsorbate has no electronic states close to the Fermi energy and the contributions of the adsorbate result from the broad tails of (usually several) molecular orbital (MO) resonances. For simple atomic adsorbates the apparent height was found to correlate with the atomic polarizability or spatial extension of the dominant orbital, respectively, whereas the orbital energy seems to be less important [34]. STM images of more complex adsorbates, for example, (poly)aromatic molecules, have often been interpreted in terms of the molecules HOMO and LUMO. This interpretation seems too simplistic in the light of modern calculations, which indicate contributions of a large number of MOs to the image and a dependence of the observed shape of the molecule on the adsorption site [4, 34]. In general, only a small influence of tunnel bias on the images is expected for the range of U_{bias} accessible in *in situ* measurements in electrochemical environment. However, if electronic states of the adsorbate are located in the energy window defined by U_{bias} strong resonant tunneling through these states may occur, resulting in pronounced contrast (see Section 2.5). For close-packed adlayers additional complications arise. Here the tip may interact simultaneously with several neighboring adsorbates, resulting in electronic interferences that may strongly affect the image. For this reason, images of dense adlayers depend more strongly on the atomic-scale tip shape than on images of isolated molecules. Due to these ambiguities in the interpretation of STM images, unambiguous clarification of the complex structures at electrochemical interfaces is often not possible by STM alone, but requires complementary techniques.

As a characteristic example, *in situ* STM images of an ordered "($\sqrt{7} \times \sqrt{3}$)" sulfate adlayer on Au(111) are presented in Fig. 5 [47]. This type of adlayer structure exists on many (111) surfaces of fcc metals at potentials positive of a critical value, where it is formed in a sharp first-order phase transition [44]. In the images a prominent maximum, corresponding to the sulfate species, and a second, weaker maximum, which is nowadays attributed to a water species bound via hydrogen bonds, is found per unit cell (see Fig. 5a). The images of the sulfate adlayer depend strongly on the tunnel resistance. At low resistance, that is, small tip-sample distance, the adlayer becomes even largely transparent, and the STM images the underlying substrate lattice (Fig. 5b). This shift in image contrast is gradual, reversible, and has been observed for a number of anionic [29, 47–50] and organic [51] adsorbate systems. It can be employed to determine the positions of the maxima in the superstructure relative to the substrate lattice, allowing clarification of the adsorption site in

Fig. 5 High-resolution *in situ* STM images of Au(111) in H_2SO_4 solution. (**a**) Characteristic image of the ordered sulfate adlayer (3.6×4.4 nm^2). (**b**) Effect of tip-sample distance on STM image, illustrated by stepping the tunnel resistance twice from 15 MΩ (*top, center, bottom*) to 2.5 MΩ and back (10 × 8.5 nm^2). (**c**) Change in STM image upon stepping the potential from 0.68 V_{SCE} (*upper part*), where the substrate lattice is observed, to the potential regime of the ordered adlayer phase (*lower part*) at 0.69 V_{SCE} (21 × 12 nm^2, after ref. [47]. Reproduced by permission of The Royal Society of Chemistry)

simple adlayer systems [49]. The same is possible by stepping the potential across the phase transition (Fig. 5c). Since the sulfate adsorbate is disordered and highly mobile in the potential range negative of the ordered phase, only the periodicity of the substrate lattice remains visible in the STM image. The position of the adlattice relative to the substrate atoms can then be obtained by extrapolating the substrate lattice into the range of the image where the superstructure is observed. This and other "potentiodynamic" STM experiments utilize the fact that the slow scanning direction (by convention chosen along the vertical direction in the images) also provides a time axis.

Modern studies utilize *in situ* STM to characterize the structure of complex supramolecular assemblies on electrode surfaces with submolecular resolution. Two recent examples of this rapidly progressing field are presented in Fig. 6. The first illustrates lateral organization of molecular species for a bimolecular adsorbate system, a mixed monolayer of a Co phthalocyanine and a Cu tetraphenyl-porphyrin on reconstructed Au(100) in $HClO_4$ solution (Fig. 6a) [52]. The STM images reveal ordering in alternating rows of clearly distinct molecules, which can be attributed to the phthalocyanine (I) and the porphyrin (II) species. An example for vertical as well as lateral organization is shown in Fig. 6b [53, 54]. Here a $c(2 \times 2)$ Cl adlayer on a Cu(100) surface provides a template for the ordered adsorption of supramolecular squares (as well as co-adsorbed chainlike oligomers of these molecular units, also visible in the image), formed from bipyridine and Pt ions. Specifically, the squares are oriented parallel to the underlying anion adlattice and can, under optimized preparation conditions, exhibit long-range order. Studies of this type can play an important role in bottom-up approaches for nanostructure formation. Additional control over the molecular organization can in many cases be gained by the applied potential (see below), another advantage of the electrochemical environment. The adlayers shown in Fig. 6 should only be considered as representative examples of the large number of organic adlayer systems that have been studied by *in situ* STM, which even include biomolecular adsorbates such as functional, self-assembled proteins [55]. The structural characterization of functionalized electrodes afforded by STM greatly assists the interpretation of the electrochemical response of these complex systems.

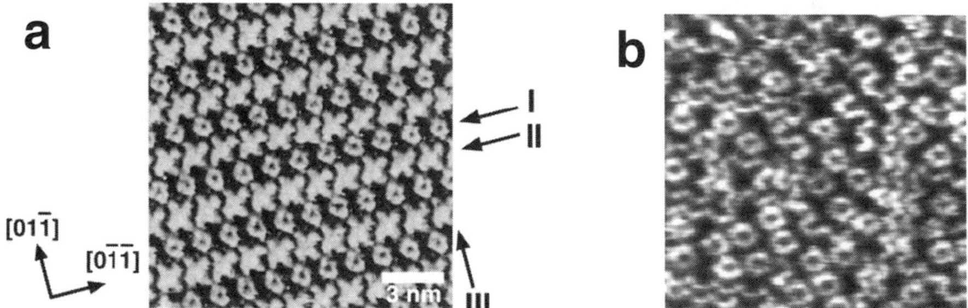

Fig. 6 High-resolution *in situ* STM images of supramolecular adsorbate structures. (**a**) Mixed Co phthalocyanine/Cu tetraphenyl-porphyrin adlayer on reconstructed Au(100) in 0.1 M $HClO_4$ solution ($15 \times 15\,nm^2$. Reprinted from ref. [52] with permission by the author. Copyright by American Chemical Society). (**b**) Square supramolecular aggregates of bipyridine and Pt ions on a $c(2 \times 2)$ Cl adlayer covered Cu(100) surface in HCl solution ($20.5 \times 20.5\,nm^2$. Reprinted from ref. [53] with permission by the author. Copyright by Wiley-VCH)

2.3 *Studies of Phase Transitions and Transport on Electrode Surfaces*

STM not only can image static surface structures, but is also a valuable tool to study surface-dynamic phenomena. Electrochemical interfaces, where phenomena such as adsorption/desorption, surface phase transitions, and surface transport processes can be controlled in a very easy way by the potential, are a particularly rich field for this type of studies. The accessible temporal resolution depends on the image-acquisition rates, which for most STMs usually do not exceed some 10 s. Most *in situ* STM studies therefore focus on structural changes on the timescale of minutes.

An early application of time-resolved *in situ* STM studies was the investigation of the potential-induced reconstruction of Au single-crystal electrodes, a well-studied phenomenon in electrochemical surface science [36]. According to studies by variety of *ex situ* and *in situ* methods, the Au(111), Au(100), and Au(110) surface reconstructions, found on those surfaces under UHV conditions, are only stable negative of a critical potential that depends on crystallographic orientation and electrolyte composition, whereas at more positive potentials a reversible phase transition to the unreconstructed (1×1) surfaces occurs. *In situ* STM studies have contributed significantly to the understanding of the complex mechanisms of the transitions between reconstructed and unreconstructed surfaces [9, 36, 56–62]. In particular, they have revealed the nucleation and growth mechanisms involved in the formation of the (1×1) or the reconstructed phase, respectively, and showed the pronounced influence of long-range mass transport on the kinetics of these transitions. A detailed description of these studies can be found in Ref. [36].

A more recent example of potential-induced structural phase transitions can be found on Pt(111) electrodes covered with electrodeposited mono- and bilayer Au films (Fig. 7) [63]. These films are pseudomorphic to the underlying Pt substrate lattice (i.e., significantly compressed with respect to the Au bulk lattice) at potentials $<0.35\ V_{Ag/AgCl}$ (Fig. 7a), but undergo a sequence of phase transitions to dislocation network structures with uniaxial and two-dimensionally expanded lattices. As in the case of the Au(111) surface reconstruction (see Fig. 3a), the dislocation lines can be directly observed in the STM images (Fig. 7a and b, dark lines). Specifically, the structural transitions start with the formation of a unidirectionally relaxed lattice ("striped phase") in the Au bilayer islands (marked as (ii) in Fig. 7a), manifesting as a pattern of dislocation lines (Fig. 7b). With increasing potential, first the formation of a similar "striped phase" in the Au monolayer (marked as (i) in Fig. 7a) and then a transition from the "striped" into a "hex" dislocation network phase in the Au bilayer is found. In all these phases the dislocation line spacing increases continuously with potential, that is, the Au surface density gradually decreases (as also revealed by changes in the island morphology), approaching values close to that in bulk Au(111) planes at the most positive potentials. Upon decreasing the potential, the Au layers are compressed again and a transition back into the pseudomorphic (1×1) phase occurs (Fig. 7d). This complex phase behavior can be explained by the reduction of

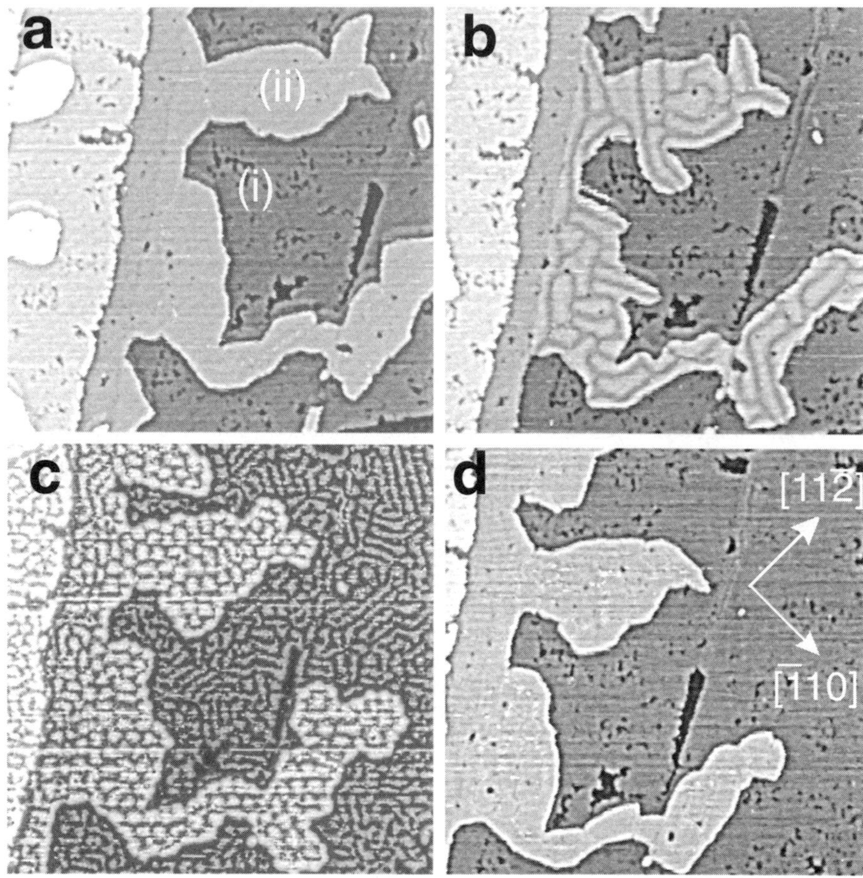

Fig. 7 Selected *in situ* STM images of a series recorded on a 1.3 ML Au deposit on Pt(111) in 0.1 M H_2SO_4 solution in the same surface area during a full potential cycle between 0.3 and 0.85 $V_{Ag/AgCl}$. (**a**) Initial pseudomorphic Au film at 0.3 V, Au monolayer, and bilayer islands are indicated by (i) and (ii), respectively. (**b, c**) Au film with dislocation network structures (*dark lines*) at (**b**) 0.6 V and (**c**) 0.85 V. (**d**) Pseudomorphic Au film after returning the potential to 0.30 V (20.5 × 20.5 nm^2. Reprinted from ref. [63]. Copyright (2003) by American Physical Society)

the in-plane stress in the Au surface layer(s) due to the adsorption of anions and the minimization of strain energy in the Au film. The possibility to generate and control a wide range of dislocation networks by potential is of interest for the formation of ordered-metal nanostructures, for which such networks may serve as templates (see Section 2.4).

Likewise, STM has been used to study phase transitions in adsorbate layers on electrode surfaces. Although these transitions often occur on timescales of milliseconds, that is, too rapidly for conventional STM studies (see, for example, Fig. 5c), details on the mechanisms of the transition can be observed in selected systems. An example is bipyridine on Au(111) electrodes, which forms various ordered phases

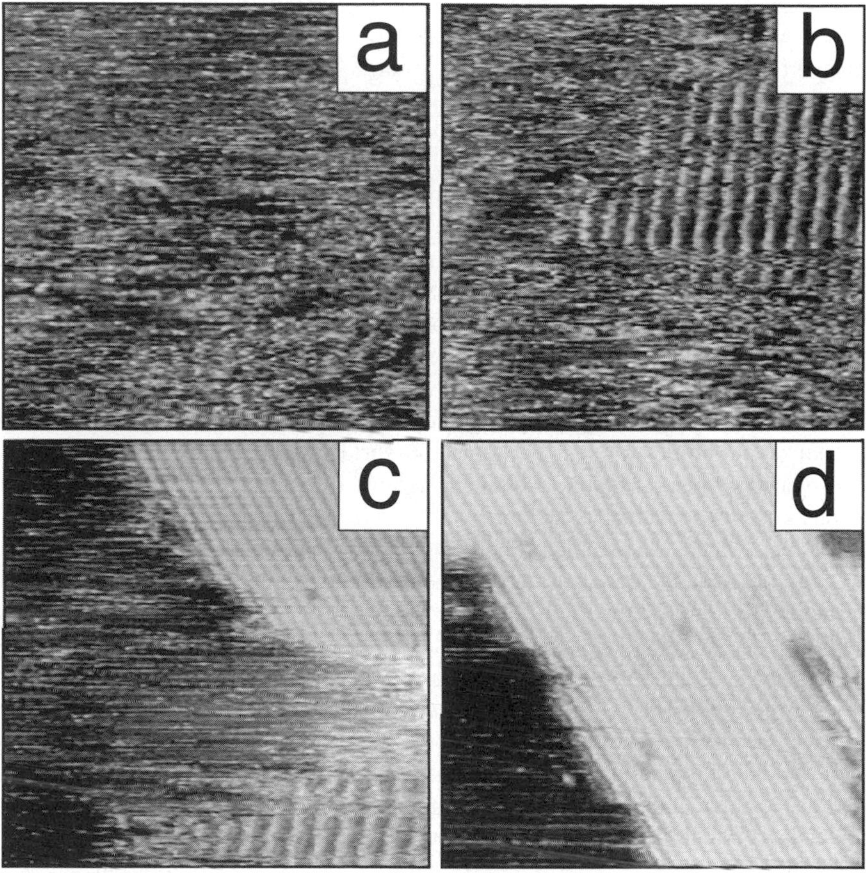

Fig. 8 Selected *in situ* STM images of a series showing the potential-induced formation of the ordered bipyridine phase on Au(111) in 0.1 M H_2SO_4 solution. The images were recorded (**a**) at 0.10 and (**b–d**) at 0.5 V_{SCE} (3×35 nm^2. Reprinted from ref. [64] with permission by the author. Copyright by Elsevier Science)

composed of stacked molecular chains at positive-charge densities. Potential- and temperature-induced transitions in this adsorbate system could be followed directly by *in situ* STM in a study by Dretschkow et al. [64]. In Fig. 8 selected STM images from a series recorded during the nucleation and growth of the ordered bipyridine phase after an increase in potential from 0.10 (Fig. 8a) to 0.5 V_{SCE} (Fig. 8b–d) are shown, indicating an anisotropic growth along the stacking direction and annealing of defects within the ordered phase. Furthermore, a metastable, low-density adlayer with a different chain direction is found prior to the formation of the stable bipyridine phase (Fig. 8b,c). These microscopic observations provide important details on the nanoscale mechanisms of the structural transitions in the adlayer and are a central prerequisite for a realistic modeling of its kinetic behavior.

A second major topic of *in situ* STM studies, related to the surface dynamics of electrochemical interfaces in the double-layer regime, is the investigation of mass transport on electrode surfaces. In early studies, morphological changes, such as the decay of adatom or vacancy islands, were qualitatively studied, revealing a pronounced dependence of transport rates on electrolyte composition and potential [65, 66]. If the transport is limited by surface diffusion the decay rates depend in general on the local environment. Nevertheless, quantitative measurements are possible for special geometric configurations and have been reported for electrochemical systems [67]. A comprehensive, quantitative analysis of the diffusion of adatoms of the electrode metal on the electrode surface has been performed for single crystalline Au, Ag, and Cu in elegant STM studies by Giesen and co-workers as well as other groups, employing two different approaches (see Ref. [68] for a detailed overview). The first method is based on STM observations of monolayer islands (or, alternatively, vacancy islands). By analyzing the island shape, the free energies of steps and kinks can be determined. Furthermore, surface-transport mechanisms can be investigated by measuring the Brownian motion or decay of such islands, observed in long series of sequential STM images. In the second method, fluctuations of steps on the metal surface are studied. These fluctuations manifest as a frizzy appearance of the steps in the STM images and can be attributed to the movement of kinks due to rapid emission, capture, and step edge diffusion of metal adatoms. Quantitative studies employ a one-dimensional (1D) scan mode rather than 2D imaging, where the tip scans continuously along a line across the step(s), resulting in an improved time resolution. Such 1D scanning ("x–t scans") is a frequently applied trick to increase the time resolution on cost of the spatial information and has also been used to study adsorption and metal-deposition processes [37, 69]. From x–t scans of the step fluctuations one can obtain time correlation functions $F(t)$ that exhibit power laws which are characteristic for the dominant mass-transport mechanism. For example, studies of the close-packed steps of Ag(111) in sulfuric acid solutions indicated that the step fluctuations diverge close to the potential regime of Ag dissolution (Fig. 9) [70]. Based on the scaling behavior of $F(t)$, diffusion along the step edge was suggested as the dominant mechanism at negative potentials, whereas close to dissolution rapid adatom exchange between steps and terraces as well as between terraces and the electrolyte seems to prevail. Furthermore, the corresponding activation energies may be evaluated by temperature-dependent measurements.

Direct observations with high time resolution are possible using specialized STMs, capable of image-acquisition rates in the video-frequency range or higher. Crucial elements of such Video-STMs are a very stable mechanical setup and a large bandwidth of the preamplifier for the tunnel current ($\geq$500 kHz). Video-STMs not only allow studies of faster processes, which is particularly important in liquid environment where the available temperature range is limited to a regime, where the thermal motion of most adsorbates is high, but also to obtain large data sets in a few minutes, allowing statistically relevant, quantitative studies. As an example, Video-STM studies of adsorbate diffusion on electrode surfaces – an important, but only marginally investigated elementary step in many electrochemical reactions – are illustrated in Fig. 10 for adsorbed sulfide on Cu(100) in HCl solution [10]. In

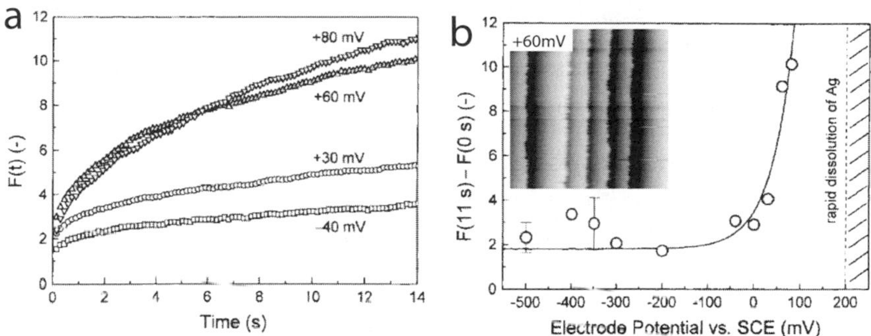

Fig. 9 (**a**) Time correlation functions $F(t)$ at several potentials, obtained from 1D *in situ* STM scans on Ag(111) in 0.1 M H_2SO_4 solution, and (**b**) resulting potential dependence of $F(t)$. An example of a 1D scan across five parallel steps, where only the x-direction is a spatial axis (50 nm) and the y-direction represents a time axis (60 s), is shown in the inset (Reprinted from ref. [70] with permission by the author. Copyright by Elsevier Science)

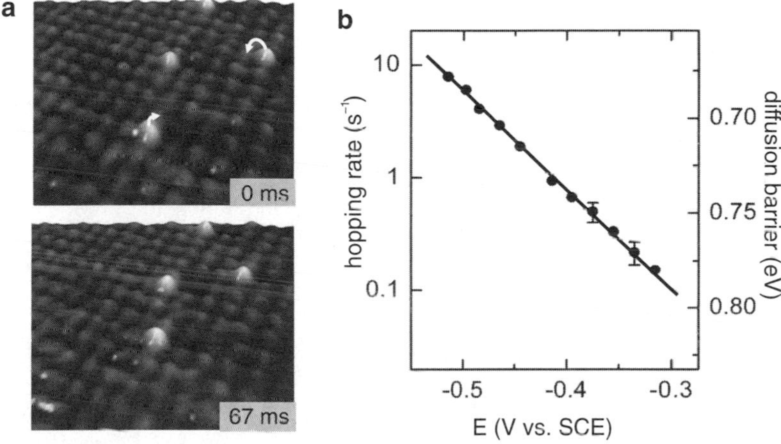

Fig. 10 *In situ* Video-STM studies of the diffusion of adsorbed sulfur on Cu(100) in 0.01 M HCl solution. (**a**) two successive images from a video sequence recorded at $-0.32\ V_{SCE}$ at 15 images/s, showing S_{ad} hopping diffusion. (**b**) Potential dependence of sulfur hopping rates at 296 K and corresponding diffusion barrier, obtained from a statistical analysis of video sequences (Reprinted from ref. [10]. Copyright (2006) by American Physical Society)

this system, isolated S_{ad} occupy positions of the $c(2 \times 2)$ Cl adlattice, replacing one of the Cl_{ad} species. The S_{ad} hopping diffusion to neighboring sites in the Cl_{ad} lattice is directly observed in the recorded video sequences (Fig. 10a, indicated by arrows). From a detailed statistical analysis of the S_{ad} positions in the frames of a video sequence, the probability $p(d)$ that S_{ad} is displaced by a distance d (in units of the Cl_{ad} lattice) during the time between successive frames is obtained. By fits of these jump-distribution functions to a simple random-walk model, the S_{ad} hopping rate was determined as a function of temperature and electrode potential. Specifically, a strong exponential decrease by an order of magnitude per 100 mV potential increase was found (Fig. 10b). Based on the temperature-dependent data, this could be clearly attributed to a linear potential dependence of the activation energy for S_{ad} hopping ("diffusion barrier"), which in turn can be rationalized by an electrostatic contribution to the S_{ad} adsorption energy caused by the electric field at the interface. The latter is in agreement with thermodynamic considerations and data on Au(100) surface diffusion [67] and of general validity for surface transport at electrochemical interfaces. Hence, it is of considerable importance for understanding the kinetics of electrochemical reactions.

Video-STM also allows to directly study the dynamics of nanoscale structures, a topic of obvious relevance to nanoscience. A recent example is the observation of a quasi-collective diffusion of the basic elements of the Au(100) surface reconstruction, five atom wide strings of hexagonally ordered Au, within the surrounding square lattice of the Au surface layer (Fig. 11). Video-STM observations revealed a surprisingly high surface mobility of these "hex" stripes, with rapid displacements perpendicular and along the stripe direction [9, 62]. As an example, the mobile free "hex" stripe visible near the upper, left corner of the first image (marked by arrows; the other stripes in these images are pinned to surface defects) moves by more than 10 nm in only 66 ms. Based on similar quantitative studies as described above [71], the sideway hopping of strings can be attributed to the formation and rapid 1D propagation of lateral distortions ("kinks") within the strings (also manifest in form of a wiggly appearance of the mobile string). Equivalently, this mechanism of motion can be interpreted as a gliding of surface dislocations through the "hex" string. A similar quasi-collective mechanism as well as 1D diffusion of Au

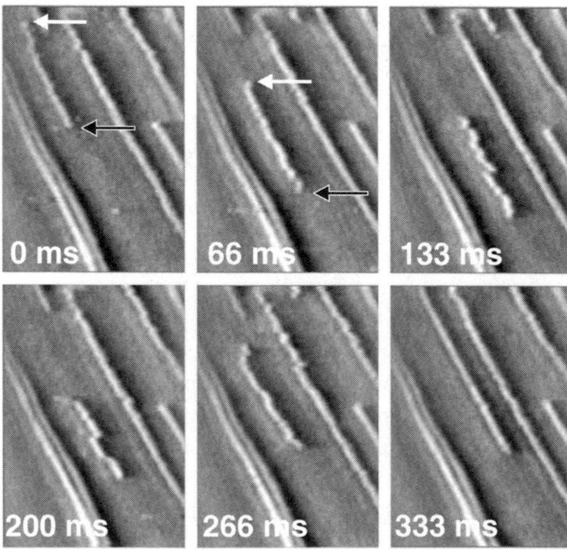

Fig. 11 Sequence of successive *in situ* Video-STM images showing the quasi-collective motion of nanoscale "hex" strings on Au(100) in Cl-containing 0.01 M Na_2SO_4 solution ($18 \times 29\,nm^2$, 15 frame/s video, from ref. [9])

adatoms along the string edges [9] may contribute to the mass transport along the string, necessary for the motion in string direction.

2.4 Studies of Electrochemical Phase Formation Processes

The modification of electrode surfaces by faradayic processes, such as electrochemical deposition, etching, and oxidation, is an area of particular significance to nanoscale electrochemistry. As described in Chapters 6, 7, 8, 9 of this book, these processes are utilized extensively in current scientific and technological applications. *In situ* studies by scanning probe methods such as STM have helped significantly to elucidate the microscopic mechanism of electrochemical phase formation processes. The main emphasis in this area has been on the initial stages of heteroepitaxial metal electrodeposition, which has been investigated for a large number of different metal deposits and substrates [14, 15, 37, 69, 72–74]. Typical studies focus on clarifying the lattice structure and the surface morphology of the deposit and to rationalize those in terms of the thermodynamic growth mode and kinetic effects (e.g., the rates of nucleation as well as intra- and interlayer transport). In many cases, not only the admetal and the substrate, but also parameters such as overpotential or deposition rate, film thickness, and electrolyte composition were found to influence the growth behavior substantially.

As a characteristic example, again ultrathin Au films on Pt(111), discussed already above (see Fig. 7), are chosen. STM studies of the electrochemical growth of these films at large overpotentials reveal nucleation and 2D growth of Au monolayer islands (Fig. 12a), followed by a layer-by-layer growth of the first two Au layers (Fig. 12b and c) and subsequent 3D growth (Fig. 12d), that is, a classical Stranski–Krastanov behavior [73]. While the mono- and bilayer deposit is pseudomorphic to the Pt substrate, the presence of a *moiré* pattern on top of the 3D Au islands (see Fig. 12d) indicates a structurally relaxed Au lattice at larger film thickness, again in accordance with a Stranski–Krastanov growth mode. In addition, kinetic effects were found to rule the nucleation of Au islands in the submonolayer regime, determining the equilibrium island density on the surface. Specifically, for (diffusion-limited) deposition from sulfuric acid solution, the nucleation density continuously increases to more positive deposition potentials, resulting in a change by a factor of 40 for a potential increase from 0.2 to 0.5 V_{SCE}. Since at the employed, high overpotentials the deposition rate does not depend on potential and Au desorption can be neglected, these STM observations can be understood using the same concepts of kinetic growth theory that have successfully explained UHV–STM data on MBE metal nucleation and growth. In particular, they point toward a pronounced blocking of Au surface transport by co-adsorbed sulfate. Studies of this type can help to evaluate fundamental similarities as well as differences between electrodeposition and deposition under vacuum conditions, clarifying the influence of the potential and co-adsorbed species (anions, water, etc.) on the growth behavior.

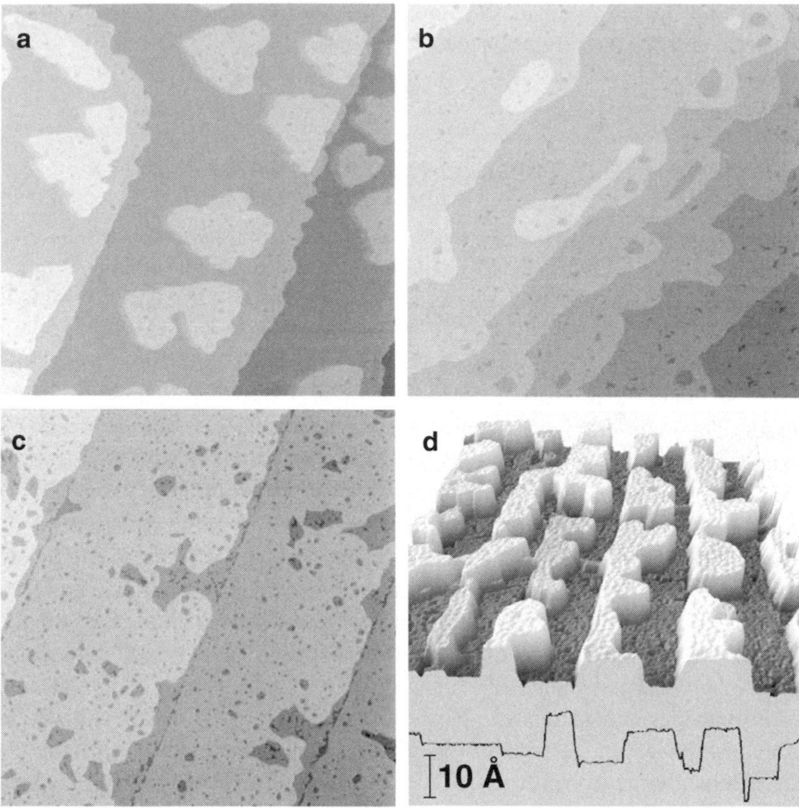

Fig. 12 *In situ* STM images showing (**a**) submonolayer islands (0.3 ML coverage), (**b**) a monolayer (1.2 ML coverage), (**c**) a bilayer (1.8 ML coverage), and (**d**) a multilayer (3.7 ML coverage) Au deposit on Pt(111). The Au films were formed by electrodeposition in $KAuCl_4$-containing 0.1 M H_2SO_4 + 10^{-3} M HCl at 0 V_{SCE} (370 × 370 nm^2. Reprinted from ref. [73]. Copyright by Elsevier Science)

At low growth rates, electrodeposition processes can be directly followed by *in situ* STM. This is illustrated for Ni electrodeposition on Au(111) in Fig. 13 a–d, where nucleation and subsequent 2D growth of second-layer islands on a Ni monolayer deposit are observed, revealing the annealing of structural defects (manifesting as a better-ordered *moiré* pattern) during the growth [75]. The possibility of such *in situ* studies is a clear advantage over UHV–STM studies of MBE processes, where the imaged area is geometrically shaded by the STM tip. As shown explicitly in recent finite-element calculations [76], the local electrodeposition rate under the tip is only slightly lower than that far away from the tip position, provided the diffusion of the metal ions in the electrolyte occurs substantially faster than the charge transfer at the electrode surface. The latter is not a very restrictive condition for electrodeposition processes with slow charge-transfer kinetics, such as Ni, but is difficult to fulfill for systems where the charge transfer is fast. For example,

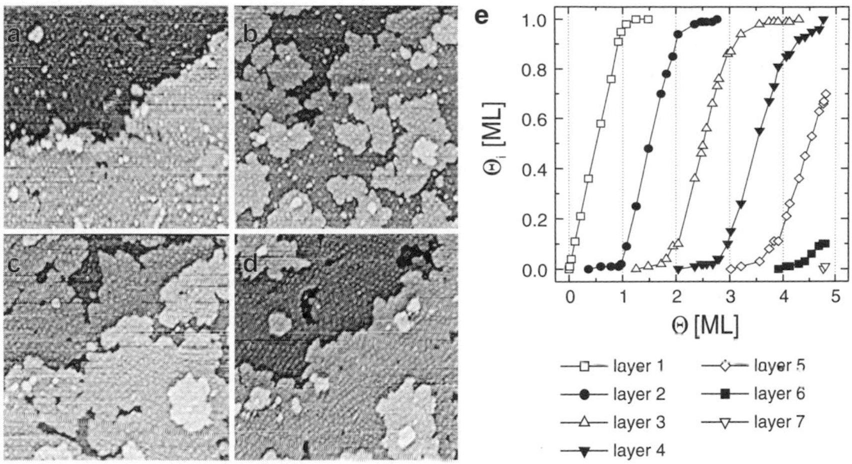

Fig. 13 (**a–d**) Sequence of *in situ* STM images recorded on Au(111) in 0.1 M Ni sulfate solution in the same surface area ($90 \times 100\,\mathrm{nm}^2$). The images show the growth of the Ni deposit at $-0.66\,V_{AgAgCl}$ and correspond to a Ni coverage of (**a**) 1 ML, (**b**) 1.5 ML, (**c**) 2 ML, and (**d**) 2.3 ML. (**e**) Occupation θ_i of the first seven Ni layers as a function of total Ni coverage θ, obtained by quantitative evaluation of series of STM images during Ni electrodeposition from Watts electrolyte (Reprinted from ref. [75]. Copyright (1997) by American Physical Society)

the growth rates observed by *in situ* STM during Cu electrodeposition from sulfuric acid solution are orders of magnitudes smaller than those expected from electrochemical measurements [69]. This as well as other effects of the tip on the observed process can be probed by significantly expanding the scan range and comparing the previously scanned area with the surrounding surface. From series of STM images recorded during deposition, a comprehensive picture of the growth process can be obtained, including nucleation sites, preferred growth directions, and roughness evolution of the deposit. As a simple example, a quantitative evaluation of the deposit morphology in the series of images obtained for Ni on Au(111) clearly demonstrates the almost-perfect layer-by-layer growth of the first five layers (Fig. 13e). Other *in situ* STM studies have shown that, similarly as for MBE hetero-epitaxial growth, also in electrochemical environment the structure of ultrathin deposited films can deviate significantly from that of the bulk admetal, in particular on strongly corrugated substrates such as fcc(100) surfaces. A well-known example are Cu deposits on Au(100), where the entire deposit changes at a film thickness >10 layers from a pseudomorphic Cu lattice, resembling a bcc copper phase, to a buckled structure with orthorhombically distorted domains that manifest as a pronounced unidirectional modulation in STM images of these deposits [77, 78].

Of special interest to nanoscience are *in situ* STM studies that focus on the electrodeposition of nanoscale structures. Electrochemical processes involve a charge-transfer process, which may exhibit a rather different kinetics at different surface sites. Under favorable conditions the discharge of the admetal species may even be limited to specific sites, which then act as a template for the formation of admetal nuclei. This offers attractive possibilities for the preparation of nanostructured surfaces. The most popular method and a very common phenomenon in metal-on-metal systems (Fig. 12a) is the decoration of the steps of the substrate on the lower terrace side, resulting in quasi-1D monolayer deposits (i.e., nanowires) [79, 80]. In some cases also 3D nanowires can be formed, for example, for Ni electrodeposited on Ag(111), where *in situ* STM studies revealed a very selective step decoration at intermediate overpotentials (Fig. 14a) [81]. Also, at higher overpotentials 3D islands on the terraces are formed, whereas at very small overpotentials a 2D film growth is observed. *In situ* observations of the nanowire growth process indicated a mechanism involving heterogeneous nucleation at a defect in the Ni deposit, created by the underlying step, and a higher growth rate for Ni on Ni as compared to Ni on Ag. Also, other heterogeneities on the electrode surface, such as surface reconstructions, were found to act as templates for the nucleation and growth of the admetal. This is illustrated for submonolayer electrodeposits of Ni and Ru, formed at low overpotentials, which decorate selected sites on the reconstructed Au(111) surface (Fig. 14b,c). For Ru, this selective growth occurs in the fcc areas between the dislocation lines, resulting in a replication of the reconstruction domain pattern (Fig. 14b) [82]. In addition, the opposite phenomenon, decoration of the hcp areas, can be observed, as shown in recent STM studies for Co [83] and for Au–Cd surface alloy growth [84]. In the case of Ni electrodeposition, islands nucleate at bending points ("elbows") of the dislocation lines, resulting in arrays of monolayer islands on surfaces exhibiting a well-ordered herringbone reconstruction (Fig. 14c) [75, 85, 86]. This nucleation behavior was also found for MBE growth of Ni and other transition metals under UHV conditions [87–89] and attributed to place exchange of Ni adatoms with Au surface atoms at the elbow sites, followed by Ni island nucleation on top [88]. Other possible templates for the electrodeposition of regular nanoscale structures may be dislocation networks on ultrathin metal films (as in Fig. 7) [90]

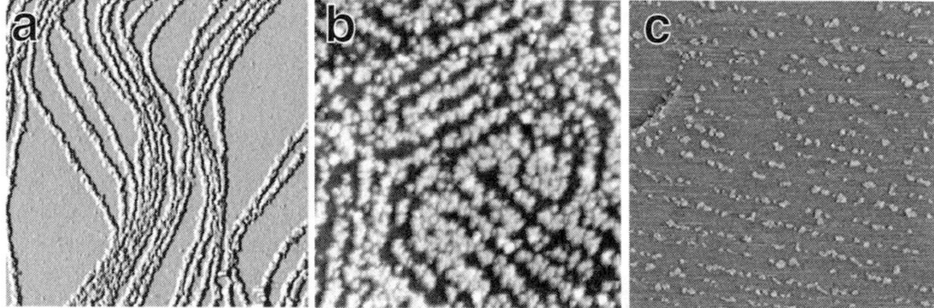

Fig. 14 *In situ* STM images of electrodeposited metal nanostructures, employing surface inhomogeneities as a template. (**a**) 3D Ni nanowires deposited along steps of the Ag(111) substrate ($850 \times 850\,\mathrm{nm}^2$. Reprinted from ref. [81]. Copyright (1999) by American Physical Society). (**b**) Submonolayer Ru monolayer islands deposited selectively in the fcc areas of the reconstructed Au(111) surface ($100 \times 100\,\mathrm{nm}^2$. Reprinted from ref. [82]. Copyright (1999) by American Physical Society). (**c**) Decoration of the elbows of the Au(111) reconstruction by electrodeposited Ni ($230 \times 230\,\mathrm{nm}^2$. Reprinted from ref. [86]. Copyright (1996) by American Physical Society)

and molecular adlayers [91]. Due to its high resolution *in situ* STM has been an unrivaled tool for the characterization of such nanostructured metal electrodes.

Even more complex electrochemical phase-formation processes, specifically also the deposition of nonmetallic materials, have been studied by *in situ* STM. Stickney and co-workers have developed electrochemical atomic layer epitaxy (ECALE) of compound semiconductors, where a deposit is formed layer by layer using repetitive, successive adsorption steps (e.g., alternating metal and anion adsorption), and investigated the formation of the first layers by a combination of STM, electrochemical, and *ex situ* methods [92, 93]. As an example, the structure of CdTe bilayers on Au(111) was found to depend on the sequence of the individual growth steps as well as on the potential [93]. For layers where Te was deposited first, resulting in a (13 × 13) adlayer structure, the successively deposited Cd layer exhibited a $(\sqrt{7} \times \sqrt{7})R19^{\circ}$ (Fig. 15a) and a (3 × 3) structure (Fig. 15b) for Cd deposition at −0.5 and $-0.6\,V_{Ag/AgCl}$, respectively. Clarification of such complex deposits usually requires a multimethod approach where STM is complemented by other structure-sensitive or spectroscopic techniques. Furthermore, *in situ* STM studies in nonaqueous electrolytes have been performed, where a much wider range of materials can

Fig. 15 *In situ* STM images of electrodeposited semiconductors. (**a,b**) CdTe bilayers deposited by the ECALE method on Au(111) from Cd and Te containing H_2SO_4 solutions at (**a**) −0.5 ($20 \times 20\,nm^2$) and (**b**) $-0.6\,V_{Ag/AgCl}$ ($10 \times 10\,nm^2$). Details are given in the text (Reprinted from ref. [93] with permission by the author. Copyright by Elsevier Science). (**c,d**) Electrodeposits of (**c**) Ge bilayer islands ($150 \times 150\,nm^2$. Reprinted from ref. [95, 96] with permission by the author. Reproduced by permission of the PCCP Owner Societies) and (**d**) Si mono- and bilayer islands ($500 \times 500\,m^2$. Reprinted from ref. [97] with permission by the author. Copyright by American Chemical Society), formed by electrodeposition on Au(111) from ionic liquids

be electrodeposited, including less-noble metals and semiconductors. Endres et al. have constructed an STM, capable of working in rigorously controlled nonaqueous environments (i.e., at water and O_2 contaminations <2 ppm) and applied it to studies of metal and semiconductor deposition from ionic liquids [94–97]. Characteristic images of Ge [95, 96] and Si [97] deposits on Au(111), obtained *in situ* during deposition, are shown in Fig. 15c and d, respectively, demonstrating that high-quality STM images can also be obtained under these conditions. Dependent on the deposition conditions, the formation of thin epitaxial films starting by 2D island nucleation and growth (Fig. 15c,d) or of 3D nanocrystalline aggregates was observed [95, 96]. Such measurements are very demanding due to the experimental efforts involved in maintaining the required purity of the electrolyte.

Studies of electrochemical dissolution and oxidation processes have employed *in situ* STM in a similar way, that is, have investigated the resulting surface structure on the atomic and mesoscopic scale as well as have directly followed the structural changes during these processes. For example, the microscopic mechanisms underlying the pronounced influence of anionic or organic adsorbates on these processes were revealed by such measurements (see, e.g., ref. [15, 44, 98]). Often, STM studies of the electrodeposition of ultrathin metal films include observations of the dissolution of the admetal, which can provide data on the dissolution mechanisms, the defect density in the deposit, and structural changes in the substrate morphology due to surface alloying. Details on metal dissolution and oxidation studies are given in the chapter on nanoscale corrosion (Chapter 10), where these processes play a central role.

More complex phase formation processes, such as ECALE, often involve two or more consecutive steps, which often require a change in electrolyte composition (e.g., introduction/removal of electroactive species or pH changes). Although this can, in principle, be accomplished in an *in situ* STM experiment by exchanging the electrolyte, the latter often leads to changes in the lateral tip position, making it difficult to directly observe the effect of the exchange by comparing the structural changes in the same surface area. To solve this problem Ammann et al. developed complex microfabricated probes with an integrated STM tip and a surrounding "generator electrode", whose potential could be independently controlled [99, 100]. This allowed to rapidly generate species for an electrochemical reaction (e.g., metal ions or H_3O^+) in the electrolyte close to the sample. Since the species are delivered to the electrolyte volume in the immediate vicinity of the tip, they have to diffuse only $\approx$50 μm, resulting in almost immediate initiation of the reaction in the imaged surface area (response times below a second).

STM studies of the (nonequilibrium) dynamics of electrodeposition/-dissolution with high temporal and spatial resolution can be performed using similar approaches as those described in the previous chapter for studies of surface transport in the double-layer regime. Using Video-STM fundamental atomic-scale processes during metal dissolution and (re)deposition can be directly observed, as illustrated in Fig. 16 for Cu(100) in HCl solution [7, 101, 102]. In this system removal and attachment of Cu along monoatomically high steps occurs strongly anisotropically (e.g., in Fig. 16 only along the step direction denoted as [010]), which was attributed

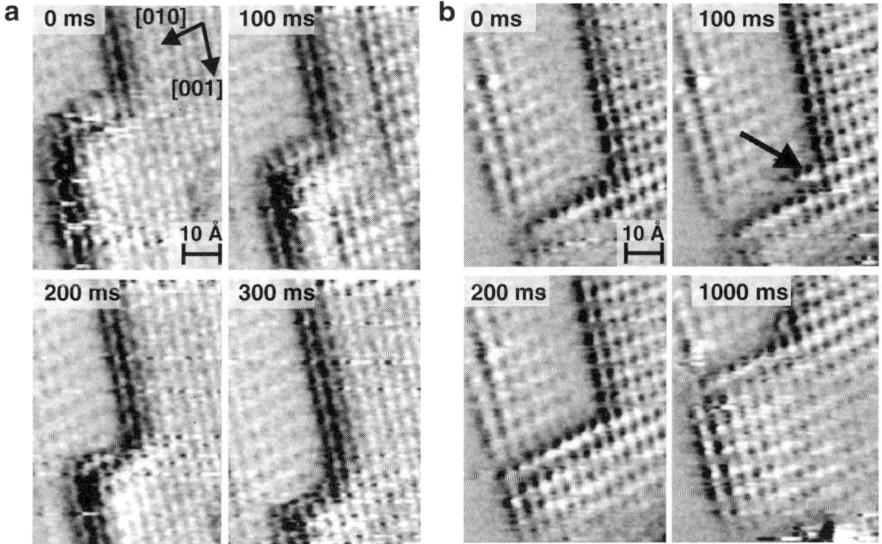

Fig. 16 Series of successive *in situ* Video-STM images showing (**a**) local dissolution and (**b**) local growth of a terrace on Cu(100) in 0.01 M HCl solution. In both cases a strongly anisotropic process is found, caused by the presence of the c(2 × 2) Cl adlayer (5 × 8 nm^2, 10 frames/s video. Reprinted from ref. [7]. Copyright by Elsevier)

to the influence on co-adsorbed Cl anions. Specifically, according to these studies, the active sites of the deposition or dissolution reaction, kinks in the Cu steps (Fig. 16b, arrow), are determined by the ordered c(2 × 2) Cl adlattice, which exhibits a very different coordination at Cu steps running at 90° angles, resulting in a different reactivity of kinks in these steps. This is supported by Video-STM observations of isotropic growth and dissolution at the Cu(100) bilayer steps, which do not exhibit this adsorbate-induced structural anisotropy. Quantitative measurements of the kink propagation during Cu(100) dissolution in HCl solution, which allowed conclusions on the local reaction rates, were performed using a variation of the *x*–*t* scans described in Section 2.3 [103]. Although only line scan rates of 20 s^{-1} were used, kink propagation rates of the order of 10^3 s^{-1} could be measured by this technique. Even higher time resolution could, in principle, be obtained by combining this method with Video-STM scan rates. Despite these attempts, detailed quantitative studies of the nonequilibrium atomic-scale dynamics of dissolution and growth processes as have been performed for surfaces under equilibrium conditions are still largely lacking.

2.5 Scanning Tunneling Spectroscopy

Beyond microscopic images, STM techniques can also provide local spectroscopic data, including data on electronic or vibronic excitations of single molecules and nanostructures. This feature is extensively used in nanoscience and has produced many spectacular results. Scanning tunneling spectroscopy (STS) is based on the dependence of the tunnel rate on the electronic structure of tip and sample and the distance-dependent tunnel probability (see refs. [1–4] for a detailed description). Typically, the tunnel current I_t is measured as a function of the tunnel bias U_{bias} or tip-sample distance Δz by sweeping U_{bias} or Δz over a wider range at a fixed sample position (or in some more elaborate experiments at every point of an STM image). Alternatively, the slopes dI_t/dU_{bias} or $dI_t/d\Delta z$ are acquired parallel to the surface topography, which is a simple method to detect spatial variations in the spectroscopic signal. Since the signal also depends on the electronic structure of the tip, well-defined, stable tip states are an important requirement for STS measurements. The best data are obtained in UHV at cryogenic temperatures using tips that are prepared by well-defined methods and characterized by known spectroscopic features (e.g., surface states) of clean samples. At room temperature, especially in electrochemical environment, the tip stability is considerably reduced and more difficult to characterize. Under these conditions, the most reproducible data are obtained by "blunt" tips, which average the STM signal over larger surface areas, while tips capable of resolving atomic-scale features are usually less suited. Furthermore, at room temperature, spectroscopic features are thermally broadened, and thermal drift motion makes maintaining a constant position on top of a surface feature of interest difficult. Despite these problems, STS is gaining increasing attention for studies of electrochemical interfaces.

The first *in situ* spectroscopic data in electrochemical environment were measurements of the tunnel barrier, that is, of the decay of I_t with increasing tip-sample distance. Early studies by several groups [104–106] revealed effective barrier heights $\bar{\Phi}$ of 1–2 eV, which is noticeable smaller than typical tunnel barriers obtained under UHV conditions. The origin of this decrease in barrier height in the presence of the electrolyte solution, specifically of liquid water, was addressed in several theoretical studies, where it was attributed to the lowering of the metal work functions due to the electronic polarizability of the solvent [107] or to resonant tunneling via intermediate states formed by solvent fluctuations [108]. More recent models have taken the molecular structure of the solvent explicitly into account [109–113]. The most extensive of these calculations indicate that both effects indeed play a role [112, 113]. However, the resonant states in the solvent are found several eV above the Fermi level, that is, they act as virtual intermediate states. In the energy range where true resonant tunneling via water states would occur strong electron scattering, resulting in a loss of spatial resolution, was predicted. More detailed barrier height measurements by Vaught et al. indicated significant deviations from a true exponential decay at gap resistances below $10^8\,\Omega$, which were attributed to changes of the water structure in the thin tunnel gap [114]. More recently, Hugelmann and Schindler reported measurements of $I_t(\Delta z)$ over 6 orders of magnitude that exhibited clear (potential-independent) oscillations of $\bar{\Phi}$ with a period of 3.5 Å (Fig. 17) [115]. The latter were associated with the formation of discrete water layers in the tunnel gap. Similar phenomena were also observed in *in situ* STS studies on semiconductor samples [116]. Furthermore, very recent STS studies in pure water and

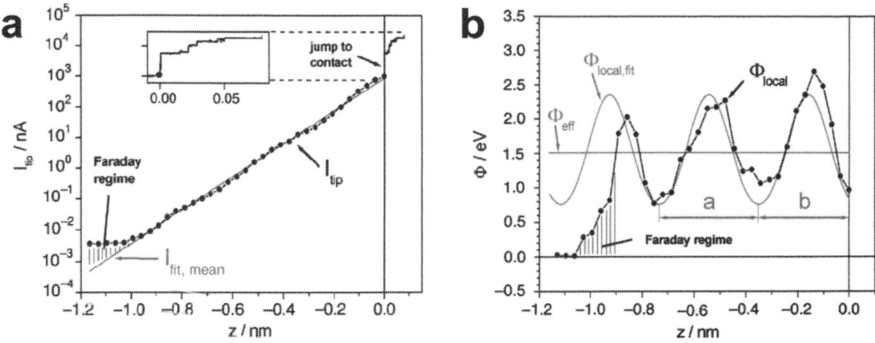

Fig. 17 Local barrier height of the tunneling contact between a Au tip and a Au(111) sample in 0.02 M $HClO_4$, showing oscillations in the local barrier height $\bar{\Phi}$ as a function of tip-sample distance. At distances >0.9 nm reliable measurement of I_t is prevented by the increasing contribution of faradayic tip currents (Reprinted from ref. [115] with permission by the author. Copyright by Elsevier)

1 mM $HClO_4$ where I_t and $dI_t/d\Delta z$ were simultaneously acquired, even report a drop of $dI_t/d\Delta z$ to zero below a critical distance, which was attributed to field-induced freezing of water in the tunnel gap at room temperature [117]. A general problem in all of these studies is to determine the absolute tip-sample distance. Usually, the position where the abrupt formation of a metal nanowire contact occurs (resulting in a jump in I_t) is chosen as $\Delta z = 0$. However, according to a recent theoretical study, this assumption may result in an underestimation of the actual gap width by as much as 3 Å [35]. Nevertheless, studies of this type are not only important for a fundamental understanding of *in situ* STM operation in electrochemical environment, but also may shed light on the electrolyte structure near the interface, a problem of considerable importance in interfacial electrochemistry.

Applying the more powerful *I*/*U* tunneling spectroscopy to *in situ* studies of electrochemical interfaces is considerably more difficult, not only due to problems with tip stability and reproducibility (see above), but also due to the limited potential range. The available range of U_{bias} is at most $\approx$1 V in aqueous electrolytes, which is too small for most systems to show clear spectroscopic signatures, for example, of molecular adsorbates. Wider ranges of up to 4 V can be obtained in nonaqueous electrolytes and have been employed in *in situ* STS measurements to demonstrate semiconducting behavior of electrodeposited Ge films in ionic liquids [96]. However, as discussed above, the technical difficulties of studies in such electrolytes are substantial. In addition, very well isolated tips are required to keep the faradayic and capacitive currents at the tip significantly lower than I_t during sweeps of U_{bias}. For these experimental reasons *in situ I*/*U* spectroscopy has not gained wide usage till date.

A variation of tunneling spectroscopy that is specific to the electrochemical environment has been first proposed on theoretical grounds by Schmickler and co-workers [118] and later demonstrated experimentally by Tao [119]. It is based

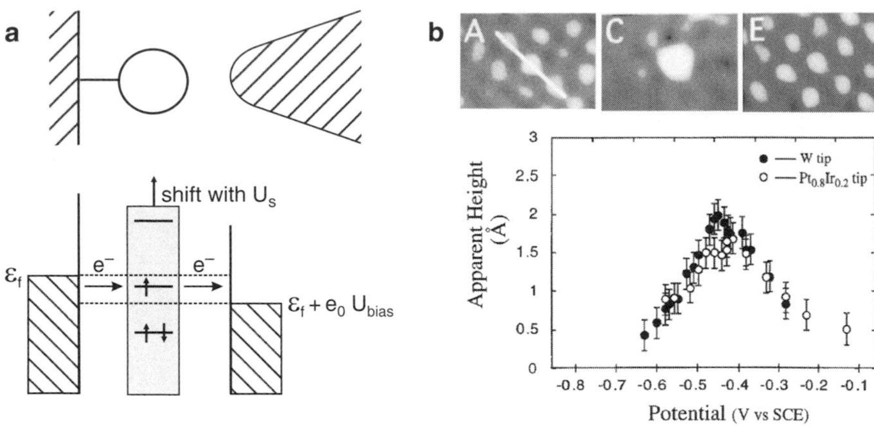

Fig. 18 (**a**) Schematic illustration of resonant tunneling via molecular states near the Fermi level of an adsorbed, redox-active molecule. (**b**) *In situ* STM images of a redox-active Fe(III) protoporphyrin and surrounding redox-inactive protoporphyrins on graphite at sample potentials U_s of −0.15 (A), −0.42 (B), and −0.65 V_{SCE} (*top*) and apparent height of the Fe(III) protoporphyrin as a function of U_s, showing the increase close to the redox potential (Reprinted from ref. [119] with permission by the author. Copyright (1996) by American Physical Society)

on resonant tunneling via molecular states near the Fermi level of an adsorbed redox-active molecule, resulting in an increase in I_t (see Fig. 18a). A prerequisite for a significant effect is a fast and reversible electron transfer, which typically occurs if the reduced and oxidized molecular forms are similar and not strongly coupled to solvent modes. By changing the sample potential U_s at constant U_{bias} the molecular states are shifted relative to the tunnel gap, leading to a resonant contribution if empty states of the molecules are situated in the energy window defined by U_{bias}. This configuration may be considered as a molecular field effect transistor with the electrolyte acting as the gate (i.e., U_s acts as a gate voltage). First direct observations of this effect were reported by Tao for mixed protoporphyrin adlayers on HOPG, where the redox-active Fe(III) protoporphyrins appeared in the STM images up to 2 Å higher relative to the surrounding (redox-inactive) protoporphyrins for sample potentials close to the redox potential (Fig. 18b). For this system, the shift in the molecular energy levels seems to be approximately equal to the change in electrode potential, which can be rationalized by the low density of states in the graphite substrate [120]. More recently, this spectroscopic approach has been applied to other systems such as metal nanoclusters attached to electrode surfaces via redox-active molecules [121] and molecular junctions, where the molecule is covalently bound to both sample and tip [122]. The possibility to modify the electron-transport properties of such nanoscale assemblies by employing the potential as a gate is a unique advantage of the electrochemical environment. *In situ* STS studies of this type are currently gaining increasing interest.

3 Electrochemical AFM

3.1 Experimental Realization and Operation Modes

In scanning force microscopy (also called atomic force microscopy, AFM) the force between the tip of the probe and the sample surface, which typically is in the range 10^{-12}–10^{-9} N, is used to control the tip-sample distance. Since its invention in 1986 [123] AFM has become the most popular scanning probe method. This is related to (a) its applicability to a wide variety of samples, including insulators and biological materials, (b) the large number of different force interactions (e.g., hard–core repulsive, attractive van der Waals, electrostatic, magnetic) that can be employed, (c) the possibility to easily measure local surface properties (e.g., adhesion, stiffness, and friction) in parallel to the surface topography, (d) the possibility to combine the force measurements with other local probe measurements, such as conductivity measurements using conducting tips or near-field spectroscopy studies (see Section 4), and (e) the more robust and user-friendly operation (as compared to STM) for studies on lateral length scales of micrometers. In the following chapter, the focus will be on the AFM operation modes commonly used in electrochemical environment. A more detailed, general description of AFM can be found elsewhere [2, 4].

For a basic understanding of scanning force microscopy first the different contributions to the force between an uncharged tip and a sample separated by a homogeneous medium (e.g., vacuum or a simple dielectric liquid) have to be considered in more detail (Fig. 19). The force between the outmost atoms of tip and sample (Fig. 19a and b, dotted lines) consists of the short-range repulsive interaction between the ion cores and the long-range attractive van der Waals interactions. However, the latter also are nonnegligible for atoms in the bulk of tip and sample separated by a larger distance. The sum of these long-range attractive forces (Fig. 19, dashed lines) usually is considerably larger than the local forces between the outmost atoms, which has important implications for the operation modes and the resulting lateral resolution of the AFM. For example, at small distances z the local repulsive forces may be largely compensated by the long-range attraction between tip and sample, that is, strong local forces may act on the tip atoms although

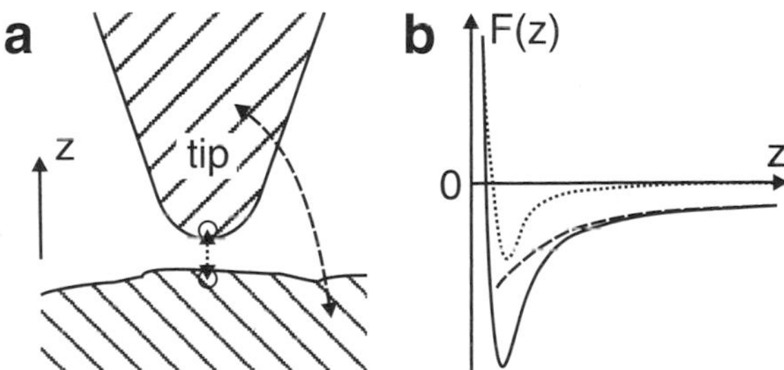

Fig. 19 Schematic illustration of (**a**) the tip-sample contact in an AFM and (**b**) the corresponding force distance curve $F(z)$, showing short-range (*dotted lines*), long-range (*dashed lines*), and total forces (*solid line*) acting between the AFM tip and the sample

the total, externally measurable force (Fig. 19b, solid line) is only weakly repulsive or even attractive. However, the long-range van der Waals interactions can be substantially reduced in water, which allows AFM operation at forces down to 10^{-11} N [124].

To measure the force the AFM tip is attached to a micromechanical cantilever, whose mechanical response provides the signal. Upon approach of the tip to the sample the attractive forces result in a small downward deflection of the cantilever until the force gradient exceeds the force constant of the cantilever and the tip snaps toward the surface ("jump-to-contact"). Further approach results in a continuous opposite deflection of the cantilever with a slope that depends on the stiffness of the sample.

For stable feedback the AFM has to be operated in a part of the force-distance curve where the signal depends monotonically on z. In the simplest mode of operation, "contact-mode" AFM, the measurements are performed in the range of the repulsive interactions, where the force gradient is large, and the static deflection of the cantilever is used to determine the force. Since the local forces are large in this regime, contact-mode studies are restricted to sufficiently hard samples that are not mechanically modified during imaging. This problem is less severe in *in situ* AFM studies due to the screening of the van der Waals forces by the electrolyte solution (see above). The electrochemical environment therefore allows in general a gentler contact-mode imaging of surfaces than AFM studies in air or vacuum. In addition, contact-mode AFM allows to measure in parallel the friction force acting on the moving tip by monitoring the torsion of the cantilever.

Studies of softer, mechanically less-stable surfaces require measurements in the attractive branch of the $F(z)$ curve via a dynamic AFM mode, where mechanical oscillations of the cantilever along the z direction are induced. Typically, the cantilever is driven at a constant frequency close to the resonance frequency. The interaction of the tip with the surface causes a change in the resonance frequency, resulting in turn in a change in the oscillation amplitude, which provides the feedback signal. If the tip remains in the range of the attractive interactions over the entire oscillation period ("noncontact mode") the influence of the tip on the surface is very small, however, also due to the long-range nature of the interactions the resolution is rather low. For this reason, an intermittent mode ("tapping-mode") is used more frequently, where the tip is in brief contact with the surface at the outmost point of the oscillation. This results in a substantial increase in resolution as compared to noncontact mode, without exerting much larger lateral forces on the surface. Even higher resolution can be achieved using frequency-modulated AFM where the cantilever oscillates at a fixed, small amplitude and frequency is used as the signal [125–127] (see also Section 3.2).

Almost all scanning force microscopes for operation in liquid environment are based on the same experimental setup, where the cantilever deflection is measured via the deflection of a laser beam, reflected from the back of the cantilever. The cantilever oscillation in dynamic AFM modes is generated either indirectly by inducing acoustic waves in the liquid via a piezoelectric actuator in the cantilever holder or directly by employing cantilevers with a magnetic coating on the backside that are excited electromagnetically. The latter in general allows more easy identification of the true cantilever resonance frequency, but the magnetic coating may be incompatible with the electrochemical environment. Both drive methods seem to afford a similar cantilever excitation and microscopic resolution [128]. AFM usually does not require electric measurements (apart from specialized experiments with conducting tips) and hence is easy to combine with potentiostatic or galvanostatic electrochemical control of the sample. Furthermore, commercially available AFM cantilevers (typically made from silicon or silicon nitride) feature tips that are (on scales of $\geq$10 nm) much better structurally defined than typical STM tips. Together with the ability of AFM to provide stable images on nonconductive (e.g., passivated) surface areas, these properties make AFM often superior to STM in studies of technical electrode surfaces and applied electrochemical processes. Earlier reviews of AFM in electrochemical environment can be found in ref. [14, 18].

3.2 In Situ AFM Imaging of Electrode Surfaces

Since the samples in electrochemical systems usually are electronically conducting, the key advantage of AFM over STM, that is, the possibility to study isolator surfaces, is of lesser importance. For this reason STM, which in general provides higher microscopic resolution, is still more frequently employed than AFM for *in situ* studies of electrochemical systems. Nevertheless, *in situ* AFM has been widely used for the microscopic characterization of electrode surfaces, ranging from atomic-resolution images to studies of the surface morphology on the μm scale.

The possibility to resolve the atomic lattice of bare metal electrode surfaces and metal adlayers was demonstrated by Mane et al. in a study of Cu UPD on Au(111) by contact-mode AFM [129]. Since then, a large number of atomic resolution *in situ* AFM studies have been reported, including studies of various metal and salt-like adlayers, oxide films, and organic adsorbates (see ref. [14] for an overview). As an example, contact-mode AFM observations of Bi UPD on Au(111) are shown in Fig. 20, illustrating resolution of the Au substrate lattice (Fig. 20a), as well as of a (2×2) (Fig. 20b) and an uniaxially incommensurate (Fig. 20c) Bi-adlayer [130], in good agreement with a complementary surface X-ray diffraction study [131]. Although such high-resolution contact-mode AFM images can clearly resolve the surface lattice structure, the forces are typically so large that a wider area of the tip is in contact with the surface rather than only the apex atom of the AFM tip. Atomic resolution can in this case result if the atomic lattice at the tip locks into the sample lattice during scanning, leading to a corrugation with the periodicity of the lattice spacing. Experimental evidence for this is the characteristic absence of deviations from a periodic lattice, that is, of surface defects, such as vacancies, domain boundaries, or substrate steps, in the AFM images.

Since AFM and STM employ different interactions, comparative studies by both techniques may give complementary data, which may help to clarify the surface structure. As shown in Fig. 21 for a guanine monolayer on graphite electrodes, the

Fig. 20 *In situ* contact-mode AFM images of Au(111) in 0.1 M $HClO_4$ solution (**a**) positive of the Bi UPD range, showing the Au substrate lattice, (**b**) in the range of the (2×2) Bi UPD adlayer phase, and (**c**) in the range of the uniaxially incommensurate Bi UPD adlayer phase ($5 \times 5\,nm^2$. Reprinted from ref. [130] with permission by the author. Copyright by American Chemical Society)

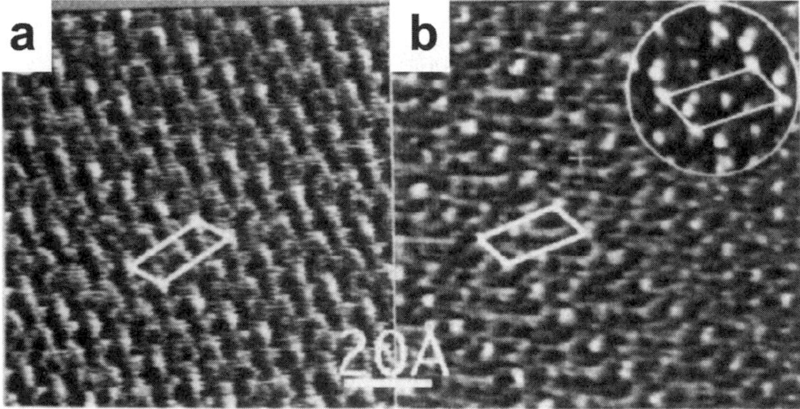

Fig. 21 (**a**) *In situ* contact-mode AFM and (**b**) *in situ* STM images of a guanine monolayer on graphite in 0.1 M NaCl solution ($5 \times 5\,nm^2$. Reprinted from ref. [132] with permission by the author. Copyright by American Chemical Society)

in situ AFM (Fig. 21a) and STM (Fig. 21b) images both resolve a molecular structure with identical lattice parameters, but clearly differ with respect to details of the image within the unit cell [132]. In principle, noncontact AFM probes the ion cores (i.e., the total electron density) and AFM images consequently should reflect more closely the geometric surface structure than STM images, which are (within the Tersoff–Haman approximation) determined primarily by the electron density at the Fermi level (see Section 2.2). However, the influence of the tip structure on both STM and AFM images makes a detailed analysis of the differences in the images difficult if not outright impossible. This problem may be overcome by frequency-modulated AFM, where the interaction between the outmost tip and the surface is dominant. This technique has been recently shown to be capable of true atomic-resolution imaging with an image quality equal or even surpassing high-resolution STM [126, 127]. Up to now frequency-modulated AFM has been performed almost exclusively under UHV conditions. Very recently, the adaptation of this method to measurements in liquid environment has been demonstrated [133, 134], suggesting that frequency-modulated AFM may develop into a promising future method for electrochemical surface science.

Similar to scanning tunneling microscopy, scanning force microscopy has been extensively used for microscopic studies of a wide range of electrode processes, for example, electrodeposition or dissolution processes. Compared to STM, local surface features such as steps are less well resolved by AFM. On the other hand, the AFM tip is usually more stable and the imaging is less affected by changing electrochemical conditions. For these reasons, AFM is generally preferred for studies on the mesoscopic scale (i.e., for scan ranges from 0.5 to 100 μm). In particular, AFM is superior to STM for studies on strongly heterogeneous samples, where materials with different electronic properties (e.g., oxides and metals) coexist in the studied surface area. For example, AFM has been successfully employed to study the electrodeposition of Pb [135] and Au [136] on Si(111) electrode surfaces, whereas stable STM imaging proved difficult due to the strongly different imaging conditions required for the metal and the semiconductor surface. Another import field is corrosion, where the samples are often covered by oxide films with poor electronic conductivity and where AFM is the method of choice, particularly in applied corrosion studies (see also Chapter 10). Apart from qualitative observations of the surface morphology larger-scale AFM images can also be analyzed quantitatively. Specifically, systematic quantitative studies of the surface roughness can provide insights into the mechanisms of growth and dissolution processes. An elegant example is the evaluation of the morphology of electrodeposited Cu films (Fig. 22a) by scaling analysis, performed by Hou and Schwarzacher [137]. In this type of studies, the dependence of the root-mean-square surface width $w(l)$ on the lateral length scale l over which $w(l)$ is measured and on the deposition time is analyzed (Fig. 22b). For deposition from pure, additive-free sulfate electrolyte, the surface roughness could only at low current density (4 mA cm^{-2} at 0.3 M Cu^{2+} concentration) be described

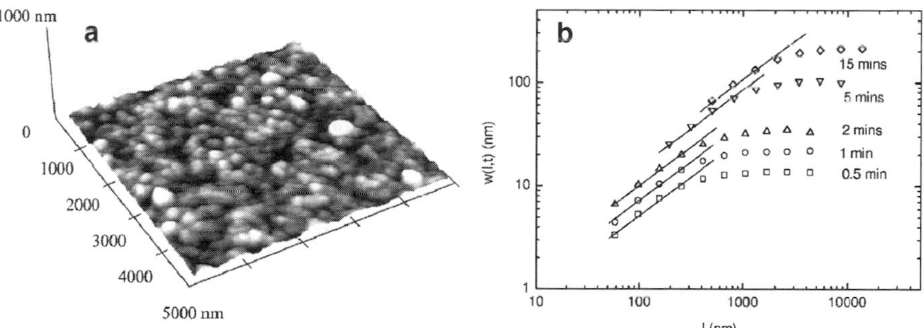

Fig. 22 AFM studies of the morphology of Cu deposits formed by electrodeposition from acidic 0.3 M $CuSO_4$ solution. (**a**) AFM image for 1 min deposition time. (**b**) Scaling analysis, showing the surface width $w(l)$ over regions of length l for different deposition times (Reprinted from ref. [137] with permission by the author. Copyright (2001) by American Physical Society)

by normal dynamic scaling, where w is independent of deposition time at low l. In contrast, at higher current density ($\geq$16 mA cm^{-2}) anomalous scaling was observed, where an extra scaling exponent β_{loc} is required to describe the time evolution of the roughness at short-length scales ($\beta_{loc} = 0.30$ for the data shown in Fig. 22b). The origin of this behavior, although not yet completely understood, may be related to surface transport during deposition. Studies of this type may help to unravel the complex mechanisms of polycrystalline film growth.

In situ dynamic AFM in liquid environment is primarily used to study soft samples, for example, biological systems or polymers, where it has become a major technique for high-resolution imaging and characterization of local material properties. While numerous dynamic AFM studies on nonconducting substrates (e.g., mica) exist, only a few *in situ* studies of electrochemical systems have been reported, primarily in the field of bioelectrochemistry. An early example was a tapping-mode AFM study by Boussaad and Tao of myoglobin on bare graphite electrodes and graphite modified by a didodecyldimethylammonium bromide self-assembled monolayer [138], an important model system for studying electron transfer reactions in proteins. More recently, it has been shown that dynamic AFM can be combined with scanning electrochemical microscopy (see Section 4) to study in parallel the structure and electrochemical reactivity of soft biological samples [139]. Although the preparation of the conducting, partly isolated AFM tips required for these measurements is substantial, modern microstructuring methods may help to build suitable probes on a routine base [139, 140]. Such combined SPM studies could develop into a valuable tool to unravel complex bioprocesses on the nanometer scale.

3.3 Forces at Electrochemical Interfaces

The quantitative determination of forces at interfaces is another important area of application for AFM. Often specialized instruments with improved force detection and very linear tip movement along the surface normal direction are employed in these experiments. For precise force measurements, probes with cantilevers of known spring constant and defined tip geometry are required. The latter is usually difficult with common microfabricated silicon or silicon nitride tips, which provide high lateral resolution, but usually have an ill-defined apex geometry. In an alternative approach, special probes have been employed, where the tip consists of a spherical microparticle (typically silica spheres of 2–20 μm diameter), glued to the cantilever. The low surface roughness and well-defined curvature of these probes, leading to a similar configuration as in the older surface-force apparatus, allows to quantitatively determine the interactions with the sample and to compare those experimental results with quantitative theories.

The forces between two solids immersed in electrolyte solutions contain contributions from van der Waals forces and electrostatic double-layer forces, resulting from the mutual perturbations of the double layers at both solid surfaces at small separations. Most of these experiments were performed on insulators, where the surface charge is mainly determined by the pH. Some of the first AFM studies of metal electrodes used Au electrodes and silicon nitride tips [141] or silica spheres [142] as probes. As expected, the force on the negatively charged probes depended on the surface charge of the Au electrode: while repulsive interactions were found in the potential regime of the negatively charged surface (i.e., at potentials below the potential of zero charge), the probe is attracted to the surface at positive surface charges. Furthermore, the force curves were in good agreement with predictions from Derjaguin–Landau–Verwey–Overbeek (DLVO) theory at large separations. A detailed treatment of the investigation of double-layer forces by AFM and its theoretical basis is beyond the scope of this chapter, but can be found elsewhere [16, 143].

In addition, AFM-based force measurements can be used to probe more short-range, chemical interactions, including adhesive forces and elastic properties of adsorbed molecules. Spectacular examples of this are force spectroscopy studies of single molecules that allowed to directly obtain mechanical molecular properties [144–147]. Studies at electrochemical interfaces have not yet reached the single molecule level. However, studies of the adhesion forces between tips and metal electrodes functionalized by organic molecules become increasingly popular and permitted to measure quantitatively properties, such as surface and interface free energy [148, 149]. An overview of the earlier work in this rapidly growing field can be found in ref. [150].

4 Other SPM Techniques

Apart from STM and AFM scanning electrochemical microscopy (SECM) and scanning near-field optical microscopy (SNOM) are of relevance to electrochemical nanoscience. SECM is a genuine electrochemical SPM technique, where the sample is scanned by an ultramicroelectrode (UME), whose current or potential provides the signal (see also Chapter 11) [17, 151–154]. A common mode of operation is the feedback mode, where a redox-active mediator species is consumed at the UME. If the mediator is regenerated at the sample the probe current increases with decreasing probe-sample distance ("positive feedback"), otherwise it decreases ("negative feedback"). Although the spatial resolution of SECM is usually only in the micrometer range, it provides a valuable technique for measuring local (electro) chemical reactivity. Furthermore, SECM-based approaches can also be used to modify surfaces locally via electrochemical reactions (e.g., deposition, etching, and oxidation) at the probe position, enabling the formation of micro- and nanoscale structures (see Section 5). SECM has been successfully combined with STM as well as contact and noncontact AFM, allowing parallel characterization of surface structure and electrochemical properties [27, 139, 140, 155] (see also Section 3.2).

Although SNOM has not been used extensively till date in electrochemical environment, it offers good prospects for future *in situ* studies. SNOM allows optical near-field studies with lateral resolution far below the wavelength of the employed light [2]. It employs local electromagnetic fields generated either via submicrometer apertures or metal tips. The latter "apertureless SNOM" is based on the huge enhancement of optical fields in nanoscale gaps. A particularly exciting new development is tip-enhanced Raman or infrared spectroscopy, which offers prospects to combine the high spatial resolution of SPM methods with the chemical sensitivity of vibrational spectroscopy. Recent experiments indicate that optical signals from areas of less than 10 nm extension can be obtained, enabling studies of small ensembles of molecular adsorbates and individual nanoparticles [156, 157].

5 Nanostructuring by Electrochemical SPM

5.1 Nanostructuring by Local Mechanical Interactions

SPMs can not only be employed for imaging, but also for controlled modification of surfaces on the nanoscale (see, e.g., ref. [158, 159] for a detailed overview). The most spectacular example of this is the positioning of individual atoms on surfaces by low-temperature UHV-STM [160], which opens up opportunities to directly study quantum physical phenomena (e.g., standing electron waves in quantum corrals [5, 161, 162]) or chemical processes (e.g., tip-induced catalytical reactions [163]). For practical purposes larger-scale structures with dimensions of a few nanometers, which are stable at room temperature, are of considerable interest, for example, for catalysis or nanoscale electronics. SPM-based nanostructuring methods also allow the preparation of this type of structures. In the simplest of approaches, this is done by mechanical contact between tip and sample, leading to a mechanical deformation of the surface or to the transfer of material between tip and sample, as has been demonstrated already in early STM studies by Gimzewski et al. [164].

SPM in electrochemical environment allows to develop the latter type of methods into more sophisticated techniques, where local mechanical modification by the tip is combined with an electrochemical process, in general, electrodeposition or dissolution. Typically, the mechanical interaction in these approaches defines the nanostructure dimension, whereas the electrochemical process provides the material for the formation of the structure. A very simple method is to use the tip as a nanomechanical tool for the creation of defects in the sample surface, which subsequently act as a preferred nucleation sites, resulting in local electrochemical deposition or dissolution at this site. One of the earliest examples of this method was reported by the group of Penner for STM-induced nucleation of metal electrodeposition on graphite [158, 165, 166]. In this study, a defect in the graphite surface was generated by applying a voltage pulse to the STM tip, which subsequently could be decorated selectively by electrodeposition. Obviously, this technique works best if the overpotential for nucleation on the original sample is much higher than that for nucleation at the induced defect. The latter can be easily achieved on samples covered by a passivating layer, which is locally removed by the tip (Fig. 23a). Examples

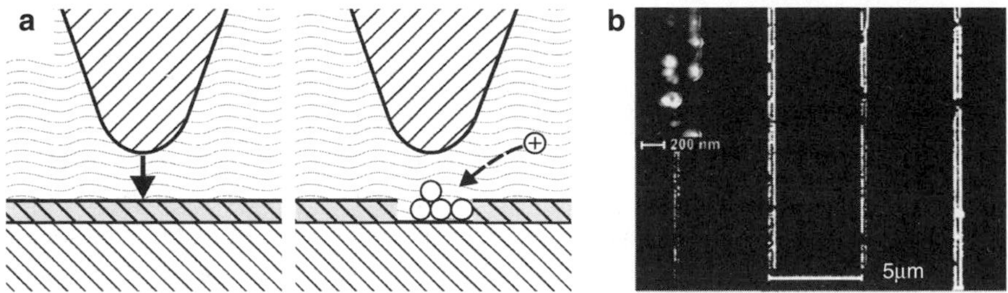

Fig. 23 (**a**) Schematic illustration of nanostructuring by electrochemical decoration of SPM-induced defects. (**b**) Cu lines formed by Cu electrodeposition on AFM-scratched p-type Si(100) (Reprinted from ref. [168] with permission by the author. Copyright 2001 by American Institute of Physics)

of this approach include submicron structures formed by local electrodeposition of Cu on AFM-induced defects in oxidized Cu surfaces [167] or of Cu and Pd onto AFM-scratched silicon surfaces [168, 169] (Fig. 23b).

An alternative method, where material is first electrodeposited on an STM tip and then mechanically transferred to the surface (Fig. 24a), was developed and extensively employed by Kolb and co-workers [170–174]. In this method, the tip potential is kept slightly negative of the Nernst potential, resulting in continuous metal loading of the tip, whereas the sample potential is kept $\approx 10\,\text{mV}$ positive of the Nernst potential to avoid bulk deposition. The mass transfer from the tip to the sample is initiated by approaching the sample close enough that a metal bridge between tip and sample is formed ("jump to contact"; Fig. 24a, right). This bridge breaks during subsequent retraction of the tip, leaving a small metal cluster on the sample surface (Fig. 24b). The cluster height of 1–4 monolayers can be controlled by the tip displacement Δz during approach. For fast electrodeposition processes (e.g., of Cu) cluster formation at kHz rates has been demonstrated, allowing preparation of large, highly uniform cluster arrays in a few minutes. Transfer of material from the tip to the sample can only occur if the adhesion of the deposit to the sample metal exceeds

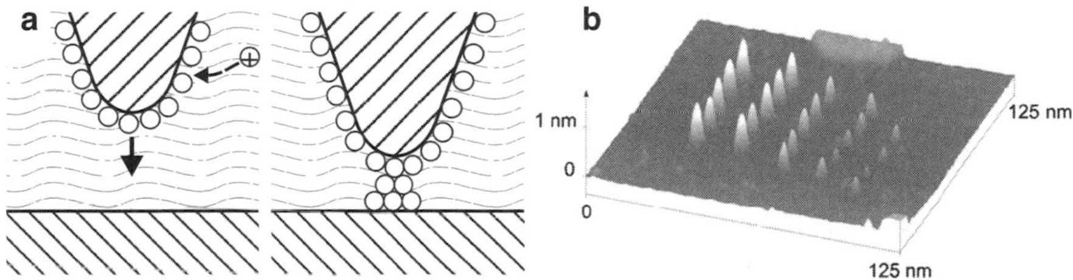

Fig. 24 (**a**) Schematic illustration of nanostructuring via electrodeposition on the tip and subsequent mechanical transfer to the sample. (**b**) Five rows of five Cu clusters each formed on Au(111) by this method. The tip displacement Δz increases from right to left, resulting in an increased cluster size (Reprinted from ref. [172] with permission by the author. Copyright by Elsevier)

the cohesion within the deposit, that is, in systems where underpotential deposition of the admetal is observed and the sample is consequently covered by a UPD adlayer. This is confirmed by the experimental observation that well-defined clusters can only be generated in UPD systems (e.g., Cu [170–172] or Pd [173, 175] on Au(111)), whereas in non-UPD systems cluster formation does not (e.g., for Ni on Au(111) [174]) or only erratically (e.g., for Cu on Ag(111) [172]) occur. In addition, computer simulations of this nanostructuring process indicated substantial alloying with the substrate during cluster formation, which may explain the exceptionally high stability of these clusters against electrochemical dissolution [175, 176]. However, this point is still under debate [174]. Overall, this tip-induced cluster formation is, till date, in terms of reproducibility and achievable feature size, one of the most successful examples of nanostructure formation by *in situ* SPM and has, despite its limitation to certain material combinations, greatly stimulated interest in electrochemical nanostructuring methods.

5.2 Nanostructuring by Modification of the Local Electrochemistry

Rather than using the tip of an *in situ* SPM as a nanomechanical tool, it also can be employed to change the electrochemical conditions at the interface. In the simplest case, the composition of the solution is locally changed by an electrochemical process at the tip, which is kept a few nanometers away from the surface. This composition change triggers a change in the local rates of an electrochemical reaction at the sample surface in the vicinity of the tip position. Conceptually, this approach is identical to operating the SPM in an SECM mode, differing mainly in the tip geometry: SPM tips are usually much sharper than SECM tips, but feature less well-defined tip geometry, making a quantitative description of the involved electrochemical and transport processes difficult. The formation of metal nanostructures on metal and semiconductor electrodes by this method was reported by Lorenz, Schindler, and co-workers [177–179]. In these studies, the STM was operated in a metal ion-containing electrolyte with the sample potential being kept slightly more positive than the nucleation threshold (Fig. 25a). The nanostructure was formed by first loading the tip with a metal by electrodeposition from solution, followed by stepping the tip potential to a very positive value, resulting in rapid redissolution of the deposit. Due to the resulting supersaturation electrochemical nucleation and growth of a metal cluster with typical diameters of 10–30 nm occurs. Examples of nanoclusters prepared by this technique include Co clusters on Au(111) and Pb on Si(111) (Fig. 25b) [177–179].

Local changes in the electrochemical behavior of the sample can also be induced by directly influencing the electrochemical reactions at the underlying electrode surface by the tip. Since the tip-sample distance in SPM is very small (e.g., typically $\approx$1 nm in STM), the electrochemical double layers of tip and sample can overlap, which may affect the electrostatic conditions, the electron transfer, and the mass transport in the gap between tip and sample. For example, locally enhanced

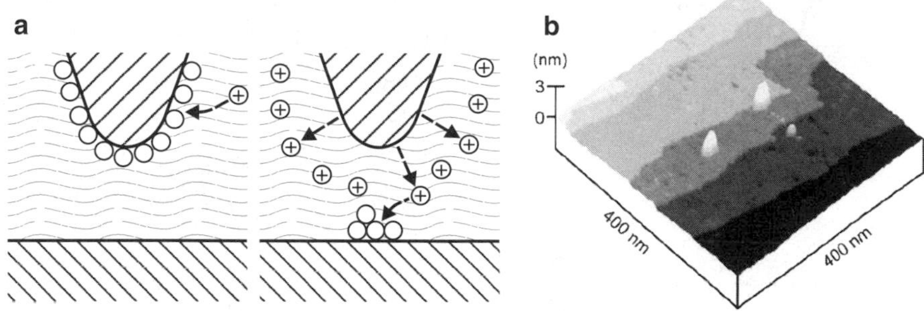

Fig. 25 (**a**) Schematic illustration of SPM-based nanostructuring, based on changing the local electrolyte composition via burst-like dissolution of previously electrodeposited material from the tip. (**b**) Two 22-nm diameter Pb clusters on n-type Si(111), prepared subsequently by local electrolyte supersaturation via Pb desorption from the STM tip (Reprinted from ref. [179] with permission by the author. Copyright by Elsevier)

metal dissolution rates of Cu [180] and Ag [181] electrodes at the tip position were observed in *in situ* STM studies if the tip potential was kept positive of the sample potential (i.e., at negative U_{bias}). This phenomenon was tentatively attributed to an oxidation of metal surface atoms induced by the tunneling of electrons from the sample to the tip [180]. Due to the high local-current density at the tip position such a process may cause a noticeably increased dissolution rate even if its efficiency is very low.

Schuster et al. introduced an interesting technique, where nanosecond pulses are employed to spatially confine electrochemical reactions, allowing the electrochemical machining of the sample with submicrometer precision [182, 183]. The characteristic timescale for double-layer charging, required for establishing the potential on the solution side of the interface, depends on the electrolyte resistance between tip and sample, which is orders of magnitudes smaller at sample areas close to the tip position (see Fig. 26a). By applying suitable chosen nanosecond pulses to the tip (or an arbitrary-shaped metal tool) a deposition or dissolution reaction may be limited to areas within submicrometer distances of the tip. As illustrated by the 5-μm deep spiral trough shown in Fig. 26b features with dimensions approaching 100 nm and a surface roughness of a few 10 nm can be etched by this method. Furthermore, not only etching of various metals and silicon surfaces, but also local electrodeposition has been demonstrated [182], making this technique a versatile tool for electrochemical micro- and possibly also nanostructuring.

5.3 Nanostructuring by Electrochemical Nanocells

An alternative to performing the nanostructuring in a bulk electrolyte is to confine the electrochemical environment to a droplet of nanometer dimensions (i.e., an electrochemical "nanocell") between tip apex and sample (Fig. 27a). The most popular

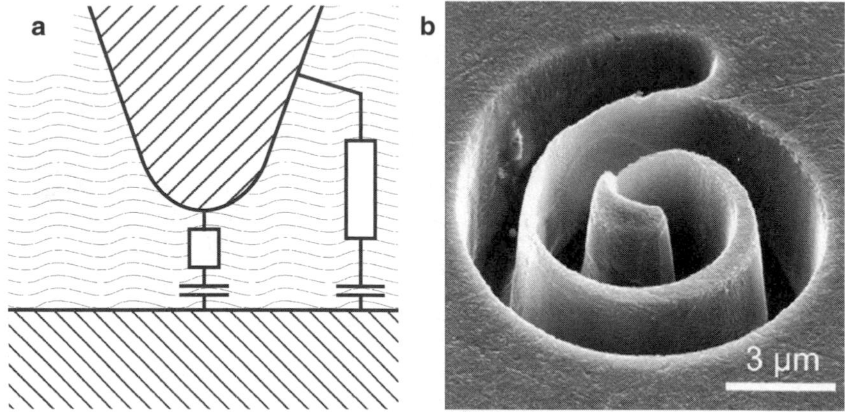

Fig. 26 (**a**) Principle of electrochemical micromachining by locally confined double-layer charging via ns-pulses, illustrating schematically the differences in the electrolyte resistance at different sample locations. (**b**) Spiral trough etched into Ni sheet using 3 ns pulses of 2 V (Reprinted from ref. [183] with permission by the author. Copyright by Elsevier)

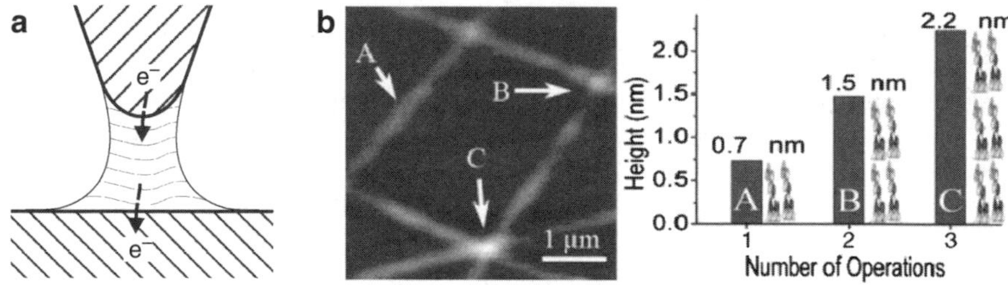

Fig. 27 (**a**) Schematic illustration of SPM-based nanostructuring methods, utilizing the tip-sample gap as an electrochemical nano cell. (**b**) Multiple mercaptopropyltrimethoxysilane lines prepared by dip-pen nanolithography combined with electro-oxidation via the AFM tip (Reprinted from ref. [191] with permission by the author. Copyright by American Chemical Society)

method based on this principle is a local oxidation of the sample by a conducting AFM tip. Although a number of metals and semiconductors that form stable oxides have been similarly nanostructured, most studies have focused on the local oxidation of Al and Si. The lateral dimensions and oxide thickness can be controlled by the ambient humidity, the applied voltage, and the tip-scanning speed. In addition, apart from test structures, simple devices, such as a single-electron transistor [184], have been fabricated. An overview of the extensive work done in this area is given in ref. [159, 185]. An interesting variation of the "nanocell" method was reported by Maoz et al., who employed it to electrochemical-pattern organic monolayers by tip-induced oxidation or reduction [186, 187].

Local electrochemistry in the liquid meniscus between tip and sample can also be combined with "dip-pen" nanolithography, where molecules are applied to the tip and then transferred via capillary transport through the meniscus onto the sample [188]. Liu and co-workers have introduced an electrochemical "dip-pen" method that utilizes local electrochemical reduction or oxidation reactions involving the delivered molecules, which allowed direct writing of metal or oxide lines and selective local etching of the surface [189, 190]. Cai and Ocko have combined "dip-pen" nanolithography with the approach of Maoz et al. demonstrating local conversion of an organic monolayer into a reactive state and "dip-pen" delivery of molecules (e.g., silanes) binding to these activated surface areas in a single sweep of the AFM tip [191]. As shown in Fig. 27b (right graph), subsequent application of this method even allows the formation of defined multilayer structures built by these molecules. Altogether, SPM nanostructuring methods employing an electrochemical nanocell are among the most versatile techniques to create structural or chemical pattern on surface with dimensions from a few 10 nm to the μm range and consequently have become very popular in the last years.

6 Conclusions

SPMs, in particular, scanning tunneling and scanning force microscopy, had and still have a significant part in ushering in a paradigm change toward true science and technology on the nanoscale. The possibility to view, analyze, and to manipulate surfaces down to atomic dimensions afforded by these methods makes them indispensable tools in a wide variety of fields, including electrochemistry. Because only few techniques allow *in situ* structure-sensitive studies of electrochemical interfaces, SPM methods were strongly embraced and have significantly contributed to a true microscopic picture of electrode surfaces and electrochemical processes since their introduction at the end of the 1980s of the last century. To the achievements of *in situ* SPM techniques belongs the clarification of the adlayer structure in a large number of adsorbate systems, the study of the mechanisms of phase-formation processes, such as electrochemical metal deposition, dissolution, and oxidation, the investigation of mass transport at electrochemical surfaces, and the study of double-layer structure and electron transfer by SPM spectroscopy. Furthermore, various nanostructuring methods which are based on the combination of SPM with electrochemical methods have been developed. Emerging trends are to apply SPM to even more complex systems and to employ it as a tool for the quantitative determination of local interface properties and elementary steps in electrochemical reactions. This will, in part, require multifunctional probes that allow to simultaneously measure different physical quantities or approaches that provide new information, for example, local chemical sensitivity. As has been discussed in this chapter, exciting new developments in this area are currently emerging.

References

1. H.-J. Güntherodt, R. Wiesendanger (eds.), *Scanning tunneling microscopy I-III, Springer Series in Surface Sciences* (Springer, Berlin, 1992).

2. R. Wiesendanger, *Scanning Probe Microscopy and Spectroscopy – Methods and Applications* (Cambridge University Press, Cambridge, 1994).
3. C.J. Chen, *Introduction to Scanning Tunneling Microscopy* (Oxford University Press, New York, 1993).
4. A. Foster, W. Hofer, *Scanning Probe Microscopy – Atomic Scale Engineering by Forces and Currents* (Springer, Berlin, 2006).
5. F.C. Crommie, C.P. Lutz, D.M. Eigler, *Science* **262**, 218 (1993).
6. J.W.M. Frenken, T.H. Oosterkamp, B.L.M. Hendriksen, M.J. Rost, *Mater. today* **8**, 20 (2005).
7. O.M. Magnussen, L. Zitzler, B. Gleich, M.R. Vogt, R.J. Behm, *Electrochim. Acta* **46**, 3725 (2001).
8. L. Zitzler, B. Gleich, O.M. Magnussen, R.J. Behm, *Proc. Electrochem. Soc.* **99–28**, 29 (2000).
9. M. Labayen, C. Ramirez, W. Schattke, O.M. Magnussen, *Nat. Mater.* **2**, 783 (2003).
10. T. Tansel, O.M. Magnussen, *Phys. Rev. Lett.* **96**, 026101-1 (2006).
11. H. Siegenthaler, R. Christoph, in: *Scanning tunneling microscopy and related methods, NATO ASI Series E. Vol. 184*, R.J. Behm, N. Garcia and H. Rohrer (eds.), Kluwer, Dordrecht (1990) p. 315.
12. H. Siegenthaler, in: *Springer series in surface sciences. Vol. 28. Scanning tunneling microscopy I*, H.-J. Güntherodt and R. Wiesendanger (eds.), Springer, Berlin-Heidelberg (1992) p. 7.
13. *Nanoscale probes of the solid/Liquid interface* A.A. Gewirth and H. Siegenthaler (eds.) (Kluwer Academic Publishers, Dordrecht-Boston-London, 1995).
14. A.A. Gewirth, B.K. Niece, *Chem. Rev.* **97**, 1129 (1997).
15. K. Itaya, *Prog. Surf. Sci.* **58**, 121 (1998).
16. H.-J. Butt, in: *Thermodynamics and Electrified Interfaces, Encyclopedia of Electrochemistry Vol. 1*, A.J. Bard and M. Stratmann (eds.), Wiley-VCH, Weinheim (2002) p. 225.
17. B.R. Horrocks, in: *Instrumentation and Electroanalytical Chemistry, Encyclopedia of Electrochemistry Vol. 3*, A.J. Bard and M. Stratmann (eds.), Wiley-VCH, (2001) p. 444.
18. J.V. Macpherson, in: *Instrumentation and Electroanalytical Chemisty, Encyclopedia of Electrochemistry Vol. 3*, A.J. Bard and M. Stratmann (eds.), Wiley-VCH, (2001) p. 415.
19. T.P. Moffat, in: *Instrumentation and Electroanalytical Chemistry, Encyclopedia of Electrochemistry Vol. 3*, A.J. Bard and M. Stratmann (eds.), Wiley-VCH, (2001) p. 393.
20. G.K. Binnig, H. Rohrer, *Helv. Phys. Acta* **55**, 726 (1982).
21. R. Sonnenfeld, P.K. Hansma, *Science* **232**, 211 (1986).
22. L.A. Nagahara, T. Thundat, S.M. Lindsay, *Rev. Sci. Instrum.* **60**, 3128 (1989).
23. C.E. Bach, R.J. Nichols, W. Beckmann, H. Meyer, A. Schulte, J.O. Besenhard, P.D. Jannakoudakis, *J. Electrochem. Soc.* **140**, 1281 (1993).
24. P. Allongue, V. Costa-Kieling, H. Gerischer, *J. Electrochem. Soc.* **140**, 1009 (1993).
25. P. Allongue, *Phys. Rev. Lett.* **77**, 1986 (1996).
26. J. Meier, J. Schiotz, P. Liu, J.K. Norskov, U. Stimming, *Chem. Phys. Lett.* **390**, 440 (2004).
27. T.H. Treutler, G. Wittstock, *Electrochim. Acta* **48**, 2923 (2003).
28. M. Szklarczyk, J.O. Bockris, *J. Electrochem. Soc.* **137**, 452 (1990).
29. S.-L. Yau, C.M. Vitus, B.C. Schardt, *J. Am. Chem. Soc.* **112**, 3677 (1990).
30. O.M. Magnussen, J. Hotlos, R.J. Nichols, D.M. Kolb, R.J. Behm, *Phys. Rev. Lett.* **64**, 2929 (1990).
31. J.V. Barth, H. Brune, G. Ertl, R.J. Behm, *Phys. Rev. B* **42**, 9307 (1990).
32. J. Tersoff, D.R. Hamann, *Phys. Rev. Lett.* **50**, 1998 (1983).
33. W.A. Hofer, A.S. Foster, A.L. Shluger, *Rev. Mod. Phys.* **75**, 1287 (2003).
34. P. Sautet, *Chem. Rev.* **97**, 1097 (1997).
35. W.A. Hofer, A.J. Fisher, R.A. Wolkow, P. Grütter, *Phys. Rev. Lett.* **87**, 236104-1 (2001).
36. D.M. Kolb, *Prog. Surf. Sci.* **51**, 109 (1996).

37. E. Budevski, G. Staikov, W.J. Lorenz, *Electrochemical phase formation and growth* (VCH, Weinheim, 1996).
38. E. Ammann, M.Sc. thesis, University of Bern (1995).
39. U. Müller, D. Carnal, H. Siegenthaler, E. Schmidt, W.J. Lorenz, W. Obretenov, U. Schmidt, G. Staikov, et al., *Phys. Rev. B* **46**, 12899 (1992).
40. N. Batina, T. Yamada, K. Itaya, *Langmuir* **11**, 4568 (1995).
41. O.M. Magnussen, J.X. Wang, R.R. Adzic, B.M. Ocko, *J. Phys. Chem.* **100**, 5500 (1996).
42. M. Schmid, H. Stadler, P. Varga, *Phys. Rev. Lett.* **70**, 1441 (1993).
43. F. Maroun, F. Ozanam, O.M. Magnussen, R.J. Behm, *Science* **293**, 1811 (2001).
44. O.M. Magnussen, *Chem. Rev.* **102**, 679 (2002).
45. T. Wandlowski, in: *Thermodynamics and electrified interfaces, Encyclopedia of Electrochemistry Vol. 1*, A.J. Bard and M. Stratmann (eds.), WILEY-VCH, Weinheim (2003) p. 383.
46. J. Wintterlin, R.J. Behm, in: *Springer Series in Surfaces Sciences, Vol. 20. Scanning Tunneling Microscopy I*, H.-J. Güntherodt and R. Wiesendanger (eds.), Springer, Berlin-Heidelberg (1992) p. 40.
47. O.M. Magnussen, J. Hageböck, J. Hotlos, R.J. Behm, *Faraday Discuss.* **94**, 329 (1992).
48. N.J. Tao, S.M. Lindsay, *J. Phys. Chem.* **96**, 5213 (1992).
49. P. Broekmann, M. Wilms, M. Kruft, C. Stuhlmann, K. Wandelt, *J. Electroanal. Chem.* **467**, 307 (1999).
50. P. Broekmann, M. Wilms, K. Wandelt, *Surf. Rev. Lett.* **6**, 907 (1999).
51. M. Kunitake, N. Batina, K. Itaya, *Langmuir* **11**, 2337 (1995).
52. K. Suto, S. Yoshimoto, K. Itaya, *J. Am. Chem. Soc.* **125**, 14976 (2003).
53. C. Safarowsky, L. Merz, A. Rang, P. Broekmann, B.A. Hermann, C.A. Schalley, *Angewandte Chemie-International Edition* **43**, 1291 (2004).
54. C. Safarowsky, A. Rang, C.A. Schalley, K. Wandelt, P. Broekmann, *Electrochim. Acta* **50**, 4257 (2005).
55. J. Zhang, Q. Chi, A.M. Kuznetsov, A.G. Hansen, H. Wackerbarth, H.E.M. Christensen, J.E.T. Andersen, J. Ulstrup, *J Phys Chem B* **106**, 1131 (2002).
56. X. Gao, A. Hamelin, M.J. Weaver, *Phys. Rev. Lett.* **67**, 618 (1991).
57. X. Gao, A. Hamelin, M.J. Weaver, *Phys. Rev. B* **44**, 10983 (1991).
58. N.J. Tao, S.M. Lindsay, *Surf. Sci.* **274**, L546 (1992).
59. X. Gao, G.J. Edens, A. Hamelin, M.J. Weaver, *Surf. Sci.* **296**, 333 (1993).
60. O.M. Magnussen, J. Hotlos, R.J. Behm, N. Batina, D.M. Kolb, *Surf. Sci.* **296**, 310 (1993).
61. O.M. Magnussen, J. Wiechers, R.J. Behm, *Surf. Sci.* **289**, 139 (1993).
62. M. Labayen, O.M. Magnussen, *Surf. Sci.* **573**, 128 (2004).
63. E. Sibert, F. Ozanam, F. Maroun, O.M. Magnussen, R.J. Behm, *Phys. Rev. Lett.* **90**, 056102 (2003).
64. Th. Dretschkow, D. Lampner, Th. Wandlowski, *J. Electroanal. Chem.* **458**, 121 (1998).
65. D.J. Trevor, C.E.D. Chidsey, D.N. Loiacono, *Phys. Rev. Lett.* **62**, 929 (1989).
66. R.J. Nichols, O.M. Magnussen, J. Hotlos, T. Twomey, R.J. Behm, D.M. Kolb, *J. Electroanal. Chem.* **290**, 21 (1990).
67. M. Giesen, G. Beltramo, S. Dieluweit, J. Müller, H. Ibach, W. Schmickler, *Surf. Sci.* **595**, 127 (2005).
68. M. Giesen, *Prog. Surf. Sci.* **68**, 1 (2001).
69. T. Will, M. Dietterle, D.M. Kolb, in: *Nanoscale probes of the solid/liquid interface*, Vol. E288. A.A. Gewirth and H. Siegenthaler (eds.), Kluwer Academic Publishers, Dordrecht-Boston-London (1995) p. 137.
70. M. Giesen, M. Dietterle, D. Stapel, H. Ibach, D.M. Kolb, *Surf. Sci.* **384**, 168 (1997).
71. M. Labayen, C. Haak, O.M. Magnussen, *Phys. Rev. B* **71**, 241409-1 (2005).
72. R.J. Nichols, in: *Imaging of surfaces and interfaces, Vol. IV*. R.N. Ross and J. Lipkowski (eds.), VCH, New York (1999)
73. E. Sibert, F. Ozanam, F. Maroun, R.J. Behm, O.M. Magnussen, *Surf. Sci.* **572**, 115 (2004).
74. R.J. Randler, D.M. Kolb, B.M. Ocko, G.N. Robinson, *Surf. Sci.* **447**, 187 (2000).

75. F. Möller, J. Kintrup, A. Lachenwitzer, O.M. Magnussen, R.J. Behm, *Phys. Rev. B* **56**, 12506 (1997).
76. O. Skylar, T.H. Treutler, N. Vlachopoulos, G. Wittstock, *Surf. Sci.* **597**, 181 (2005).
77. B.M. Ocko, I.K. Robinson, M. Weinert, R.J. Randler, D.M. Kolb, *Phys. Rev. Lett.* **83**, 780 (1999).
78. R. Randler, M. Dietterle, D.M. Kolb, *Z. Phys. Chem.* **208**, 43 (1999).
79. D.Y. Petrovykh, F.J. Himpsel, T. Jung, *Surf. Sci.* **407**, 189 (1998).
80. P. Berenz, S. Tillmann, H. Massong, H. Baltruschat, *Electrochim. Acta* **43**, 3035 (1998).
81. S. Morin, A. Lachenwitzer, O.M. Magnussen, R.J. Behm, *Phys. Rev. Lett.* **83**, 5066 (1999).
82. S. Strbac, O.M. Magnussen, R.J. Behm, *Phys. Rev. Lett.* **83**, 3246 (1999).
83. P. Allongue, L. Cagnon, C. Gomes, A. Gundel, V. Costa, *Surf. Sci.* **557**, 41 (2004).
84. M.D. Lay, J.L. Stickney, *J. Am. Chem. Soc.* **125**, 1352 (2003).
85. F. Möller, O.M. Magnussen, R.J. Behm, *Phys. Rev. Lett.* **77**, 5249 (1996).
86. A. Lachenwitzer, PhD thesis, University Ulm (2000).
87. D. D.Chambliss, R.J. Wilson, S. Chiang, *Phys. Rev. Lett.* **66**, 1721 (1991).
88. J.A. Meyer, I.D. Baikie, E. Kopatzki, R.J. Behm, *Surf. Sci.* **365**, L647 (1996).
89. A.W. Stephenson, Ch.J. Baddeley, M.S. Tikhov, R.M. Lambert, *Surf. Sci.* **398**, 172 (1998).
90. H. Brune, M. Giovanni, K. Bromann, K. Kern, *Nature* **394**, 451 (1998).
91. Y.-C. Yang, S.-L. Yau, Y.-L. Lee, *J. Am. Chem. Soc.* **128**, 3677 (2006).
92. B.W. Gregory, J.L. Stickney, *J. Electroanal. Chem.* **300**, 543 (1991).
93. K. Varazo, M.D. Lay, T.A. Sorenson, J.L. Stickney, *J. Electroanal. Chem.* **522**, 104 (2002).
94. F. Endres, W. Freyland, *J. Phys. Chem. B* **102**, 10229 (1998).
95. F. Endres, *Phys. Chem. Chem. Phys.* **3**, 3165 (2001).
96. F. Endres, S. Zein, El Abedin, *Phys. Chem. Chem. Phys.* **4**, 1640 (2002).
97. N. Borisenko, S. Zein, El Abedin, F. Endres, *J. Phys. Chem. B* **110**, 6250 (2006).
98. O.M. Magnussen, R.J. Behm, *Mat. Res. Bull.* **24**, 16 (1999).
99. E. Ammann, C. Beuret, P.-F. Indermühle, R. Kötz, N.F. de Rooij, H. Siegenthaler, *Electrochim. Acta* **47**, 327 (2001).
100. H. Siegenthaler, E. Ammann, P.-F. Indermühle, G. Repphun, in: *Nanoscale Science and Technology*, N. Garcia and et al. (eds.), Kluwer Academic Publishers, (1998) p. 297.
101. O.M. Magnussen, W. Polewska, L. Zitzler, R.J. Behm, *Faraday Discuss.* **121**, 43 (2002).
102. W. Polewska, R.J. Behm, O.M. Magnussen, *Electrochim. Acta* **48**, 2915 (2003).
103. O.M. Magnussen, M.R. Vogt, *Phys. Rev. Lett.* **84**, 357 (2000).
104. J. Wiechers, T. Twomey, D.M. Kolb, R.J. Behm, *J. Electroanal. Chem.* **248**, 451 (1988).
105. R. Christoph, H. Siegenthaler, H. Rohrer, H. Wiese, *Electrochim. Acta* **34**, 1011 (1989).
106. M. Binggeli, D. Carnal, R. Nyffenegger, H. Siegenthaler, R. Christoph, H. Rohrer, *J. Vac. Sci. Technol. B* **9**, 1985 (1991).
107. W. Schmickler, D. Henderson, *J. Electroanal. Chem.* **290**, 283 (1990).
108. J.K. Gimzewski, J.K. Sass, *J. Electroanal. Chem.* **308**, 333 (1991).
109. O. Pecina, W. Schmickler, K.Y. Chan, D.J. Henderson, *J. Electroanal. Chem.* **396**, 303 (1995).
110. W. Schmickler, *Surf. Sci.* **335**, 416 (1995).
111. G. Nagy, G. Denuault, *J. Electroanal. Chem.* **437**, 37 (1997).
112. A. Mosyak, P. Graf, I. Benjamin, A. Nitzan, *J. Phys. Chem. A* **101**, 429 (1997).
113. M. Galperin, A. Nitzan, *J. Phys. Chem. A* **106**, 10790 (2002).
114. A. Vaught, T.W. Jing, S.M. Lindsay, *Chem. Phys. Lett.* **236**, 306 (1995).
115. M. Hugelmann, W. Schindler, *Surf. Sci. Lett.* **541**, L643 (2003).
116. R. Hiesgen, D. Eberhardt, D. Meissner, *Surf. Sci.* **597**, 80 (2005).
117. E.-M. Choi, Y.-H. Yoon, S. Lee, H. Kang, *Phys. Rev. Lett.* **95**, 085701-1 (2005).
118. W. Schmickler, C. Widrig, *J. Electroanal. Chem.* **336**, 213 (1992).
119. N.J. Tao, *Phys. Rev. Lett.* **76**, 4066 (1996).
120. W. Schmickler, in: *Imaging of surfaces and interfaces*, Vol. 5. J. Lipkowski and P.N. Ross (eds.), Wiley-VCH, NY (1999) p. 305.
121. D.I. Gittins, D. Bethell, D.J. Schiffrin, R.J. Nichols, *Nature* **408**, 67 (2000).

122. B. Xu, X. Xiao, X. Yang, L. Zang, N. Tao, *J. Am. Chem. Soc.* **127**, 2386 (2005).
123. G.K. Binnig, C.F. Quate, C. Gerber, *Phys. Rev. Lett.* **56**, 930 (1986).
124. F. Ohnesorge, G.K. Binnig, *Science* **260**, 1451 (1993).
125. T.R. Albrecht, P. Grütter, D. Horne, D. Rugar, *J. Appl. Phys.* **69**, 668 (1991).
126. F.J. Giessibl, *Science* **267**, 68 (1995).
127. F.J. Giessibl, *Rev. Mod. Phys.* **75**, 949 (2003).
128. I. Revenko, R. Proksch, *J. Appl. Phys.* **87**, 526 (2000).
129. S. Manne, P.K. Hansma, J. Massie, V.B. Elings, A.A. Gewirth, *Science* **251**, 133 (1991).
130. C.-H. Chen, A.A. Gewirth, *J. Am. Chem. Soc.* **114**, 5439 (1992).
131. J. Wang, G.M. Watson, B.M. Ocko, *Physica A* **200**, 679 (1993).
132. N. Tao, Z. Shi, *J. Phys. Chem.* **98**, 7422 (1994).
133. T. Fukuma, K. Kobayashi, K. Matsushige, H. Yamada, *Appl. Phys. Lett.* **87**, 034101-1 (2005).
134. B.W. Hoogenboom, H.J. Hug, Y. Pelimont, S. Martin, P.L.T.M. Frederix, D. Fotiadis, A. Engel, *Appl. Phys. Lett.* **88**, 193109-1 (2006).
135. D.M. Kolb, R.J. Randler, R.I. Wielgosz, J.C. Ziegler, *Mat. Res. Soc. Symp. Proc.* **451**, 19 (1997)
136. M.L. Munford, F. Maroun, R. Cortès, P. Allongue, A A. Pasa, *Surf. Sci.* **537**, 95 (2003).
137. S. Huo, W. Schwarzacher, *Phys. Rev. Lett.* **86**, 256 (2001).
138. S. Boussaad, N.J. Tao, ., *J. Am. Chem. Soc.* **121**, 4510 (1999).
139. A. Kueng, C. Kranz, A. Lugstein, E. Bertagnolli, B. Mizaikoff, *Angew. Chem.* **42**, 3238 (2003).
140. P.S. Dobson, J.M.R. Weaver, *Anal. Chem.* **77**, 424 (2005).
141. R. Raiteri, M. Grattarola, H.-J. Butt, *J. Phys. Chem.* **100**, 16700 (1996).
142. A.C. Hillier, S. Kim, A.J. Bard, *J. Phys. Chem.* **100**, 18808 (1996).
143. J. Israelachvili, *Intermolecular and Surface Forces* (Academic Press, 1991).
144. M. Rief, F. Oesterhelt, B. Heymann, H.E. Gaub, *Science* **275**, 1295 (1997).
145. M. Rief, M. Gautel, F. Oesterhelt, J.M. Fernandez, H.E. Gaub, *Science* **276**, 1109 (1997).
146. A.F. Oberhauser, P.E. Marszalek, H.P. Erickson, J.M. Fernandez, *Nature* **393**, 181 (1998).
147. M. Grandbois, M. Beyer, M. Rief, H. Clausen-Schaumann, H.E. Gaub, *Science* **283**, 1727 (1999).
148. S.K. Sinniah, A.B. Steel, C.J. Miller, J.E. Reutt-Robey, *J. Am. Chem. Soc.* **118**, 8925 (1996).
149. H.-C. Kwon, A.A. Gewirth, *J. Phys. Chem. B* **109**, 10213 (2005).
150. H. Takano, J.R. Kenseth, S.-S. Wong, J.C. O'Brien, M.D. Porter, *Chem. Rev.* **99**, 2845 (1999).
151. A.J. Bard, F.-R. Fan, M.V. Mirkin, in: *Electroanalytical Chemistry*, A.J. Bard and M. Dekker (ed.) New York (1994) p. 244.
152. A.J. Bard, F. R. Fan, M. Mirkin, in: *Physical Electrochemistry: Principles, Methods, and Applications*, I. Rubenstein and M. Dekker (ed.) New York (1995) p. 209.
153. *Scanning Electrochemical Microscopy* A.J. Bard and M.V. Mirkin (eds.) (John Wiley & Sons, New York, 2001).
154. M.V. Mirkin, *Anal. Chem.* **68**, 177A (1996).
155. J.V. Macpherson, P.R. Unwin, *Anal. Chem.* **73**, 550 (2001).
156. B. Ren, G. Picardi, B. Pettinger, R. Schuster, G. Ertl, *Angew. Chem.* **117**, 141 (2005).
157. A. Cvitkovic, N. Ocelic, J. Aizpurua, R. Guckenberger, R. Hillenbrand, *Phys. Rev. Lett.* **97**, 060801-1 (2006).
158. R.M. Nyffenegger, R.M. Penner, *Chem. Rev.* **97**, 1195 (1997).
159. A.A. Tseng, A. Notargiacomo, T.P. Chen, *J. Vac. Sci. Technol. B* **23**, 877 (2005).
160. D.M. Eigler, E.K. Schweizer, *Nature* **344**, 524 (1990).
161. F.C. Crommie, C.P. Lutz, D.M. Eigler, *Nature* **363**, 524 (1993).
162. H.C. Manoharan, C.P. Lutz, D.M. Eigler, *Nature* **403**, 512 (2000).
163. S.-W. Hla, L. Bartels, G. Meyer, K.-H. Rieder, *Phys. Rev. Lett.* **85**, 2777 (2000).
164. J.K. Gimzewski, R. Möller, D.W. Pohl, R.R. Schlittler, *Surf. Sci.* **189/190**, 15 (1987).
165. W. Li, J.A. Virtanen, R.M. Penner, *J. Phys. Chem.* **96**, 6529 (1992).

166. J.V. Zoval, R.M. Stiger, R.P. Biernacki, R.M. Penner, *J. Phys. Chem.* **100**, 837 (1996).
167. J.R. LaGraff, A.A. Gewirth, *J. Phys. Chem.* **98**, 11246 (1994).
168. L. Santinacci, T. Djenizian, P. Schmuki, *Appl. Phys. Lett.* **79**, 1882 (2001).
169. L. Santinacci, T. Djenizian, H. Hildebrand, S. Ecoffey, H. Mokdad, T. Campanella, P. Schmuki, *Electrochim. Acta* **48**, 3123 (2003).
170. R. Ullmann, T. Will, D.M. Kolb, *Chem. Phys. Lett.* **209**, 238 (1993).
171. D.M. Kolb, R. Ullmann, T. Will, *Science* **275**, 1097 (1997).
172. D.M. Kolb, R. Ullmann, J.C. Ziegler, *Electrochim. Acta* **43**, 2751 (1998).
173. G.E. Engelmann, J.C. Ziegler, D.M. Kolb, *J. Electrochem. Soc.* **145**, L33 (1998).
174. D.M. Kolb, F.C. Simeone, *Electrochim. Acta* **50**, 2989 (2005).
175. M.G. Del Pópolo, E.P.M. Leiva, H. Kleine, J. Meier, U. Stimming, M. Mariscal, W. Schmickler, *Electrochim. Acta* **48**, 1287 (2003).
176. M.G. Del Pópolo, E.P. M.Leiva, W. Schmickler, *Angew. Chem.* **113**, 4807 (2001).
177. R.T. Pötzschke, G. Staikov, W.J. Lorenz, W. Wiesbeck, *J. Electrochem. Soc.* **146**, 141 (1999).
178. W. Schindler, D. Hofmann, J. Kirschner, *J. Electrochem. Soc.* **148**, C124 (2001).
179. W. Schindler, P. Hugelmann, M. Hugelmann, F.X. Kärtner, *J. Electroanal. Chem.* **522**, 49 (2002).
180. Z.-X. Xie, D.M. Kolb, *J. Electroanal. Chem.* **481**, 177 (2000).
181. S.G. García, D.R. Salinas, C.E. Mayer, W.J. Lorenz, G. Staikov, *Electrochim. Acta* **48**, 1279 (2003).
182. R. Schuster, V. Kirchner, P. Allongue, G. Ertl, *Science* **289**, 98 (2000).
183. M. Kock, V. Kirchner, R. Schuster, *Electrochim. Acta* **48**, 3213 (2003).
184. K. Matsumoto, M. Ishii, K. Segawa, Y. Oka, B.J. Vartanian, J.S. Harris, *Appl. Phys. Lett.* **68**, 34 (1996).
185. Ph.Avouris, R. Martel, T. Hertel, R. Sandstrom, *Applied Physics A* **66**, S659 (1998).
186. R. Maoz, S.R. Cohen, J. Sagiv, *Adv. Mater.* **11**, 55 (1999).
187. R. Maoz, E. Frydman, S.R. Cohen, J. Sagiv, *Adv. Mater.* **12**, 725 (2000).
188. R.D. Piner, J. Zhu, F. Xu, S. Hong, C.A. Mirkin, *Science* **283**, 661 (1999).
189. Y. Li, B.W. Maynor, J. Liu, *J. Am. Chem. Soc.* **123**, 2105 (2001).
190. B.W. Maynor, J. Li, C. Lu, J. Liu, *J. Am. Chem. Soc.* **126**, 6409 (2004).
191. Y. Cai, B.M. Ocko, *J. Am. Chem. Soc.* **127**, 16287 (2005).
192. F. Möller, diploma thesis, University Ulm (1993).

X-ray Lithography Techniques, LIGA-Based Microsystem Manufacturing: The Electrochemistry of Through-Mold Deposition and Material Properties

James J. Kelly and S.H. Goods

1 Introduction to LIGA Fabrication and Its Applications

Certain microsystem fabrication techniques are critically dependent on the electrochemistry of metal deposition into lithographically defined features that are developed in insulating molding materials. One such technique, developed originally at the Forschungzentrum Karlsruhe, Germany, is known as LIGA, the German acronym for lithography, electroplating, and replication (*Lithographie, Galvanoformung*, and *Abformung*) [1–3]. An example of typical miniature structures formed by plating through thick photoresist (the insulating molding materials) is shown in Fig. 1. Since its inception in Germany in the 1980s, LIGA research activities have expanded throughout Europe, as well as in Asia and North America.

Thanks to the intense and collimated X-ray synchrotron radiation used in the lithography step, the definition of high aspect ratio features in thick photoresist with good-dimensional fidelity is a unique attribute of the technique [4]. For example, high aspect ratio features (i.e., the height-to-width ratio) as high as 100 and feature depths up to $\approx$1 mm have been realized. These features are then filled with a metal via electrodeposition through the X-ray exposed photoresist mold that is bonded to a conductive substrate (the details of the LIGA process and other LIGA-like processes will be discussed later in this chapter). The unique geometries encountered in LIGA through-mask electrodeposition present distinct challenges. This is particularly the case with respect to the resultant materials that are required to have special properties (e.g., mechanical performance characteristics or magnetic properties). Consequently, a substantial portion of LIGA research is dedicated to first understand the electrochemistry of the LIGA through-mask electrodeposition process, and second, the subsequent material properties for electrodeposited metals, alloys, and metal-matrix composites [5–6].

James J. Kelly (✉)
IBM, Electrochemical Processes, 255 Fuller Rd., Albany, NY 12203, USA
e-mail: mjkelly@us.ibm.com

P. Schmuki, S. Virtanen (eds.), *Electrochemistry at the Nanoscale,* Nanostructure Science and Technology, DOI 10.1007/978-0-387-73582-5_3,

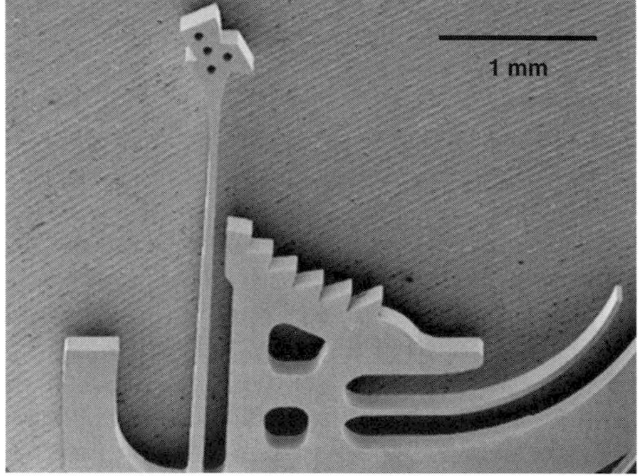

Fig. 1 Example of Ni structures fabricated by electroplating through PMMA resist that had been patterned by X-ray lithography. This technique of using electrodeposition to deposit metal into molds that have been lithographically formed with X-ray synchrotron radiation is referred to as the LIGA technique (German for *lithographie, galvanoformung*, and *abformung*; in English, lithography, electroplating, and replication). The top optical shows a variety of structures still attached to the metallized substrate. The lower image is an scanning electron micrograph showing a portion of a released structure. Images courtesy of Georg Aigeldinger, Sandia National Laboratories, Livermore, California

Most structures fabricated with the LIGA process generally have critical lithographic dimensions on the micron scale [2]. In principle, submicron features are possible, but such fine features are not routinely fabricated [2, 7]. Although the physical dimensions of LIGA-fabricated components are rarely on the nanoscale, important microstructural aspects of the electrodeposit often are. For example, electrodeposited materials are well known for exhibiting fine microstructures, with grain sizes in the nm range [8]. In fact, this is somewhat fortuitous, as a grain size on a scale much smaller than that of the physical component dimensions is usually a requisite for that component to have a spatially uniform microstructure. But, as will be discussed later, the degree of this uniformity usually depends on the presence of other species (other metal ions, additives, or adsorbed species, for example) that may affect the deposit structure, particularly when the concentration of these species is sensitive to transport.

Our objective here is not a detailed review of the LIGA process or its technological applications; several reviews of this type are available [9–15]. Instead, focus is on the unique aspects of the electrochemistry of LIGA through-mask metal deposition and the generation of the fine and uniform microstructures necessary to ensure proper functionality of LIGA components. Before addressing these issues, brief descriptions of the LIGA process, its history, and applications are given for those unfamiliar with the technique.

1.1 LIGA Process Flow

The general flow of the process steps necessary to create LIGA microparts is shown in Fig. 2. The process consists of using synchrotron X-rays and an X-ray absorbing mask to transfer a desired pattern to a metallized wafer substrate on to which is bonded a thick, X-ray photoresist blank. This photoresist is most commonly high molecular weight polymethylmethacrylate (PMMA) which is rendered more susceptible to chemical dissolution by X-ray exposure. The resist is then "developed" using a commercial solvent to create prismatic cavities into which metal is electrodeposited. The remaining PMMA is then dissolved, and the prismatic metallic structures can be released by chemically dissolving the original metallization layer. What we have described here and shown in Fig. 1 is, in fact, often referred to as "direct" LIGA, wherein the final released parts are those intended to be used. Often, perhaps more typically, the final part is a monolith consisting of the electrodeposited structures attached to a backing plate (either the plating substrate itself or a top surface blank formed by overplating the mold). This "parent" tool is then used to replicate multiple generations of microparts in a variety of ways.

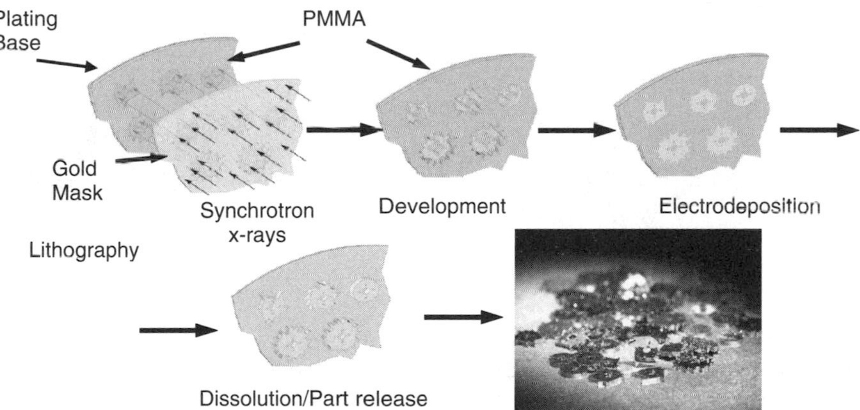

Fig. 2 Schematic diagram of the LIGA process flow. A LIGA mold consisting of a plating base (usually a metallized wafer) with thick PMMA resist is patterned with synchrotron X-rays by using a gold X-ray-absorbing mask. The features are developed and then electroplated, after which dissolution of the PMMA may take place. In the case shown here, the parts are released from the plating base to yield separate piece parts

Pattern Layout – Before starting the LIGA process shown in Fig. 2, a pattern layout must be created in order to fabricate the X-ray-absorbing mask that will be used during exposure. This entails taking a desired two-dimensional device geometry from a drawing (from CAD software, for example) and distributing it across an area corresponding to the surface of a wafer substrate. An example of a pattern layout of a spring prototype is shown in Fig. 3a. Unless special exposure techniques are employed [2], LIGA-fabricated devices are prismatic, that is, the part geometry does not vary in the z dimension, which is perpendicular to the substrate. Figure 3b shows the final-plated LIGA component from the layout shown in Fig. 3a.

A fair amount of experience is involved in arranging the devices judiciously across the wafer so as to facilitate subsequent processing. For example, a certain minimum separation distance between features may be desired to avoid thin, fragile areas of resist. Also, since large regions of uninterrupted resist tend to lead to significant physical distortions due to swelling and thermal expansion, ancillary structures may be added to improve the dimensional fidelity of features in critical regions. Where dimensional tolerances are of particular concern, these ancillary structures

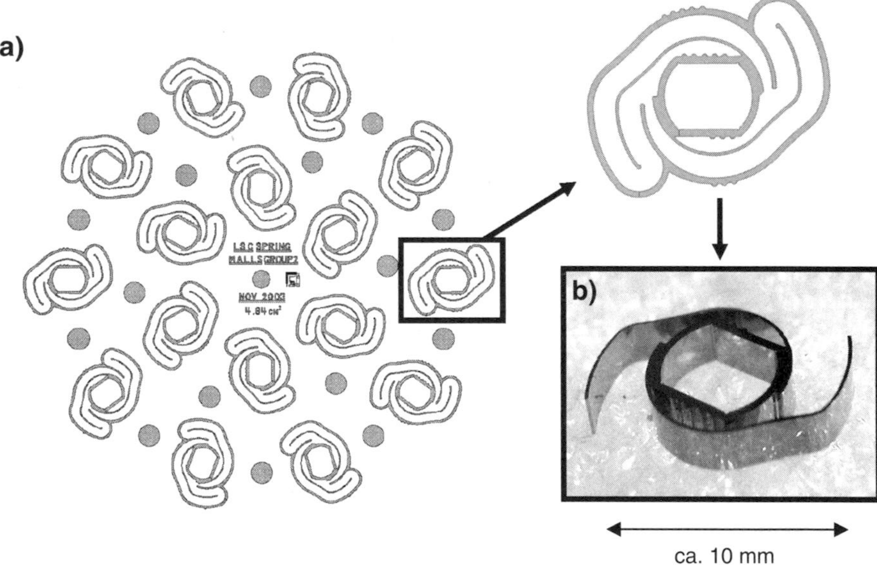

Fig. 3 (**a**) A layout of spring prototypes used in the photolithography necessary to fabricate the gold X-ray absorbing mask. The colored areas correspond to regions where thin UV resist remains after patterning for the X-ray mask. Since thick gold (several to tens of microns) is electroplated around and through these thin resist features to form the X-ray-absorbing mask, the pattern is transferred into the PMMA resist as its negative (the lack of thick gold in this region allows X-rays to impinge on the PMMA, sensitizing it for development to form open features in the resist). These features are in turn filled with electroplated metal, producing a final part after mechanical planarization and release from the substrate as shown in (**b**). Layout image courtesy of M. Hekmati and L. Hunter of Sandia National Laboratories, Livermore CA

are placed close to and conformal to part features. The intent is to minimize the uninterrupted length of resist immediately adjacent to these critically dimensioned structures, reducing the overall swelling and CTE-driven distortions. This also has the beneficial effect with respect to the electrodeposition process itself. By placing openings in large unpatterned areas of the resist, these ancillary features improve the homogeneity of the overall wafer with respect to the "active area density"[16], making the pattern-scale current distribution more uniform. This improves the overall uniformity with respect to deposition rate, which in turn, may favorably influence material properties and structure. In Fig. 3a, the moat-like features surrounding the spring geometry are an example of a structure serving these purposes, and as an added benefit, they can be used as sacrificial samples for materials analysis. One other benefit derived from these ancillary structures is the fact that they help support critical features during planarization and lapping, avoiding the tendency for fine features to lap faster (resulting in thinner in-plane dimensions) than neighboring large features. Once an acceptable layout of desired parts is obtained, one may proceed with the production of the X-ray-absorbing mask.

Mask Substrate – The fabrication procedure for an X-ray mask mainly depends on the particular X-ray spectrum at the synchrotron facility. Thicker absorbers are typically needed when the X-ray energies used are relatively high, and the exact absorber thickness required for a given X-ray spectrum has been worked out by a number of groups [2, 17–18]. Electrodeposited gold is usually used as the absorbing material, although other materials and deposition processes have been investigated [2]. We limit our discussion to the more common method that involves the use of electroplated gold.

For low absorber thicknesses ($<$10 μm), thin metallic membrane substrates can be used; this is usually the situation at synchrotron beamlines delivering a soft (low-energy) X-ray spectrum. At such soft sources, the substrate must be sufficiently "transparent" to X-rays to allow them to pass through the unmasked regions. Titanium has been used as an X-ray-transparent membrane for this purpose (as well as other materials) [2]. In this instance, after patterning and deposition of the absorber gold, the titanium is partially etched away from the area of interest (usually the center of a wafer) resulting in an X-ray-transparent membrane that is supported by thicker annular ring or "picture frame" of Ti. For harder X-ray sources, thicker absorbers ($>$10 μm) must be used, and the substrates are required to have greater mechanical integrity to support the patterned absorber gold. For this mask type, silicon and beryllium are often used as sheet substrates in thickness that can range from 100 to 500 μm. If the substrate is not conductive, a thin metallization layer is applied. Since the X-ray spectrum employed is fairly hard in this situation, X-ray transparency in the unmasked regions is generally of less concern.

Mask Photoresist – The next step, the patterning of photoresist, is the same for either approach. Commercially available photoresist of a desired thickness is patterned by standard contact UV lithography to define the pattern-layout geometry on the mask. Where the gold is deposited (whether as the positive or negative of the feature to be fabricated) depends on the development characteristics of the mold photoresist used to pattern the structures to be electroformed. This is the so-called tone of the mask. Regardless, only the active (electrically conductive) areas receive the plated gold. The thickness of the resist must be about 25% higher than the target thickness of the deposited gold so as to not overplate the resist (which would reduce the dimensional fidelity of the final electrodeposited structures). Sulfite baths are often used for the gold deposition to avoid damaging the UV mask photoresist. After gold deposition, the resist is usually removed from the mask by dissolution in a solvent.

Since high-precision patterning of UV photoresist becomes more difficult with increasing resist thickness [2], thin absorbers on membrane masks yield the greatest precision for small features. Sheet-substrate masks with thicker absorber gold may be mechanically more robust, but tend to be used for less demanding, more routine LIGA prototyping where high tolerances are less critical. These sheet-substrate masks are convenient as the entire wafer area is available for processing, whereas for membrane masks, only the thinned region may be used. Generally speaking, for mask resist measuring a few microns in thickness, features may be reliably defined down to about a micron in width, although process limits will vary with lithography infrastructure at a given laboratory [2]. More advanced techniques for defining finer features for submicron LIGA will be discussed at a later point [2, 7].

Mold Exposure – After an X-ray mask is in hand, the next step in the LIGA process is the exposure at a synchrotron to produce molds for the subsequent electrodeposition step. The molds, consisting of wafers bearing PMMA photoresist of a desired thickness, must be prepared in advance. This entails the bonding of a disk of PMMA to the wafer that has been metallized with a conductive seed layer for subsequent electrodeposition. The exposure of these molds necessitates a substantial level of infrastructure. At a minimum, an apparatus for holding and translating the mold so as to expose the entire wafer to the X-ray beam is required. More sophisticated equipment is necessary for more advanced exposure techniques, such as angled exposures to produce graded side walls, or alignment tools for multilevel processing that require multiple masks and X-ray exposures [2]. The mask is placed between the X-ray source and the mold, allowing the X-rays to pass through the areas where there is no gold absorber. The X-rays that pass through the mask impinge on the PMMA, scissioning the PMMA polymer chains, thereby sensitizing it to dissolution in the subsequent development step. The exposure times are dictated by the characteristics of the X-ray source, beamline parameters (mirrors and filters), mask substrate material and thickness, and finally, the thickness of the PMMA resist [17].

Mold Development and Seed Layer Preparation – After exposure, the X-ray-exposed regions of the PMMA may be dissolved by immersion of the mold in an appropriate chemical developer. Developers are mixtures of various organic solvents formulated to selectively dissolve the exposed PMMA, while leaving behind undissolved, unexposed PMMA. In this way, the geometry of the discrete structures on the mold is defined. Dissolution times depend on the mold thickness and specific development conditions, but typically range from a few minutes to several hours.

When development is complete (confirmation may necessitate observing features under a microscope to check for residual PMMA at the bottom of the exposed mold features), the mold is rinsed with deionized water and dried.

Depending on the metallization applied to the wafer, certain dry or wet etching processes may be necessary after development to fully reveal the desired metal seed layer. A number of materials have been used as conductive seed surfaces in LIGA [19–20]. Since the X-rays may interact with the seed layer material during exposure, some consideration must be given to the choice of metal used. These interactions often involve the emission of secondary radiation that may damage (overexpose) the PMMA/wafer interface. These aspects will be discussed in more detail in the electrodeposition section.

Electrodeposition – In order for the electrodeposition reaction to start, the seed surface must be first wetted by the electrolyte. This is usually accomplished by immersing the mold in water that has been de-aerated, as condensed gases in aerated water may form bubbles within the developed cavities of the mold. The container of water with the immersed mold is then placed in a chamber where a vacuum is used to facilitate the penetration of water into the mold features. A surfactant may be added to the water to lower its surface tension, further facilitating the wetting of the features. After the features are filled with water, the mold is then placed in the desired bath for electrodeposition, where the water in the features mixes with the bulk electrolyte. Depending on the feature geometry, some time is given for this mixing to occur before the electrodeposition of the mold is started.

Planarization and Release – Electrodeposition is complete when all of the patterned features exhibit some degree of overplating. Due to wafer- and pattern-scale effects, the local deposition rate varies from point to point on the wafer, so the extent of overplating varies across the mold. This overplating must be removed mechanically, by grinding and/or lapping off the excess metal. Grinding is followed by a planarization step, yielding a mold with both the final desired thickness and an appropriate surface finish. After planarization, the PMMA is dissolved in a solvent, such as acetone, if the parts are to be freed from the wafer and used individually (direct LIGA). However, in some cases, a new, thick metal layer is deposited on top of the planarized surface to form a structural backing on top of the plated features in order to use the LIGA mold as an embossing tool for replication. Another way to accomplish the same end is by simply allowing the electrodeposited metal to grow completely out of the features, coalesce, and form into a monolithic metal disk. These latter two approaches are illustrated schematically in Fig. 4.

After dissolution of the PMMA, if the plated features are going to be used as individual microparts themselves, the plating base layer must be dissolved to liberate the individual structures. Since copper is often used as a plating base and the LIGA structures themselves are often Ni or Ni alloys, chemical solutions selective to Cu may be used to dissolve the plating base, leaving the Ni unaffected. The microparts may then be collected and sorted for postprocessing metrology or testing required for the desired application.

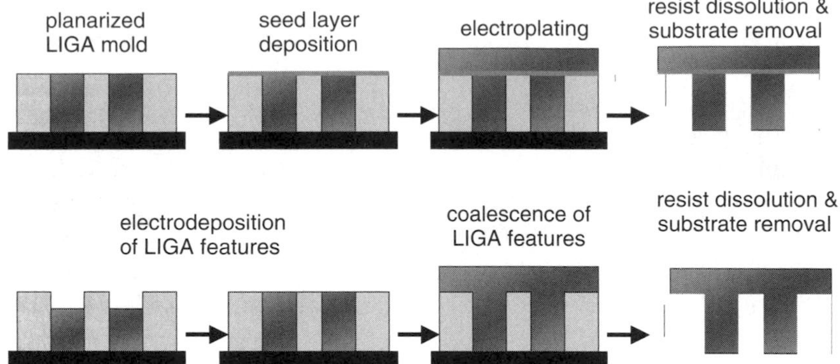

Fig. 4 Process schemes for producing a thick tool bearing LIGA features for subsequent embossing or injection molding. (*top*) The LIGA mold is planarized so a new metal seed layer may be applied to the top of the mold, typically by vacuum deposition. A layer of thick Ni is then electroplated over the seed layer, after which the resist is dissolved and the substrate removed (either chemically or mechanically). (*bottom*) Electrodeposition is allowed to continue up and out of the resist until the overplated regions coalesce into a solid Ni piece. The substrate is removed as described in the prior approach. For this technique to work, it is necessary for the parts to be located close to one another so the overplated regions can join together

1.2 LIGA History

Electrodeposition into features defined in resist by electron beam and X-ray lithography was first explored in the late 1960s and early 1970s by Romankiw and coworkers at IBM [21]. Also, the through-mask plating technology for producing gold X-ray masks necessary for X-ray exposures was developed by the same workers during this period. At the same time, through-mask electroplating was already becoming a preferred technique for the realization of patterned metal structures in the electronics industry for packaging and memory applications. It was thought possible that X-ray lithography would become necessary to replace optical lithography for features less than one micron. Romankiw gives a thorough review of the role of through-mask electroplating in the early stages of e-beam and X-ray lithography [21]. Figure 5 shows one of the earliest X-ray masks fabricated by Romankiw and co-workers by through-mask electroplating of gold using first an e-beam fabricated mask and subsequent X-ray lithography. The X-ray sources used up to 1973 by the group at IBM were laboratory sources and hence of low power. That year, high-power X-rays from a synchrotron in Hamburg, Germany, were employed by the IBM group to fabricate features in PMMA resist 12 μm thick and 1 μm wide. The separation between the features was 0.1 μm. Gold was electrodeposited to a thickness of 6 μm to fill the features, resulting in the first through-mask deposited structures in PMMA patterned with synchrotron radiation. These structures are shown in Fig. 6. Since such high aspect ratio features were not required for electronic devices, subsequent work by the group at IBM focused on thin structures having small lateral dimensions.

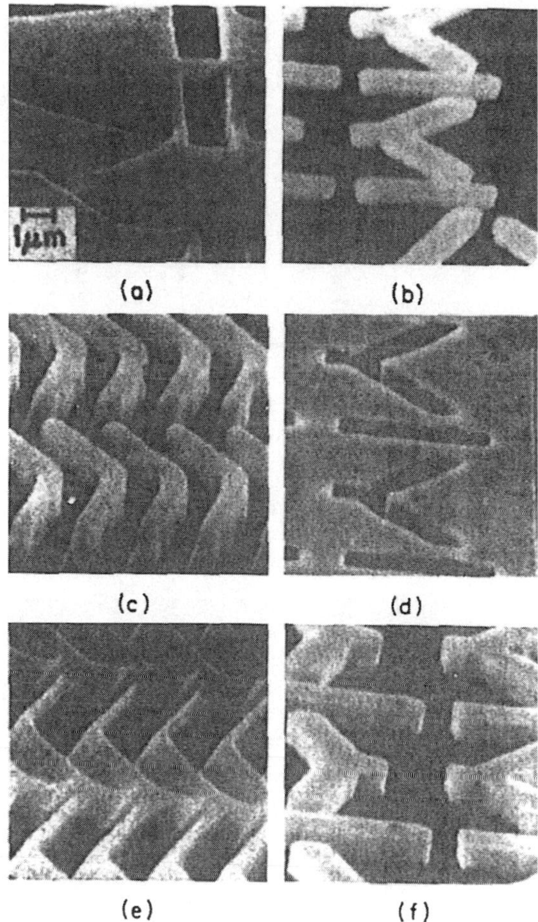

Fig. 5 X-ray absorbing masks produced by Romankiw et al. by gold electrodeposition [21]. Photoresist patterns are on the *left*, while electroplated gold patterns are shown on the *right*. The *top* two images are from the original e-beam fabricated mask, resulting in an X-ray negative copy in the middle. The pattern was inverted another time to return to the original tone in the *bottom* two images. Reprinted from Electrochimica Acta, Vol. 43, L. T. Romankiw, A Path: From Electroplating Through Lithographic Masks in Electronics to LIGA in MEMS, pp. 2985–3005, Copyright (1997), with permission from Elsevier

By the early 1980s, investigations involving the fabrication of structures using features defined in thick PMMA resist by X-ray synchrotron radiation and subsequent electrodeposition were in progress at the Forschungzentrum Karlsruhe (FZK), in Karlsruhe, Germany. In this case, the researchers were not interested in the fabrication of electronic or magnetic devices, but instead in structures meant to be used for the isotopic separation of uranium [1]. Apparently, the critical features of the device needed to be not only very narrow but also very tall (i.e., high aspect ratio) for the separation process to work. X-ray lithography, with its capability to precisely form features having very small lateral dimensions in thick resist, was thus understandably an attractive fabrication technique for such structures. An example of one of the first devices fabricated at FZK is shown in Fig. 7. The potential of the technique relative to the fabrication of structures for microsystems became apparent,

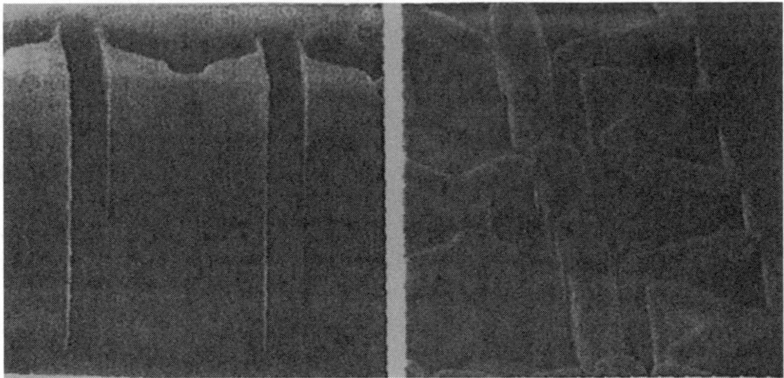

Fig. 6 Patterned PMMA 12 μm in thickness (*left*) exposed at the DESY accelerator in Hamburg and corresponding electroplated gold structures (*right*) as shown by Romankiw in ref. [21]. The resist-pattern line widths are 1 μm with 100-nm spacers, resulting in electroplated gold structures spaced 100 nm apart. Reprinted from Electrochimical Acta, Vol. 43, by L. T. Romankiw, "A Path: From Electroplating Through Lithographic Masks in Electronics to LIGA in MEMS", pp. 2985–3005, Copyright (1997), with permission from Elsevier

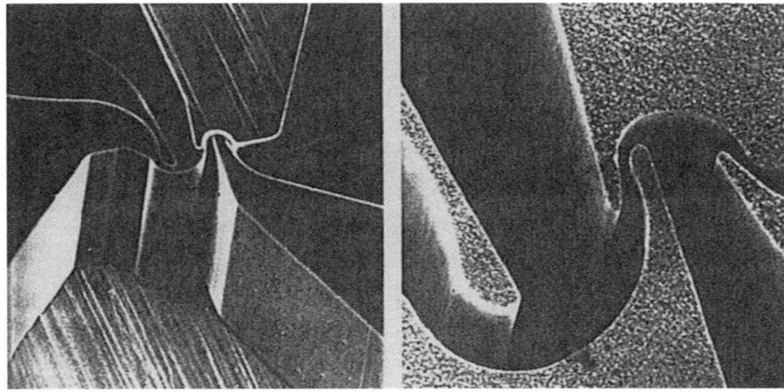

Fig. 7 Patterned PMMA (*left*) and corresponding through-mold plated Ni nozzle structures for gas isotope separation (*right*) fabricated at the Forschungzentrum Karlsruhe [1]. The openings in the Ni of closest approach are about 3 μm. Reprinted from Naturwissenschaften, "Production of Separation-Nozzle Systems for Uranium Enrichment by a Combination of X-ray Lithography and Galvanoplastics", by E. W. Becker, W. Ehrfeld, D. Muenchmeyer, H. Betz, A. Heuberger, S. Pongratz, W. Glashauser, H. J. Michel, and R. v. Siemens, Vol. 69, p. 520, Fig. 3, (1982, with permission of Springer-Verlag

and work continued on electroplating through PMMA resist to produce structures for other applications [3, 4, 22].

Furthermore, the researchers at FZK were interested in using X-ray lithography to form a metal master tool that would be subsequently used to mass-produce structures (presumably plastic) *via* embossing or injection-molding processes. Thus, the German acronym, LIGA, used generally to refer to electroplating through thick resist to form metal structures includes the word *abformung*, indicating the intention of using the metal structures as a tool to mass produce or replicate patterns.

Subsequent to the establishment of the LIGA process at FZK in the early 1980s, X-ray lithography efforts elsewhere in the world began to appear, naturally centered at synchrotron facilities in Europe, Asia, and North America. Since then, a large body of LIGA research has been published in the literature and presented at international conferences. An attempt to comprehensively summarize the LIGA work that has been done over two decades would be beyond the scope of this chapter; thus, only a relatively small number of articles and presentations are referenced where appropriate to the topic at hand.

It must be noted that many researchers worldwide (the current authors and their colleagues included) use X-ray lithography as a tool for fabricating a mold of structures, with the intention of using the electroplated parts separately in some sort of assembly (usually a microsystem). Although such a process does not involve the replication (*abformung*) step, practitioners still typically use the term LIGA to describe their work (occasionally this is referred to as "direct LIGA" as we have previously indicated). Since synchrotron radiation is usually necessary for the thick resist-exposure step, a large fraction of LIGA research groups are located in academic institutions or government-sponsored national laboratories. However, commercial companies have established their presence at such facilities, and some notable examples will be mentioned in the next section discussing applications of LIGA technology.

1.3 LIGA Applications

In surveying the numerous articles and presentations involving the use of the LIGA technique, it is evident that an equally large number of applications for LIGA-enabled structures have been proposed. However, over the course of the technology's 20-year or so history, only a few of these proposed applications have been developed into commercial products. The need for an X-ray synchrotron source, no doubt an issue for those wishing to adapt the technology for larger-scale production, has been alleviated somewhat by the fact that most synchrotron facilities have active programs for allowing access by researchers and industry [14]. In this section, we discuss only the most popularly reported applications, with some emphasis on those that have been commercially realized to give the reader a general idea of the types of applications of interest. Good reviews are available for those interested in more detailed treatments of LIGA applications [2, 11, 13–15].

One of the most commonly discussed applications of structures fabricated *via* the LIGA process involves mechanical devices. Since lateral geometries in LIGA are defined lithographically, very high dimensional fidelity is theoretically possible across large areas, enabling the batch fabrication of large numbers of precision structures simultaneously. Moreover, deep, high-aspect-ratio structures are possible if thick resist is used, desirable for structures to be used as active devices transmitting useful amounts of mechanical energy.[1] For example, the amount of torque that a micromotor can deliver is directly proportional to the height of the motor itself. Further, the use of electrodeposition as an *additive* materials-fabrication technique offers the possibility of a large materials set with interesting mechanical, magnetic, and electrical properties as compared to *subtractive* materials-fabrication techniques, such as silicon surface micromachining (some challenges in taking advantage of this potential materials set will be discussed in detail later).

Gears, actuators, rotors, and drivers have been fabricated *via* the LIGA method for a large variety of purposes (see ref. [15] and references cited therein). Figures 1 and 8 show examples of LIGA structures fabricated from Ni-based electrodeposits at Sandia National Laboratories, Livermore (SNL), CA. Mechanical structures designed as sensors (motion, magnetic, pressure) have also been fabricated *via* the LIGA process. An SNL Ni accelerometer prototype device is shown in Fig. 9. Other LIGA researchers have fabricated similar structures (see, e.g., ref. [23] by Qu et al.). Christenson and Guckel have demonstrated several different types of mechanical devices, including linear and rotational actuators and micromotors [24]. An example of a minimotor assembly is shown in Fig. 10 [25]. This device was fabricated by the group at IBM using optical lithography and through-mask plating. More examples of such devices can be found in Madou [2].

Another area of application for LIGA-fabricated devices is in the field of optical systems and communications. The ability to produce thick structures that are highly parallel, with optical-quality side-wall roughness is useful in the fabrication of gratings, waveguides, and lenses [15]. These structures can be made from polymers that are directly patterned with X-rays, developed and used immediately

Fig. 8 Example of Ni test gear device formed from electrodeposition through a patterned PMMA mold. Micrograph courtesy of Georg Aigeldinger, Sandia National Laboratories, Livermore, California

[1] This is an important characteristic that distinguishes LIGA-fabricated microdevices from silicon MEMS technology.

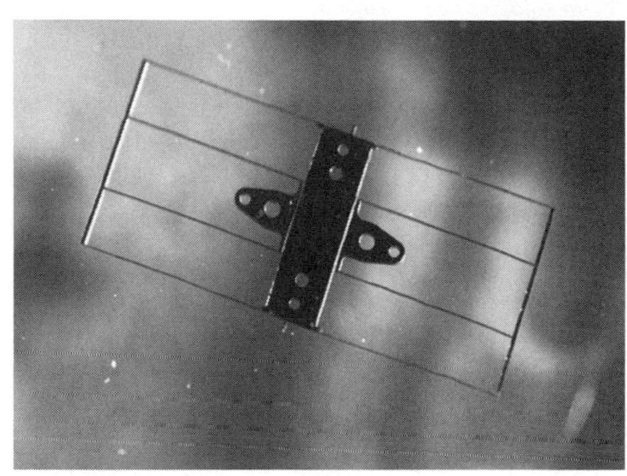

Fig. 9 Example of an acceleration-sensing prototype device fabricated by the LIGA process at Sandia National Laboratories, Livermore, California. The long, slender flexures are 25-μm wide and 250-μm tall; the entire width of the device is approximately 3 cm. A portion of the device is anchored in place, while another is free to move by the bending of the flexures, allowing for the sensing of acceleration. The device above is made of high-strength Ni

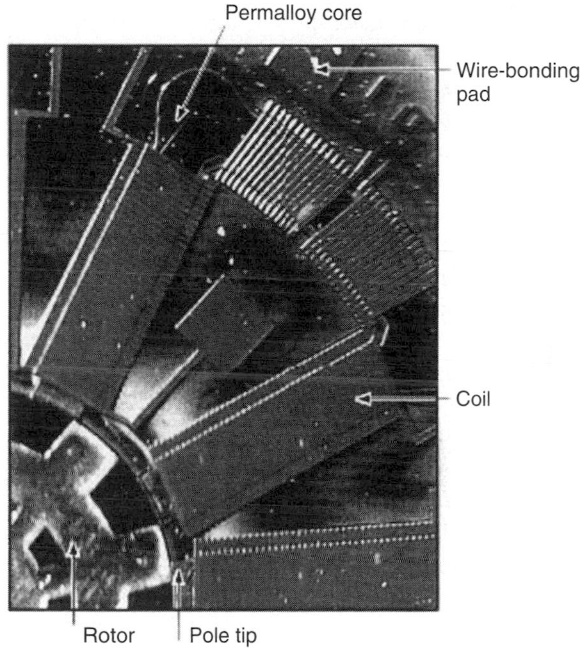

Fig. 10 Minimotor assembly composed of piece parts formed by through-mask plating of thick and thin UV-photoresist by workers at IBM [25]. Electroplated copper and a nickel–iron alloy was necessary for its fabrication. The plated Cu coils are 60-μm wide. Photograph reproduced with permission of the IBM Journal of Research and Development

afterward. LIGA-fabricated structures have been developed for collimating grids for high resolution solar-imaging missions [26] and adaptive optics systems in astronomy [27]. LIGA has enabled various connecting and mounting systems for optical fibers and components [2, 15]. The LIGA technique has been used to fabricate metal transmission lines with applications in the field of microwave- and millimeter-wave integrated circuits [28].

The last area that can be broadly grouped as a LIGA application field is biological applications and microfluidics [15, 29] LIGA tools for embossing patterns into plastic structures to be used as platforms in disposable biological or medical applications ("biochips" or "lab-on-a-chip" devices) are of special interest, since such approaches are low in cost as compared to the direct LIGA route [30]. LIGA tools have also been employed for the microstructuring of composite materials used as artificial bone analogs to improve bone cell-growth conditions [31]. A range of microfluidic devices have been proposed, such as microreactors, micropumps, nozzles, and microvalves [32–34]. Both direct LIGA and the molding of plastic with a LIGA tool have been used for microfluidic device fabrication [35].

A few more applications that do not fit into the above categories are worth mentioning due to their unique use of LIGA technology. Spinnerets for synthetic microfiber production have been fabricated by LIGA to make high aspect ratio, narrow features of arbitrary prismatic cross-section [2, 13]. These are essentially nozzles for the extrusion of polymer material; the LIGA technique allows for finer features with smooth side walls as compared to conventional fabrication methods. The simultaneous batch production of large numbers of nozzle structures is more rapid than conventional machining, where one structure is formed at a time.

Microscale cross-flow heat exchangers have been fabricated using a LIGA embossing tool and subsequent replication steps to form Ni metal microstructures with higher heat-transfer/volume ratios than conventional devices [36]. The possibility of using LIGA to form small Ni structures with microscale barbs for tagging has been explored and could be useful for military and law-enforcement applications [37]. In the area of defense, industry is studying the use of LIGA technology to fabricate fusing devices for munitions [38].

Some progress has been made in the last few years in the development of LIGA-fabricated devices for commercial products. A number of optical, mechanical, and fluidic products are available from STEAG microParts, GmbH [39]. This company uses advanced injection-molding techniques in combination with LIGA technology to make inhalers, microspectrometers, and lab-on-a-chip reaction platforms, as well as other microfluidic products. Zero-backlash gears and other precision positioning systems are fabricated by Micromotion GmbH using a LIGA-related technique (see Fig. 11) [40]. LIGA technology has been employed to make optical fiber connectors by Spinner GmbH in collaboration with FZK [15]. In the United States, Axsun Technologies, Inc., uses the LIGA technique to make alignment structures for photonics systems [38].

As evidenced by this wide range of applications, the unique capability of the LIGA and LIGA-related techniques to fabricate high-aspect-ratio structures with a range of useful geometries has been known for several years. More recently, groups working on LIGA technology have started to focus on the fabrication of structures with special material properties. This often necessitates the consideration of materials other than electrodeposited Ni, in contrast to most of the examples discussed in this section. The next section discusses the unique challenges posed by the geometries encountered in the LIGA process during the electrodeposition of Ni and Ni alloys with special properties.

Fig. 11 Precision position components fabricated by a LIGA-related process commercially available from Micromotion GmbH. *Bottom* image is a closeup of the gear assembly in the *top* photo. Photographs courtesy of Micromotion, GmbH

2 Electrodeposition into Deep High-Aspect-Ratio Features for LIGA

2.1 Introduction

The science and process sensitivities of electrodeposition through thin resist[2] used to produce metallic-conducting interconnects, packaging, and contacts for the microelectronics industry have been much discussed [16]. In contrast, electrodeposition through thick resist for deep, high aspect ratio LIGA features has received relatively little attention (although we note that one study involving design rules for the LIGA electrodeposition step has recently become available [41]). In surveying the literature where electrodeposition through patterned photoresist is discussed, it is evident that the microelectronics industry has devoted a significant amount of experimental and modeling resources to understanding through-mask electrodeposition (more

[2]We define "thin" resist loosely as less than several microns, while "thick" resist is considered to be hundreds of microns to several millimeters.

detailed reviews of these developments are available) [16, 42]. A central theme of the prior literature is the classification of three length scales over which electrochemical processes occur; the first, *the workpiece scale*, involves the distribution of current over the scale of the substrate or wafer; the second, *the pattern scale*, involves the distribution of current between features defined in the insulating photoresist; the third, *the feature scale*, involves the distribution of current to and within an individual opening or feature in the resist. These three scales are depicted in Fig. 12 after DeBecker and West [43]. As will be discussed later, the principles governing the deposition of a metal structure or film having uniform thickness and properties over all these scales (usually desired for all electrodeposition processes) are largely the same, regardless of the final application. In terms of attaining uniformity *across* a substrate, the prior literature regarding the workpiece and pattern scales for electrodeposition through thin resist is directly applicable to electrodeposition through thick resist.

The important difference regarding metal deposition through thin and thick resist comes when *through-thickness* uniformity is considered; as the electrodeposited metal films for microelectronics are usually only a few to several microns in thickness, appreciable through-thickness variations in film structure do not usually occur since the thickness is so small. Moreover, any small microstructural deviation that may be present through the film thickness does not usually critically impact the function of the electrodeposited metal film in microelectronics. But in contrast to the thin films employed in microelectronics, large variations in deposit microstructure may occur over the hundreds of microns that comprise the through-mask electrodeposited thick films in LIGA. Since most of the issues in attaining uniformity across the substrate are similar to those in microelectronics, this discussion focuses rather on attaining through-thickness uniformity in structures electrodeposited through thick resist, a challenging aspect unique to LIGA.

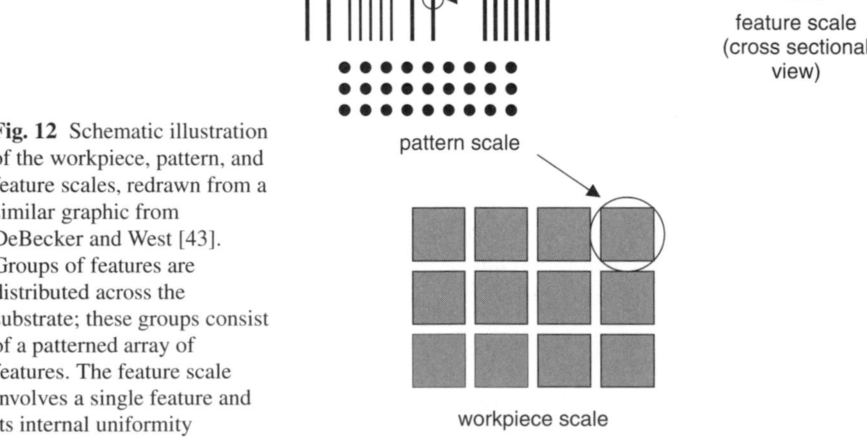

Fig. 12 Schematic illustration of the workpiece, pattern, and feature scales, redrawn from a similar graphic from DeBecker and West [43]. Groups of features are distributed across the substrate; these groups consist of a patterned array of features. The feature scale involves a single feature and its internal uniformity

Besides differences in resist and metal thicknesses, another major difference between films electrodeposited for microelectronics and microsystem applications lies in the choice of materials. In microelectronics, the metal films are serving as conductors or contacts; hence, materials such as copper, gold, and lead–tin alloys (more recently, lead-free alternatives are under development) have attractive physical properties and are widely employed. In microsystems, the thick metal structures are typically serving as components in some mechanical device or as a replication mold; the mechanical properties are thus of most interest in these situations. Electrodeposited nickel has a long history of use as a structural material, since its mechanical properties may be tailored by the use of alloying elements and electrolyte bath additives (both organic and inorganic) [8, 44]. Depending on deposition conditions and chemistries, Ni may be electrodeposited with low stress, making thick films possible. The generally good corrosion resistance of electrodeposited Ni also makes it attractive for mechanical applications in a wide variety of environments. The material properties of Ni and Ni alloys will be discussed in more detail later in the chapter. In comparing electrodeposition processes for thin and thick resists, it is to be kept in mind that the materials that are deposited into these lithographic molds may have very different deposition characteristics as well.

2.2 *Workpiece- and Pattern-Scale Effects*

In prior treatments of through-mask electrodeposition, the first length scale often identified is the workpiece scale; this is usually the characteristic length of the substrate (or wafer) itself onto which resist defining the desired structures has been applied. Variations in the local current density typically occur over the workpiece scale. The cell geometry, the size of the substrate, and other deposition parameters dictate the uniformity across the substrate; the degree of this uniformity may be described by the Wagner number, a dimensionless parameter that has been discussed in more detail previously [16]. Generally speaking, if no measures are taken to improve the current distribution across the wafer, regions close to the edge will have a locally higher current density than areas near the center of the substrate.

A few different approaches may be used to render the current distribution across the wafer more uniform. The appropriate placement of insulating shields with respect to the wafer may result in a more uniform current distribution [45]. The use of auxiliary electrodes (also known as thief electrodes) in minimizing variations across the workpiece has been discussed previously [46]. For example, another cathode placed around the workpiece (as a type of surrounding "frame") may improve the workpiece current distribution significantly. In industrial plating tools, all these aspects are carefully considered in optimizing the wafer-scale uniformity. Obviously, since a large number of wafers are typically run under fixed conditions in the microelectronics industry, a fair amount of time and effort is spent in finding the optimal cell geometry and deposition conditions. These approaches for homogenizing the wafer-scale uniformity are directly applicable to electrodeposition through thick resist in LIGA and have in some cases been adopted [45].

The second length scale is referred to as the pattern scale. The relative amounts of exposed active metal and insulating photoresist may vary across the substrate due to the geometry and layout of the pattern defining the structures to be electrodeposited. Thus, the "active area density" varies depending on essentially the distribution of the insulating resist material. This leads to variations in the local current density (the deposition rate) depending on the local pattern geometry in the resist. Mehdizadeh et al. studied this problem for electrodeposition through photoresist having various pattern densities [47]. West et al. considered the effect of patterning on multiple electrodes of disks and lines [48]. These and other studies [16, 49] may be drawn upon in designing layouts where pattern-scale effects are minimized.

In our experience, LIGA-pattern layouts that consist of an array of uniformly spaced features having similar dimensions are not commonly encountered; instead, patterns tend to consist of fine features (dense resist areas) located in close proximity to large, open areas of exposed seed metal (defining a wide, low aspect ratio feature). Thus, pattern-scale effects must be managed on a fairly routine basis. Although it may not be possible to alter the particular part geometry and size (necessary for the device function), some discretion may be exercised in distributing the features in the resist defining the part geometry so as to mitigate pattern-scale effects. Moreover, sacrificial features (openings in the resist) may be added to the pattern to break up areas that have long runs of uninterrupted insulating resist, resulting in a more uniform active area-density distribution [46–47]. This practice of inserting sacrificial "moats" or "frames" around features where dimensional fidelity is critical was previously mentioned with regard to minimizing dimensional errors due to PMMA swelling and thermal expansion after immersion in the electroplating bath [50]. It is somewhat fortuitous that the same measures that aid in minimizing these dimensional errors in LIGA also may help in rendering the pattern-scale current distribution more uniform. Mehdizadeh et al. have discussed the possibility of carefully considering the wafer- and pattern-scale current distribution and arranging features in the resist across the workpiece so as to attain a more balanced current distribution [47]. Even when detailed current distribution models are unavailable, past experience can be a useful guide in judiciously arranging the part layout to avoid uneven distributions of current at the pattern scale.

2.3 The Feature Scale

As mentioned previously, the similarity between microelectronics processing and LIGA through-mask electrodeposition diminishes as one moves from the workpiece and pattern to the feature scale. The principal reason for this is the large resist thicknesses typically employed for X-ray lithography. As pointed out by Dukovic [16], at the feature scale, the dimensions of the lithographic features are usually small enough (microns) that electrical-field effects become less important, and the influence of the concentration field of the reacting species grows in determining the feature scale current distribution. Since mixing is limited inside the feature due to the presence of the surrounding resist material, the concentration of a reacting species may vary with position within the feature. Another aspect that must be considered for both thin and thick resists is that as deposition proceeds, the feature geometry changes as the deposited metal fills the feature. Generally speaking, obtaining good mixing inside lithographic features becomes more difficult with increasing resist thickness.

When the electrodepostion reaction involves just a single metal cation, variations in its concentration inside the feature typically result in variations in the local rate of deposition and therefore deposit thickness [51]. The deposition characteristics of copper, gold, and tin alloys through relatively thin resists (less than 20 microns) for bump structures have been considered with attention to the thickness distribution of the deposit within a single feature [52–57]. It is desirable that these bump structures have good as-plated uniformity for subsequent packaging processing. Another example of an important feature-scale electrodeposition problem is the case of copper Damascene electrodeposition for copper interconnects [58]; for this process, proprietary electrolyte bath additives and active feature side walls lead to more complicated feature-filling behavior, despite the fact that only a single elemental species is being deposited [59, 60] The understanding and control of through-mask electrodeposition at the feature scale has undoubtedly been important in enabling the use of these processes for the large-scale manufacturing of microelectronics packaging and copper interconnects [16, 42]

In the case of electrodeposition through thick resist for LIGA, the importance of the feature scale is not its influence on the local deposit-thickness uniformity (the wafer is planarized after plating, removing thickness variations across the wafer at all scales). Rather, it is the more serious issue of poor mixing in the mold features themselves due to their small widths and large heights. This has broad implications with respect to the potential for variations in the concentrations of electrolyte species that may, in turn, impact the local deposit composition, morphology, and properties. For example, if additives or other metal species are present as alloying elements, profound changes may occur through the thickness owing to the constantly changing feature geometry that accompanies the metal growth. If the concentration of one of these species is close to zero at the feature bottom (due to its rapid incorporation into the deposit, for example), large changes in the flux of this species will occur as the deposit grows, and the diffusion length from the bulk electrolyte shortens. The resulting nonuniform deposit morphology or alloy composition may be unacceptable for some applications. The electrodeposition of the NiCo and NiMn systems and the resulting feature-scale uniformity will be contrasted to demonstrate the importance of transport into LIGA features later in the chapter.

Interesting experiments involving the feature scale were carried out by Leyendecker et al. [61]. These authors studied the effect of thick photoresist on mass transport and on the pH within a LIGA feature using special electrodes. In the first series of experiments, a rotating disk electrode masked by an insulating patterned polymer layer of various thicknesses and feature sizes is employed to study a model redox reaction occurring under transport control. The aqueous electrolyte consists of 2 mM potassium ferro- and ferri-cyanide each in 1 M potassium chloride at 25°C.

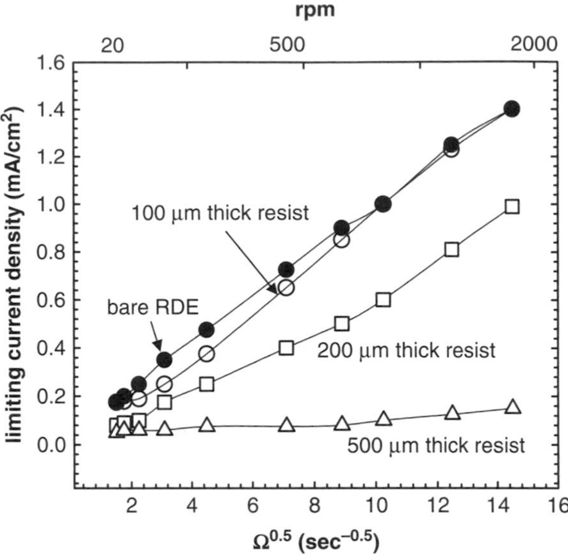

Fig. 13 Limiting current densities for ferri–ferro cyanide redox couple on bare and patterned rotating-disk electrodes replotted from the results of Leyendecker et al. [61]. Feature hole diameter is 200 μm. The presence of the resist has a large effect on the transport of species to the feature bottom, even for an aspect ratio of one. Solid points are the bare rotating disk electrode, while the hollow points are electrodes patterned with resists of various thicknesses as indicated

Results of their experiments are reproduced in Fig. 13; although various geometries were investigated, they present results for a constant hole diameter of 200 μm and resist heights of 100, 200, and 500 μm (in addition to an unmasked electrode).

In the case of an unpatterned, bare rotating disk, good linearity is obtained as expected by plotting the limiting current density as a function of the square root of the electrode rotation rate $1/\Omega^{0.5}$. When the electrode is covered with photoresist, the difficulty in generating mixing into the features becomes evident. For a hole diameter of 200 μm and a resist height of 100 μm (an aspect ratio of 0.5) the measured limiting current densities are close to those on the bare electrode. But for a resist thickness of 200 μm (aspect ratio 1), the measured limiting current densities are much lower than for unpatterned electrodes and the electrode with 100 μm of resist. Even at this modest aspect ratio, increasing the electrode-rotation rate becomes less effective at transporting reactant to the electrode surface, resulting in a flatter curve for the 200-μm thick resist in Fig. 13. When the resist thickness becomes 500 μm (aspect ratio 2.5), almost no increase is obtained in the limiting current density with increasing rotation rate. This is somewhat surprising, since if metal deposition through thick resist occurred under transport control, one would expect impractically low deposition rates to be necessary for the deep, high-aspect-ratio features used in LIGA process. In fact, structures with aspect ratios of 10 having appreciable thicknesses (hundreds of microns) are not uncommon, and, while on the low end of the current-density operating range, these structures are typically deposited at rates that are not drastically lower than those conventionally used in electroplating and electroforming. The reasons for this seeming discrepancy will become apparent when other modes of transport are discussed.

In the same paper [61], Leyendecker et al. perform a second series of experiments investigating the pH behavior within a deep, high aspect ratio feature as Ni is electroplated into the feature. As the current efficiency during Ni electrodeposition is not always 100%, the pH may rise close to the cathode surface (especially in a feature in thick resist), resulting in the presence of adsorbed hydroxide species that may affect the deposit morphology. Since Ni is often electrodeposited for LIGA applications, these experiments are highly relevant.

Special PMMA structures were fabricated by laminating three 10-μm thick antimony metal foils in between 240-μm thick PMMA layers; the laminated structure was then bonded to a metal substrate. Microholes were drilled through the layers until the conducting substrate was reached. The final structure thus consists of a metal plating base with four levels of PMMA separated by the antimony sheets, through which arrays of microholes are drilled, the entire piece is approximately 1000 μm in thickness. Figure 14 is a schematic diagram of the laminated structure taken after Leyendecker et al. [61]. The antimony metal acts as a pH sensor due to the pH-dependent equilibrium with its oxide. The Sb/Sb_2O_3 system has been used previously as a pH probe. By monitoring the potential between the antimony sheets and a reference antimony electrode in the bulk electrolyte, the authors could study *in situ* the pH in the LIGA feature as Ni electrodeposition occurred. This is the only study of its kind in the literature of which the current authors are aware.

Results of filling these features were given in this paper for the 200 μm diameter microholes. They are reproduced in Fig. 15. Nickel is deposited into the features at 50 mA/cm^2 from a sulfamate bath (76 g/L Ni and 40 g/L boric acid) at 52°C and a bulk pH of 3.8. Since the exact depth of the drilled features was not known (the drilling proceeded some distance into the bulk metal piece used as the plating base), deposition occurs until the first Sb layer is overgrown. At that point, plating is stopped (the position of the Ni deposit surface within the features is then known) and the experiment paused until the pH in the feature returns to the bulk pH. Then, plating resumes, with the potential of the second Sb layer monitored until the growing Ni deposit reaches this Sb layer. Ni deposition continues from that point with

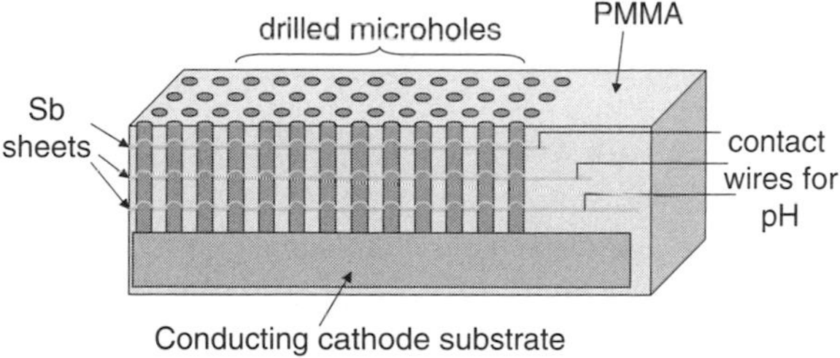

Fig. 14 Diagram of special microstructured pH electrode used by Leyendecker et al. [61], redrawn after a similar sketch that was reported [61]. Feature hole diameters of 200, 100, and 50 μm were considered. The Sb sheets embedded in the PMMA resist function as *in situ* pH electrodes, allowing for the monitoring of local pH within the features as electrodeposition proceeds from the cathode at the bottom of the features

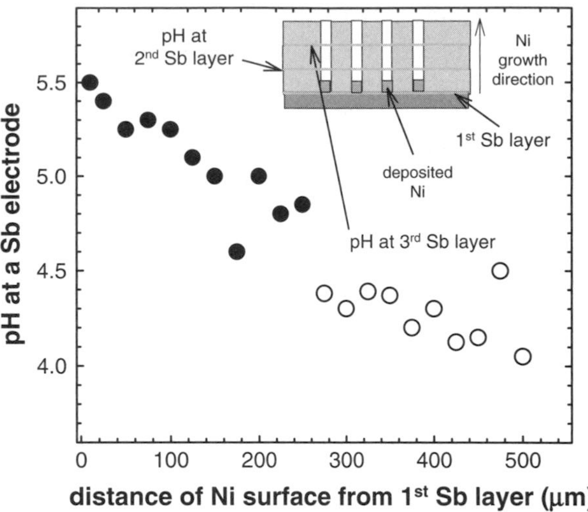

Fig. 15 *In situ* pH measurements from the microstructured pH electrode shown in the previous figure, replotted from Leyndecker et al. [61]. In this case, the hole diameter of all features is 200 μm. The inset helps one visualize the position of the Sb pH electrode layers with respect to the growing Ni deposit. A Ni sulfamate bath was used operating at 50°C and a current density of 50 mA/cm^2

the potential of the third Sb layer then being monitored until it too is overgrown by the Ni deposit. Thus, the zero value on the *x* ordinate in Fig. 15 represents the point when the Ni is even with the first Sb layer (about 750 μm from the feature mouth into the bulk electrolyte). The plotted values represent the pH measured at the respective Sb layers (fixed in position) as the Ni fills the features in the laminated structure. The pH at the second Sb layer (500 μm from the feature mouth) is as high as 5.5 when the Ni deposition begins. As it grows, the second Sb layer pH value drops toward 4.5. After the Ni deposit grows past the second Sb layer, the third Sb layer (itself 250 μm from the feature mouth) is monitored. The measured pH values at the third Sb layer are just slightly higher than the bulk value of 3.8. Finally, when the Ni deposit has grown to a point just below the depth of the third Sb layer, the pH at that position is close to that in the bulk.

These results are important for a number of reasons. First of all, although the PMMA feature thickness corresponding to the zero position on the *x* axis is 750 μm, the aspect ratio of this feature is only 3.75, which is not considered to be very aggressive for LIGA electrodeposition (aspect ratios of 10–15 approach the high end of what is commonly fabricated). Despite this modest aspect ratio, the measured pH 500 μm from the feature mouth is as high as 5.5; if the pH near the growing Ni surface is this high or higher (a likely case, since protons are consumed at the interface) the deposit morphology could be profoundly influenced. Deposition at high pH values has been typically associated with high hardnesses and the possible incorporation of oxygen into the deposit, presumably from hydroxide species adsorbed at the deposit surface [8, 44, 62–63]. As the deposit grows, the microstructure of the deposit may therefore change as the feature opening is approached, and the pH drops toward its value in the bulk electrolyte. If the pH value reaches extremely high values, the impurities in the deposit from hydroxide species may compromise its material properties. Another issue is that the increasing pH is indicative of hydrogen generation within the feature; if the hydrogen trapped within the feature accumulates and forms a bubble, it may be occluded into the deposit as a defect that can lead to failure. These issues are of real concern for structures that will be used in mechanical applications, especially for structures having aspect ratios more extreme than those considered in Fig. 15.

Another series of experiments relevant to the feature scale in LIGA involving transport limitations is reported by Griffiths et al. [64]. They use a commercial Ni sulfamate plating electrolyte to deposit Ni into features drilled into PMMA. A single piece of Cu foil attached to one side of the PMMA acts as the substrate, from which the deposited Ni structures could be removed for weighing in the determination of the current efficiency. Current densities range from 1.1 to 108 mA/cm^2, and an electrolyte temperature of 38°C is employed. The drilled microholes have diameters of 1.7, 3.2, and 6.4 mm and depths from 17 to 42 mm. They also calculate the Sherwood number (the ratio between the measured current to the diffusion-limited current) for each case.

The experimental results are surprising in that the Sherwood numbers are all between 10 and 100, indicating a much larger deposition current than expected based upon the diffusion-limited case. The reported data for the experimental conditions are reproduced in Table 1 [64]. Measured current efficiencies indicate that hydrogen evolution accounts for only a few percent of the total current (except in one case, where the current efficiency was only 89%), suggesting that the large current densities are associated with the metal deposition reaction itself. They also indicate their results were reproducible, further suggesting that the observed behavior was not due to hydrogen evolution. The authors explain the unexpectedly high currents as resulting from enhanced transport of the reacting metal ion due to natural convection in the feature driven by its depletion at the substrate. This hypothesis is further supported by the fact that when the experimental cell was inverted, the measured currents were an order of magnitude less than with the deposition surface facing up.

Table 1 Experimental data reproduced from Griffiths et al. [62]. Experimentally measured currents during through-mask electrodeposition into microfeatures drilled into PMMA show current densities larger than those expected from pure diffusion as indicated by the Sherwood number. The authors explain that buoyancy-driven convection is an important mode of transport in LIGA features

Cavity diameter (mm)	Feature aspect ratio	Current density (mA/cm^2)	Current efficiency	Sherwood number
1.7	10	11	99.7	48.0
1.7	15	86	96.4	82.2
1.7	25	86	89.6	137.5
3.18	8	86	99.5	82.2
6.35	1.6	108	100.9	41.1

These authors go on to demonstrate by numerical modeling that flow across the top of the features cannot support the transport of the reacting metal ions at the observed rates, and that instead buoyancy-driven flows are probably important in allowing such high deposition rates [64]. Nilson and Griffiths present more detailed calculations of natural convection in high aspect ratio trenches in a subsequent article [65]. They find that the relative importance of natural convection depends on the feature geometry and the inclination of the substrate. The authors conclude that some mixing from natural convection is expected to occur in features having aspect ratios less than 1.2, 6.6, and 37 for resist thicknesses of 0.1, 1.0, and 10 mm, respectively. Thus, natural convection becomes important for a wider range of feature aspect ratios as the resist thickness increases. Since resist thicknesses in microelectronics are usually only tens of microns or less, the importance of convective flow had not been widely discussed until electrodeposition into deep LIGA features was considered. It is interesting to contrast the findings of Leyendecker et al. [61], discussed at the beginning of this section, with those of Griffiths et al. [64]. In the former case, the insulating PMMA mask was shown to have strongly diminished mixing, resulting in lower transport rates of the reacting ions, while in the latter case, buoyancy-driven transport rates into deep high-aspect-ratio features higher than those expected from pure diffusion control were observed. Since Leyendecker et al. used a redox couple to measure transport rates through microstructured electrodes, they were not able to benefit from potential buoyancy-driven flows induced by the consumption of metal ions at the feature bottom.

Another series of studies focusing on feature-scale uniformity in LIGA was carried out by Schwartz and co-workers [66–68]. Leith et al. considered the electrodeposition of NiFe at 23°C from sulfamate-chloride chemistries having various ratios of Ni to Fe electrolyte loadings [67]. They pointed out that the plating characteristics of an electrolyte must be carefully considered before employing it for through-mask electrodeposition using thick resist. They found that electrolytes having less dissolved iron (e.g., 20:1 Ni^{2+}:Fe^{2+} as opposed to 10:1 and 5:1 also considered in their study) are less sensitive to mixing variations at relatively high current densities, suggesting that they would be more suitable for the electrodeposition of LIGA structures. Leith and Schwartz also describe a cell designed to maximize workpiece- and pattern-scale uniformity, using the NiFe system (since it is a challenging system for uniformity) to demonstrate good uniformity across the wafer [67]. The through-thickness uniformity was not investigated in this study [67]. In another article, Wang et al. used an electrolyte with a 10:1 Ni^{2+}:Fe^{2+} loading (where the deposited NiFe alloy composition is more sensitive to mixing) to study convective-diffusion conditions during the through-mask deposition of LIGA structures [68]. They term the approach "process archaeology" since the plated structure was planarized to half its deposited thickness for compositional analysis *via* energy dispersive X-ray analysis [68]. By observing in plan view the planarized structures, regions in the feature where locally high degrees of mixing occur could be easily identified since the Fe content of the alloy increases with mixing. In this case, the authors were using the transport-sensitive deposition kinetics of the Fe as a type of "tracer" tool to study the mixing within small cavities such as LIGA resist features.

Another group of authors also considered the NiFe system to focus on the feature-scale uniformity of through-mask-deposited LIGA structures. Thommes et al. used a sulfate-based NiFe electrolyte at 50°C with a Ni^{2+}:Fe^{2+} ratio of about 12:1 to deposit NiFe structures at 5 mA/cm^2 with good through-thickness uniformity [69]. Figure 16, reproduced from their data, shows the compositional behavior of the deposition system using a rotating-disk electrode. It is clear from Fig. 16 that for low current densities, the electrode-rotation rate does not influence the deposited alloy composition, suggesting that under these conditions the alloy composition is not greatly influenced by the amount of mixing. This relative insensitivity to mixing would appear promising for the feature-scale uniformity of LIGA NiFe structures.

The suitability of the authors' deposition conditions for the electrodeposition of uniform high aspect ratio NiFe microstructures is evidenced by the through-thickness compositional data shown in Fig. 17. For a resist thickness of 180 μm and a feature line width of 8 μm (aspect ratio 22.5), the authors obtained very good NiFe alloy uniformity throughout the entire structure thickness using a deposition current density of 5 mA/cm^2. Through-thickness uniformity suffers at higher deposition current densities. Figure 17 also shows experimental results for deposition at 14 mA/cm^2 using the same feature geometry. The Fe content increases through the deposit as the top PMMA surface is approached, varying from about 20% to 50% across the thickness. Though not shown here, the authors filled cylindrical features with a diameter of 5 μm and a height of 90 μm depositing NiFe at 5 mA/cm^2. They

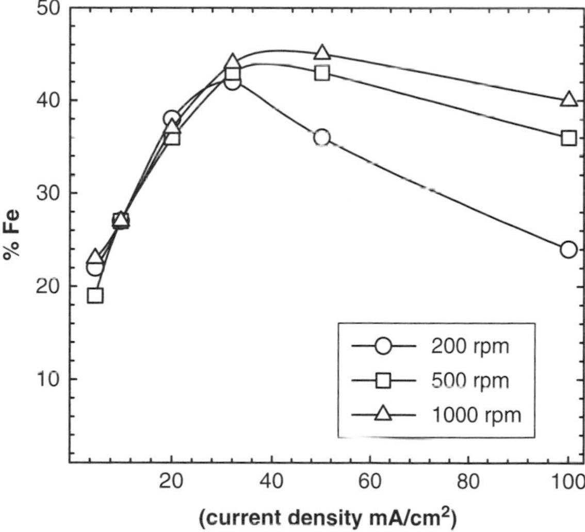

Fig. 16 Electrodeposited NiFe alloy composition dependence on deposition conditions, for conditions given by and replotted from Thommes et al. [69]. The relative compositional insensitivity at low current densities suggests that the system may be appropriate for the electrodeposition of uniform, high-aspect-ratio structures

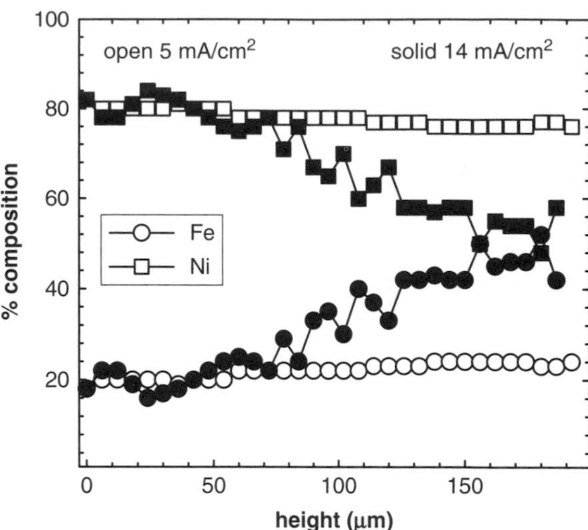

Fig. 17 Electrodeposited NiFe alloy composition through-thickness uniformity replotted from Thommes et al. [69]. The structure width is 8 μm and its height is 180 μm. Open points are for deposition at 5 mA/cm^2, while the solid points are at 14 mA/cm^2. Deposition at the lower current density (where composition was shown to be less sensitive to mixing) leads to uniform alloy content through the structure thickness, as compared to 14 mA/cm^2

show similarly good results as for the structure in Fig. 17 deposited at the same current density.

The electrodeposition process outlined by Thommes et al. has been shown to be capable of producing high-aspect-ratio structures with good alloy through-thickness compositional uniformity [69]. To the authors' knowledge, few other studies exist of alloy deposition through thick resist where the feature-scale uniformity is considered in such detail. An aspect that is likely important in explaining the good results obtained by these authors is the bath temperature; Brenner has discussed the fact that NiFe electrodeposition becomes less anomalous (and hence less transport sensitive) with increasing bath temperature [70]. Andricacos and Romankiw also discuss the importance of bath temperature as well as other process variables on electrodeposited magnetic alloy composition [71]. Given the limitations in obtaining good mixing in small LIGA features, it is not surprising that achieving good feature-scale uniformity for anomalous deposition systems (such as NiFe) is challenging for geometries encountered in LIGA electrodeposition [70–71].

New techniques are being employed for studying the electrodeposition reaction inside of small features like those fabricated by the LIGA process; for example, special microelectrodes and chronoamperometric methods have been developed to measure the local metal cation concentrations during alloy electrodeposition inside small cavities [72]. Kuepper and Schultze study concentration variations inside of a cavity where a NiCu alloy is being electrodeposited [72]; they also observe the formation of a Ni hydroxide and nickel oxide at the deposition surface during electrodeposition of the NiCu alloy from a sulfate electrolyte with potassium nitrate as supporting electrolyte. It is likely that specialized, *in situ* techniques such as those used by Kuepper and Schultze [72] and others [61] will yield new insights into electrodeposition into deep high-aspect-ratio features.

3 Summary: Uniformity at the Workpiece, Pattern, and Feature Scales in LIGA

The detailed study of Thommes et al. is a good example of the considerable effort required to develop an electrodeposition process capable of producing uniform structures at the feature scale for geometries encountered in LIGA electrodeposition. When mixing effects are important (as often the case in Ni-alloy deposition) [70–71], this is especially difficult. After some confidence in the feature-scale uniformity is attained for a given chemistry and set of deposition conditions, the uniformity at the pattern and workpiece scales must still be addressed as well. This makes LIGA through-mask electrodeposition in some ways more challenging than electrodeposition through thin resist, as the feature scale adds literally another dimension to the problem of obtaining uniformity across multiple length scales. Considerable time and effort may be saved by first characterizing the electrodeposition kinetics and the importance of mass transport for an electrodeposition process before attempting to apply it to LIGA fabrication (especially in the case of an alloy system). If one has a good understanding of kinetics, transport, and alloy composition, electrolytes and deposition parameters suitable for deposition through thick resist may be chosen more prudently [67, 69]. Examples of such situations will be given when the mechanical properties of electrodeposits are discussed.

4 Electrodeposition in LIGA: Materials and Other Aspects

Up to this point, the range and types of materials electrodeposited for LIGA applications have not been discussed in depth. In the next section, we will go into more detail on these topics in regard to requirements for LIGA fabricated structures to be used in mechanical applications. Before doing so, we briefly mention materials studied for other types of LIGA applications, as well as other aspects related to the integration of the electrodeposition step into the LIGA process.

Besides the study of Thommes et al. discussed above as an example of feature-scale uniformity [69], the possibility of using electrodeposited NiFe structures as magnetic components in microsystems has been discussed or demonstrated in a number of articles (a sampling is given in the references) [25, 73–79]; but as shown by Thommes et al. [69], some care must be taken in insuring that a given deposition process will result in acceptable alloy uniformity. Chin reviews a variety of different types of magnet films, besides the traditional plated iron-group alloys for applications in microsystems, discussing the potential of electrodeposition in fabricating some more exotic magnetic alloys [80]. Despite the challenges posed by the thick resist often used in LIGA, the possibility of electroplating magnetic alloys for LIGA microstructures with useful magnetic properties makes LIGA and LIGA-like processes attractive to microsystem designers. In general, such LIGA-fabricated NiFe microstructures appear to have magnetic properties that are comparable to similar bulk NiFe materials [75–76, 78].

Compared with Ni and Ni alloys, investigations of other electroplated materials for LIGA processing are relatively few in number; this is most likely due to the fact that Ni may be electroplated with low stress in thick films with reasonable mechanical properties, as discussed in the next section. However, copper [28] and gold [21] (both of which can be plated as thick films) have been used with the LIGA process to fabricate thick-plated LIGA structures. Copper is of interest of course where a high electrical conductivity is necessary, and gold is attractive when structures having a high mass are required due to its high density (~19.3 g/cm^3). One study reports the electrodeposition of a Bi–Te alloy into a LIGA mold to form BiTe microposts for microcooling probes [81].

Though not discussed at length here, another important area of LIGA electrodeposition research involves the nucleation and adhesion of the plated structures to the seed layer material used as the substrate. Since the substrate is exposed to X-rays during the exposure of the PMMA resist, metallization layers between the substrate and resist (such as those commonly used as plating bases, like copper) can reemit the X-ray radiation (e.g., by fluorescence), damaging the bond at the resist-metallization interface and provoking adhesion failures during subsequent processing [82]. Since this damage can lead to the failure of the entire LIGA mold, alternative plating bases that do not behave in such a way are thus of interest [82]. At the Forschungzentrum Karlsruhe in Germany, an approach using TiO_2 as a plating base has been developed and successfully employed [83]. Other suitable plating base materials have been recommended, but electrodeposition processes where good nucleation and adhesion to the base still need to be developed for these materials [82].

In this section, the challenges in attaining good uniformity in deposit morphology and/or deposit alloy composition over various length scales during the LIGA electrodeposition process step was emphasized. In practice, an equally important aspect in electrodeposition for LIGA is the subject of electrolyte ageing and maintenance. Although it is not the focus of this chapter, a few comments on these topics are appropriate, especially since they are little discussed in the LIGA electrodeposition literature. These aspects are particularly important if the LIGA process is to be used for the manufacturing of a product with some specified, reproducible material properties that depend on the deposit morphology and/or alloy composition. For example, sulfamate baths are very popular for the electrodeposition of Ni LIGA microstructures. But as the electrolyte is used and ages, ammonium ions (originating from the sulfamate ion) may accumulate and adsorb at the deposition surface, acting as an additive and changing the deposit morphology and hardness [84]. Moreover, it has been known from the traditional electroforming literature that sulfur-containing breakdown species (originating presumably from the sulfamate ion) may also accumulate over time, leading to changes in film stress and morphology [85]. Both of these aging effects complicate the deposition of a film, with consistent morphology and stress during manufacturing. We have observed complications arising from sulfamate electrolyte aging during the deposition of a NiMn alloy from this chemistry [86]. Since the aging of electrolytes often leads to species that induce unwanted changes in the plated film microstructure and material properties that an electrodeposition process is engineered to obtain, an understanding of electrolyte ageing and maintenance is critical for the use of LIGA as a manufacturing process. Before addressing these topics in more detail, we first discuss the development of electrodeposited materials for LIGA mechanical applications in the next section.

5 Properties and Structure of Electrodeposited Materials for LIGA-Based Microsystem Applications

5.1 Introduction

We describe here the material characteristics necessary to successfully fabricate precision structures for microsystem applications. In particular, we focus on the electrodeposited (ED) component structures that are required to have mechanical functionality: for example, springs or flexures. In such instances, mechanical strength, ductility, fatigue life, and a well-characterized modulus are critical material properties that must be well understood and reproducible. These structures typically have micron-scale tolerances with respect to critical dimensions and insofar as their mechanical and physical behavior are governed by their submicron grain size, crystallographic texture, and the very high defect densities typical of electrodeposits, we may consider these materials to be nanostructured.

In this section we restrict ourselves largely to the material properties and performance of structures that are fabricated *via* the LIGA process, which we described at the beginning of this chapter. In some instances, though, we refer to work based on the deposition of blanket films where LIGA-prototyping processes were not employed. A number of constraints limit the range of practical electrolytes and therefore electrodeposited material systems. Some of these constraints include: (1) low bath temperatures, which often must be well below those recommended by common electrodeposition practice (necessary to reduce thermal distortions of the resist material that define the lateral dimensions of precision structures) [50, 87]; (2) transport limitations – particularly important in deep, high-aspect-ratio features where circulation of the bulk electrolyte is impeded, being typically minimized by the use of reduced deposition rates (i.e., the use of atypically low current densities) [86]; (3) low intrinsic film stresses to minimize piece-part distortions and survivability of the mold during plating; and (4) pH compatibility with LIGA-fabricated molds and substrates and pH control (both in the bulk electrolyte as well as within narrow, stagnant cavities as described in the previous section). These various factors are summarized in Table 2.

With regard to the properties of electrodeposits themselves, where structures have an intended mechanical functionality, the most critical material property considerations are (1) strength – as section size of microscale components decrease, the strength must increase to avoid plastic deformation; (2) ductility – structures within mechanical assemblies cannot fail in a brittle, catastrophic fashion; (3) modulus – the material stiffness is a first order input parameter in the design of a spring or flexure; and (4) fatigue – the fatigue response of an electrodeposit can be an important

Table 2 Deposition-process characteristics and requirements

Characteristic	Requirement
Near-ambient temperature deposition	Dimensional tolerances
Transport: Cation/impurity/additive	Composition, microstructure, property uniformity in high aspect ratio features
Intrinsic film stress	Contour tolerances Thick-section deposit
pH restrictions	Substrate/resist compatibility, effects on ED structure and properties

Table 3 Electrodeposit properties and requirements

Characteristic	Requirement
Strength	Elastic response of mechanically functional structures
Ductility	Preclude catastrophic brittle fracture
Well-characterized modulus	Predictable flexural response
Fatigue	Predictable response to dynamic environments

material property depending on the ambient or functional environment. Table 3 summarizes these requirements.

The practical choice of ED materials compatible with all aspects of LIGA fabrication is restricted to gold, copper, and nickel (and a limited number of nickel-based alloys). Not withstanding this limited family of materials, the range of microstructures and resulting material properties is quite extensive, and it is the electrochemistry of metal deposition that is principally responsible for defining the microstructure and properties of these materials which form the free-standing structures from which microsystem component parts are fabricated. However, with respect to the fabrication of mechanically functional structures, neither gold nor copper has suitable strength (yield strength of ED copper is typically less than 500 MPa, while even "hard" gold will rarely exhibit yield strengths that exceed 250–300 MPa) [8].

We focus then on ED nickel and Ni-based alloys since these materials can exhibit yield strengths that range between 350 MPa and 1200 MPa, depending on electrolyte and the particular plating conditions employed. The main challenge lies in adapting approaches that have been used for years to deposit high-strength Ni for industrial electroforming and electroplating to the through-mask electrodeposition of LIGA structures.

5.2 Measurement Techniques for Strength and Ductility

The strength of electrodeposits can be characterized either indirectly through surface indentation hardness testing or directly by testing net shape deposited mechanical test specimens (tensile, compression, fatigue, etc.). In the former instance, yield strength, the strength parameter most meaningful with respect to the design of mechanically functional structures, can only be roughly estimated even if the general work hardening characteristics of the material are known (more typically, attempts are made to correlate "flow stress" with indentation hardness measurements). The principal difficulty is the inability to reliably characterize the "constraint" factor that relates the contact pressure under an indenter tip to an equivalent uniaxial compressive stress. This constraint factor is dependent on a host of considerations, including the geometrical shape of the indenter tip, the rate of indenter loading, and the work hardening and relaxation characteristics of the material of interest [88]. Thus, hardness measurements can only be related to the strength of a material if an explicit correlation has been derived between indentation measurements and strength measurements for each individual electrodeposit of interest.

For example, Fig. 18 shows the correlation between indentation hardness and the yield and the ultimate tensile strengths (UTS) of an ED NiMn alloy, where we have adjusted the deposition parameters to yield a series of deposits whose strengths vary over the wide range indicated. For the data shown here, the yield strength (in conventional SI units of MPa) follows the relationship:

$$\sigma_y \approx -300 + 3\mathrm{H}_{25\mathrm{gm}}$$

where $\boldsymbol{\sigma_y}$ is the yield strength and $\mathbf{H}_{25\mathrm{gm}}$ is the Vickers microhardness (using a 25-g load). The UTS of these NiMn electrodeposits has a somewhat higher correlation slope. With such correlation curves, one can, with some confidence, use hardness measurements as a surrogate for yield strength, where the direct measurement of mechanical properties is impractical (due to structure geometry or size, for example). As the work hardening characteristics of each ED material system

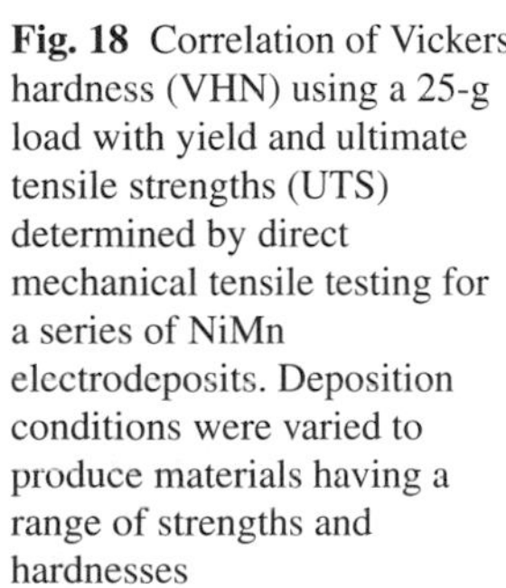

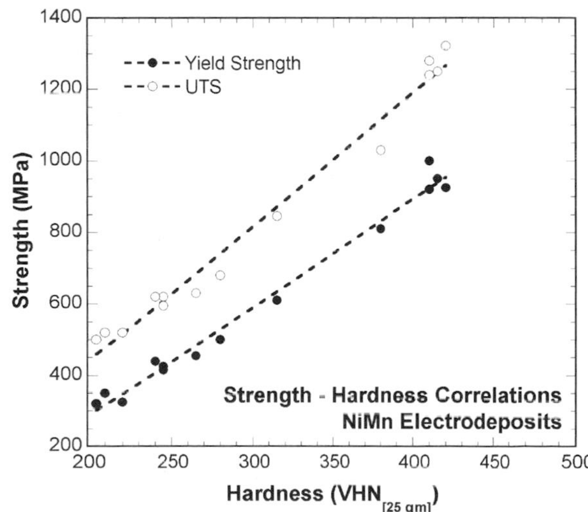

Fig. 18 Correlation of Vickers hardness (VHN) using a 25-g load with yield and ultimate tensile strengths (UTS) determined by direct mechanical tensile testing for a series of NiMn electrodeposits. Deposition conditions were varied to produce materials having a range of strengths and hardnesses

is different, in principle, a similar correlation curve would be necessary for each family of electrodeposit. Furthermore, the intrinsic ductility of an ED (or any other) material remains uncharacterized by a hardness measurement.

With respect to the direct characterization of strength and ductility, these measurements are made directly using standard, albeit, much higher resolution, testing methodologies on test instruments that are usually appropriate for the size scale of the test specimens. Because specimen sizes are small, noncontacting, optical-strain measurement systems are often employed [89–91]. As will be described later, it is these strain measurements that are most problematic when specimen sizes approach those of actual microsystem component structures.

5.3 Grain Refinement for Improved Electrodeposit Strength

Alloying and Organic Additives – Strength in engineering materials is derived from several sources. Common among these sources is solution strengthening, strengthening from the presence of rigid second phases or inert particle additives, and, lastly, grain refinement. Generally speaking, for Ni-based electrodeposits, grain size is the principal microstructural feature that governs strength; the wide range of strength that can be realized reflects the equally wide range in grain size that is achievable. For example, Fig. 19 shows tensile curves for an ultra-fine-grained Ni deposited from a standard Watts bath electrolyte (nickel sulfate, nickel chloride, and boric acid) as well as that for a lower-strength ED nickel deposited from a standard sulfamate bath chemistry (nickel sulfamate and boric acid). In the example shown in Fig. 19, saccharin is included in the electrolyte as a stress reliever, and the resulting Watts bath-deposited Ni has a grain size on the order of 10–20 nm as shown in the TEM insert in Fig. 19. The yield and ultimate tensile strengths (σ_y and UTS) are $\approx$1200 and 1800 MPa, respectively. This can be contrasted to the sulfamate bath ED Ni, where the grain size measured transverse to the columnar axis is on the order of 1–2 μm and the resulting yield and ultimate tensile strengths are $\approx$350 and 600 MPa. Indeed, Ebrahimi et al. [92] and Thompson [93] have shown that over a very wide range, nickel electrodeposits follow the Hall–Petch relationship [94] with respect to the effect of grain size on strength as illustrated in Fig. 20.

The importance of grain size is further illustrated in Fig. 21 where the Vickers hardness of ED NiCo (deposited from a 1.25 M nickel sulfamate and 0.025 M cobalt sulfamate electrolyte) is plotted against the concentration of incorporated cobalt. It is clear that the hardness, and therefore the strength, rises rapidly with cobalt concentration to a maximum of 450 VHN at approximately 40 wt% cobalt. One is tempted to view this increased hardness as the result of solid solution strengthening, since cobalt is substitutional in nickel and completely miscible at these concentrations [95]. However, the SEM micrograph inserts in Fig. 21 reveal that the principal effect of cobalt is to dramatically reduce the grain size from 1–3 μm at low Co concentrations to 50–100 nm at the maximum hardness [96]. The absence of a significant solution strengthening effect (even in this relatively high alloy content

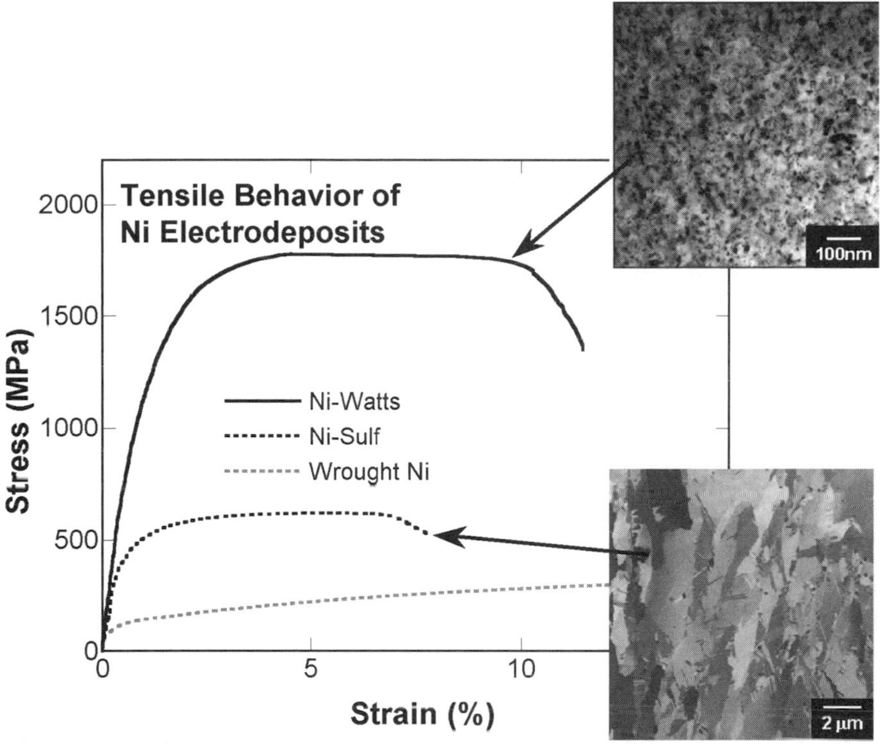

Fig. 19 Stress–strain behavior of electrodeposited and wrought Ni. When a Watts bath is used with saccharin (a common organic additive in Ni plating) a fine grain size is obtained (see inset TEM plan view micrograph), resulting in a high-strength Ni deposit. In contrast to the fine-grained Watts deposit made with saccharin, Ni from an additive-free sulfamate bath has larger grains (see inset focused ion beam cross-sectional micrograph), resulting in a lower strength deposit. Compared to commercial purity, annealed Ni, both electrodeposits have high strength due to the fine as-deposited grain size

material) is further illustrated in Fig. 22, where the tensile behavior of this alloy electrodeposit is compared to that of unalloyed ED Ni plated from a sulfamate bath. The as-deposited yield strength of the NiCo alloy is $\approx$1.1 GPa, approximately 800 MPa greater than the yield strength of the pure nickel deposit. In contrast, after annealing at 700°C for 1 h, the yield strength of the alloy electrodeposit is only about 150 MPa greater than that of the nickel deposit. Thus, any contribution to the increase in mechanical properties due to solution strengthening is marginal. Even this small difference in the annealed strength most likely reflects slight differences in the final grain size of the tested material.

Figure 21 indicates that it takes tens of percent of Co addition to significantly increase the strength and hardness of ED nickel. However, metallic solutes can be potent grain refiners and therefore material strengtheners in nickel electrodeposits even at very low concentrations. The work of Stephenson and others has demonstrated this for nickel–manganese electrodeposits [70, 91, 97–99]. Small

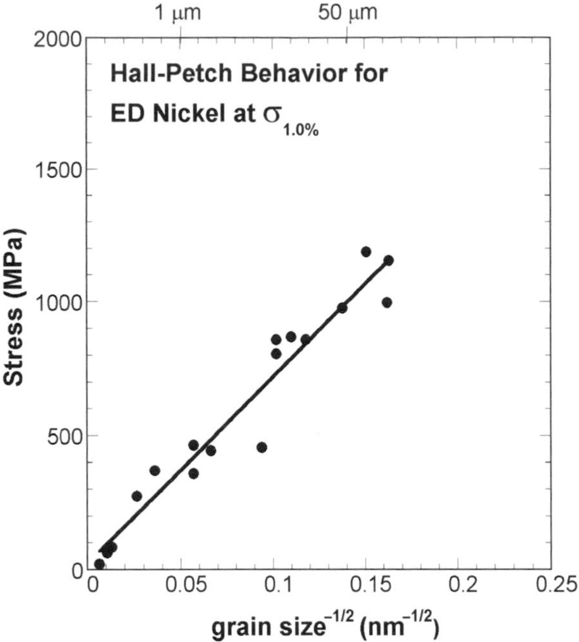

Fig. 20 Hall–Petch behavior for electrodeposited Ni, replotted from the data of Ebrahimi et al. [92] and Thompson [93]. The grain size of electrodeposited Ni plays a large role in determining its strength

additions of a Mn salt to a Ni-sulfamate electrolyte result in exceptionally high-strength ED Ni alloys having Mn concentrations that are typically less than 1 wt.%. In these reports, the principal effect of Mn incorporation was to reduce the grain size of the Ni-sulfamate electrodeposit by approximately 100-fold. Figure 23 illustrates effectiveness of small Mn additions to ED Ni using a sulfamate chemistry. This series of tensile curves shows the progressive increase in strength with increasing Mn concentration. As Mn is not a potent solution strengthening species, the more than twofold increase in strength is derived from the continuing grain refinement as the solute concentration increases. Generally speaking, therefore, nickel-alloy electrodeposits derive their strength from the effect of the solute on suppressing grain growth during deposition (or alternatively, increasing grain nucleation) rather than from the more traditional metallurgical effects associated with solution strengthening.

Employing the grain-refining attributes of solutes in nickel-alloy electrodeposits must be approached with caution for deep, high aspect ratio, through-mold structures. The principal reason for concern arises from diffusive transport limitations of the co-depositing solute that can limit compositional and microstructural uniformity. The nickel–cobalt system affords a useful example of this limitation. Cobalt is easily deposited in uniform concentrations in bulk films over a wide range of concentration. Indeed, the hardness data shown in Fig. 21 is an example of the synthesis of bulk films over a wide range of uniform, through-thickness Co concentration. However, when filling a deep, high-aspect-ratio feature on a LIGA mold,

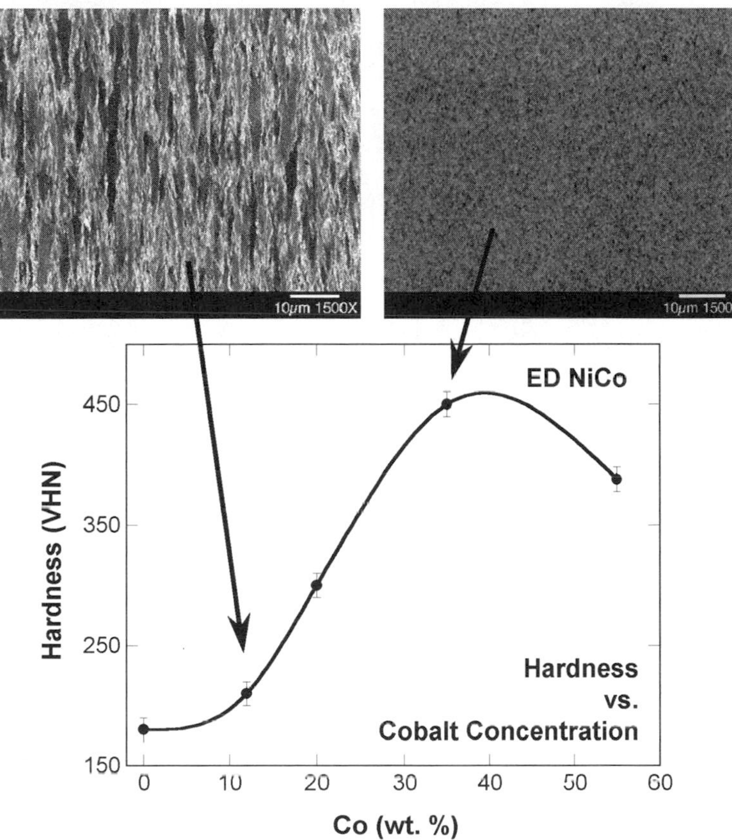

Fig. 21 Hardness dependence on Co content for an electrodeposited NiCo alloy. The SEM cross-sectional micrographs show that the increased hardness obtained at higher Co contents arises from a refinement in the alloy grain size. (Reproduced from Yang, et al. [96]

the consumption of cobalt ions at the substrate surface may lead to concentration gradients in the electrolyte within the feature for this species (this was discussed previously for the NiCu system) [72] as mixing of the electrolyte deep within high-aspect-ratio features is minimal.

As we discussed for the case of NiFe, the NiCo system may also exhibit what is known as anomalous co-deposition, which exacerbates the problem. For anomalous systems, under certain conditions (e.g., low bath temperatures), the alloying element deposits preferentially; that is, the fraction of the alloying element in the deposit is larger than that expected based on the ratios of the two metal cations in the electrolyte [70]. In the particular case of NiCo, the bath is dilute in cobalt ions, but they are deposited and are thus consumed at a high rate because of the anomalously high cobalt-deposition current density. Under such circumstances, the electrolyte close to the deposition surface is rapidly depleted of cobalt ions, leading to a lower-than-expected concentration of cobalt in the deposit. However, as the metal grows

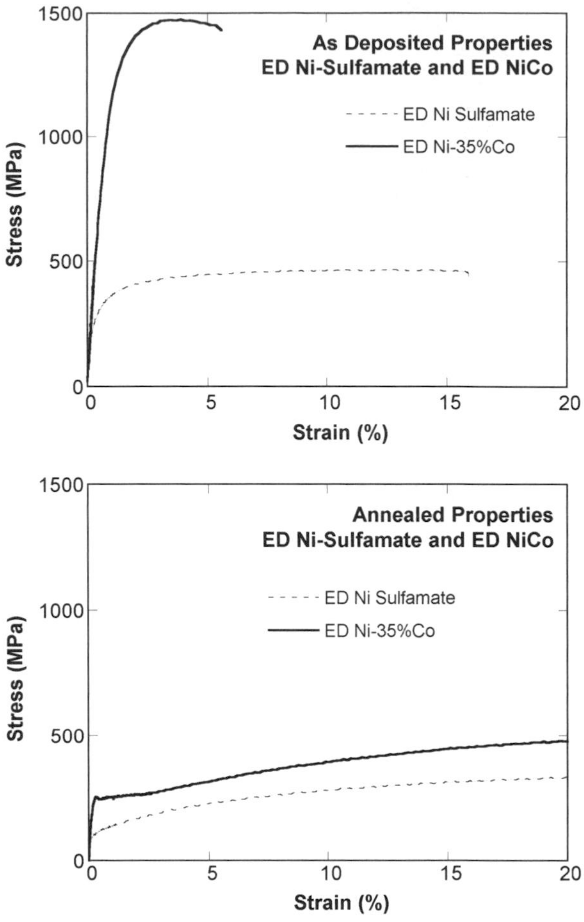

Fig. 22 (*top*) Stress–strain behavior for as-deposited NiCo and pure Ni, both from sulfamate electrolytes. The as-deposited NiCo is much stronger than the pure Ni. (*bottom*) Both materials soften after annealing at 700°C for 1 h, but the NiCo alloy loses almost all of its strength advantage over the pure Ni after the anneal, resulting in similar stress–strain behavior for both materials

toward the top opening in the resist, the geometry of the feature changes, leading to a constantly changing set of local electrolyte conditions. For example, as the aspect ratio is reduced, the diffusive transport of cobalt ions increases, allowing for their more rapid replenishment to the deposit–electrolyte interface. The net result is a progressive increase in the concentration of solute in the alloy as the high-aspect-ratio feature is filled.

As an example, Fig. 24 a shows the evolution in concentration of cobalt in a 25-μm wide by 250-μm tall LIGA-fabricated NiCo beam. The plating conditions were set so as to yield a deposit consisting of 30 wt% Co–70 wt% Ni. The cobalt concentration is characterized *via* microprobe-based, wavelength-dispersive spectroscopy (WDS); and in this figure the seed surface (the surface at the "bottom" of the lithographically defined feature) is to the left axis, while the top of the feature is to the right. Transport limitations due to lack of mixing within the LIGA feature, as described above, result in a deposit that is severely depleted in cobalt at the

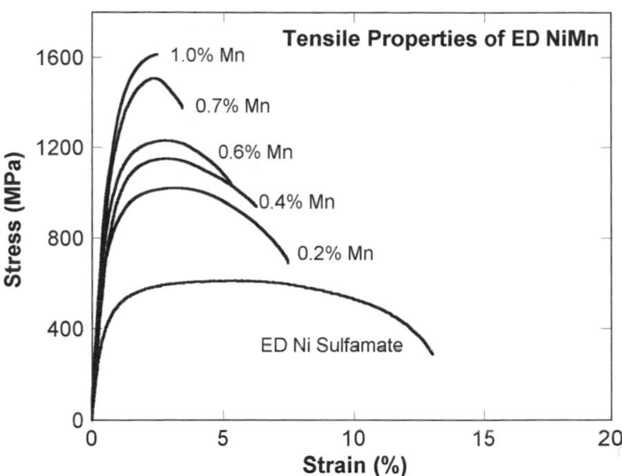

Fig. 23 Stress–strain behavior for a series of NiMn electrodeposits. The alloy is deposited from a Ni-sulfamate electrolyte containing 1–5 g/L of Mn added as Mn chloride. The reason the alloy strength increases with Mn content is the increasingly fine grain size induced by the Mn codeposition [98–99]

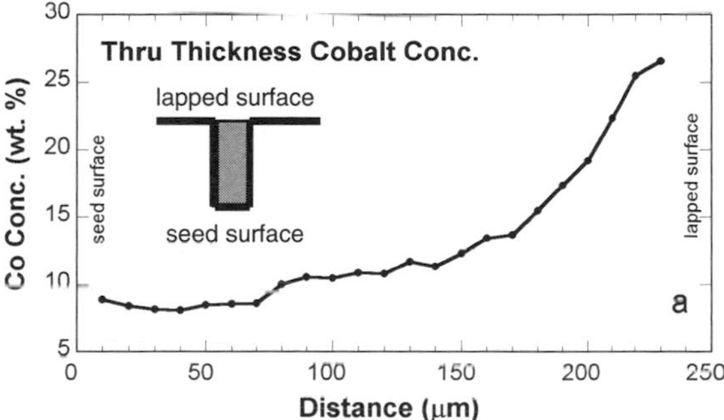

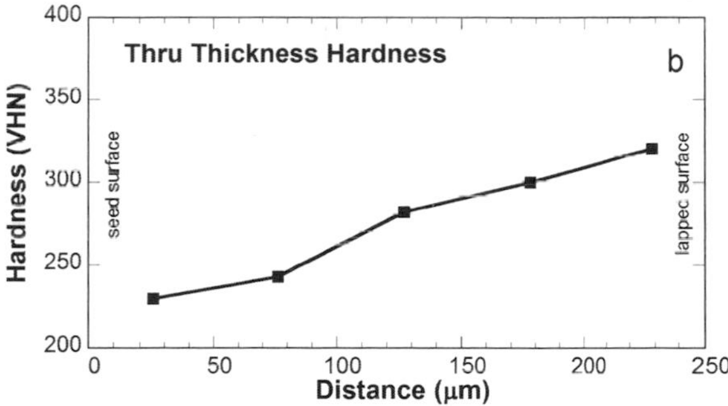

Fig. 24 (**a**) Co content of an electrodeposited NiCo alloy within a LIGA feature having an aspect ratio 10 (25-μm wide and 250-μm tall). Poor mixing within the feature results in cobalt ion gradients in the electrolyte and hence in the deposited metal. (**b**) The varying Co content manifests itself in a hardness gradient across the feature thickness

base of the feature. Only as the cavity fills and the deposit-solution surface rises toward the top-third of the high-aspect-ratio feature (where convective mixing can aid in the replenishment of the cobalt cations) does the bulk concentration of Co approach its target concentration. Since solute content determines the grain size of the deposit, the mechanical properties of such a compositionally graded structure are correspondingly compromised as shown in Fig. 24b. Here, the indentation hardness is profiled from the seed surface to the top of the feature, and it is clear that the hardness rises in direct relationship to the increase in Co concentration in close agreement to the data shown in Fig. 21.

The NiCo system is particularly sensitive to these transport-limitation-induced effects on composition and structure, since this system exhibits anomalous co-deposition. Figure 25 compares its feature-filling behavior to that of the NiMn system, which is not anomalous in its deposition characteristics, for a geometry similar to that shown in Fig. 24 (specifically, 32-μm wide and 350-μm tall). Since the target concentration for the NiMn alloy was approximately 1% as compared to 30% for the NiCo alloy, we plot the concentration profile in Fig. 25 normalized to the WDS-measured concentration near the seed surface for each alloy. Similarly, location from the substrate is normalized to the absolute feature height for both deposits. While the concentration of Mn solute in the NiMn deposit exhibits a measurable gradient, the Co concentration gradient in the NiCo deposit is tenfold greater, a direct consequence of the sensitivity of the anomalous NiCo system to mixing. The increased Mn uniformity in the NiMn alloys is reflected in the enhanced uniformity of the hardness of the electrodeposit along the thickness of the deposit as shown in

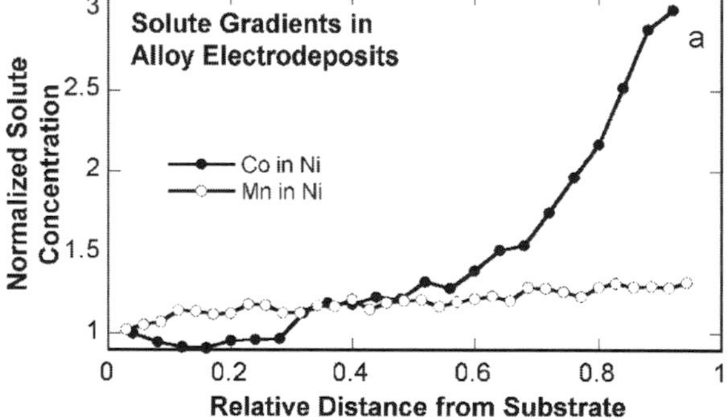

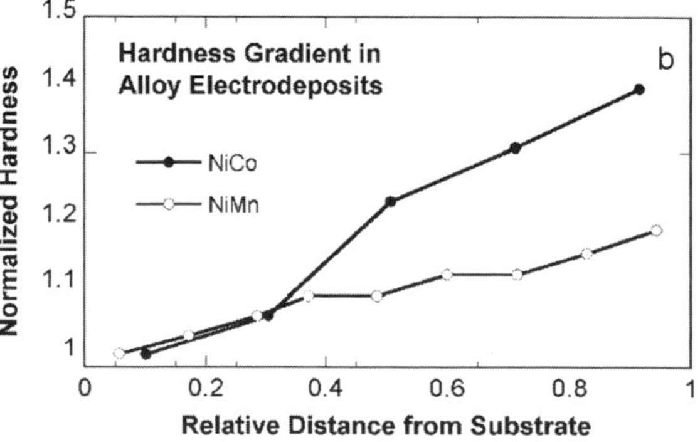

Fig. 25 (**a**) Solute concentration for NiCo and NiMn alloys, normalized to the solute concentration at the bottom of the feature, for alloys deposited into a LIGA feature with an aspect ratio of about 10. The distance from the substrate is also normalized (for NiCo, the feature width and height is 25 μm and 250 μm, respectively, while for NiMn it is 32 and 350 μm, respectively). The Mn solute concentration is more uniform since the Mn deposition rate is not sensitive to mixing. (**b**) Through-thickness hardnesses for the two-alloy systems normalized to the hardness at the feature bottom. The more uniform composition of the NiMn alloy results in a more uniform through-thickness hardness

Fig. 25. The behavior illustrated in Fig. 25 demonstrates the importance of understanding how the electrochemistry of the deposition reaction influences the resulting material properties of LIGA-fabricated structures.

Other Additives – Other electroplating bath additives can have similarly significant effects on the grain structure of nickel electrodeposits. We have already discussed the example of the organic electrolyte additive saccharin. At concentrations of about 1 g/L and higher, saccharin reduces high film stresses obtained when depositing nickel from a Ni–Watts plating bath (nickel sulfate, nickel chloride, and boric acid) [8]. It has the additional effect of reducing the grain size of the as-deposited material into the tens of nanometer regime as a result of its adsorption at the deposit-solution interface. The impact of this refined grain size on the deposit's mechanical properties has already been illustrated in Fig. 19, where it was shown that the as-deposited strength of the deposit exceeds that of ED Ni–sulfamate by nearly fourfold, while retaining significant ductility. However, the adsorption of the saccharin molecule and subsequent incorporation of sulfur into the deposit (the resulting sulfur concentration in a Ni–Watts deposit is on the order of 500–700 wt. ppm) have severe consequences with respect to the thermal response of the deposit. Sulfur is a well-known embrittling agent of nickel and nickel-based alloys at concentrations far lower than that present in the electrodeposit. The consequences of this sulfur incorporation is made obvious in Fig. 26, which shows the effect of vacuum-annealing temperature on the room temperature mechanical properties of Ni deposited from a Watts electrolyte with a saccharin additive. Ductility loss is evident in specimens annealed at temperatures as low as 200°C. At temperatures above 200°C, embrittlement is so severe that there is no measurable ductility. Tensile specimens simply deform elastically and then fail in a catastrophic brittle fracture. Because of this complete lack of ductility, the individual tensile curves cannot be discerned in Fig. 26. As an aid to visualizing the degree to which the material is embrittled, the insert in this figure shows the tensile curve for a specimen annealed at 500°C and then tested at room temperature. In this instance, fracture occurred at approximately 50 MPa and at a total plastic strain of less than 0.025%. Figure 26b shows the fracture surface of a tensile specimen tested to failure at room temperature after an elevated temperature anneal. The intergranular nature of the fracture is evident in this micrograph. Similar effects are found in electrodeposited Ni–Fe alloys, where the saccharin addition to the NiFe electrolyte is again required to reduced plating stress [100]. In this instance, there is direct evidence of the segregation of sulfur to grain boundaries after annealing at elevated temperatures. The results of these studies lead to the conclusion that, insofar as ED Ni is concerned, no sulfur-bearing electrodeposit can be expected to exhibit any significant amount of ductility, after even modest elevated temperature exposure.

Even unintentional additives to a plating bath can result in significant changes to the resulting structure and properties of electrodeposited materials. Deposition of through-mold structures from Ni-sulfamate baths affords an example of this. Ni-sulfamate is an attractive electrolyte for use in LIGA fabrication for a variety of reasons. From a properly conditioned bath, deposition can occur at relatively

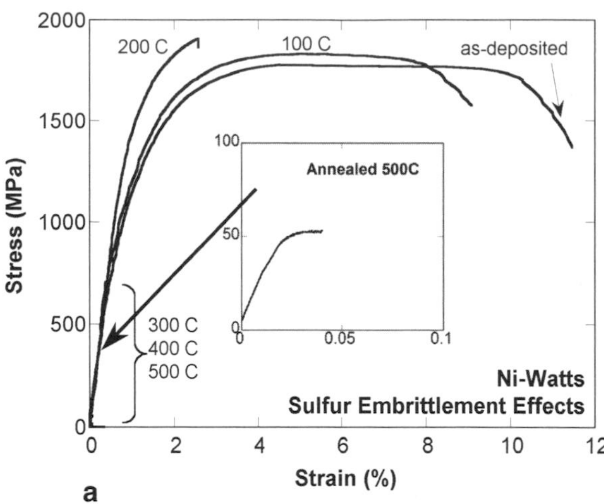

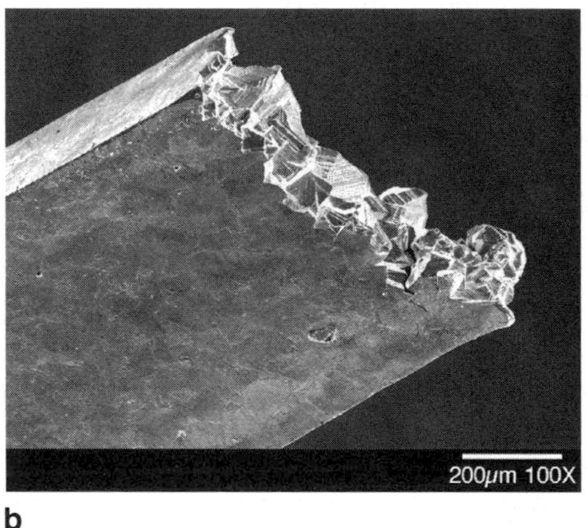

Fig. 26 (**a**) Stress–strain behavior for various anneals of Ni electrodeposited from a Watts chemistry containing 1 g/L saccharin. The incorporation of sulfur from the saccharin additive leads to the embrittlement of the electrodeposit after thermal exposures. (**b**) Fracture surface of an annealed Ni test sample plated from a saccharin-containing Watts bath. The intergranular nature of the fracture surface is easily discernable *via* SEM

high rates due to its high current efficiency, and the resulting nickel deposits exhibit good ductility and low internal film stress (an important consideration when fabricating high-precision microcomponent structures, where elastic distortions can compromise dimensional fidelity). However, hydrolysis of the sulfamate ion may result in the build-up of sulfate and ammonium ion [85, 101–102], the latter of which acts effectively as an unintended additive species. This may occur over the natural age of the plating bath, or if the temperature of the nickel-sulfamate concentrate makeup solution or Ni-sulfamate plating bath exceeds 60°C for extended periods.

The ammonium decomposition product in nickel-sulfamate plating baths affects the intrinsic film stress, the crystallographic film texture, as well as the film grain structure and mechanical properties. Lin et al. [84] reported that film stress rises from about 22 MPa in the absence of ammonium to over 45 MPa at ammonium ion concentrations of $\approx$250 wt ppm. At the same time, the <001> texture, characteristic of the free-growth mode of Ni-sulfamate deposits, is suppressed in favor of an "inhibited" growth mode, giving rise to a <110> characteristic texture. Such texture changes can affect the elastic response of through-mold structures and will be discussed later in this section. With respect to mechanical strength, though, the presence of ammonium ion has important implications. Figure 27 compares the microstructure and tensile behavior of LIG-fabricated mechanical-test specimens plated from a freshly prepared Ni-sulfamate electrolyte to that of a specimen fabricated from an "aged" electrolyte (i.e., after several hundred A hours of prior use). In both cases, the deposition conditions are identical, with respect to plating temperature, current density, pH, and plating-cell geometry. In the former instance, the microstructure and tensile behavior is as expected; a coarse, columnar grain structure giving rise to a yield strength of $\approx$300 MPa. In the later instance, the presence of $\approx$150 wt ppm of ammonium results in drastically refined grain size, and as a result, a nearly threefold increase in yield strength to 800 MPa.

The mechanism by which additives and alloying species can influence the structure and properties of nickel electrodeposits is a subject of much discussion. Some

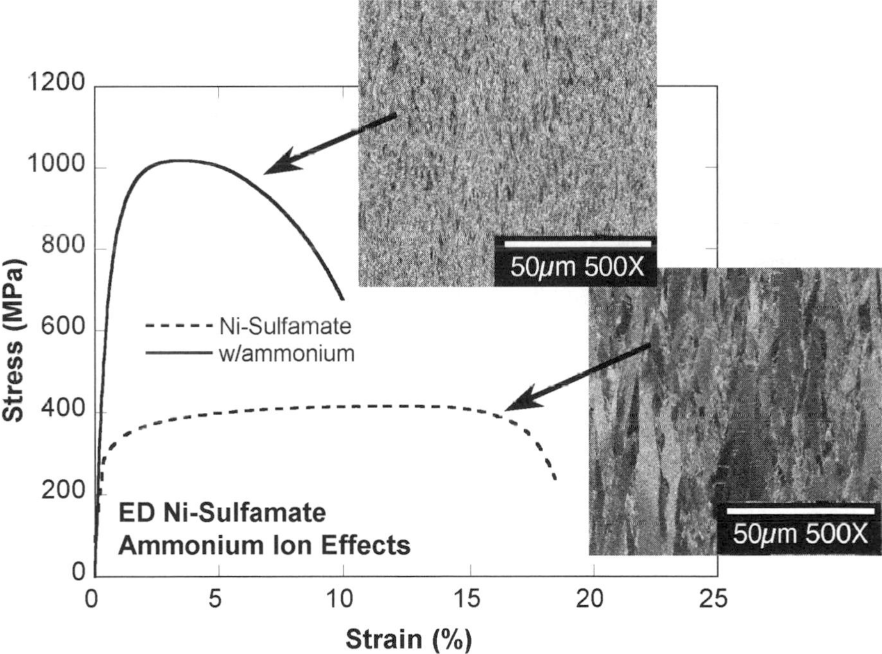

Fig. 27 Stress–strain behavior for Ni electrodeposited from a fresh sulfamate plating bath (*dashed line*) and an identical bath that has $\approx$150 ppm ammonium ion due to several hundred A hours of use. The accompanying SEM cross-sectional micrographs show that the higher strength of the Ni from the ammonium-containing bath is due to its finer grain structure

qualitative understanding of the phenomenon can be drawn from the proposition that any adsorbed species, be it a foreign co-depositing metal cation or a chemical constituent of the plating bath (inorganic or organic) may interfere with the usual nucleation and growth of the electrodeposited film, resulting in a grain structure much different than that obtained in the absence of such species [103–105]. Sometimes, the potency of the interfering species may be ascertained by studying its effects on the metal-deposition reaction kinetics; a species that strongly prevents the deposition reaction from occurring is thought to have a strong inhibitive effect on the film growth, resulting in a finer-grained film [105]. Thus, additives are sometimes classified by their inhibitive effects on film growth, but it is clear that their impact on the nucleation process can strongly influence the final film microstructure as well [103–105].

Current Density Control –Beyond the intentional or unintentional incorporation of alloy or chemical additives, plating-process parameters can have a significant effect on the structure and properties of nickel electrodeposits. For example, the imposed current density during deposition can have a profound impact on grain size. Zentner et al. [106] have shown that, for bulk ED Watts nickel (nickel sulfate + nickel chloride) deposited at pH 3.0 and at 55°C, grain size first increases as current density increases to intermediate levels. Under these conditions, the deposit develops an exaggerated coarse columnar grain structure. However, as current density continues to increase, the trend reverses, such that the grain size becomes finer overall. This behavior is sometimes rationalized as being governed first by the inhibiting effect of weakly adsorbed species in the electrolyte at very low current densities (such species may be effective inhibitors despite being weakly adsorbed since the driving force for film growth is low) that are overwhelmed as current densities and overpotentials increase to intermediate levels where the metal deposition rate is correspondingly higher. At these intermediate current densities, deposition is said to be "uninhibited", and coarser-grained microstructures are favored. Ultimately, at the very highest current densities, nucleation once again dominates; the net result is a reversal in the trend to favor more crystallite (grain) nucleation at very high deposition rates.

The response to applied current density (in terms of strength) can be seen in other reports in the literature, although the trends are not always consistent. For example, Sharpe reported that both the yield and UTS decreased monotonically with increasing current density across the range of 3–50 mA/cm^2 for through-mold ED Ni tensile specimens plated from a sulfamate chemistry, as shown in Fig. 28a [89]. Such behavior is easily explained based on conditions that promote uninhibited growth of the deposit with a corresponding increase in grain size of ED Ni at these intermediate current densities. However, the work of Kim and Weil run counter to this, as shown in Fig. 28b, which suggests no strong trend at applied current densities between 3 and 25 mA/cm^2 for ED Ni plated from a sulfate chemistry [107]. A clear increase in yield strength is noted at a current density of 50 mA/cm^2, however, and pulse plated deposits exhibit higher strengths than direct-current (DC) deposits. Pulse plating has long been considered a method for encouraging nucleation and thus finer-grained deposits as we will discuss shortly. Safranek reports similarly

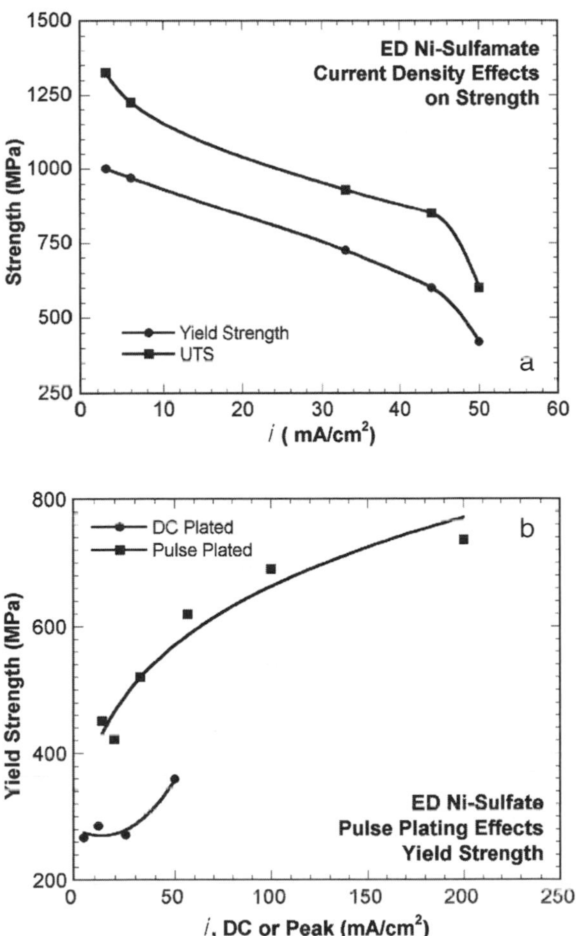

Fig. 28 (**a**) Yield and ultimate strengths for Ni electrodeposited from a sulfamate electrolyte reproduced from data given by Sharpe [89]. Strength decreases with increasing current density under these conditions. (**b**) Yield strength data reproduced from Weil and Kim who used a sulfate chemistry for Ni electrodeposition [107]. Their results for DC deposits show no strong dependence of yield strength on current density for low deposition rates, with an increase occurring at 50 mA/cm^2. Their pulse-plated deposits are stronger than those plated using DC, with strength trending higher with higher peak current densities

contradictory results for ED Ni-sulfamate, with the strength trending in opposite directions with increasing current densities between 200 and 600 mA/cm^2 [8].

Such inconsistencies are not uncommon when reviewing the literature and illustrate the fact that broad assumptions as to particular effects of current density may not be drawn between different electrolyte systems (sulfamate vs. sulfate, for example). Indeed, even for a single electrolyte, the potential effects of even the smallest variation in other deposition process variables such as bath age (discussed above), chemistry, pH, and temperature may overwhelm the effect of current density. A graphic example of the impact of even minor changes in deposition-process parameters is shown in Fig. 29, where the tensile behavior of ED Ni is shown for through-mold test specimen plated from a sulfamate chemistry at 28°C and at 3 mA/cm^2. The only difference in the plating processes for the two results shown is the circulation

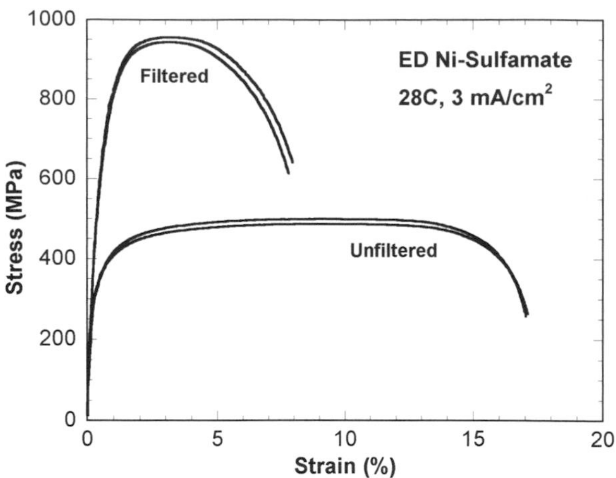

Fig. 29 Effect of electrolyte filtering on stress–strain behavior for electrodeposited Ni from a sulfamate bath at 28°C and 3 mA/cm^2. The passage of the bath through the filter at lower temperatures is thought to lead to the formation of polyborate species that refine the grain structure at low deposition rates, resulting in higher strength Ni [86]

pump-driven passage of the electrolyte through a 5-μm debris filter in one instance but not the other.

Filtration is a common industrial practice and has not been thought to influence the *electrochemistry* of deposition. However, it is evident that the properties and microstructure of the deposits are profoundly affected by the presence or absence of such filtration under certain conditions. ED Ni specimens plated from a filtered sulfamate bath exhibit a relatively high yield and ultimate tensile strength (700 MPa and 950 MPa respectively), while specimens made from an unfiltered bath exhibit corresponding strengths equal to about half of these values; these lower strengths seen for the unfiltered electrolyte are more in line with expectations from the literature for the sulfamate chemistry [8]. The differences observed in the mechanical properties are reflected in the microstructures shown in Fig. 30a and b, where the deposit from the filtered bath exhibits about a five- to tenfold decrease in grain size. The inverse-pole figure inserts in Fig. 30 reveal even more unexpectedly a change in the characteristic texture of ED Ni-sulfamate from <001> in Fig. 30b to <011> in Fig. 30a. These effects are dependent on the boric-acid content and occur most profoundly at lower bath temperatures ($<40°$C) and current densities (<5 mA/cm^2); we thus postulate that precipitation of the boric acid during passage through the filter leads to the formation of polyborates, which in turn act as weak inhibitors at low deposition rates [86].

As mentioned previously, pulsed deposition schemes are also used to influence deposit properties and structure through current-density manipulation. Square-wave pulses, in which the deposition current is on for a given time followed by an interval in which no current flows, are most often used. Additional control over the properties of deposits relative to constant, DC deposition results, because it is possible to independently vary the current density, the pulse frequency, and the duty cycle (the ratio of the on-time to the pulse period, i.e., the sum of the on-time and the off-time). The properties of such pulsed-plated (PP) deposits can be manipu-

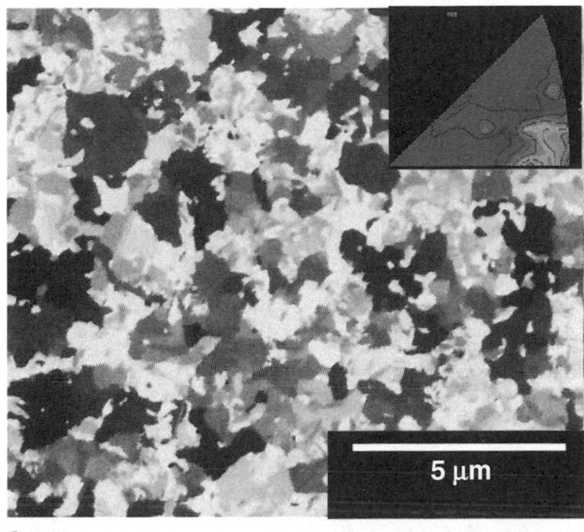

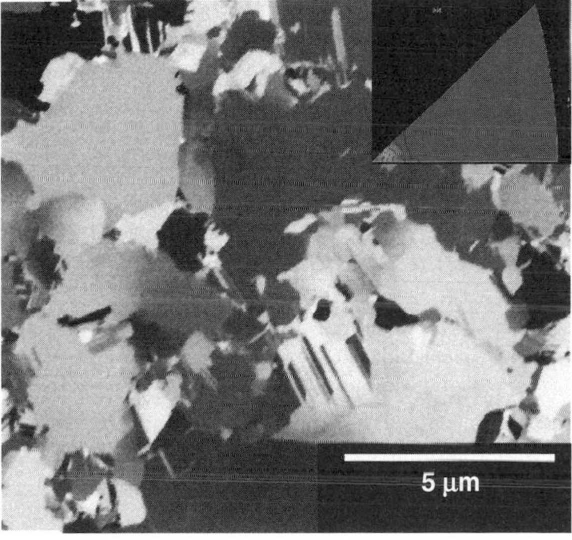

Fig. 30 (**a**) Planview focused ion beam image of a nickel deposit from a filtered sulfamate electrolyte at 3 mA/cm^2 and 28°C reveals a fine-grain structure with a predominantly <011> texture (inset inverse pole figure). (**b**) Same as in (**a**), but from an unfiltered electrolyte. The nickel deposit instead has a coarse grain structure with a <001> texture. (Microscopy and texture courtesy of J. Michael and A. Talin, Sandia National Laboratories)

lated in theory, because the accompanying higher peak-current densities result in higher nucleation rates and thus in grain refinement, as we and others have discussed [108–111]. Brighter deposits, typically an indication of a fine-grained deposit, frequently result [112]. Figure 28b shows additional work of Kim and Weil illustrating the effect of current density-induced grain refinement for pulse-deposited Ni-sulfate at peak current densities up to 200 mA/cm^2. At lower mean current densities, it is still possible to find reports where the reverse trend is observed. Fritz and co-workers [113] reported continuously decreasing indentation hardness for pulse-deposited ED Ni plated from a sulfamate electrolyte at 40°C and a pH of 3.2 between mean current densities of 2 mA/cm^2 and 20 mA/cm^2; as expected, the lower observed hardness values are accompanied by an increase in the grain size.

pH and Temperature Control – Other processing parameters can affect the structure and properties of ED nickel. Safranek has compiled results from many sources and shown that for ED nickel chloride-, sulfate-, sulfamate-, and Watts-type deposits, hardness generally rises with increasing pH [8]. The effects tend to be small between pH values of 1–4, but then rise rapidly as pH is increased to values of about 5 or higher [106, 114]. Since hydroxide species start to precipitate at about this pH for most Ni plating baths, the increases in hardness observed at pH values of 4.5 and higher are likely due to adsorbed metal hydroxide species acting as inhibitors [8, 115]. Conversely, increasing deposition temperature generally results in decreased strength of electrodeposits. Phillips and Clifton reported a decrease of almost 50% in the UTS of Watts nickel deposits plated at 60°C relative to deposits plated at 25°C [116]. Such changes may be due to the adsorption–desorption kinetics of a potentially inhibiting electrolyte species; for example, the adsorption and subsequent effects of the electrolyte bath additive saccharin have been shown to be very temperature dependent [117]. However, as with the earlier discussion regarding current-density effects on material properties, identifying consistent trends from the literature with respect to temperature and pH effects is often problematic, since the final characteristics of any electrodeposit are the result of the interactions of many processing variables.

5.4 Particulate Additive Effects

A path toward enhanced mechanical strength that is independent of deposition parameters and electrolyte chemistry is found through the physical incorporation of an inert, particulate dispersoid, typically a hard oxide such as alumina (Al_2O_3), into the host nickel electrodeposit, creating an oxide-dispersion-strengthened (ODS) particulate composite. The source of the strength increase is the Orowan hardening phenomena, wherein the distributed particulates act as barriers to dislocation motion within the metal matrix [118–119]. Additional benefits can be realized if the particulate microstructure acts to pin the grain boundaries, thereby rendering the deposit less sensitive to anneal softening. The ability to successfully co-deposit a particulate in both thin and thick electrodeposited films has been demonstrated previously by a number of groups [120–125]. More recently, we have demonstrated the ability to uniformly incorporate a 10-nm (average diameter) Al_2O_3 particulate into through-mold electrodeposited structures [126]. The use of such a small-diameter particulate is necessary in order to insure that the transport of this hardening constituent is not impeded within the narrow, high-aspect-ratio channels that are typical of LIGA-fabricated structures. Figure 31 shows the influence of a uniform distribution of oxide particles co-deposited with nickel from a nickel sulfamate bath. In the

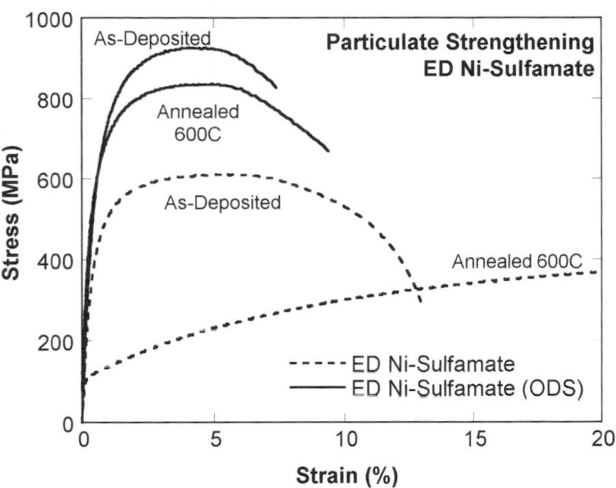

Fig. 31 Effect of alumina particulate incorporation on stress–strain behavior of ED Ni plated from a sulfamate chemistry [126]. The incorporated particulate strengthens the plated Ni; ED Ni plated under the same conditions but without the particulate is shown for comparison. The particulate also allows the plated Ni to retain much of its strength after a 600°C anneal as compared to the particulate-free Ni

as-deposited condition, the ODS electrodeposit exhibits a 50% increase in both the yield and UTSs. Because the oxide particulates also act to pin grain boundaries as suggested above, resistance to anneal softening is also realized. The two traces in Fig. 31 labeled as "annealed 600C" represent the tensile behavior after a 1-hr vacuum-annealing treatment, and it is evident that the ODS deposit suffered only negligible loss in strength, while the baseline ED Ni plated from sulfamate bath suffered a 75% loss in yield strength.

While dispersion strengthening is clearly an effective means to increase the mechanical performance of Ni-based electrodeposits, significant hurdles remain with respect to implementing the appropriate plating processes in LIGA-based fabrication. Principal among these would be the potentially problematic effects of routine particle filtration of the plating bath, a routine procedure. As commented earlier, this filtering is employed to remove unintended species resulting from accidental contamination, which would lead to internal defects in the plated material (e.g., dissolved anode particulate). Such filtering media would have to be carefully sized to minimize any tendency for it to strip the intentionally added particulate out of the bath, compromising the original intent of the process.

5.5 *Modulus of Electrodeposits*

Because many LIGA-based microsystem components have a mechanical functionality, the elastic modulus of electrodeposits is often a first-order design parameter. For example, the elastic modulus is required to predict the elastic displacement of a spring-type structure. Handbook or literature values for modulus cannot be assumed, since these are typically based on randomly oriented polycrystalline materials. In contrast, most ED nickel materials exhibit out-of-plane (parallel to the growth direction) crystallographic texture. For very thin films (<10 microns), there can be significant in-plane texture as well (in-plane being normal to the growth direction of the electrodeposit), depending on the character of the seed surface substrate [127]. For thicker deposits (>50 microns) any in-plane texture usually disappears, leaving a material that is randomly oriented within the plane. However, the out-of-plane texture can remain strong, and the particular crystallographic orientation and texture strength depends on the ED material system, the deposition parameters and the chemistry (chemical purity) of the electrolyte. As a result, it is often necessary to explicitly characterize the in-plane modulus of through-mold electrodeposited structures not only for each material system, but also when there are even minor changes in deposition conditions.

Some work has been reported in the literature where modulus has been determined through the characterization of the resonance behavior electroformed structures [128–129]. In the work of Shi et al. [128], a modulus of 110 GPa is reported for electroformed Ni resonators, which is compared to a literature value of 140 GPa for electroplated Ni. This data appears to be insupportably low (see the discussion below) and undoubtedly suffers from the inability to determine critical specimen dimensions with sufficient accuracy. Beyond this, the authors do not indicate the ED Ni system employed or the particular deposition parameters. In another study by Majjad et al. [129], the authors do not indicate the ED Ni system studied, but the values they report (≈195 GPa) are in agreement with companion studies employing apparently the same material and are more in line with expected values for a range of likely Ni deposit systems.

Most often, the modulus of a material is determined in a straightforward fashion from the initial linear-loading slope of a mechanical test specimen. From such testing, the modulus of ED Ni-sulfamate has been variously reported from low values of ≈90 GPa to as high as 208 GPa [90, 130]. Accurate characterization of the elastic modulus of LIGA-fabricated test specimens in this fashion is made difficult by two principal factors, the specimen size and alignment considerations. With respect to the former, material test specimens fabricated using LIGA techniques characteristically have small dimensions. For example, gauge sections of lithographically patterned tensile specimens range from a few hundred microns to a few millimeters in length. As a result, the elastic deformation range over which the modulus is measured is quite small. This point is illustrated in Fig. 32, which shows a typical tensile test curve for ED-Ni deposited from a sulfamate electrolyte at 50°C [126]. While the overall curve is unremarkable, for this relatively soft electrodeposit (yield strength on the order of 400 MPa), the insert in Fig. 32 shows that the total elastic elongation of a test specimen having a 3-mm gauge length is only ≈4 μm. If we presume that 25–50 unique data points are necessary to adequately define the initial loading slope, then the required displacement resolution is in the order of 200 nm. Accurate and reproducible measurements with noncontacting extensometry (necessary because of the inability to mechanically attach displacement transducers to these very small test specimens) at this resolution are quite difficult. Some groups have had good success with custom-built instruments using interferometric techniques (Sharpe and co-workers [89], for example), but there are few commercial devices that have sufficient accuracy for these types of measurements.

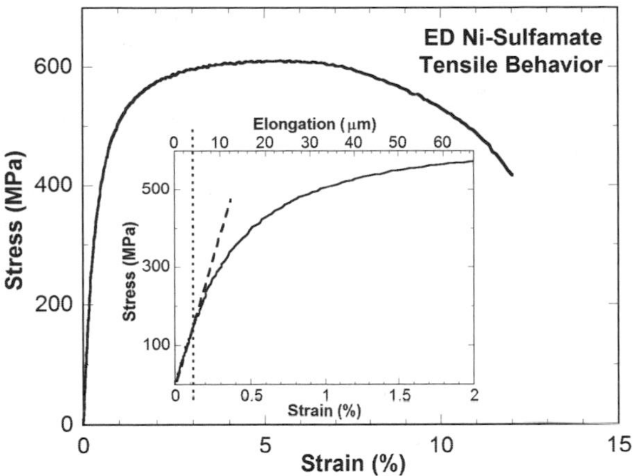

Fig. 32 Stress–strain data for Ni electrodeposited from a sulfamate chemistry [126]. The measured displacements in the elastic region, necessary for the determination of the elastic modulus by this method, are very small, thus limiting the accuracy of moduli determined in this fashion

With respect to the latter consideration, the very small loads required to deform these small specimens preclude the use of alignment fixtures. The resulting off-axis bending can mask the true linear loading slope and can result in significant uncertainty in the calculation of the modulus. Thus, even when displacement-measuring instruments with sufficient resolution are available, one must approach the use of the initial tensile loading slope with great caution.

We have instituted a three-point bending test procedure that obviates these problems. These tests are performed on LIGA-fabricated, rectangular beams of uniform cross-section, and the modulus is extracted from the centerline deflection and the deflection load according to:

$$\mathrm{E} = \frac{P \cdot L^3}{4\Delta x \cdot w \cdot h^3}$$

where E is the modulus, P is the centerline flexure load, L is the support span length, x is the centerline displacement, w is the specimen width, and h is the thickness of the beam parallel to the displacement line. Because the applied loads are small, the system and fixture compliance (with respect to displacement measurement) are negligible. Therefore, centerline displacement can be monitored by the optical encoder of the cross-head drive mechanism that is present in any modern mechanical test instrument. Since the specimen is point loaded, alignment requirements are far less critical than is the case for conventional uniaxial mechanical testing. Results, presented below, generally show less than about 1% standard deviation in the computed modulus when 6–10 replicate measurements are performed on individual specimens; far less variability than that typically found in the literature.

For the majority of microsystem applications where the mechanical response of component structures is required, design considerations are most often concerned with the in-plane modulus. Hemker and Last showed that the in-plane modulus of an electrodeposit that exhibits out-of-plane texture can be modeled by averaging over all possible directions in the plane of the electrodeposit (assuming that the plane of the electrodeposit is the plane of the texture) [131]. The calculation is simplified by the fact that the in-plane texture is typically random as described above.

For ED nickel deposited from sulfamate and sulfate solutions under conventional deposition conditions, the as-deposited structures most typically have a <001> out-of-plane texture and exhibit either no or only weak in-plane texture. Following Hemker and Last [131], the in-plane directional dependence of E on the (001) plane is:

$$E(\theta) = \{S_{11}(\cos^4\theta + \sin^4\theta) + (2S_{12} + S_{44})\cos^2\theta\sin^2\theta\}^{-1}$$

where S_{11}=0.00734 GPa^{-1}, S_{12}=-0.00274 GPa^{-1} and S_{44}=0.00802 GPa^{-1} are the relevant compliances [132] and θ is the angle from the [001] direction on the (001) plane. E for a <001> textured material is shown as a function of θ in Fig. 33 along with that for a <011> textured deposit since these are the most relevant texture considerations for ED Ni-sulfamate. For the <001> preferred texture, the modulus reaches a maximum of 323 GPa along the <011> directions and exhibits minimum values of 137 GPa along the <001> directions. Thus, it can be concluded that in no instance can the modulus of an electrodeposit with a preferred <001> out-of-plane texture *ever* exhibit a modulus either greater than 323 GPa or less than 137 GPa regardless of the extent of in-plane texture. Hemker and Last further indicate that, since bulk film Ni-sulfamate is randomly oriented in-plane, it is appropriate to average the curves in Fig. 33. Such averaging yields a value for E for <001> textured deposits of 178 GPa assuming a Voigt solid model [133] or a somewhat lower value of 172 GPa assuming a Reuss solid model [134]. For a deposit with a preferred <011> texture, Fig. 33 shows that E(θ) exhibits a different periodicity with respect to the in-plane direction, with average moduli of 230 GPa and 210 GPa, respectively, depending on the solid model used.

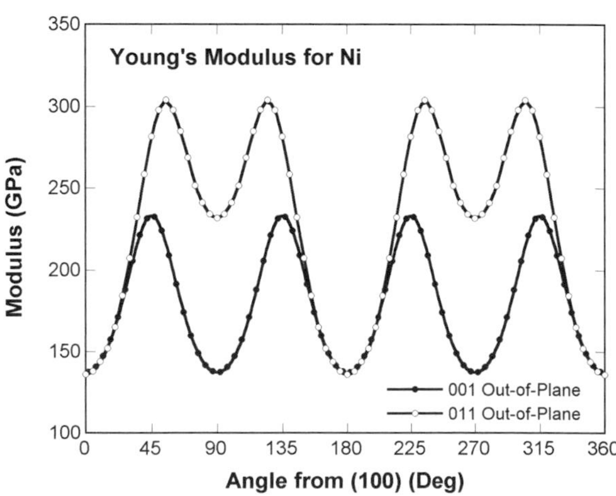

Fig. 33 Calculated elastic moduli for Ni, taken after Hemker and Last [131]

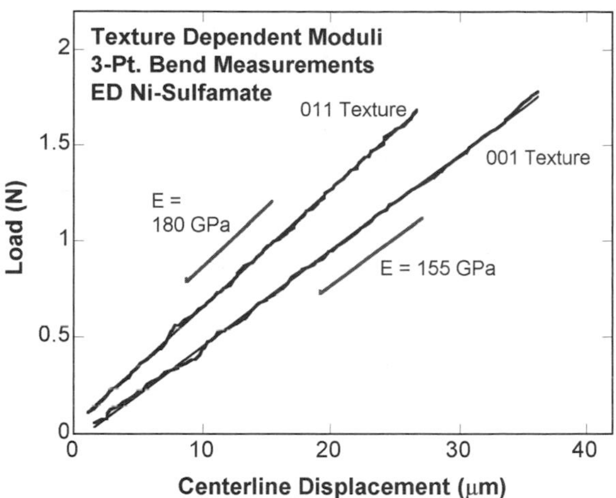

Fig. 34 Three-point bend tests for <001> and <011> textured ED Ni. The difference in film texture is reflected in the moduli determined from this measurement

We have shown that the crystallographic texture developed in an electrodeposit is influenced by deposition-processing parameters. Based on Fig. 33, the modulus of differently textured ED Ni structures should be measurable. Figure 34 shows representative results for modulus measurements performed as described earlier. The slope of each curve represents the modulus for two different ED Ni-sulfamate films deposited using deposition conditions such that in one instance the preferred texture is <001>, while in the other instance the preferred texture is <011>. In the former case, the <001> textured material exhibits a modulus of 153.2 ± 1.4 GPa, while in the later case, the modulus is 186.9 ± 1.8 GPa. It is evident that, while the trends agree with the modulus calculations presented in Fig. 33, the absolute values in each case are significantly lower than expected. On possible explanation for these discrepancies may be associated with the assumption that these deposits have no in-plane texture. Any significant in-plane texture component may skew the measured modulus of these materials to values that are lower or higher than the predicted Reuss or Voigt averages. More likely, the above consideration of modulus does not account for the high fraction of material associated with grain boundaries in these very fine-grained electrodeposits. A high "volume fraction" of disordered grain boundary may lead to a material that is overall, more compliant. Studies like those of Hemker, and the experimental work presented here, suggest that additional work is appropriate with respect to understanding the relationship between the structure of electrodeposits and physical properties such as modulus.

5.6 Summary: Electrodeposited Materials for LIGA-Based Microsystems

We have attempted to summarize and discuss the important characteristics of electrodeposited materials used for LIGA-based applications and the ways in which these characteristics may be manipulated to obtain material properties desirable for such applications. For the mechanical applications discussed here, a material with a uniformly nano-sized grain structure is ideal; while this can be done in a fairly straightforward fashion for blanket electrodeposited films, it is quite challenging to do so when electrodepositing into deep, high-aspect-ratio features formed by X-ray lithography. A variety of processing parameters, ranging from the bulk chemistry of the electrolyte, the effect of adsorbed species on the nucleation and growth of the depositing film, to mass transport limitations in deep cavities, all can have a strong effect on the microstructure and material properties of the through-mold deposited structures. We have only discussed Ni and Ni alloys here, but it is certain that similar challenges exist in integrating electrodeposition processes for other materials with LIGA processing. As interest develops for new materials necessary for new applications of LIGA technology, such materials and process development for LIGA will ensue [11, 15].

6 Other Aspects of LIGA Technology

In this final section, we address some miscellaneous aspects of LIGA technology that do not explicitly involve electrodeposition but are worth mentioning for completeness. They entail the areas of replication, alternative resists for LIGA, and X-ray lithography at the nanoscale.

As mentioned in the introductory section, when the LIGA process was developed, a major interest was the use of the lithographically patterned mold as a means to make a thick, electroplated monolithic metal master tool for subsequent replication. This plated Ni tool, bearing the lithographically defined pattern, would in turn be used to inexpensively mass produce plastic parts *via* embossing or injection molding [135–137]. As discussed in the LIGA applications section, replication is an important technique in LIGA technology that is used to this day in the low-cost fabrication of commercial products [39]. Other variants of the replication technique, such as the electroforming of lost plastic molds (the plating of molded-plastic substrates) and the casting of metal powders into ceramic molds (obtained from plastic parts made from a metal LIGA mold) to form solid microparts have been discussed by Ruprecht et al. [138]. The lost plastic-mold approach is very attractive from a cost standpoint, but structures electroformed from these substrates typically have small internal voids (the plastic mold side walls become electrically active after the necessary metallization layer is vacuum deposited; electrically conductive plastic molds present the same problem) [138–139]. More recent work has shown greater success with respect to obtaining defect-free material with mechanical properties comparable to direct LIGA-fabricated test structures [140]. Ruprecht et al. [138] also describe the challenges in obtaining fully dense metal structures with minimal shrinkage by casting metal powders into ceramic molds derived from LIGA masters.

While replicating large numbers of parts with a single master tool lowers the overall cost of the LIGA process, another way to achieve this end is to use a resist that may be prepared using standard UV exposure equipment, obviating the need for

X-ray synchrotron access. Since collimated X-rays are not used, the lithographic dimensional fidelity is not as high as that obtained with the true LIGA process. Wafers bearing patterned resist may be electroplated as in the normal LIGA process to produce through-mask plated structures [141]. Another complication is the removal of some of the thick UV resists typically employed, such as SU8; whereas PMMA may be dissolved in standard organic solvents like acetone, for example, resists such as SU8 usually do not dissolve in such solvents due to its high degree of cross-linking [142]. Dentinger et al. reviews some of the prior methods for SU8 removal and presents some new routes for the removal of this resist [142]. Although thick resist such as SU8 have been widely used as a process tool for fabricating LIGA-like structures, there are limitations with regard to dimensional precision at large resist thicknesses that must be mitigated by careful resist process control [143–146].

Most LIGA research has focused on structures whose lateral critical dimensions are on the micron scale. In principle, it is possible to fabricate structures that have lateral critical dimensions that are less than a micron, but this has not been much explored. A possible reason why the LIGA process has not been employed frequently at the submicron scale involves the typical PMMA resist thickness; since resist is usually hundreds of microns in thickness, submicron features are difficult to fabricate and process. Nonetheless, some work is being done at a submicron feature scale using X-ray lithography, typically using fairly thin PMMA resist [147–148]. The gold-absorbing masks must be carefully fabricated, and synchrotron radiation with the appropriate characteristics must be used to insure high resolution and fidelity at submicron scales [149]. Recently, X-ray and nanoimprint lithography have been used together to fabricate three-dimensional structures patterned over different length scales [148].

Acknowledgments The authors thank the Sandia/California LIGA prototyping team and metallography group for their assistance in the fabrication of the masks, molds, sample preparation, and analyses essential to the work described here. Colleagues at Sandia/New Mexico are acknowledged as well for their contribution to some of the mechanical testing and microscopy. Georg Aigeldinger and Sam McFadden are acknowledged for their comprehensive review of this work.

Sandia is a multiprogram laboratory operated by Sandia Corporation, a Lockheed Martin Company, for the United States Department of Energy under contract DE-AC04-94AL85000.

References

1. E. W. Becker, W. Ehrfeld, D. Muenchmeyer, H. Betz, A. Heuberger, S. Pongratz, W. Glashauser, H. J. Michel, and R. v. Siemens, "Production of Separation-Nozzle Systems for Uranium Enrichment by a Combination of X-ray Lithography and Galvanoplastics", *Naturwissenschaften*, **69**, 520 (1982).
2. M. Madou, *Fundamentals of Microfabrication*, CRC Press, New York, p. 275 (1997).
3. E. W. Becker, W. Ehrfeld, P. Hagmann, A. Maner, and D. Muenchmeyer, "Fabrication of Microstructures With High Aspect Ratios and Great Structural Heights by Synchrotron Radiation, Lithography, Galvanoforming, and Plastic Moulding (LIGA Process), *Microelectronic Engineering*, **4**, 35 (1986).

4. J. Mohr, W. Ehrfeld, and D. Munchmeier, "Requirements on Resists Layers in Deep-etch Synchrotron Radiation Lithography", *Journal of Vacuum Science and Technology B*, **6**(6), 2264 (1988).
5. W. Ehrfeld, V. Hessel, H. Loewe, Ch. Schulz, and L. Weber, "Materials of LIGA Technology", *Microsystems Technologies*, **5**, 105 (1999).
6. J. Hormes, J. Goettert, K. Lian, Y. Desta, and L. Jian, "Materials for LIGA and LIGA-based Microsystems", *Nuclear Instruments and Methods in Physics Research B*, **199**, 332 (2003).
7. M. W. Boerner, M. Kohl, F. J. Pantenburg, W. Bacher, H. Hein, and W. K. Schomburg, "Submicron LIGA Process for Movable Microstructures", *Microelectronic Engineering*, **30**, 505 (1996).
8. W. H. Safranek, *The Properties of Electrodeposited Metals and Alloys* (2nd Ed.), American Electroplaters and Surface Finishers Society, U.S.A. (1986).
9. A. Rogner, J. Eicher, D. Muenchmeyer, R.-P. Peters, and J. Mohr, "The LIGA Technique-What Are the New Opportunities", *Journal of Micromechanics and Microengineering*, **2**, 133 (1992).
10. D. W. L. Tolfree, "Microfabrication Using Synchrotron Radiation", *Report on Progress Physics*, **61**, 313 (1998).
11. R. K. Kupka, F. Bouamrane, C. Cremers, and S. Megtert, "Microfabrication: LIGA-X and Applications", *Applied Surface Science*, **164**, 97 (2000).
12. R. A. Lawes, G. Arthur, and A. Schneider, "LIGA- A Fabrication Technology for Industry?", *Proceedings of SPIE*, **4593**, 145 (2001).
13. Y. Cheng, B.-Y. Shew, M. K. Chyu, and P. H. Chen, "Ultra-deep LIGA Process and its Applications", *Nuclear Instruments and Methods in Physics Research A*, **467–468**, 1192 (2001).
14. J. Hruby, "LIGA Technologies and Applications", *MRS Bulletin*, **26**(4), 337 (2001).
15. C. K. Malek and V. Saile, "Applications of LIGA Technology to Precision Manufacturing of High-Aspect-Ratio Micro-components and –systems: a Review", *Microelectronics Journal*, **35**, 131 (2004).
16. John O. Dukovic, "Current Distribution and Shape Change in Electrodeposition of Thin Films for Microelectronic Fabrication", in *Advances in Electrochemical Science and Engineering*, H. Gerischer and C. W. Tobias (Eds.), Vol. 3, VCH, Weinheim, p. 117 (1994).
17. S. K. Griffiths, A. Ting, and J. M. Hruby, "The Influence of Mask Substrate Thickness on Exposure and Development Times for the LIGA Process", *Microsystem Technologies*, **6**, 99 (2000).
18. S. K. Griffiths, J. Hruby, and A. Ting, "Optimum Doses and Mask Thickness for Synchrotron Exposure of PMMA Resists", *Proceedings of the SPIE,* **3680**, 498 (1999).
19. H. M. Manohara, C. Khan Malek, A. S. Dewa, and K. Deng, "Low Z Substrates for Cost Effective High-Energy, Stacked Exposures", *Microsystem Technologies*, **4**, 17 (1997).
20. A. El-Kholi, K. Bade, J. Mohr, F. J. Pantenburg, X.-M. Tang, "Alternative Resist Adhesion and Electroplating Layers for LIGA Process", *Microsystem Technologies*, **6**, 161 (2000).
21. L. T. Romankiw, "A Path: From Electroplating Through Lithographic Masks in Electronics to LIGA in MEMS", *Electrochimica Acta*, **42**, 2985 (1997).
22. W. Bacher, W. Menz, J. Mohr, C. Muller, and W. K. Schomburg, "The LIGA Process and its Potential for Microsystems", *Naturwissenschaften*, **81**(12), 536 (1994).
23. W. Qu, C. Wenzel, and K. Drescher, "A Vertically Sensitive Accelerometer and its Realization by Depth UV Lithography Supported Electroplating", *Microelectronics Journal*, **31**, 569 (2000).
24. T. R. Christenson and H. Guckel, "Deep X-ray Lithography for Micromechanics", *Proceedings of the SPIE*, **2639**, 134 (1995).
25. E. J. O'Sullivan, E. I. Cooper, L. T. Romankiw, K. T. Kwietniak, P. L. Trouilloud, J. Horkans, C. V. Jahnes, I. V. Babich, S. Krongelb, S. G. Hegde, J. A. Tornello, N. C. LaBianca, J. M. Cotte, and T. J. Chainer, "Integrated, Variable-reluctance Magnetic Minimotor", *IBM Journal of Research and Develop*ment, **42**,681 (1998).

26. R. A. Brennen, M. H. Hecht, D. V. Wiberg, S. J. Manion, W. D. Bonivert, J. M. Hruby, M. L. Scholz, T. D. Stowe, T. W. Kenny, K. H. Jackson, and C. K. Malek, "Fabricating Sub-collimating Grids for an X-ray Solar Imaging Spectrometer Using LIGA Techniques", *Proceedings of the SPIE*, **2640**, 214 (1995).
27. M. Ghigo, E. Diolaiti, F. Perennes, and R. Ragazzoni, "Use of the LIGA Process for the Production of Pyramid Wavefront Sensors for Adaptive Optics in Astronomy", *Proceedings of the SPIE*, **5169**, 55 (2003).
28. T. L. Willke and S. S. Gearhart, "LIGA Micromachined Planar Transmission Lines and Filters", *IEEE Transactions on Microwave Theory and Techniques*, **45**, 1681 (1997).
29. F. Aristone, P. Datta, Y. Desta, A. M. Espindola, and J. Goettert, "Molded Multilevel Modular Micro-fluidic Devices", *Proceedings of the SPIE*, **4982**, 65 (2003).
30. L. James Lee, M. J. Madou, K. W. Koelling, S. Daunert, S. Lai, C. G. Koh, Y.-J. Juang, Y. Lu, and L. Yu, "Design and Fabrication of CD-Like Microfluidic Platforms for Diagnostics: Polymer-Based Microfabrication", *Biomedical Microdevices*, **3:4**, 339 (2001).
31. A. Schneider, S. Rea, E. Huq, and W. Bonfield, "Surface Microstructuring of Biocompatible Bone Analogue Material HAPEX^{TM} Using LIGA Technique and Embossing", *Proceedings of the SPIE*, **5116**, 57 (2003).
32. A. Ruzzu, J. Fahrenberg, M. Heckele, Th. Schaller, "Multi-functional Valve Components Fabricated by Combination of LIGA Processes and High Precision Mechanical Engineering", *Microsystems Technologies*, **4**, 128 (1998).
33. L. Huang, W. Wang, M. C. Murphy, K. Lian, and Z.-G. Ling, "LIGA Fabrication and Test of a DC Type Magnetohydrodynamic (MHD) Micropump", *Microsystems Technologies*, **6**, 235 (2000).
34. S. Baik, J. P. Blanchard, and M. L. Corradini, "Development of Micro-Diesel Injector Nozzles via Microelectromechanical Systems Technology and Effects on Spray Characteristics", *Journal of Engineering for Gas Turbines and Power*, **125**, 427 (2003).
35. A. Morales, J. Brazzle, R. Crocker, L. Domeier, E. Goods, J. Hachman, Jr., C. Harnett, M Hunter, S. Mani, B. Mosier, and B. Simmons, "Fabrication and Characterization of Polymer Microfluidic Devices for Bio-agent Detection", Proceedings of the SPIE, 5716, 89 (2005).
36. C. Harris, K. Kelly, T. Wang, A. McCandless, and S. Motakef, *Journal of Microelectromechanical Systems*, **11**, 726 (2002).
37. A. Cox, E. Garcia, "Three-Dimensional LIGA Structures for Use in Tagging", *Proceedings of the SPIE*, **3673**, 122 (1999).
38. Axsun Technologies Homepage, www.axsun.com, accessed 12/20/04.
39. SEAG microParts GmbH, www.microparts.de, accessed 12/20/04.
40. Micromotion GmbH, www.mikrogetriebe.de, accessed 12/20/04.
41. Klaus Stefan Drese, "Design Rules for Electroforming in the LIGA Process", *J. Electrochem. Soc.*, **151**, D39 (2004).
42. M. Datta and D. Landolt, "Fundamental Aspects and Applications of Electrochemical Microfabrication", *Electrochimica Acta*, **45**, 2535 (2000).
43. B. DeBecker and A. C. West, "Workpiece, Pattern, and Feature Scale Current Distributions", *J. Electrochem. Soc.*, **143**, 486 (1996).
44. J. L. Marti and G. P. Lanza, "Hardness of Sulfamate Nickel Deposits", *Plating*, **56**, 377 (1969).
45. J. M. Lee, J. T. Hachman, J. J. Kelly, and A. C. West, "Improvement of Current Distribution Uniformity on Substrates for MEMS," *Journal of Microlithography, Microfabrication, and Microsystems*, **3**(1), 146 (2004).
46. S. Mehdizadeh, J. Dukovic, P. C. Andricacos, L. T. Romankiw, and H. Y. Cheh, "Optimization of Electrodeposit Uniformity by the Use of Auxiliary Electrodes", *Journal of the Electrochemical Society*, **137**, 110 (1990).
47. S. Mehdizadeh, J. O. Dukovic, P. C. Andricacos, L. T. Romankiw, and H. Y. Cheh, "The Influence of Lithographic Patterning on Current Distribution: A Model for Microfabrication by Electrodeposition", *Journal of the Electrochemical Society*, **139**, 78 (1992).

48. A. C. West, M. Matlosz, and D. Landolt, "Normalized and Average Current Distributions on Unevenly Spaced Patterns", *Journal of the Electrochemical Society*, **138**, 728 (1991).
49. A. C. West, "Ohmic Interactions Within Electrode Ensembles", *Journal of the Electrochemical Society*, **140**, 134 (1993).
50. S. K. Griffiths, J. A. W. Crowell, B. L. Kistler, and A. S. Dryden, "Dimensional Errors in LIGA-Produced Metal Structures due to Thermal Expansion and Swelling of PMMA". *Journal of Micromechanics and Microengineering*, **14**, 1548 (2004).
51. S. Mehdizadeh, J. Dukovic, P. C. Andricacos, L. T. Romankiw, and H. T. Cheh, "The Influence of Lithographic Patterning on Current Distribution in Electrodeposition: Experimental Study and Mass-Transfer Effects", *Journal of the Electrochemical Society*, **140**, 3497 (1993).
52. K. Kondo, K. Fukui, K. Uno, and K. Shinohara, "Shape Evolution of Electrodeposited Copper Bumps", *Journal of the Electrochemical Society*, **143**, 1880 (1996).
53. K. Kondo, K. Fukui, M. Yokoyama, and K. Shinohara, "Shape Evolution of Electrodeposited Copper Bumps with High Peclet Numbers", *Journal of the Electrochemical Society*, **144**, 466 (1997).
54. K. Kondo and K. Fukui, "Current Evolution of Electrodeposited Copper Bumps with Photoresist Angle", *Journal of the Electrochemical Society*, **145**, 840 (1998).
55. H. Watanabe, S. Hayashi, and H. Honma, "Microbump Formation by Noncyanide Gold Electroplating", *Journal of the Electrochemical Society*, **146**, 574 (1999).
56. K. Hayashi, K. Fukui, Z. Tanaka, and K. Kondo, "Shape Evolution of Electrodeposited Bumps into Deep Cavities", *Journal of the Electrochemical Society*, **148**, C145 (2001).
57. B. Kim and T. Ritzdorf, "Electrodeposition of Near Eutectic SnAg Solders for Wafer Level Packaging", *Journal of the Electrochemical Society*, **150**, C577 (2003).
58. P. C. Andricacos, C. Uzoh, J. Dukovic, J. Horkans, L. Deligianni, *IBM Journal of Research and Development*, **42**, 567 (1998).
59. P. Taephaisitphongse, Y. Cao, and A. C. West, "Electrochemical and Fill Studies of a Multicomponent Additive Package for Copper Deposition", *Journal of the Electrochemical Society*, **148**, C492 (2001).
60. D. Josell, B. Baker, C. Witt, D. Wheeler, and T. P. Moffat, "Via Filling by Electrodeposition. Superconformal Silver and Copper and Conformal Nickel", *Journal of the Electrochemical Society*, **149**, C637 (2002).
61. K. Leyendecker, W. Bacher, W. Stark, and A. Thommes, "New Microelectrodes for the Investigation of the Electroforming of LIGA Microstructures", *Electrochimica Acta*, **39**, 1139 (1994).
62. J. Ji, W. C. Cooper, D. B. Dreisinger, and E. Peters, "Surface pH Measurements During Nickel Electrodeposition", *Journal of Applied Electrochemistry*, **25**, 642 (1995).
63. D. R. Gabe, "The role of Hydrogen in Metal Electrodeposition Processes", *Journal of Applied Electrochemistry*, **27**, 908 (1997).
64. S. K. Griffiths, R. H. Nilson, R. W. Bradshaw, A. Ting, W. D. Bonivert, J. T. Hachman, and J. M. Hruby, "Transport Limitations in Electrodeposition for LIGA Microdevice Fabrication", *Proceedings of the SPIE*, **3511**, 364 (1998).
65. R. H. Nilson and S. K. Griffiths, "Natural Convection in Trenches of High Aspect Ratio", *J. Electrochem. Soc.*, **150**, C401 (2003).
66. S. D. Leith and D. T. Schwartz, "Through-mold Electrodeposition Using the Uniform Injection Cell (UIC): Workpiece and Pattern Scale Uniformity", *Electrochimica Acta*, **44**, 4017 (1999).
67. S. D. Leith, S. Ramli, and D. T. Schwartz, "Characterization of Ni_xFe_{1-x} ($0.10 < x < 0.95$) Electrodeposition from a Family of Sulfamate-Chloride Electrolytes", *Journal of the Electrochemical Society*, **146**, 1431 (1999).
68. W. Wang, S. D. Leith, and D. T. Schwartz, "Convective-Diffusive Mass Transfer Inside Complex Micro-molds During Electrodeposition", *Journal of Microelectromechanical Systems*, **11**, 118 (2002).

69. A. Thommes, W. Stark, K. Leyendecker, W. Bacher, H. Liebscher, and Ch. Ilmenau, "LIGA Microstructures From a NiFe Alloy: Preparation by Electroforming and Their Magnetic Properties", in *Proceedings of the 3rd International Symposium on Magnetic Materials, Processes, and Devices PV 94-6*, L. T. Romankiw and D. A. Herman, Jr. (Eds.), The Electrochemical Society, U.S.A., p. 89 (1994).
70. A. Brenner, *Electrodeposition of Alloys*, Academic Press, New York (1963).
71. P. C. Andricacos and L. T. Romankiw, in *Advances in Electrochemical Science and Engineering*, H. Gerischer and C. W. Tobias (Eds.), vol. 3, VCH Verlagsgesellschaft, Weinheim, Germany, p. 227 (1993).
72. M. Kuepper and J. W. Schultze, "Spatially Resolved Concentration Measurements During Cathodic Alloy Deposition in Microstructures", *Electrochimica Acta*, **42**, 3023 (1997).
73. H. Guckel, T. Christenson, and K. Skrobis, "Metal Micromechanisms via Deep X-ray Lithography, Electroplating, and Assembly", *Journal of Micromechanics and Microengineering*, **2**, 225 (1992).
74. H. Guckel, "Progress in Magnetic Microactuators", *Microsystem Technologies*, **5**, 59 (1998).
75. T M. Liakopoulos and C. H. Ahn, "A Micro-fluxgate Magnetic Sensor Using Micromachined Planar Solenoid Coils", *Sensors and Actuators*, **77**, 66 (1999).
76. D. J. Sadler, T. M. Liakopoulos, and C. H. Ahn, "A Universal Electromagnetic Microactuator Using Magnetic Interconnection Concepts", *Journal of Microelectromechanical Systems*, **9**, 460 (2000).
77. F. Yi, L. Peng, J. Zhang, and Y. Han, "A New Process to Fabricate the Electromagnetic Stepping Micromotor Using LIGA Process and Surface Sacrificial Layer Technology", *Microsystem Technologies*, **7**, 103 (2001).
78. J.-W. Park, J. Y. Park, Y.-H. Joung, and M. G. Allen, "Fabrication of High Current and Low Profile Micromachined Inductor With Laminated Ni/Fe Core", *IEEE Transactions on Components and Packaging Technologies*, **25**, 106 (2002).
79. T. W., Andrew, B. McCandless, Sean Ford, Kevin W. Kelly, Richard Lienau, Dale Hensley, Yohannes Desta, and Zhong G. Ling, "High-Aspect-Ratio Microstructures for Magnetoelectronic Applications", *Proceedings of the SPIE*, **4979**, 464 (2003).
80. Tsung-Shune Chin, "Permanent Magnet Films for Applications in Microelectromechanical Systems", *Journal of Magnetism and Magnetic Materials*, **209**, 75 (2000).
81. L. Huang, W. Wang, and M. C. Murphy, "Microfabrication of High Aspect Ratio Bi-Te Alloy Microposts and Applications in Micro-sized Cooling Probes", *Microsystem Technologies*, **6**, 1, (1999).
82. S. K. Griffiths and A. Ting, "The Influence of X-ray Fluorescence on LIGA Sidewall Tolerances", *Microsystem Technologies*, **8**, 120 (2002).
83. M. Strobel, U. Schmidt, K. Bade, and J. Halbritter, "Nucleation and Growth of Ni-LIGA Layers", *Microsystem Technologies*, **3**, 10 (1996).
84. C. S. Lin, P. C. Hsu, L. Chang, and C. H. Chen, "Properties and Microstructure of Nickel Electrodeposited From a Sulfamate Bath Containing Ammonium Ions", *Journal of Applied Electrochemistry*, **31**, 925 (2001).
85. N. V. Mandich and D. W. Baudrand, "Troubleshooting Ni Sulfamate Plating Installations",*Plating and Surface Finishing*, **89** (9), 68 (2002).
86. J. J. Kelly, S. H. Goods, and A. A. Talin, "Ageing of Nickel Sulfamate Electrolytes During the Electrodeposition of MEMS Structures", in *Electrochemical Processing in ULSI and MEMS: Proceedings of the 205th Meeting of the Electrochemical Society, San Antonio*, H. Deligianni, S. T. Mayer, T. P. Moffat, and G. R. Stafford (Eds.), The Electrochemical Society, U.S.A. PV 2004–17, pp. 432–447 (2004).
87. A. Ruzzu and B. Matthis, "Swelling of PMMA in Aqueous Solutions and Room Temperature Ni-Electroforming", *Microsystem Technologies*, **8**, 116 (2002).
88. A. C. Fischer-Cripps, *Nanoindentation*, Springer, New York (2002).
89. W. N. Sharpe, "Murray Lecture Tensile Testing at the Micrometer Scale: Opportunities in Experimental Mechanics", *Experimental Mechanics*, 43 (3), 228 (2003).

90. T. E. Buchheit, D. A. LaVan, J. R. Michael, T. R. Christenson, and S. D. Leith, "Microstructural and Mechanical Properties Investigation of Electrodeposited and Annealed LIGA Nickel Structures", *Metallurgical and Materials Transactions A*, **33A**, 539 (2002).
91. S. H. Goods, J. J. Kelly, and N. Y. C. Yang, "Electrodeposited Nickel-Manganese: an Alloy for Microsystem Applications", *Microsystem Technologies*, **10**(6–7), 498 (2004).
92. F. Ebrahimi, G. R. Bourne, M. S. Kelly, and T. E. Matthews, "Mechanical Properties of Nanocrystalline Nickel Produced by Electrodeposition", *Nanostructured Materials*, **11**, 343 (1999).
93. A. W. Thompson, "Effect Of Grain Size On Work Hardening In Nickel", *Acta Metallurgica*, **25**, 83 (1977).
94. E. O. Hall, "The Luders Deformation Of Mild Steel", *Proceedings of Royal Society London*, **64**, 747 (1951).
95. H. Baker and H. Okamoto (Eds.), ASM Handbook: Volume 3, Alloy Phase Diagrams, ASM International, U.S.A., p. 145, (1992).
96. N. Y. C Yang, C. H. Cadden, C. W. San Marchi, *LIGA Microsystems Aging: Evaluation and Mitigation*, SAND2003-8800, Sandia National Laboratories, Livermore, CA, (2003).
97. W. B. Stephenson, Jr., "Development and Utilization of a High Strength Alloy for Electroforming", *Plating*, **53**, 183 (1966).
98. G. A. Malone, "New Developments in Electroformed Nickel-Based Structural Alloys", *Plating and Surface Finishing*, **74**(1), 50 (1987).
99. J. J. Kelly, S. H. Goods, and N. Y. C. Yang, "High Performance Nanostructured NiMn Alloys for Microsystem Applications", *Electrochemical and Solid State Letters*, **6**, C88 (2003).
100. T. E. Buchheit and S. H. Goods, unpublished results.
101. S. A. Watson, *Nickel Sulphamate Solutions*, NiDI Technical Series No. 10 052, Nickel Development Institute, Toronto, Canada (1989).
102. Don Baudrand, "Nickel Sulfamate Plating, Its Mystique and Practicality", *Metal Finishing*, **94**(7), 15 (1996).
103. H. Fischer, "Aspects of Inhibition in Electrodeposition of Compact Metals", *Electrodeposition and Surface Treatment*, **1**, 319 (1972/73).
104. D. Landolt, "Electrochemical and Materials Science Aspects of Alloy Deposition", *Electrochimica Acta*, **39**, 1075 (1994).
105. R. Winand, "Electrodeposition of Metals and Alloys- New Results and Perspectives", *Electrochimica Acta*, **39**, 1091 (1994).
106. V. Zentner, A. Brenner, and C. W. Jennings, *Plating*, 39, 865 (1952).
107. W. Kim and R. Weil, "Pulse Plating Effects in Nickel Electrodeposition", *Surface and Coatings Technology*, **38**(3), 289
108. N. Ibl, "Some Theoretical Aspects of Pulse Electrolysis", *Suface Technology*, **10**(2), 81 (1980).
109. M. Viswanathan and Ch. J. Raub, "Effect of Pulsed Direct Current (Pulsed Plating) on the Properties of Electrodeposited Coatings", *Galvanotechnik*, **66**(4), 277 (1975).
110. D. L. Rehrig, H. Leidheiser, and M. R. Notis, "Influence of the Current Waveform on the Morphology of Pulse Electrodeposited Gold", *Plating and Surface Finishing*, **64**(12), 40 (1977).
111. T. L. Lam, I. Ohno, and T. Saji, *Journal of the Metal Finishing Society of Japan*, **33**, 29 (1982).
112. L. G. Holmbom and B. E. Jacobson, "Effects of Bath Temperature and Pulse-Plating Frequency on Growth Morphology of High-Purity Gold", *Plating and Surface Finishing*, **74**(9), 74 (1987).
113. T. Fritz, H. S. Cho, K. J. Hemker, W. Mokwa, and U. Schnakenberg, "Characterization of Electroplated Nickel", *Microsystem Technologies*, **9**(1–2), 87 (2002).
114. E. J. Roehl, *Plating*, **35**, 452 (1948).
115. J. Amblard, I. Epelboin, M. Froment, and G. Maurin, "Inhibition and Nickel Electrocrystallization", *Journal of Applied Electrochemistry*, **9**, 233 (1979).

116. W. M. Phillips and F. L. Clifton, *Proceedings of the American Electroplaters' Society*, **35**, 87 (1948).
117. J. Edwards, "Radiotracer Study of Addition Agent Behavior: 4 Mechanism of Incorporation", *Transactions of the Institute of Metal Finishing*, **41**, 140 (1964).
118. E. Orowan, *Dislocations in Metals*, AIME, Warrendale, PA, p. 69 (1954).
119. E. Dieter, *Mechanical Metallurgy*, McGraw-Hill, NY, p. 144 (1961).
120. V. P. Greco, "Review of Fabrication and Properties of Electrocomposites", *Plating and Surface Finishing*, **76**(10), 68 (1989).
121. C. A. Addison and E. C. Kedward, *Transactions of the Institute of Metal Finishing*, **55**, 1 (1977).
122. M. Thoma, Plating and Surface Finishing, "Cobalt/Chromic Oxide Composite Coating for High-Temperature Wear Resistance", *Plating and Surface Finishing*, **71**(9), 51 (1984).
123. F. K. Sautter, *J. Electrochem. Soc.*, **110**, 557 (1963).
124. X. M. Ding, N. Merk, and B. Ilschner, "Mechanical Behaviour of Metal-Matrix Composite Deposits", *Journal of Materials Science*, **33**, 803 (1998).
125. X. M. Ding, N. Merk, and B. Ilschner, "Functional Behaviour of Particle-Volume-Graded Electrodeposited Composite Coatings", *Journal of the Chinese Society of Mechanical Engineers*, **183**, 145 (1997).
126. S. H. Goods, T. E. Buchheit, R. P. Janek, J. R. Michael, and P. G. Kotula, "Oxide Dispersion Strengthening of Nickel Electrodeposits for Microsystem Applications", *Metallurgical and Materials Transactions A- Physical Metallurgy and Materials Science*, **35A**, 2351 (2004).
127. A. Talin, Sandia National Laboratories, unpublished results.
128. Q. Shi, S. C. Chang, M. W. Putty, and D. B. Hicks, "Characterization of Electroformed Nickel Microstructures", *Proceedings of the SPIE*, **2639**, 191 (1995).
129. H. Majjad, S. Basrour, P. Delobelle, and M. Schmidt, "Dynamic Determination of Young's Modulus of Electroplated Nickel Used in LIGA Technique", *Sensors and Actuators A (Physical)*, **74**(1), 148 (1999).
130. E. Mazza, S. Abel, and J. Dual, "Experimental Determination of Mechanical Properties of Ni and Ni-Fe Microbars", *Microsystem Technologies*, **2**(4), 197 (1996).
131. K. J. Hemker and H. Last, "Microsample Tensile Testing of LIGA Nickel for MEMS Applications", *Materials Science and Engineering A- Structural Materials, Properties, Microstructure, and Processing*, **319**, 882 (2001).
132. J. P. Hirth and J. Lothe, *Theory of Dislocations*, McGraw-Hill, New York, (1968).
133. W. Voigt, *Lehrbuch der Krystallphysik*, B. G. Teubner, Leipzig, (1910).
134. A. Reuss, *Zeitschrift fuer Angewandte Mathematik und Mechanik*, **9**, 49 (1929).
135. W. Bacher, K. Bade, B. Matthis, M. Saumer, and R. Schwarz, "Fabrication of LIGA Mold Inserts", *Microsystem Technologies*, **4**, 117 (1998).
136. M. Heckele, W. Bacher, and K. D. Mueller, "Hot Embossing- The Molding Technique for Plastic Microstructures", *Microsystem Technologies*, **4**, 122 (1998).
137. M. S. Despa, K. W. Kelly, and J. R. Collier, "Injection Molding of Polymeric LIGA HARMS", *Microsystem Technologies*, **6**, 60 (1999).
138. R. Ruprecht, T. Benzler, T. Hanemann, K. Mueller, J. Konys, V. Piotter, G. Schanz, L. Schmidt, A. Thies, H. Woellmer, and J. Hausselt, "Various Replication Techniques for Manufacturing Three-Dimensional Metal Microstructures", *Microsystem Technologies*, **4**, 28 (1997).
139. K. Kim, S. Park, J.-B. Lee, H. Manohara, Y. Desta, M. Murphy, and C. H. Ahn, "Rapid Replication of Polymeric and Metallic High Aspect Ratio Microstructures Using PDMS and LIGA Technology", *Microsystem Technologies*, **9**, 5 (2002).
140. A. M. Morales, L. A. Domeier, M. Gonzales, J. Hachman, J. M. Hruby, S. H. Goods, D. E. McLean, N. Yang, and A. D. Gardea, "Microstructure and Mechanical Properties of Nickel Microparts Electroformed in Replicated LIGA Molds," Proceedings of the SPIE, 4979, 440 (2003).

141. R. Bischofberger, H. Zimmermann, and G. Staufert, “Low-cost HARMS Process”, *Sensors and Actuators A*, **61**, 392 (1997).
142. P. M. Dentinger, W. M. Clift, and S. H. Goods, “Removal of SU-8 Photoresist for Thick Film Applications”, *Microelectronic Engineering*, **61–62**, 993 (2002).
143. W. W. Flack, W. P. Fan, and S. White, “The Optimization and Characterization of Ultra-Thick Photoresist Films”, *Proceedings of the SPIE*, **3333**, 1288 (1998).
144. Bradley Todd, Warren W. Flack, and Sylvia White, “Thick Photoresist Imaging Using A Three Wavelength Exposure Stepper”, *Proceedings of the SPIE*, **3874**, 330 (1999).
145. W. W. Flack, H.-A. Nguyen, and E. Capsuto, “Process Improvements for Ultra-Thick Photoresist Using a Broadband Stepper”, *Proceedings of the SPIE*, **4336**, 956 (2001).
146. W. W. Flack, H.-A. Nguyen, and E. Capsuto, “Characterization of an Ultra-Thick Positive Photoresist for Electroplating Applications”, *Proceedings of the SPIE*, **5039**, 1257 (2003).
147. F. T. Hartley and C. K. Malek, “Nanometer X-ray Lithography”, *Proceedings of the SPIE*, **3894**, 44 (1999).
148. M. Tormen, F. Romanato, M. Altissimo, L. Businaro, P. Candeloro, and E. M. Di Fabrizio, “Three Dimensional Micro- and Nanostructuring by Combination of Nanoimprint and X-ray Lithography”, *Journal of Vacuum Science and Technology B*, **22**(2), 766 (2004).
149. M. W. Boerner, M. Kohl, F. J. Pantenburg, W. Bacher, H. Hein, and W. K. Schomburg, “Sub-Micron LIGA Process for Movable Microstructures”, *Microelectronic Engineering*, **30**, 505 (1996).

Direct Writing Techniques: Electron Beam and Focused Ion Beam

T. Djenizian and C. Lehrer

1 Introduction

Due to significant theoretical and technological new advances, nanostructuring of surfaces has attracted a great deal of scientific interest. The continuous demand for shrinking the dimensions of structures to reach the nanometer scale is mainly motivated by the discovery of new behaviors dominated by unique properties that are encountered when nanosize dimensions are approached (e.g., quantum confinement). Additionally, a major thrust for shrinking dimensions originates from the microelectronic field requiring the development of smaller devices, system integration, and system diversification. Thus, electronic materials as well as integrated materials adding new features must be structured at the micro- and nanometer scale.

Methods to locally micro- and nanostructuring surfaces have been intensively explored in recent years and have rapidly found technological applications. Most of these techniques are based on lithography and hence demand a masking process. To date, optical lithography is the main technique used for the integrated circuit (IC) industry, but the current strategies employed are blocked by optical diffraction. Thus, higher-resolution approaches such as X-Ray lithography and emerging lithography technologies have been explored to fabricate structures in the sub-100-nm range (see e.g., [1–6]).

Recently, electrochemistry at the nanoscale is strongly emerging because of significant advantages such as a low cost, simplicity, and compatibility with a wide range of micropatterning processes [7]. Especially, the ability to combine electrochemical techniques with several patterning approaches can be exploited for the micromachining [8, 9] and for shrinking the dimensions of structures below 100 nm (see, e.g., Ref. [10]). In this context, use of electrochemical approaches in conjunc-

T. Djenizian (✉)
Laboratoire MADIREL (UMR 6121), Université de Provence-CNRS, Centre Saint Jérôme, F-13397 Marseille Cedex 20, France
e-mail: thierry.djenizian@univ-provence.fr

P. Schmuki, S. Virtanen (eds.), *Electrochemistry at the Nanoscale*, Nanostructure Science and Technology, DOI 10.1007/978-0-387-73582-5_4,

tion with technologies based on charged particle beam lithography is also a convenient way for the nanostructuring of surfaces and the fabrication of ultra-small structures.

Charged particles beam lithography approaches include techniques using beam of electrons or ions, so-called electron-beam lithography (EBL) and focused ion beam lithography (FIBL), respectively. The process steps of these methods are similar to that described for photolithography, except that the transfer of the pattern is not achieved through the use of a mask but by scanning directly the focused particle beam across the surface [11, 12].

In this chapter, we report different strategies using the high-resolution potential of these direct writing approaches to achieve electrochemistry at the nanoscale.

2 Theoretical Aspects

2.1 Generality

Electron and ion beams are successfully used in a lot of different working areas. Although, especially electron beams are mainly used for microscopic imaging or material characterization by different techniques of electron microscopy, the number of applications increases where particle beams are used for material processing. Material processing includes not only structuring of sample surfaces by material removal or deposition, but also the application of energetic particle beams enables the well-defined and controlled modification of material properties in the surface near the regions of the specimen. For example, it is possible to alter the optical [13], mechanical, or electrical characteristics of solids by irradiation with energetic electrons or ions. In semiconductor technology the implantation of ions is used for doping of semiconductor crystals and is one of the fundamental processes for the fabrication of integrated circuits [14]. In the area of tribology, surface hardening by irradiation with energetic particles enables the fabrication of components showing increased wear resistance [15–17]. The alteration of the refractive index of materials due to irradiation with electrons or ions is also used for the fabrication of opto-electronic elements, for example, light guides [18].

All kinds of material processing by energetic particles do have in common that they are based on effects, which originate from the interaction of the energetic particles with the solid. This chapter briefly describes the theoretical background. For readers interested in a more detailed treatment literature quotations will be given.

2.2 Interaction of Energetic Particles with Solids

Energetic particles, that is, electrons or ions accelerated within an electric field, interact with atoms or molecules when penetrating a specimen. Due to these interactions, the particles are scattered and energy is transferred to the specimen. There are different inelastic and elastic effects, which can occur during the irradiation of solids with charged particles. Collisions can take place between the incident particles and bound electrons or between the incident particles and nuclei or whole atoms of the stopping matter. During elastic interactions, the internal structure of the atoms and the energy of the atoms remain roughly unchanged. Inelastic collisions result in energy loss due to the excitation or ionization of atoms or molecules or due to nuclear excitation or nuclear reactions. Also, X-ray *bremsstrahlung* and *Cerenkov* radiation are effects, which reduce the total energy of the system. Which kind of the effects predominates depends upon the mass and the energy of the energetic particles and also upon the mass and the atomic number of the solid. In general, within the energy range normally used for material processing, with particle beams inelastic collisions with electrons and elastic nuclear collisions are the dominating effects. Low-probability inelastic nuclear collisions, and also elastic collisions with electrons, resulting only in low-angle scattering can be neglected.

2.3 Scattering of Particles

The probability of a particle undergoing any kind of interaction with an isolated atom is determined by the interaction cross-section. On the assumption that inelastic and elastic processes are independent of each other, a total scattering cross-section σ_T can be defined, which is the sum of the elastic σ_{elastic} and the inelastic scattering cross section $\sigma_{\text{inelastic}}$.

$$\sigma_T = \sigma_{\text{elastic}} + \sigma_{\text{inelastic}} \tag{1}$$

The total cross-section Q_T for scattering of a particle from a specimen containing N atoms per unit volume can be defined as [19] and Q_T can be regarded as the number of scattering events per unit path that a particle travels through the sample.

$$Q_T = N\sigma_T \tag{2}$$

The energy transfer and kinematics in elastic collisions between two isolated particles can be solved by applying the principles of conservation of energy and momentum. Under the assumptions, usually made in the description of the scattering processes between energetic particles in solids [20]:

- binary collisions only are considered
- the loss of energy due to excitation or ionization of electrons does not influence collision dynamics
- one of the two interacting particles is initially at rest the energies of the scattered projectile and of the recoil atom can be calculated [21].

Figure 1 shows a typical scattering process between two particles with unequal masses M_1 and M_2. In the laboratory system, the incident particle (initial kinetic

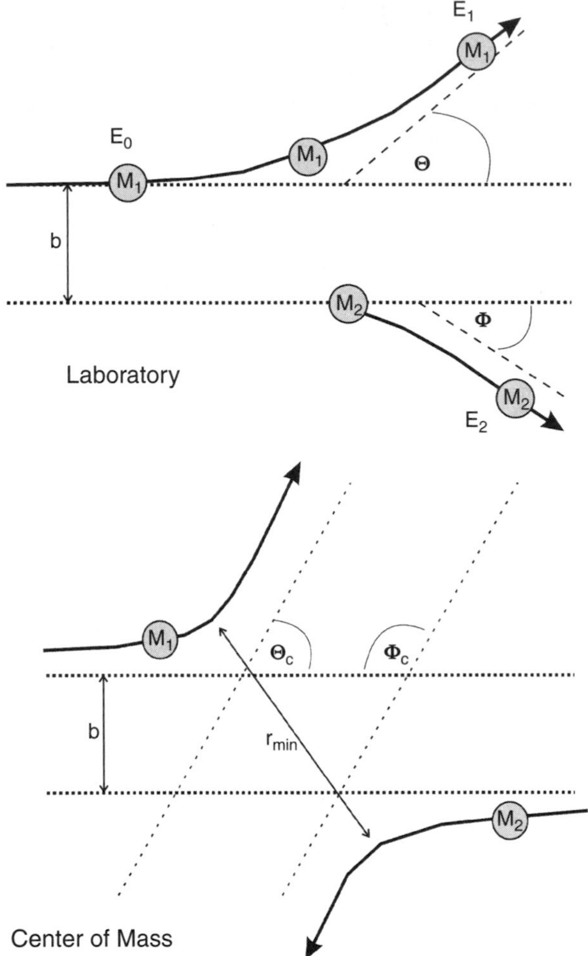

Fig. 1 Elastic collision diagram between particles of mass $M1$ and $M2$, scattering angle Θ and Φ, impact parameter b, in laboratory system and center-of-mass system

energy, E_0) is deflected by the angle Θ and transfers an energy E_2 to the recoiling atom, which leaves its original position at the angle Φ.

Energy of the scattered projectile (laboratory system) [22]:

$$K = \frac{E_1}{E_0} = \frac{\left[\mu \cos\Theta + \left(1 - \mu^2 \sin^2\Theta\right)^{1/2}\right]^2}{(1+\mu)^2} M_1 \leq M_2 \tag{3}$$

$$\frac{E_1}{E_0} = \frac{\left[\mu \cos\Theta \pm \left(1 - \mu^2 \sin^2\Theta\right)^{1/2}\right]^2}{(1+\mu)^2} M_1 > M_2 \tag{4}$$

Energy of the recoil nucleus (laboratory system):

$$\frac{E_2}{E_0} = 1 - \frac{E_1}{E_0} = \frac{4M_1M_2}{(M_1 + M_2)^2}\cos^2\Phi \tag{5}$$

E_0: energy of the incident particle
E_1: laboratory energy of the scattered particle
E_2: laboratory energy of the recoiling target
M_1, M_2: mass of incident projectile and target atom
μ: mass ratio M_1/M_2
Θ: laboratory angle of the scattered projectile
Φ: laboratory angle of the recoiling target atom

The angular distribution of projectiles scattered from an atom is described by the angular differential scattering cross-section. This quantity gives a measure of the probability of scattering an energetic particle through an angle between Θ and $\Theta + d\Theta$ into a solid angle $d\Omega$. The differential cross-section has units of area. The differential cross-section is normally calculated for center-of-mass coordinates. The energy E_2, transferred to the target atom and often referred to as T, is given by:

$$T \equiv E_2 = E_0\frac{2M_1M_2}{(M_1 + M_2)^2}(1 - \cos\Theta_C) \tag{6}$$

Conversion of scattering angles from the laboratory system to the center-of-mass system is given by:

$$\Phi = \frac{\pi - \Theta_C}{2} = \frac{\Phi_C}{2} \tag{7}$$

$$\tan\Theta = \frac{M_2\sin\Theta_C}{M_1 + M_2\cos\Theta_C} \tag{8}$$

Θ_C: center-of-mass angle of the scattered projectile
Φ_C: center-of-mass angle of the recoiling target atom

The scattering angle Θ_C is given by the classical scattering integral [22], which gives the angular trajectory information for a two-body central-force scattering:

$$\Theta_C = \pi - 2b\int_0^{u_{max}} \frac{du}{\left[1 - \frac{V(u)}{E_C} - (bu)^2\right]^{1/2}} \tag{9}$$

$u = 1/r$
$r = r_1 + r_2$, is the distance between particles in the center-of-mass system
$V(u)$: interaction potential

E_C: total kinetic energy in the center-of-mass system
u_{max}: reciprocal value of the minimum distance between the particles $1/r_{min}$
B: impact parameter

When the interaction between colliding particles is purely Coulombic and projectile and target nucleus are treated as pure nuclei, this means the interaction potential $V(u)$ can be written as the Coulomb potential, the elastic scattering in the center-of-mass system can be described by the equation of Rutherford, which was derived for the scattering of α particles [23]:

$$\frac{d\sigma(\Theta_C)}{d\Omega} = \frac{e^4}{16}\frac{Z_1^2 Z_2^2}{E_C^2}\frac{1}{\sin^4\frac{\Theta_C}{2}} \tag{10}$$

$\frac{d\sigma(\Theta)}{d\Omega}$: angular differential scattering cross section
Z_1: atomic number of the scattered particle
Z_2: atomic number of the target atom
E_C: total kinetic energy in the center-of-mass system
$e^2 = \frac{e_0^2}{4\pi\varepsilon_0}$: e_0 electron charge

The equation of Rutherford is based on the assumption that two-point charges interact with each other by Coulomb forces, and no screening by the surrounding electrons takes place. It is assumed that the kinetic energy of the incident particles is sufficiently high to penetrate the electron cloud and to approach the nucleus, so that screening of the nucleus by the electron cloud can be neglected. Deviations from the equation can occur for high and low particle energies. For high energies, the distance between the incident particle and the nucleus becomes small and nuclear forces can influence scattering. At high energies, also relativistic effects can alter the scattering. Because of their low mass, the velocity of electrons with an energy of more than 100 keV is already high enough that relativistic effects have to be considered. Within the energy range normally used for material processing by ions, the velocity of the energetic ions is significantly lower and relativistic effects can be neglected. To give a more accurate description of the scattering of high-velocity electrons, a differential cross-section corrected for relativity can be used [24]:

$$\frac{d\sigma(\Theta_C)}{d\Omega} = \frac{e^4}{16}\frac{Z^2}{E_C^2}\frac{1}{\sin^4\frac{\Theta_C}{2}}\left(\frac{m_0c^2 + E_0}{2m_0c^2 + E_0}\right)^2 \tag{11}$$

E_0: kinetic energy of the electron and
m_0c^2: rest mass energy of the electron

If the particle energy is not sufficient to get close to the nucleus, the screening effect of the surrounding electron cloud, which acts to reduce the differential cross-section, cannot be neglected. For electrons and light ions, this screening effect can be described by the modified Rutherford differential cross-section: [25]

$$\frac{d\sigma(\Theta_C)}{d\Omega} = \frac{e^4}{16}\frac{Z_1^2 Z_2^2}{E_C^2}\frac{1}{\left[\sin^2\frac{\Theta_C}{2} + \left(\frac{\Theta_0}{2}\right)^2\right]^2} \quad (12)$$

Θ_0 is the so-called screening parameter for electron scattering given by: [26]

$$\Theta_0 = \frac{0.117 Z^{1/3}}{E_C^{1/2}} \quad (13)$$

E_C in keV

It can be described by a particular scattering angle. When the scattering angle is greater than Θ_0, electron–electron interactions can be neglected and the nuclear interaction is dominant.

For heavy particles of low velocity, and in the case of relatively distant collisions, the screening effect of the electrons cannot be neglected. In order to describe the scattering of particles, the interaction potential between two atoms has to be known exactly. The detailed calculation of the potential between two atoms is only possible by numerical calculations, taking into account the electron distribution within the shell of the interacting particles. In order to enable an analytical calculation of the interaction between atoms, the potential is described by a Coulomb potential, which additionally includes a suitable screening function, taking into consideration the screening of the nuclei.

$$V(r) = \frac{Z_1 Z_2 e^2}{r}\varphi\left(\frac{r}{a}\right) \quad (14)$$

$\varphi\left(\frac{r}{a}\right)$: screening function
a: screening parameter

In literature, many different screening functions used for the analytical solution of eq. (1) can be found [27–29]. Very good accordance with numerical data can be achieved by applying the so-called ZBL screening function derived by J.F. Ziegler, J.P. Biersack, and U. Littmark [30].

2.4 Stopping of Particles in Solids

When an energetic particle penetrates a solid, it undergoes a series of collisions with atoms and electrons in the specimen. Depending on energy and mass of the incident particle as well as on the target material, the incident particle loses energy at a rate d*E*/d*x*. The kinetic energy of the projectiles, not scattered backward and leaving the specimen as the so-called "reflected" particles, is totally transferred to the target and the particle comes to rest in the solid. As the amount of energy lost per collision and the distance between collisions are random processes, the path length of identical particles impinging the surface with identical energy differs and a distribution in depth arises. The range R is determined by the energy loss along the path of the particle

$$R = \int_{E_0}^{0} \frac{1}{\mathrm{d}E/\mathrm{d}x} \mathrm{d}E \tag{15}$$

E_0 is the incident energy of the particle penetrating the specimen. The energy loss dE per traveled path dx is determined by the screened Coulomb interactions with atoms of the specimen and electrons. As already mentioned, two effects dominate the stopping of particles in solids: nuclear collisions, in which energy is transferred as kinetic energy to a target atom as a whole, and electronic collisions, in which the moving particle excites or ejects bound and free electrons. On the assumption that the two mechanisms of energy loss are independent of each other, as a good approximation, the energy-loss rate dE/dx can be expressed as:

$$\frac{\mathrm{d}E}{\mathrm{d}x} = \left.\frac{\mathrm{d}E}{\mathrm{d}x}\right|_n + \left.\frac{\mathrm{d}E}{\mathrm{d}x}\right|_e \tag{16}$$

$\left.\frac{\mathrm{d}E}{\mathrm{d}x}\right|_n$ nuclear stopping
$\left.\frac{\mathrm{d}E}{\mathrm{d}x}\right|_e$ electronic stopping

Due to the transfer of energy to target atoms, nuclear collisions can result in large, discrete energy losses and also lead to significant scattering of the projectile. Predominately for heavy ions, this process is responsible for the displacement of atoms from their lattice sites, and therefore for the production of damage. Electronic collisions involve almost no scattering of the particle, negligible lattice disorder, and much smaller energy loss per collision. The relative importance of the two effects depends on the energy, mass, and atomic number of the particles and on mass, atomic number, and structure of the specimen. Typical units for the energy-loss rate are electron-volt per nanometer. Stopping cross-sections for nuclear and electronic stopping $S_{n/e}$, are defined as:

$$S_{n,e} = -\frac{1}{N}\left(\frac{dE}{dx}\right)_{n,e} \tag{17}$$

N: atomic density

At low energies, elastic collisions with the nuclei dominate. With increasing velocity v, the energy loss due to nuclear collisions decreases and the inelastic interaction with free and bound electrons becomes the main process of interaction. Within the low-energy region, under the assumption [31] that the electrons form a free-electron gas, it can be shown that the electronic stopping cross-section is proportional to the velocity of the ions, and therefore proportional to the root of the energy. For higher energies, the electronic stopping cross-section goes through a maximum and then decreases proportionally to v^{-2} [22]. In this energy range, where the interaction potential for collisions is assumed purely Coulombic, the Bethe–Bloch theory [32, 33] is valid, which can also be applied for the stopping of electrons in solids. The electronic stopping cross-section is given by the Bethe formula

$$S_e = -\frac{1}{N}\left(\frac{dE}{dx}\right)_e = \frac{4\pi Z_1^2 e^4 Z_2}{m_e v^2}\left(\frac{M_1}{m_e}\right)\ln\frac{2m_e v^2}{I} \tag{18}$$

For most elements, the average excitation energy I can be approximated by:

$$I \cong 10Z_2 \tag{19}$$

Calculated and experimental data for the average excitation energy can be found in [34]. For high energies, the Bethe formula has to be corrected for the effect of relativity and the electronic stopping cross-section is given as:

$$S_e = -\frac{1}{N}\left(\frac{dE}{dx}\right)_e = \frac{4\pi Z_1^2 e^4 Z_2}{m_e v^2}\left(\frac{M_1}{m_e}\right)\left[\ln\frac{2m_e v^2}{I} - \ln\left(1-\beta^2\right) - \beta^2\right] \tag{20}$$

$\beta = v / c$

c: velocity of light

After transferring their kinetic energy totally to the target, the kinetic particles come to rest in the solid. Due to the collisions with the atoms of the specimen, the particles do not travel in a straight path. The net penetration into the material, that is, the projection of the range R on the incidence direction of the particle, is called the projected range.

As the interactions between particles and atoms are stochastic processes, the path length of identical particles impinging the surface with identical energy differs and a distribution in depth arises. The distribution in projected ranges is referred to as the range distribution. The most probable projected range is the so-called average or mean projected range R_p. The standard deviation of the range distribution is referred to as the projected range straggling σR_p.

Because of the high difference in mass between impinging electrons and the nuclei of the specimen, the energy loss of the electrons due to nuclear collisions can be neglected. Therefore, the deceleration of the electrons is only caused by the interaction with shell electrons. Referring to the Bethe formula, for electrons with energy between 10 keV and 100 keV the average range of the electrons can be calculated.

For ions with not too high kinetic energy, the loss of energy due to nuclear collisions cannot be neglected and in comparison to electrons, the penetration depth is reduced. For amorphous solids, at low ion doses N_{imp} and, under the assumption that all ions will be implanted, the range distribution is roughly Gaussian and can be described as:

$$N(x) = \frac{N_{imp}}{\sqrt{2\check{s}}\Delta R_P} \exp(-\frac{(x - R_P)^2}{2\Delta R_P^2}) \tag{21}$$

with a maximum doping concentration

$$N_{\max} = \frac{N_{imp}}{\sqrt{2\check{s}}\Delta R_P} \tag{22}$$

The deceleration of energetic particles in a solid by inelastic collisions leads to the excitation of target atoms and to a whole range of signals used for imaging and analytical purposes. The most important signals are X-rays and secondary electrons. Characteristic X-rays are produced if an amount of energy higher than the excitation energy is transferred to a bound electron which, as a result, is ejected and replaced by an electron from an outer shell. This transition can be accompanied by the emission of X-rays or Auger electrons. Besides characteristic X-rays, *Bremsstrahlung* X-rays can be also detected. This radiation results from kinetic particles that inelastically interact with the nucleus. The deceleration of the particle by the Coulomb field of the nucleus leads to the emission of X-rays. Depending on the strength of interaction, the particle can suffer any amount of energy loss. For electrons, the probability for the generation of *bremsstrahlung* can be described as: [35]

$$N(E) = \frac{KZ(E_0 - E)}{E} \tag{23}$$

$N(E)$: number of *bremsstrahlung* photons with energy E
E_0: energy of the electrons
K: Kramer's constant
Z: atomic number of the atom

Secondary electrons are electrons of the target material ejected due to the interaction with the penetrating particle. Besides Auger electrons, which, as already mentioned, can be ejected from an inner shell when an ionized atom returns to the ground state, there are two additional types of secondary electrons, which can be distinguished by kinetic energy. Electrons in the conduction or valence bands can be easily ejected and have energies typically below 50 eV. The emission process of this so-called "slow" electrons, which are used to form images of the specimen surface, can be quite complex, and the different production processes cannot be described by one cross-section. The number of secondary electrons is highest at about 5 eV and is close to zero for energies $\geq$50 eV. In contrast, "fast" electrons are strongly bound electrons, which are ejected from inner shells and can have up to 50% of the beam energy when they are ejected.

2.5 Radiation Damage in Solids

During the stopping of energetic particles in solids, energy is transferred to the specimen in elastic and inelastic collisions. The energy transfer responsible for the generation of a number of useful signals, for example, secondary electrons used for imaging of the sample surface, unfortunately, also causes more or less significant rearrangement of the target structure. This so-called "radiation damage" not only affects the structure of the target but can also modify many other properties of matter, for example, density, elasticity, or electrical parameters. As far as electrical parameters are concerned, they especially affect the mobility, lifetime, and, under some circumstances, the carrier concentration [14].

The resulting damage depends mainly on the energy and mass of the projectile as well as on the mass of the target atoms. Inelastic collisions between impinging particles and electrons mainly result in interband transitions and ionization and give rise to chemical-bonding changes. Elastic collisions can result in the displacement of atoms from their lattice site. This effect, which is mainly due to ions and high energy electrons, can produce collision cascades if the kinetic energy of the displaced atoms is sufficient to also displace other atoms. Collision cascades lead to an accumulation of the so-called Frenkel defects, a combination of a lattice vacancy and an interstitial atom, and also to complex lattice defects called clusters. With increasing particle dose, and thus growing density of beam-induced damage, the clusters overlap and a continuous damaged layer is formed. For a sufficiently high dose, the top layer of the specimen will be completely amorphized. It is assumed that at least 50% of the atoms have to be displaced for an amorphous layer to be formed. The extent of radiation damage created is determined by the kind and energy of the projectiles, the temperature of the specimen, the particle dose, and channeling effects, if the specimen possesses a crystalline structure. Channeling effect denotes the increased penetration of particles along major axes and planes within a monocrystalline target, because of the reduced nuclear stopping within channels formed by the symmetrical arrangement of the lattice atoms. In order to displace an atom from its lattice site, a minimum amount of energy, that is, displacement energy E_d, has to be transferred. If the transferred amount of energy is not sufficient to knock-out the atom of its atomic site, the struck atom undergoes large amplitude vibrations and no Frenkel pair will be created. The vibrational energy of the struck atom is quickly shared with nearest neighbors, and phonons will be generated. These collective oscillations of the crystal lattice result in heating of the sample and may cause damage. Specimen heating is difficult to measure experimentally because of the many variables that can effect the results, such as thermal conductivity, thickness, and surface condition of the specimen as well as the kinetic energy of the particles, and the current density of the particle beam. The effects of beam current and thermal conductivity on the specimen temperature for electrons were calculated by L.W. Hobbs [36].

Because of the crystallographic structure of a solid, the displacement energy for a lattice atom depends on the direction of the momentum of the target atom. Therefore, a range of displacement energies exists for the creation of a Frenkel pair. The average displacement energy is typically a factor one to two larger than the

Table 1 Minimum and average displacement threshold energies for some monoatomic materials [22]

Atomic number	Chemical symbol	Minimum displacement energy (eV)	Average displacement energy (eV)
6	Graphite	25	
6	Diamond	35	
12	Mg	10	
13	Al	16	27
14	Si	13	
22	Ti	19	
23	V	26	
24	Cr	28	
26	Fe (bcc)	17	
27	Co (hcp)	22	34
28	Ni	23	34
29	Cu	19	29
30	Zn	14	29
31	Ga	12	
32	Ge	15	
40	Zr	21	
41	Nb	28	
42	Mo	33	
46	Pd	26	41
47	Ag	25	39
48	Cd	19	36
49	In	15	
50	Sn (white)	22	
50	Sn (gray)	22	
71	Lu	17	
73	Ta	34	90
74	W	38	
75	Re	40	
78	Pt	33	44
79	Au	36	43
82	Pb	14	19
90	Th	35	44

minimum displacement energy. Table 1 gives the displacement energies for some monoatomic materials [22].

Based on the Kinchin–Pease formula [37], the number N_d of atoms displaced per incident particle can be calculated by:

$$N_d = \frac{E_n}{2E_d} \tag{24}$$

E_n is the total energy loss of a particle due to nuclear collisions and E_d is the displacement energy. Under the assumption that at least 50% of the atoms have to be displaced in order to create an amorphous layer, the critical dose for amorphization can be calculated by:

$$D_a = \frac{E_d N}{\left(\mathrm{d}E/\mathrm{d}x\right)_n} \tag{25}$$

N: atomic density of the target material
$(dE\,/\,dx)_n$: energy deposited in nuclear collisions per unit distance

In real terms, the critical dose for amorphization is almost always higher than the value calculated by this equation, as effects, like the annealing arising during the bombardment of the specimen or collisions with already displaced atoms, are neglected [14].

2.6 Sputtering

The erosion of a specimen by energetic particle bombardment is called sputtering. In this process, incoming particles interact with atoms in the near-surface layer of the solid, and atoms at the surface are removed. This effect, which is especially pronounced in case of heavy ions and high doses, can be regarded as a kind of damage as it (e.g., during ion implantation) unintentionally modifies the topography of the sample and leads to an alteration of the implantation profile. However, sputtering is also a very desirable effect where material has to be removed intentionally, for example, for physical vapor deposition or in the field of micro- and nanostructuring, where focused ion beams are used for high-precision material removal. The most important parameter describing sputtering is the number of sputtered atoms per incident ion. This sputtering yield Y depends on energy and mass of the ions, the angle of incidence, and the structure and composition of the specimen. Based on a model of Sigmund [38] for vertical incidence and sufficiently high energy, the sputter yield is given by:

$$Y\left(E\right) = \frac{3}{4\check{s}^2 C_0}\frac{1}{U_0}\left(\frac{M_2}{M_1}\right)\cdot S_n\left(E\right) \tag{26}$$

C_0 is a constant ($C_0 = 0.5\pi\lambda_0 a^2$; $\lambda_0 = 24$; $a = 0.0219$ nm)
(M_2/M_1) is a numerically calculable function
U_0 is the surface-binding energy

For medium mass-ion species and not too high energies, the values of Y lie between 1 and 10. Values for Y for different projectile–target combinations can be found in [39].

Sputtering removes material from the surface of the specimen. The thickness d of the removed layer is given by:

$$d = Y\frac{N_{imp}}{N} \tag{27}$$

N: atomic density of the target material
N_{imp}: ion dose

The erosion of the sample surface modifies the implantation profile. Under some simplifying assumptions, the modification of the depth profiles can be calculated by [14]:

$$N(x) = \frac{N}{2Y}\left(\operatorname{erf}\frac{x - R_P + N_{imp}\frac{Y}{N}}{\sqrt{2}\Delta R_P} - \operatorname{erf}\frac{x - R_P}{\sqrt{2}\Delta R_P}\right) \tag{28}$$

For high-ion dose, an equilibrium condition is obtained with a maximum of concentration $N_{\max}$ at the surface:

$$N_{max} = \frac{N}{2Y}\operatorname{erfc}\left(\frac{-R_P}{\sqrt{2}\Delta R_P}\right) \approx \frac{N}{Y}\text{for } R_P \geq 3\Delta R_P \tag{29}$$

3 Micro- and Nanostructuring by Electron-Beam Approaches and Electrochemical Reactions

Electrochemical technology has been combined with several electron-beam (e-beam) approaches to generate materials with features size that range from nanometers to micrometers in size. This section reports how to exploit electrochemistry with conventional EBL as well as alternative e-beam approaches to achieve the micro-and nanostructuring of the surfaces.

3.1 Microstructuring by Conventional EBL and Electrochemical Reactions

3.1.1 Basics on Electron-Beam Lithography

Electron-beam lithography (EBL) followed soon after the development of the scanning electron microscope (SEM) in 1955 [40] (see a description of a SEM in Fig. 2) and was one of the earliest processes used for IC fabrication dating back to 1957 [41]. The principle of pattern transfer based on EBL is depicted in Fig. 3. The lithographic sequence begins with coating substrates with a positive or negative resist. Then, direct e-beam exposure of the resist is achieved as it is depicted in Fig. 3a. In general, the e-beam-writing system is equipped with a lithographic tool in order to control accurately the displacement of the beam. Positive resists such as poly(methyl-methacrylate) (PMMA) become more soluble in a developing solvent after exposure, because weak radiation causes local-bond breakages and thus chain scission. As a result, the exposed regions containing material of lower mean

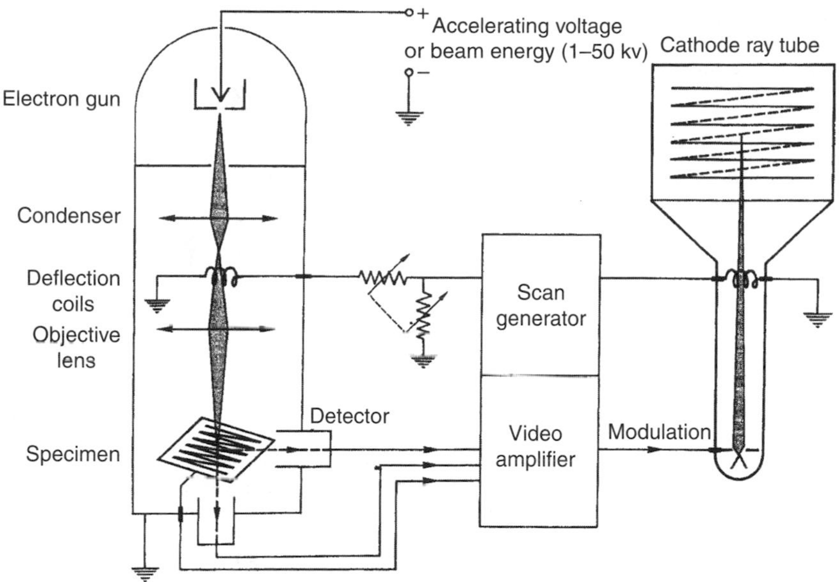

Fig. 2 Schematic drawing of a scanning electron microscope

molecular weight are dissolved after development (Fig. 3b). By contrast, negative resists become less soluble in solvent after exposure, because cross-linking of polymer chains occurs. In this case, if a region of a negative resist-covered film is exposed, only the exposed region will be covered by resist after development. Subsequently, the resist-free parts of the substrate can be selectively coated with metal – as it is shown in Fig. 3c – or etched before removal of the unexposed resist, leaving the desired patterns at the surface (Fig. 3d).

As early as 1965, sub-100-nm resolution was reported [42] and was optimized with improved electron optics and with the use of different strategies, such as membrane substrate, reactive ion etching, and lift-off process [43–52]. Compared with photolithography, a higher lateral resolution is achieved because the beam of electrons can be focused to produce probe size as small as 1 nm and electrons do not suffer from optical thin-film interference. However, several parameters other than the size of the beam determine the ultimate resolution of the process. Particularly, extent of the exposed volume in a layer of resist depends strongly on the scattering events because electrons are scattering not only within the resist but also within the substrate beneath the resist layer (see Fig. 4a). These scattered electrons slightly expose the resist in a halo around each of the exposed features. A dense array of features may contain enough scattered electrons to seriously overexpose the resist. The main characteristics of this so-called proximity effect, the range of backscattered exposure, and its relative intensity have been studied experimentally [53, 54] and by Monte Carlo calculations [55]. Chang proposed a two-Gaussian model to describe resist exposure from a point source [56]. A higher-intensity Gaussian width

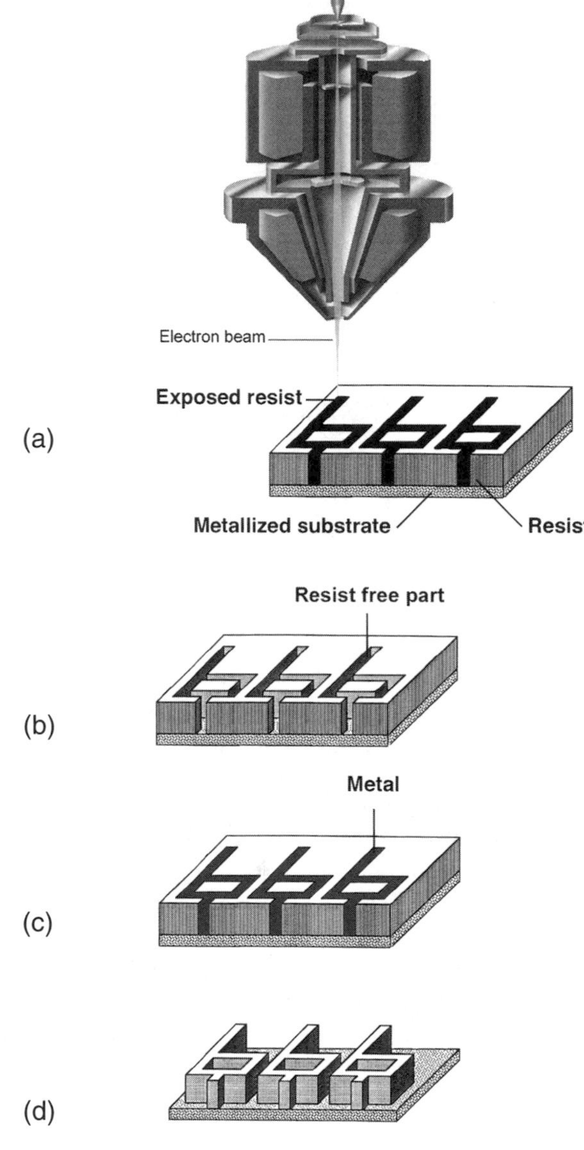

Fig. 3 Principle of the nanostructuring of surfaces using EBL technique and electroplating. E-beam exposure of a positive resist (**a**), removal of the exposed resist (**b**), filling of the resist-free locations with metal using electroplating technique (**c**), and removal of the unexposed resist leaving high-aspect-ratio metallic nanostructures at the surface (**d**)

describes the incident broadening in the resist and a less-intense broader Gaussian describes the distribution of the resist exposure due to the backscattered electrons (see Fig. 4b). The narrower distribution describes resolution and minimum feature size, while the backscattered electrons cause proximity effects. At present, these undesired effects have been minimized by using very high- and low-accelerating voltage or by correction methods. The resolution of EBL depends also on the

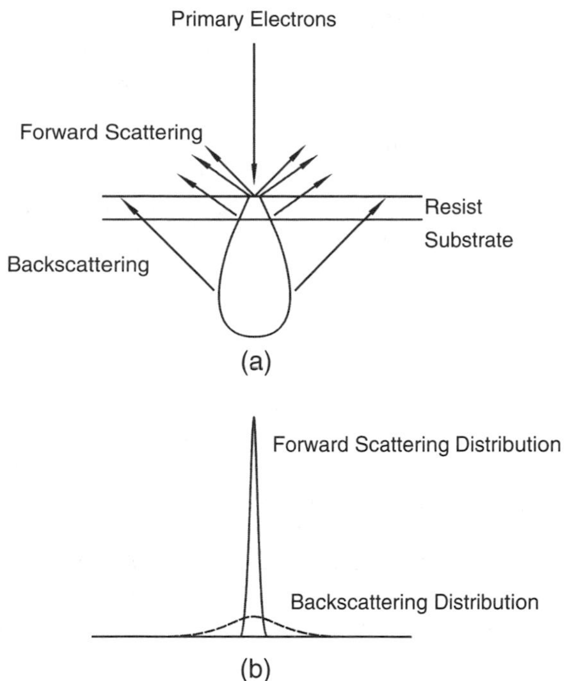

Fig. 4 Electron scattering events during exposure (**a**) and the corresponding exposure distributions (**b**)

chemical nature of the resist. PMMA was the first e-beam resist reported [57] and inspired the development of a various high-energy radiation resists. It is still commonly used because of its high resolution, but its low sensitivity and poor etch resistance under plasma conditions have forced the development of higher speed and resolution resists [58, 59]. Recently, new classes of resists, including organic self-assembled monolayers (SAMs), chemically amplified resists, and inorganic resists, have been developed to fabricate structures below 100 nm [60–65]. Thus, the electron scattering as well as the thickness and the nature of the resist determine the ultimate resolution (see, e.g., Ref. [66] for more details). Except for more recent reports of atomic resolution with a proximal probe (see, e.g., [67]), the resolution of EBL has been unsurpassed by any other form of lithography. However, the technique is far too slow for a large production and up to now is mainly used to produce masks, rapid prototyping of ICs, and specific small-volume production [63, 68].

3.1.2 Nanostructuring by EBL and Electroplating

Fabrication of metallic nanostructures has been widely explored using conventional EBL and lift-off techniques. However, this top-down approach cannot be used for the fabrication of high-aspect-ratio vertical structures, since gradual accumulation of materials at the top of the resist blocks and closes the opening of the structures during the evaporation of metal. Electroplating of metals into the holes formed in PMMA resist is a convenient alternative to circumvent this problem [48]. For example, the fabrication of dense, ultrasmall magnetic arrays by filling nanoholes with electroplated Ni has been reported [69]. Electrodeposition of Ni was performed from Ni sulfonate electrolyte by potentiostatic experiments. Depending on the electroplating time, high-aspect-ratio Ni pillars or mushroom-like structures were obtained. Figure 5 shows a SEM micrograph of mushroom-shaped micromagnetic arrays grown by overplating after removal of PMMA by oxygen plasma etching. This bottom-up approach has been also used to produce arrays of 30-nm magnets with 80-nm pitch (distance between two magnets). From the viewpoint of a practical use, this packing density translates into an equivalent memory storage capacity of over Gbit/in^2. The density of the magnetic arrays can be further increased by optimizing the EBL parameters. Under optimal conditions, the formation of 12-nm holes in 100-nm thick PMMA resist with spacing of 45 nm have been reported showing the high resolution achieved by EBL and electrodeposition of metals [69]. Combining EBL and electrochemical deposition has been also used to fabricate CdSe pillars, with diameters in the range from 180 nm to 1 μm and a fixed height of 400 nm. Depending on the size of such structures, enhanced photoluminescence (PL) has been reported, thus opening potential application for cavity resonance in the submicrometer scale [70]. The use of this template strategy has been also exploited for the fabrication of nanometer-sized metallic wires, superconducting nanowires, and magnetic multilayers [71].

3.1.3 Nanostructuring by EBL and Electrochemical Etching

A similar approach has been investigated for the formation of monocrystalline pore arrays in anodic alumina [72]. For this, a hexagonal patterns was written on the PMMA resist hole by hole with EBL. After removal of the irradiated parts, the pattern was transferred to the Al substrate by using a wet-chemical etch in phos-

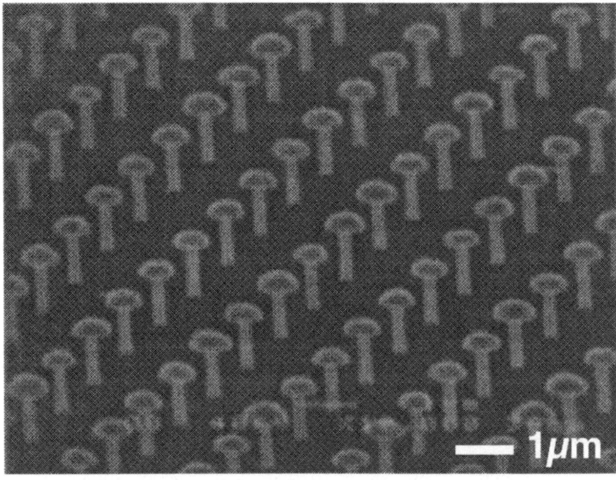

Fig. 5 SEM micrograph of an overplated micromagnet array showing the mushroom-shaped characteristic of isotropic metal deposition

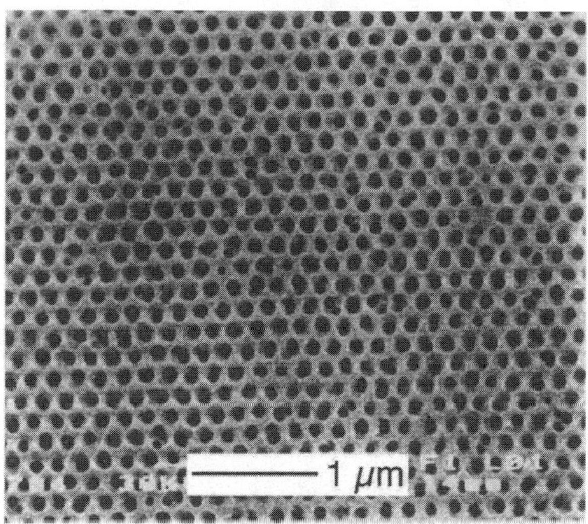

Fig. 6 Monocrystalline-pore arrays in ordered porous alumina prepared with prepattern-guided anodization. The prepattern with a pitch of 200 nm was induced by using EBL. (Reproduced by permission of the Electrochemical society, Inc.)

phoric and nitric acids. Then, PMMA was removed and the Al substrate was finally anodized in an oxalic acid solution under constant voltage. When the pore distance, which depends on the anodic voltage, matches the prepattern pitches well, the pattern can act as initiation point and guide the pore growth in the anodic film. Figure 6 shows a SEM micrograph of an ordered-pore array prepared with a 200-nm interpore distance. In this case, the anodic voltage was adjusted to 85 V based on the relationship between the pore distance and the anodic voltage. Under these conditions, very high aspect ratios (around 500) could be achieved. EBL has been also utilized for the direct microstructuring of porous silicon without the use of any sensitive resists [73]. It has been reported that the direct e-beam irradiation of electrochemically etched silicon can locally passivate the surface. This e-beam-induced enhancement of reactivity can subsequently be exploited for the selective dissolution of the exposed areas (see Fig. 7a). The ability to selectively etch silicon has been also investigated by cross-linked PMMA on porous silicon [74]. Although PMMA is usually known as a positive resist, the use of electron doses higher than 12 mC/cm^2 causes cross-links leading to a negative-tone resist behavior. Thus, this feature can be exploited to perform selective electrochemical processes (anodization) and fabricate microtips and nanomolds as shown in Fig. 7b.

3.2 Electrochemical Micropatterning Using E-beam Modification of SAMs

3.2.1 Self-Assembled Monolayers (SAMs)

SAMs are composed of organic molecules which consist of three building blocks: a head group that binds strongly to a substrate, a tail group that constitutes the

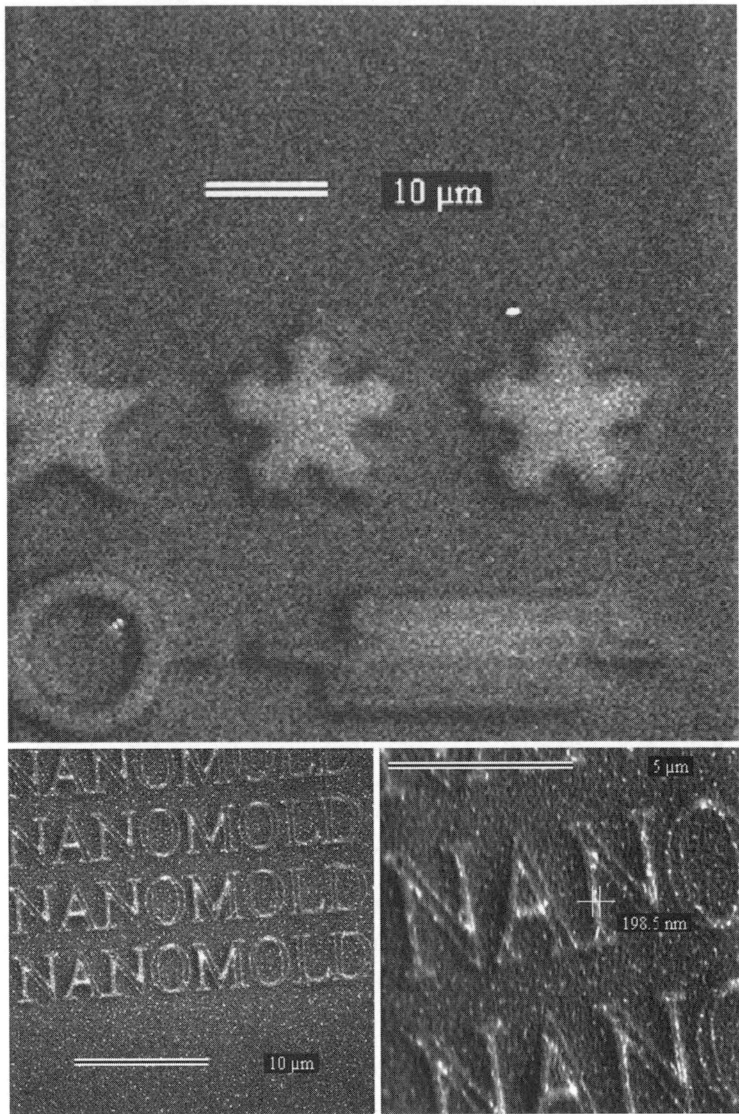

Fig. 7 SEM images of micropatterned porous silicon surfaces obtained by selective dissolution of e-beam irradiated areas (**a**) and by electrochemical etching through cross-linked PMMA mask (**b**)

outer surface of the film, and a spacer that connects head and tail. These molecules can be covalently anchored to different surfaces, such as metals, semiconductors, and oxides, with a typical thickness of 1–2 nm and an intermolecular spacing of 1–0.5 nm [75]. The interest in the general area of self-assembly, and specifically in SAMs, is mainly driven by their remarkable physical and chemical properties. In contrast to ultrathin films formed by conventional techniques, such as molecular-beam epitaxy (MBE) and chemical vapor deposition (CVD), SAMs are much more dense, homogeneous, extremely thin, highly ordered, oriented, and stable even when they are subjected to a harsh environment. For a larger overview on SAMs, see, for example, Ref. [76].

With the development of microelectronic, chemical modification of semiconductor materials, and especially Si, has been extensively studied (see, for instance, Refs. [77, 78]). Functionalization of Si using SAMs consists of replacing silicon–hydrogen (Si–H_x) bonds by more robust silicon–carbon (Si–C) bonds [79]. The formation of this covalent bonding can be achieved, for example, by reaction of unsaturated, simple, and functional alkenes, with the Si surface using several procedures, including thermal activation [80] or UV light illumination [78]. Organic monolayers covalently attached to silicon surfaces display various advantages compared to SAMs of alkanethiols on gold surface and to octadecyltrichlorisilane (OTS) monolayers on SiO_2: high resistance of the monolayer in different organic and aqueous solutions (Si–C bond is very stable), good chemical passivation of the surface, existence of a wide range of chemical functionalities compatible with the Si–H bonds terminating the silicon surface. Furthermore, the functionalization methods are easy to carry out and highly reproducible. SAMs have many potential applications for the semiconductor technology, including surface passivation, electrochemical interfaces, microsensors, etc. (see, e.g., Refs. [81–83]). In addition, highly defined functionalization of surfaces with specific interactions can be produced with fine chemical control that make SAMs ideal candidates for biochemical and biomedical applications such as biosensing [84].

Recently, patterning of SAMs has been widely investigated because the possibility to achieve selective immobilization of biological molecules on solid substrate and then to manufacture microstructures carrying specific recognition have multiple issues in diagnostics, artificial biomolecular networks, and biologically integrated systems (lab-on-a-chip). In this context, many techniques have been employed to fabricate laterally patterned SAMs, including conventional photolithography [85], microcontact printing [6], scanning probe lithography [86], ion- beam lithography [87]. Furthermore, it has been reported that many SAMs are highly e-beam sensitive [61, 88] and can be ideal candidates for the development of a new class of positive- and negative-tone e-beam resists in a wet-chemical process [60]. As SAMs are extremely thin and composed of very small subunits, the forward beam scattering in the resist is eliminated, and the electrons inelastically generated in the substrate can leave the layer before undergoing substantial lateral travel. Compared with the use of high-molecular-weight resists, such as PMMA, a significant improvement of the lateral resolution can be achieved with SAMs.

3.2.2 Selective Electrodeposition Using E-beam-induced Modification of SAMs

The selective electrochemical deposition of metals can be achieved by the e-beam writing of surfaces covered with organic SAMs (see, e.g., [89]). Figure 8 shows

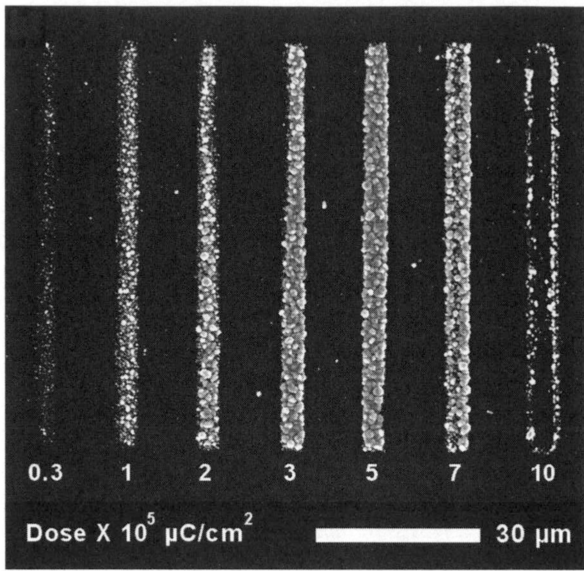

Fig. 8 SEM image of an e-beam-patterned silicon surface covered with an organic layer (1-decene) after electroplating of Cu

a SEM image of a silicon surface covered with an organic layer (1-decene) with four e-beam-modified patterns onto which copper has been electrodeposited. It has been reported that the selective copper deposition depends strongly on the electrochemical conditions as well as the electron dose. For relatively low electron doses, SAMs act as a positive-tone resist enhancing copper electrodeposition, whereas for relatively high electron doses, SAMs act a negative resist completely blocking the electrochemical process. According to surface-analysis experiments, this effect can be explained by the fact that partial removal of carbon chains occurs with low-electron doses, whereas at sufficiently high electron doses cross-linking of the molecules becomes predominant. Thus, e-beam-modified organic monolayers on Si surfaces combined with electrochemical processes is a viable alternative to achieve the fabrication of structures with features in the micrometer range.

3.3 Micro- and Nanostructuring by EBICD and Electrochemical Reactions

3.3.1 E-beam-Induced Deposition (EBID) Technique

EBID is a single-step and direct-writing technique using the beam of electrons to grow three-dimensional nanostructures. Due to the combination of high-resolution and 3D structure formation, EBID is highly appreciated in the field of exploratory nanodevice fabrication and has recently moved toward various applications for

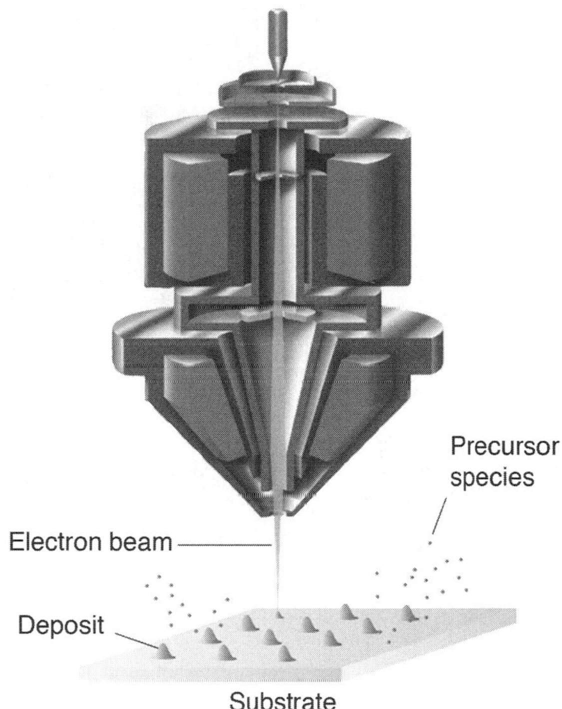

Fig. 9 Principle of EBID. The beam of electron cracks the precursor species introduced in the chamber of the SEM leading to the formation of a deposit at the point of impact of the beam

production of nanowires [90, 91], X-ray mask repair [92], photonic crystals [93], and a wide range of devices [94–98].

The principle of EBID is based on the fact that the beam of electrons decomposes adsorbed precursor molecules present in the chamber of the e-beam instrument resulting in a deposit at the point of impact of the beam as depicted in Fig. 9. The chemical composition of such deposits depends strongly on the nature of the precursor molecules introduced into the chamber of the e-beam instrument. When organometallic precursor species are injected, the e-beam-deposited materials show nanocomposite structures with metal nanocrystals of variable size embedded in an amorphous carbonaceous matrix [50, 99–102]. The resolution of the technique as well as the growth rate of such nanomaterials that is described by several models [103–107] are dependent on the vapor pressure in the chamber, the e-beam parameters, the exposure time, and the nature of the substrate (see, e.g., [92, 108–110]).

3.3.2 EBICD Technique

When decomposed precursor species are simply the residual hydrocarbon molecules issued from the pump oil, contamination writing, which consists of amorphous carbonaceous deposit is grown at the e-beam-treated locations [111–115].

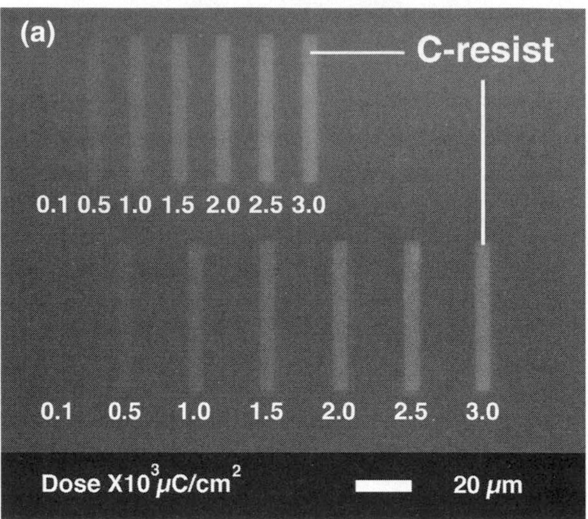

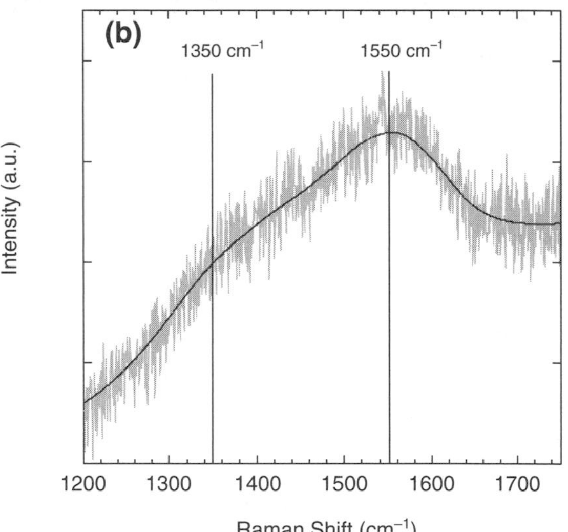

Fig. 10 SEM image of two series of seven C-patterns deposited with an increasing electron dose from the left to the right (**a**). Reprinted from Surface Science, T. Djenizian, L. Santinacci, H. Hildebrand, and P. Schmuki, vol. 524, "Electron beam induced carbon deposition used as a negative resist for selective porous silicon formation" p. 40, 2003, with permission from Elsevier. Raman spectrum obtained from a carbonaceous (**b**). (Reproduced by permission of the Electrochemical society, Inc.)

Figure 10 a shows a SEM image of ultrathin rectangular C-patterns, which have been deposited on a silicon surface using different electron doses. Figure 10b shows the Raman spectrum obtained from such C-deposited matter. In the region between 1200 cm^{-1} and 1800 cm^{-1} a broad peak centered around 1550 cm^{-1} (G-band) and the presence of the shoulder at around 1350 cm^{-1} (D-band) typical for amorphous carbonaceous films are observed [116, 117]. Examination of this spectrum confirmed that the sp^3- and sp^2-bonded carbon is incorporated in the e-beam-deposited material.

3.3.3 Contamination Lithography

As early as 1964, the fabrication of 50-nm lines ion milled into metal films using contamination resist has been reported [118]. Contamination lithography followed by ion milling of the pattern into the underlying metal has been developed [45] and exploited to produce the first functioning Aharonov–Bohm device [119]. Later, the use of carbonaceous mask has also been demonstrated for the selective growth of II–VI semiconductors by metalorganic-beam epitaxy [120]. Recently, it has been reported that e-beam-induced carbonaceous deposit can be used as a mask for electrochemical reactions, that is, it has been demonstrated that carbonaceous deposit in the nanometer-range thickness can block completely and selectively a wide range of electrochemical reactions [121]. According to the literature and surface-analysis measurements described above, e-beam-induced carbonaccous deposit is amorphous and consists mainly of a high amount of sp^3-bonded carbon [122–124] leading to chemical and physical properties very close to that observed for the diamond. Therefore, the negative-resist effect can be explained by the fact that this so-called diamond-like carbon (DLC) material is chemically inert and behaves as an excellent insulator hampering completely subsequent electrochemical reactions.

3.3.4 Micro- and Nanostructuring by EBICD Technique and Electrodeposition

Carbonaceous Masking of Electroplated Metals

Recently, it has been reported that e-beam-induced carbonaceous materials in the nanometer-range thickness can be used to block the electroplating of Au on semiconductor surfaces [125], as schematically depicted in Fig. 11. The high degree of selectivity that can be achieved by this technique is also confirmed for the selective electrodeposition of Cu.

Figure 12 shows a SEM image of n-type Si sample carrying a carbonaceous micropattern "LKO" after a cathodic potentiodynamic experiment. Clearly, the dark carbonaceous LKO micropattern surrounded by Cu crystallites corresponds to the masked area. Within this pattern, absolutely no deposited Cu particles could be detected even for very high-cathodic potentials. This result suggests the ability to exploit EBICD for the nanomasking of a wide range of electroplated metals on semiconductor surfaces.

Resolution of the Process

The fabrication of metallic nanostructures depends on the resolution achieved at the edge of the C-lines, which is governed by the morphology, size, and number of electrodeposited crystallites. Therefore, the control of electrochemical factors is a crucial step for optimizing the lateral resolution of the process. In the case of electroplating of metals onto semiconductor surfaces, the study of transients indicates that nucleation of 3D hemispherical clusters followed by diffusion-limited growth

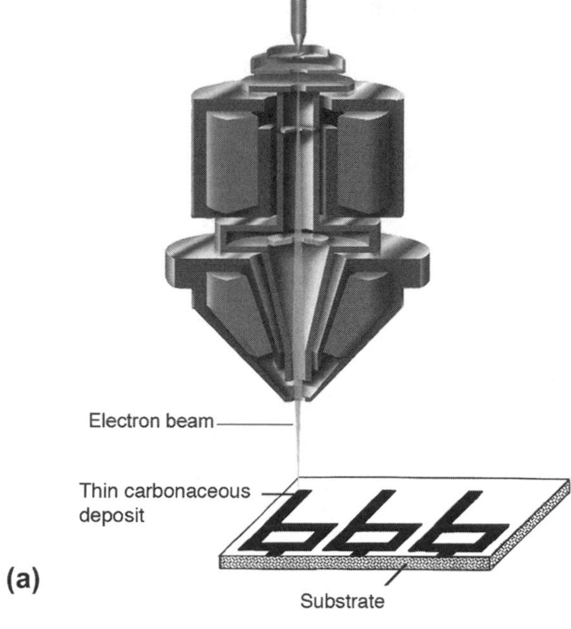

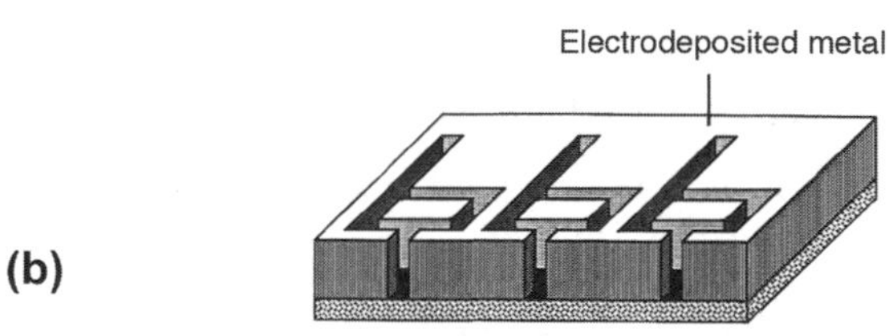

Fig. 11 Principle of the nanostructuring of surfaces using EBICD technique and electroplating. Ultrathin carbonaceous matter is deposited by direct e-beam exposure of the surface (**a**). Selective electrodeposition of metal at nonirradiated locations (**b**)

Fig. 12 Nanomasking effect of e-beam-induced carbonaceous deposits for electroplating of Cu. SEM image of n-type Si sample carrying a carbonaceous pattern "LKO" after electrochemical deposition of Cu performed by cathodic potentiodynamic experiment

occurs [126–128]. The 3D-island-growth mechanism or Volmer–Weber mechanism has already been observed for several systems and can be attributed to the relatively weak interaction energy between semiconductors and metals [129–134]. It has been observed that the density of nuclei as well as the size of the globular features size depend strongly on the applied potential, that is, the higher cathodic applied potential, the larger number of small crystallites. At relatively low cathodic potential, only few big crystallites are deposited onto Si. Using higher cathodic potential results in deposits entirely covering the Si surface with smaller mean size of particles. As the lateral resolution of the process at the edge of the C-deposit depends strongly on the formation of a smooth and continuous film, the creation of a large number of small crystallites is required because the coalescence of islands in an earlier growth stage leads to homogeneous layers. Therefore, deposits are preferentially formed at high cathodic potential.

Under optimized electrochemical conditions, it has been also shown that the resolution depends on the electron dose and the e-beam energy used for depositing the C-mask. Carbonaceous patterns written with a relatively high-beam energy (20 keV) revealed the presence of a C-background in the immediate vicinity of the predefined patterns. It is assumed that the large disk from which backscattered electrons re-emerge from the sample have enough energy to decompose organic molecules adsorbed at the Si surface. Such C-fog surrounding the patterns can block partially electrodeposition of metals and, consequently, a bad resolution is achieved at the edges of the C-lines. This effect is clearly apparent in Fig. 13a and b that show two Si samples carrying arrays of ten C-lines, equidistantly spaced (1.5 μm) and deposited using a low- and high-beam energy with increasing electron doses: $n{\cdot}\mathrm{C/cm}^2$ with $0.1 \leq n \leq 1$ after electrodeposition of Au [135].

Deposition of C-lines performed with a low-beam energy (5 keV) shows that independently of the electron dose, Au deposit between the C-lines is homogeneous and smooth leading to an excellent lateral resolution (Fig. 13a), whereas using a high-beam energy (20 keV) leads to coarse Au deposit between the C-patterns when the electron dose increases (Fig. 13b). Therefore, decreasing the spatial distribution of the backscattered electrons by decreasing the beam energy is a key step to improve the resolution of the process. In this case, Au nanowires and Au clusters in the sub-50-nm range were successfully fabricated under optimized electrochemical conditions as shown in Fig. 14a and b [135]. The linewidth of the nanowires as well as the size of the dots decrease by decreasing accurately the spacing between the carbonaceous lines. These results show clearly that combine contamination writing with electroplating of metals is a viable way to achieve the fabrication of metallic nanostructures on semiconductor surfaces in the sub-100-nm range.

3.3.5 Microstructuring by EBICD Technique and Electrochemical Etching

The masking effect of e-beam-deposited DLC has also been demonstrated for the electrochemical etching of materials in extremely aggressive environment. Particularly, carbonaceous masking has been investigated to block the porosification of Si

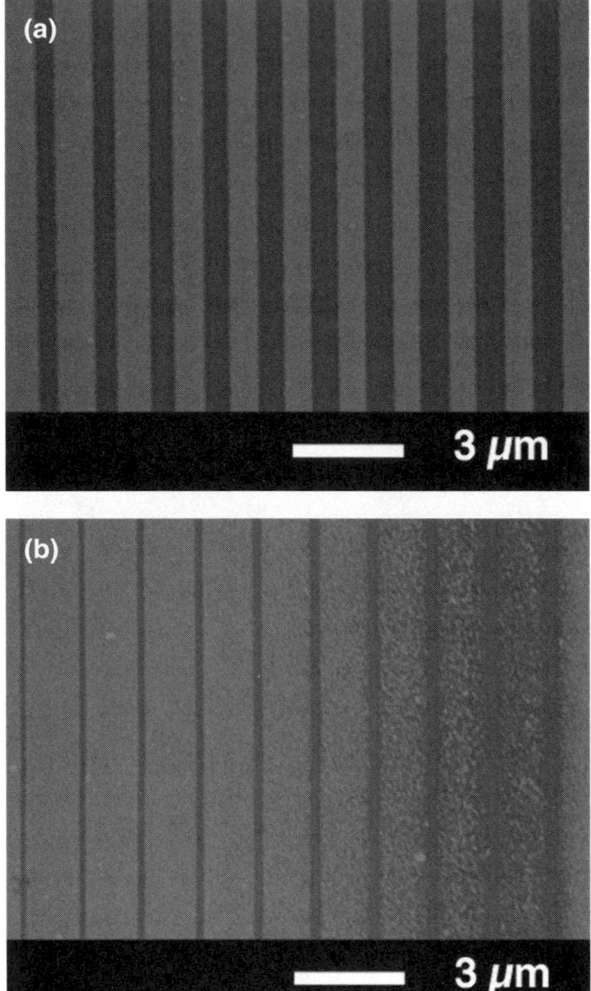

Fig. 13 Influence of the beam energy on the lateral resolution of the process. SEM images of Au electroplated on Si samples carrying arrays of ten C-lines produced with a relatively low 5 kV (**a**) and high 20 kV (**b**) accelerating voltage. (Reproduced by permission of the Electrochemical society, Inc.)

performed by galvanostatic experiments [136]. Figure 15a shows an optical image of a C-patterned p-Si sample after anodization in an HF-containing electrolyte. It is clearly apparent that the surface is modified, except at the locations carrying the C-patterns if a sufficient electron dose is used. Indeed, C-patterns deposited using electron doses higher than 1.0×10^3 μC/cm^2 do not suffer from etching. The unexposed area (surrounding the patterns) exhibits interference colors ranging from red to brown, which are typical of thin porous silicon layers. Within the rectangles the surface is intact; this region does not show signs of etching. Therefore, C-deposits produced with a dose higher than 1.0×10^3 μC/cm^2 can act as a mask for the electrochemical etching in HF solution, and thus constitute a negative resist blocking

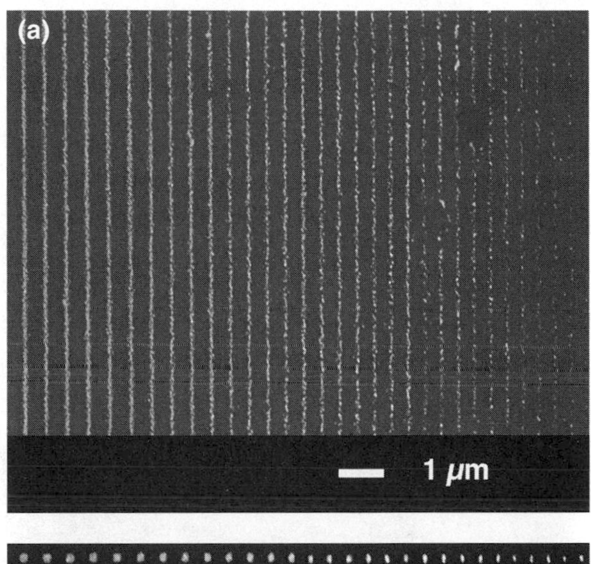

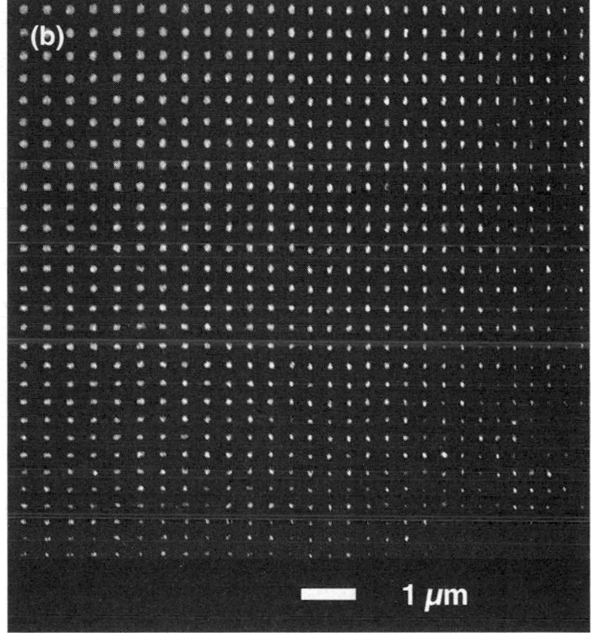

Fig. 14 SEM micrographs of Au nanowires (**a**) and Au clusters showing sizes in the sub-50-nm range (**b**). (Reproduced by permission of the Electrochemical society, Inc.)

completely and selectively the pore formation. Furthermore, selectivity in terms of optical properties of such micropatterned silicon surface has been corroborated by PL measurements. It has been reported that the PL intensity of locations protected by the C-mask deposited for electron doses higher than 1.0×10^3 μC/cm^2 is zero, and, thus, no light emitting structure is present within the C-patterned areas.

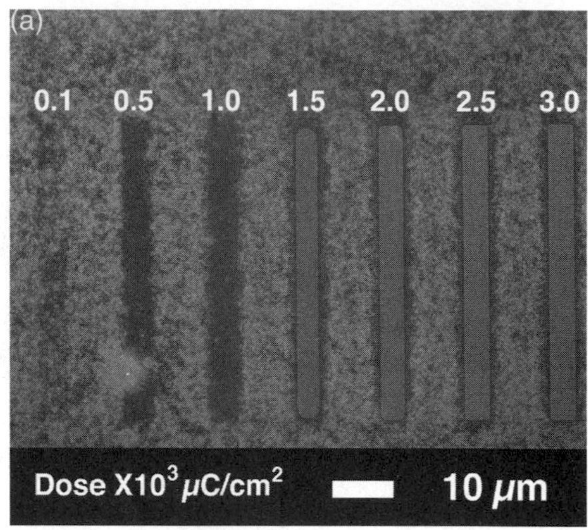

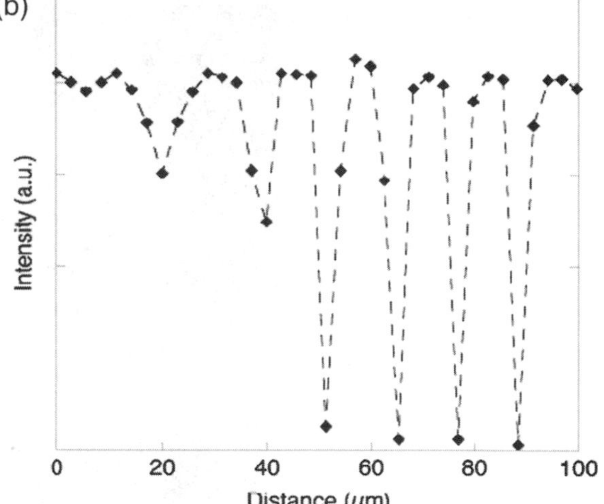

Fig. 15 SEM image of an etched Si surface carrying seven C-patterns (**a**). Luminescence scan performed across the C-patterns (**b**). Reprinted from Surface Science, T. Djenizian, L. Santinacci, H. Hildebrand, and P. Schmuki, vol. 524, "Electron beam induced carbon deposition used as a negative resist for selective porous silicon formation" p. 40, 2003, with permission from Elsevier

In contrast, the intensity between these C-deposits reveals a high value corresponding to the maximum PL response (see Fig. 15b). These assessments show the feasibility to also exploit such carbonaceous deposit in extremely aggressive chemical environment for the porous Si micropatterning and confirm that a high degree of selectivity can also be achieved in view of optical properties.

This masking effect has also been exploited to block corrosion of iron. Figure 16 shows a SEM image of a carbonaceous micropatterned iron sample after a corrosion test. Clearly, this chemical treatment leads to the selective dissolution of the iron surface resolving the grain structure of the substrate except at the e-beam treated locations. Furthermore, the high degree of protectiveness of EBICD has been

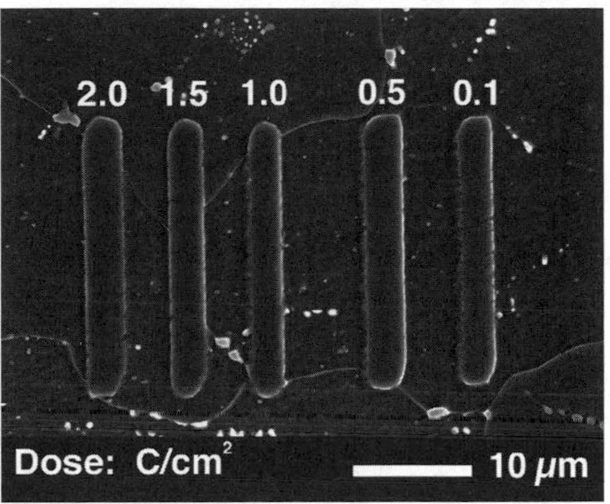

Fig. 16 SEM image showing a patterned iron sample after a corrosion test. (Reproduced by permission of the Electrochemical society, Inc.)

also demonstrated for electrochemical corrosion experiments (see, e.g., Ref. [123]). Thus, this alternative e-beam approach can be used to block a wide range of electrochemical reactions to achieve the patterning of different substrates in the nanometer range.

4 Material Processing by FIB

Focused ion beam (FIB) systems enable the generation of ion beams with diameters of a few nanometers. Similar to SEM, focused ion beam systems enable high-resolution imaging via the detection of secondary electrons or ions generated by the interaction of the energetic ions with target atoms. In comparison to SEMs, the application of heavy ions additionally enables specific material processing with high accuracy. Examples are ion implantation, physical sputtering, ion-beam-induced etching or deposition. Initially intended for maskless implantation of ions, currently focused ion beam systems are mainly used for structuring of sample surfaces by material removal or deposition. Material processing by FIBs is done in a direct writing mode where the ion beam is digitally scanned over the surface of the specimen. In contrast to other structuring techniques mainly based on optical lithography and, therefore, requiring extensive masking, this direct-writing mode makes material processing by FIBs a very flexible and versatile technique successfully applied in many different working areas.

The driving force in the development of focused ion beam systems and their applications is microelectronics. For example, defective masks made for optical lithography can be repaired using FIBs. Excessive material, forming opaque defects, can be removed and missing material can locally be replaced by material deposition [137, 138]. Material processing by focused ion beams also enables the "repair" of integrated circuits. This "rewiring" is mainly used during the development of new

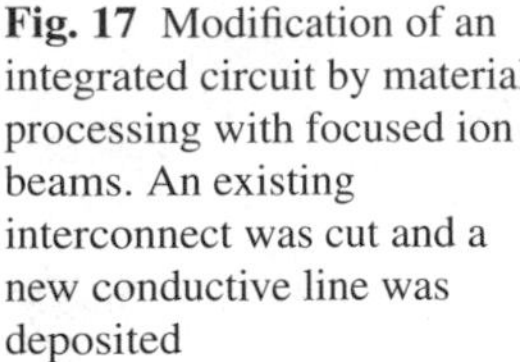

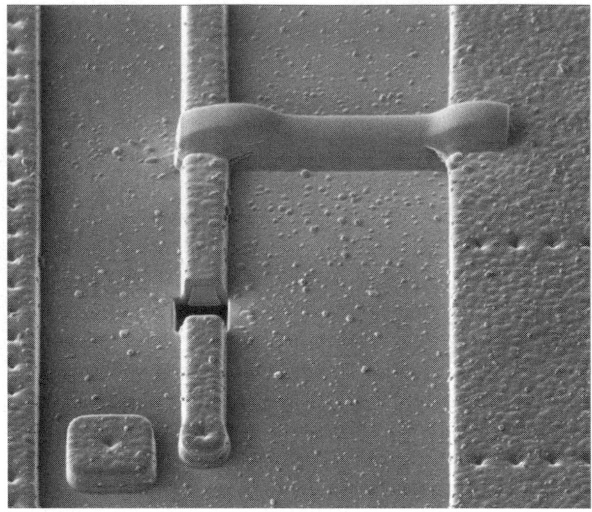

Fig. 17 Modification of an integrated circuit by material processing with focused ion beams. An existing interconnect was cut and a new conductive line was deposited

devices and enables the modification of an integrated circuit by cutting existing interconnects or by creating new conductive lines by material deposition (Fig. 17) [139].

This fast and flexible method replaces extensive and time-consuming fabrication of new prototypes for testing and allows to shorten development time and to decrease costs [140, 141].

In the area of failure analysis, material removal by FIBs is used to prepare cross-sections at definite positions of an integrated circuit and facilitates the optimization of semiconductor processing [142–144]. The capability of high-resolution imaging enables the precise alignment of the cross-section with respect to a special site of the specimen. It also enables the investigation of the cross-section within the same working process. The application of FIBs for the preparation of samples for transmission electron microscopy (TEM) also evolved from microelectronics where TEM is used for the investigation of defects and structures with dimension in the range of nanometers [145–147]. In contrast to conventional preparation techniques, the application of FIBs allows fast fabrication of TEM samples at precisely defined sites without destruction of the specimen as a whole. The extraction of the tiny TEM lamellae from the sample by modern, so-called "lift-out" techniques [148], enables the preparation of TEM lamellae from materials which cannot be processed conventionally. The application of FIBs for the preparation of TEM samples is not restricted to the characterization of semiconductor devices and is nowadays a preparation technique, which is used in many other areas by default.

The ability to fabricate structures with different size and shape in a direct writing mode has created new possibilities and applications in the area of micro- and nanotechnology. Examples are the fabrication of microtools for bio- and medical applications [149, 150] or functionalized probes for scanning probe microscopy (SPM) (Fig. 18) [151, 152].

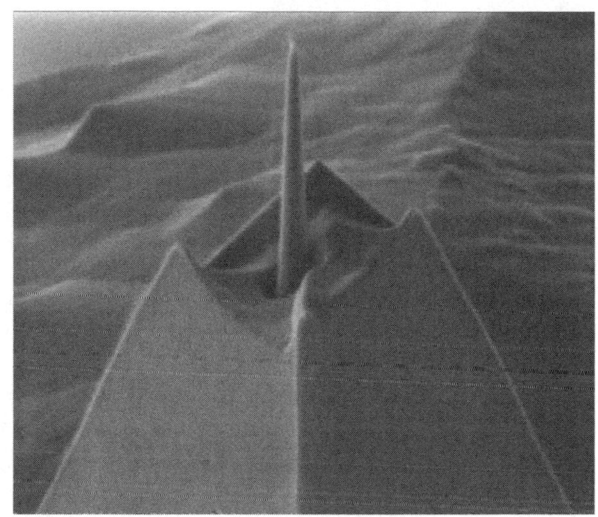

Fig. 18 SEM image of a silicon probe after FIB processing. Physical sputtering was applied to remove material and to create a tip with increased aspect ratio

4.1 Generation of Focused Ion Beams

The first ion-beam columns, dedicated for the generation of focused beams to be applied for material processing, were based on plasma ion sources. By the use of apertures and electrostatic lenses, beam diameters of less than 1 μm could be achieved [153, 154]. Further reduction of beam diameter was possible by the use of ion sources, where the generation of ions is based on the ionization of atoms within an electric field. In comparison to plasma sources, the energy distribution of ions emitted from these so-called field emission sources is clearly reduced. As a result, the beam diameter also decreases as the influence of the chromatic aberration of the ion optics on beam diameter is reduced. The first ion sources based on field ionization were realized by gas emission sources [155]. Because of some drawbacks, for example, low ion currents and technical complexity, gas emission sources nowadays only play a minor role in the generation of FIBs. Today, mainly liquid-metal ion sources are used for the generation of FIBs. This type of ion source was developed in the 1970s and comprises a needle-type emitter, which is wetted by metals or alloys [156]. By heating the emitter above the melting temperature of the metal and simultaneously applying a sufficiently high electrical field between emitter and anode, a variety of elements, for example, Al, As, Ga, Au, B, In, and Li ions can be generated [157]. The element commonly used for the generation of FIBs in commercially available FIB systems is Ga. The requirements that have to be met to generate ions of a definite element with a liquid-metal ion source are low-vapor pressure and high-surface tension of the metal at melting temperature. Additionally, the emitter has to be wetted by the metal, and no chemical reaction should occur between metal and emitter material. Ions from elements with high melting point can only by generated by liquid-metal ion sources, if the element is available in an alloy with reduced melting point in comparison to the pure element. For sources using alloys, additional mass separation has to be integrated into the ion column, as different ion species are emitted from the ion sources simultaneously.

4.2 Ion Optics

The ion column of a FIB system comprises the ion source and the ion optics. The different components of the ion optics, such as lenses, deflection units, and apertures, are dedicated for the controlled deflection and focusing of the ions after emission from the liquid-metal ion source. Figure 19 schematically shows the main parts of a FIB system.

The liquid-metal ion source is surrounded by a Wehnelt cylinder. This so-called suppressor focuses ions radially emitted from the ion source within a point lying between suppressor and extractor electrode. Between the extractor and the first lens, an electric field is applied, and the ions are accelerated. Because of the high mass of the ions, mainly electrostatic lenses are used for the generation of FIBs. The first lens is used to align the particles relatively to a beam-defining aperture. The diameter of this aperture determines the number of transmitted ions focused on the

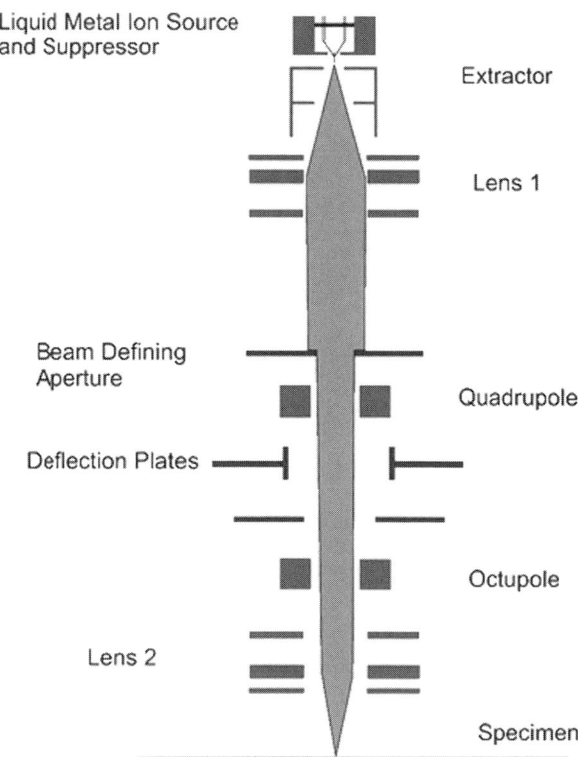

Fig. 19 Schematic showing the main parts of the ion optics of a focused ion beam system

surface of the specimen by the second lens and, therefore, controls the current of the ion beam. Additional electrical components, such as quadrupoles, octupoles, and deflection units, are applied to align the ion beam with respect to the optical axis to make corrections of astigmatism and to scan the ion beam over the sample surface. Currently, commercially available FIB systems mainly use ions with a kinetic energy between 30 keV and 100 keV. Ion-beam currents range from 1 pA to several tens of nA. Presently, the minimum achievable beam diameter (full-width half maximum) is between 5 nm and 7 nm at lowest ion-beam current.

4.3 Material Processing by Focused Ion Beams

Main areas of application of FIBs in the field of material processing are ion implantation, material removal, and material deposition. In addition to physical sputtering, gas-enhanced etching is also used for material removal by ion beams. The combination of energetic ions and etchants, such as iodine-, chlorine-, or fluorine-containing compounds, increases material-removal rates and results in high selectivity. Material deposition is based on the application of dedicated chemical compounds, so-called precursors, which are delivered to the sample surface in a gaseous phase. Due to the co-action of the impinging ions and the precursor molecules adsorbed at the surface of the specimen, different materials, for example, conducting or insulating layers, can be deposited [158].

4.4 Microstructuring by FIB and Electrochemical Reactions

FIB technology has been combined with several electrochemical processes to achieve the microstructuring of surfaces. For instance, it has been reported that aspect ratio of micropatterns can be drastically improved by using FIB etching in macroporous materials [159]. Compared with the aspect ratio obtained by FIB-etched patterns in bulk materials, which is lower than 10, it has been demonstrated that the use of porous silicon fabricated by photoelectrochemistry [160] can be used as a layered structure for the FIB in order to create holes with an overall aspect ratio of 50 (see Fig. 20). This approach has been exploited for the fabrication of 3D silicon photonic crystals and can be also used more generally for the micromachining at a submicrometer scale.

Micromachining can also be performed using the etch stoppers' property of ions-implanted substrates. Indeed, gallium- or boron-implanted areas can act as etch barriers for wet-chemical treatments making this approach useful for the microstructuring of silicon [161–163]. It has been reported that this selective etch behavior can be enhanced when photoelectrochemical etching is carried out on low-doses-implanted species [164]. The etch-barriers effect of implanted species has also been used to selectively etch III–V semiconductors, such as GaAs and InP [165, 166].

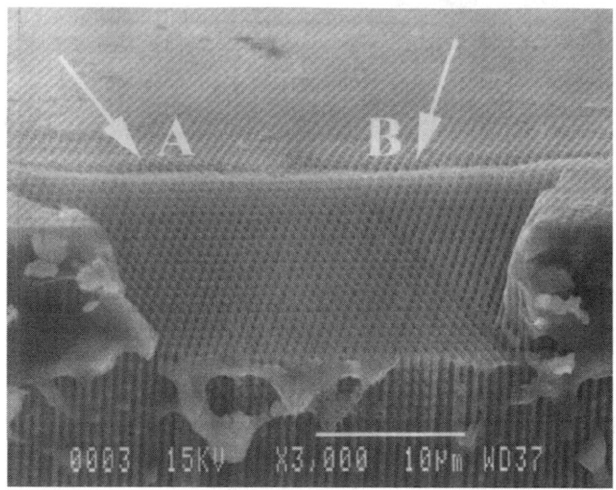

Fig. 20 Yablonovite-like crystal cross-section obtained by ion milling. The FIB etching directions are indicated by *arrows A and B*

The local electrochemical dissolution of implanted areas has also been investigated in order to establish depth profiles of implanted species. This so-called selective electrochemical delineation technique has been studied for the simultaneous electrochemical etching of As- and B-doped silicon substrate as shown in Fig. 21 [167]. This method is extremely sensitive and can be used for production and characterization of devices [168, 169].

Recently, new pathways have been explored to achieve selective electrochemical nanogrowth and nanostructuring of materials on locally sensitized single-crystal semiconductor surfaces. It has been reported how defects intentionally introduced in a surface by FIB bombardment can be used to selectively trigger electrochemical reactions [170–173]. The principle of the approach is based on semiconductor electrochemistry. When electrochemically biased, a semiconductor–electrolyte interface shows a similar electrical characteristic as a metal/semiconductor or p/n junction, that is, a current-passing (accumulation) state when forward biased, a blocking (depletion) state when reverse biased. In the blocking state, a specific "barrier breakdown" $U(\mathrm{Bd})$ potential exists that has been ascribed to the Schottky barrier breakdown of the junction. Due to the current increase at $U(\mathrm{Bd})$, electrochemical reactions are not hampered any longer by insufficient availability of charge carriers and thus can proceed at significant rate. The value of $U(\mathrm{Bd})$ is strongly affected by surface defects, that is, breakdown occurs for much lower applied voltages than for the intact surface. Thus, a processing window for local electrochemical reactions is present between the two threshold potentials $U(\mathrm{Bd})_{\mathrm{intact}}$ and $U(\mathrm{Bd})_{\mathrm{defects}}$ corresponding to the intact and defective surface, respectively. Therefore, by creating locally surface defects, the electrode can be "activated" for an electrochemical reaction only at these surface sites. Figure 22 schematically shows the electrochemical behavior of a p-type semiconductor substrate. The cathodic branch corresponds to the current blocking state, whereas in the anodic branch a passing state is observed. For n-type semiconductors, depletion and accumulation situations are reversed, that

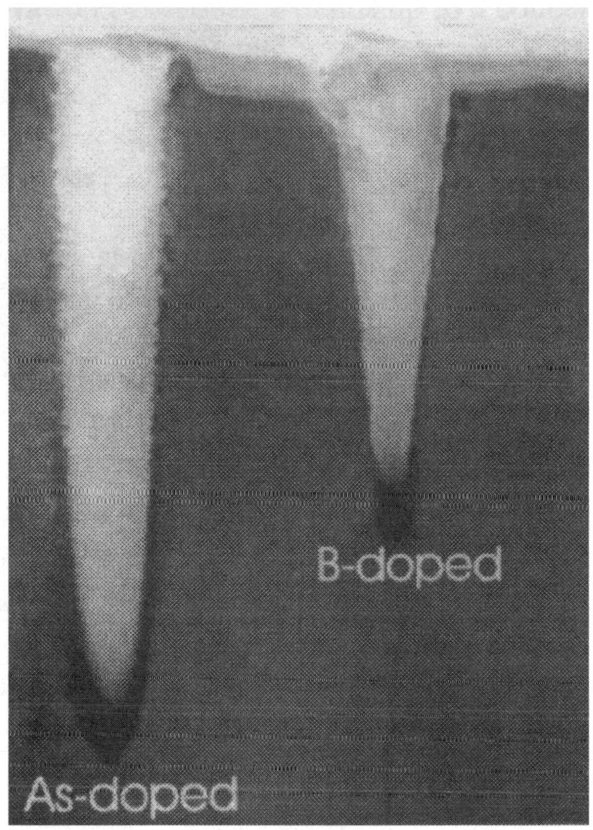

Fig. 21 TEM micrograph showing the simultaneous delineation of As- and B-doped regions after electrochemically etching

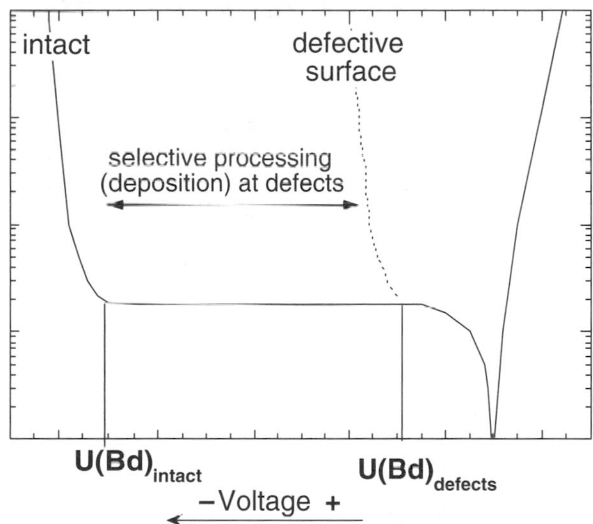

Fig. 22 Schematic current density versus voltage curve for intact and ion-implanted p-Si. Electrodeposition reactions can be initiated selectively between $U(\mathrm{Bd})_{\mathrm{intact}}$ and $U(\mathrm{Bd})_{\mathrm{defects}}$

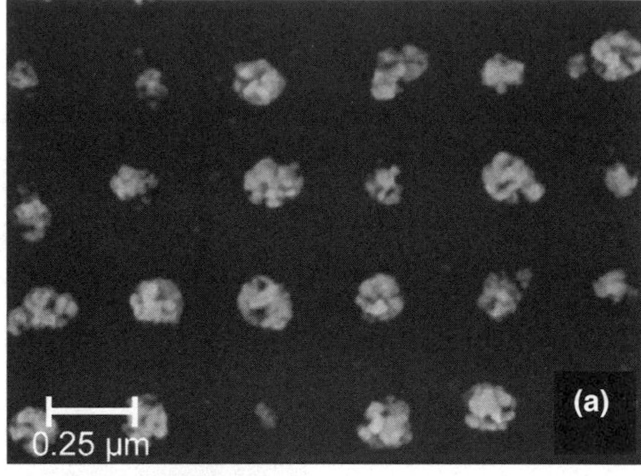

Fig. 23 Examples of selective electrodeposition of gold and electrochemical etching achieved at FIB-induced defects sites

is, current-blocking behavior is observed at anodic potentials and passing behavior at cathodic potentials. Figure 23a shows selective deposition of gold on p-type silicon surface achieved by FIB writing of defect patterns followed by electrochemical deposition of gold performed between $U(\mathrm{Bd})_{\mathrm{intact}}$ and $U(\mathrm{Bd})_{\mathrm{defects}}$. The same approach has been successfully used to form selective light-emitting porous silicon on n-type silicon surface (see Fig. 23b).

References

1. Takumi, U. Sheats, J.R. 1998, X-ray Lithography, J. R. Sheats and B. W. Smith (Ed.), Marcel Dekker, Inc., New York, 403–427.
2. Madou, M. 1997, Fundamentals of microfabrication, CRC Press, Boca Raton.
3. Stockman, L., Heyvaert, I., van Haesendonck, C., and Bruynseraede, Y. 1993, Appl. Phys. Lett., 62, 2935.
4. Matsumoto, K., Ishii, M., Segawa, K., Oka, Y., Vartanian, B.J., and Harris, J.S. 1996, Appl. Phys. Lett., 68, 34.

5. Campbell, P.M., Snow, E.S., and McMarr, P.J. 1995, Appl. Phys. Lett., 66, 1388.
6. Xia, Y. Whitesides, M.W. 1998, Angew. Chem. Int. Ed., 37, 550.
7. Schmuki, P., Maupai, S., Djenizian, T., Santinacci, L., Spiegel, A., and Schlierf, U. 2004, Techniques in Electrochemical Nanotechnology, H. S. Nalwa (Ed.), American Scientific Publishers, Stevenson Ranch, 393–410.
8. Schuster, R., Kirchner, V., Allongue, P., and Ertl, G. 2000, Science, 289, 98.
9. Allongue, P., Jiang, P., Kirchner, V., Trimmer, A.L., and Schuster, R. 2004, J. Phys. Chem. B, 108, 14434.
10. Santinacci, L., Djenizian, T., and Schmuki, P. 2001, Appl. Phys. Lett., 79, 1882.
11. Owen, G. Sheats, J.R. 1998, Electron beam lithography systems, J. R. Sheats and B. W. Smith (Ed.), Marcel Dekker, Inc., New York, 367–401.
12. Stewart, D.K. Casey, J.D.J. 1997, P. Rai-Choudry (Ed.), SPIE Optical Engineering Press, 153.
13. Townsend, P.D., Chandler, P.J., and Zhang, L. 1994, Optical effects of ion implantation, Cambridge University Press, Cambridge.
14. Ryssel, H. Ruge, I. 1986, Ion implantation, John Wiley & Sons
15. Hartley, N.E.W. 1980, Treatise of Materials Science and Tehnology, Academic Press
16. Dearnaley, G. 1982, J. Metals, 34, 18.
17. Straede, C.A. 1989, Wear, 130, 113.
18. Standley, R.D., Gibson, W.M., and Rodgers, J.W. 1972, Appl. Optics, 11, 1313.
19. Heidenreich, R.D. 1964, Fundamentals of Transmission Electron Microscopy, Interscience, New York.
20. Sigmund, P. 1972, Roumaine Phys., 17, 823.
21. Weller, R. 1995, Handbook of Modern ion-Beam Materilas Analysis, J. R. Tessmer & M. Nastasi (Ed.), Materials Research Society, Pittsburg, PA.
22. Nastasi, M.A., Mayer, J.W., and Hirvonen, J.K. 1996, Ion-Solid Interactions, Cambridge University Press.
23. Rutherford, E. 1911, Phil. Mag., 21, 669.
24. Reimer, L. Pfefferkorn, G. 1977, Rasterelektronenmikroskopie, Springer-Verlag Berlin Heidelberg, New York.
25. Werner, U. Johansen, H. 1982, Elektronenmikroskopie in der Festkörperphysik, Springer-Verlag Berlin Heidelberg New York
26. Williams, D.B. Carter, C.B. 1996, Transmission Electron Microscopy I, Plenum Press, New York.
27. Lindhard, J. Scharff, H. 1961, Phys. Rev., 124, 128.
28. Fermi, E. 1928, Z. Phys., 48, 73.
29. Thomas, L.H. 1927, Proc. Cambr. Phil. Soc., 23, 524.
30. Ziegler, J.F., Biersack, J.P., and Littmark, U. 1985, The Stopping and Range of Ions in Solids, Pergamon Press, New York.
31. Lindhard, J. Winter, A. 1964, Mat. -Fys. Medd., 34, N° 4.
32. Bethe, H. 1930, Ann. Phys., 5, 325.
33. Bethe, H. 1932, Z. Phys., 76, 293.
34. Chu, W.K. Powers, D. 1972, Phys. Lett., 40A, 23.
35. Kramers, M.A. 1923, Phil. Mag., 46, 836.
36. Hobbs, L.W. 1979, Introduction to Analytical Electron Microscopy, J. I. Goldstein & D. C. Joy J. J. Hren (Ed.), Plenum Press New York, 437.
37. Kinchin, G.H. Pease, R.S. 1955, Rep. Prog. Phys., 18, 1.
38. Sigmund, P. 1969, Phys. Rev., 184, 383.
39. Matsunami, N., Yamamura, Y., Itikawa, Y., Itoh, N., Kazumata, Y., Miyagawa, S., Morita, K., Shimizu, R., and Tawara, H. 1984, At. Data Nucl. Data Tables, 31, 1.
40. Smith, K.C.A. Oatley, C.W. 1955, Br. J. Appl. Phys., 6, 391.
41. Buck, D.A. Shoulders, K. 1957, in Proceedings Eastern Joint Computer Conference, ATEE, New York, 55.

42. Broers, A.N. 1965, Microelectron. Reliab., 4, 103.
43. Craighead, H.G., Howard, R.E., Jackel, L.D., and Mankiewich, P.M. 1983, Appl. Phys. Lett., 42, 38.
44. Sedgwick, T.O., Broers, A.N., and Agule, B.J. 1972, J. Electrochem. Soc., 119, 1769.
45. Broers, A.N., Molzen, W.W., Cuomo, J.J., and Wittels, N.D. 1976, Appl. Phys. Lett., 29, 596.
46. Rakhshandehroo, M.R. Pang, S.W. 1996, J. Vac. Sci. Technol. B, 14, 612.
47. Sung, K.T. Pang, S.W. 1992, J. Vac. Sci. Technol. B, 10, 2211.
48. Simon, G., Haghiri Gosnet, A.M., Carcenac, F., and Launois, H. 1997, Microelectron. Eng., 35, 51.
49. Allee, D.R., Umbach, C.P., and Broers, A.N. 1991, J. Vac. Sci. Technol. B, 9, 2838.
50. Matsui, S., Ichihashi, T., and Mito, M. 1989, J. Vac. Sci. Technol. B, 7, 1182.
51. Allee, D.R. Broers, A. 1990, Appl. Phys. Lett., 57, 2271.
52. Pan, X., Allee, D.R., Broers, A., Tang, Y.S., and Wilkinson, C.W. 1991, Appl. Phys. Lett., 59, 3157.
53. Jackel, L.D., Howard, R.E., Mankiewich, P.M., Craighead, H.G., and Epworth, W. 1984, Appl. Phys. Lett., 45, 698.
54. Schmoranzer, H. 1988, J. Vac. Sci. Technol. B, 6, 2053.
55. Parikh, M. Kyser, D.F. 1979, J. Appl. Phys., 50, 1004.
56. Chang, T.H.P. 1975, J. Vac. Sci. Technol., 12, 1271.
57. Haller, I., Hatzakis, M., and Srinivassan, R. 1968, IBM J. Res. Dev., 251.
58. Thomson, L.F., Stillwagon, L.E., and Doerries, E.M. 1978, J. Vac. Sci. Technol., 15, 938.
59. Dobisz, E.A., Marrian, C.R.K., and Colton, R.J. 1991, J. Appl. Phys., 70, 1793.
60. Gölzhäuser, A., Geyer, W., Stadler, V., Eck, W., Grunze, M., Edinger, K., Weimann, T., and Hinze, P. 2000, J. Vac. Sci. Technol. B, 18, 3414.
61. Lercel, M.J., Craighead, H.G., Parikh, A.N., Seshadri, K., and Allara, D.L. 1996, Appl. Phys. Lett., 68, 1504.
62. Fujita, J., Watanabe, H., Ochiai, Y., Manako, S., Tsai, J.S., and Matsui, S. 1995, J. Vac. Sci. Technol. B, 13, 2757.
63. Muray, A., Scheinfein, M., Isaacson, M., and Adesida, I. 1985, J. Vac. Sci. Technol. B, 3, 367.
64. Thackeray, J.W., Orsula, G.W., Canistro, D., and Berry, A.K. 1989, J. Photopolymer Sci. Technol., 2, 429.
65. Scherer, A. Craighead, H.G. 1987, J. Vac. Sci. Technol. B, 5, 374.
66. Broers, A.N. 1988, IBM J. Res. Dev., 32, 502.
67. Eigler, D.M. Schweizer, E.I. 1991, Nature, 344, 524.
68. Umbach, C.P., Washburn, S., Laibowitz, R.B., and Webb, R.A. 1984, Phys. Rev. B, 30, 4048.
69. Xu, W., Wong, J., Cheng, C.C., Johnson, R., and Scherer, A. 1995, J. Vac. Sci. Technol. B, 13, 2372.
70. Chen, Y.L., Chen, C.C., Jeng, J.C., and Chen, Y.F. 2004, Appl. Phys. Lett., 85, 1259.
71. Dubois, S., Duvail, J.L., and Piraux, L. 2000, Actual. Chimique, 4, 42.
72. Li, A.P., Müller, F., and Gösele, U. 2000, Electrochem. and Solid-State Lett., 3, 131.
73. Borini, S., Amato, G., Rocchia, M., Boarino, L., and Rossi, A.M. 2003, J. Appl. Phys., 93, 4439.
74. Borini, S. 2005, J. Electrochem. Soc., 152, G482.
75. Ulman, R. 1996, Chem. Rev., 96, 1533.
76. Ulman, A. 1991, An Introduction to Ultrathin Organic Films From Langmuir-Blodgett to Self-Assembly, Academic press, Inc., San Diego.
77. Effenberger, F., Goetz, G., Bidlingmaier, B., and Wezstein, M. 1998, Angew. Chem., 110/18, 2651.
78. Effenberger, F., Goetz, G., Bidlingmaier, B., and Wezstein, M. 1998, Angew. Chem. Int. Ed., 37, 2462.
79. Linford, M.R. Chidsey, C.E.D. 1993, J. Am. Chem. Soc., 115, 12631.

80. Boukherroub, R., Morin, S., Sharpe, P., Wayner, D.D.M., and Allongue, P. 2000, Langmuir, 16, 7429.
81. White, H.S., Kittlesen, G.P., and Wrighton, M.S. 1984, J. Am. Chem. Soc., 106, 5375.
82. Chidsey, C.E.D. 1991, Science, 251, 219.
83. Frisbie, C.D., Fritsch-Faules, I., Wollman, E.W., and Wrighton, M.S. 2002, Thin-Solid-Films, 210, 341.
84. Chaki, N.K. Vijayamohanan, K. 2002, Biosens. Bioelectron., 17, 1.
85. Dulcey, C.S., Georger, J.H., Krauthamer, V., Stenger, D.A., Fare, T.L., and Calver, M.J. 1991, Science, 252, 551.
86. Perkins, M.K., Dobisz, E.A., Brandow, S.L., Calvert, J.M., Kosakowski, J.E., and Marrian, C.R.K. 1996, Appl. Phys. Lett., 68, 550.
87. Ada, E.T., Hanley, L., Etchin, S., Melngailis, J., Dressick, W.J., Chen, M.S., and Calvert, J.M. 1995, J. Vac. Sci. Technol. B, 13, 2189.
88. Lercel, M.J., Whelan, C.S., Craighead, H.G., Seshadri, K., and Allara, D.L. J. Vac. Sci. Technol. B, 14, 4085.
89. Sondag Huethorst, J.A.M., van Helleputte, H.R.J., and Fokkink, L.G.J. 1994, Appl. Phys. Lett., 64, 285.
90. Koops, H.W.P., Schössler, C., Kaya, A., and Weber, M. 1996, J. Vac. Sci. Technol. B, 14, 4105.
91. Fritzsche, W., Kohler, J.M., Bohm, K.J., Unger, E., Wagner, T., Kirsch, R., Mertig, M., and Pompe, W. 1999, Nanotechnology, 10, 331.
92. Kohlmann von Platen, K.T., Thiemann, M., and Brünger, W.H. 1991, Microelectron. Eng., 13, 279.
93. Koops, H.W.P. 1996, T. C. Hale and K. L. Telschow (Ed.), Proceedings of SPIE Volume, 248.
94. Koops, H.W.P., Munro, E., Rouse, J., Kretz, J., Rudolph, M., Weber, M., and Dahm, G. 1995, Nucl. Instrum. Methods Phys. Res. B, 363, 1.
95. Weber, M., Rudolph, M., Kretz, J., and Koops, H.W.P. 1995, J. Vac. Sci. Technol. B, 13, 461.
96. Schoessler, C. Koops, H.W.P. 1998, J. Vac. Sci. Technol. B, 16, 862.
97. Koops, H.W.P., Kretz, J., Rudolph, M., and Weber, M. 1993, J. Vac. Sci. Technol. B, 11, 2386.
98. Voss, R.F., Laibowitz, R.B., and Broers, A.N. 1980, Appl. Phys. Lett., 37, 656.
99. Komuro, M. Hiroshima, H. 1997, Microelectron. Eng., 35, 273.
100. Matsui, S. Mori, K. 1986, J. Vac. Sci. Technol. B, 4, 299.
101. Utke, I., Luisier, A., Hoffmann, P., Laub, D., and Buffat, P.A. 2002, Appl. Phys. Lett., 81, 3245.
102. Hübner, U., Plontke, R., Blume, M., Reinhardt, A., and Koops, H.W.P. 2001, Microelectron. Eng., 57, 953.
103. Christy, R.W. 1960, J. Appl. Phys., 31, 1680.
104. Koops, H., Weiel, R., Kern, D.P., and Baum, T.H. 1988, J. Vac. Sci. Technol. B, 6, 477.
105. Scheuer, V., Koops, H., and Tschudi, T. 1986, Microelectron. Eng., 5, 423.
106. Hoyle, P.C., Ogasawara, M., and Cleaver, J.R.A. 1993, Appl. Phys. Lett., 62, 3043.
107. Reimer, L. 1997, Transmission Electron Microscopy, Springer-Verlag, Berlin.
108. Lee, K.L. Hatzakis, M. 1989, J. Vac. Sci. Technol. B, 7, 941.
109. Hoyle, P.C., Cleaver, J.R.A., and Ahmed, H. 1994, Appl. Phys. Lett., 64, 1448.
110. Kuntz, R.R. Mayer, T.M. 1986, J. Vac. Sci. Technol. B, 5, 427.
111. Miura, N., Ishii, H., Shirakashi, J., Yamada, A., and Konagai, M. 1997, Appl. Surf. Sci., 113/114, 269.
112. Miura, N., Numaguchi, T., Yamada, A., Konagai, M., and Shirakashi, J.I. 1997, Jpn. J. Appl. Phys., 36, 1619.
113. Miura, N., Ishii, H., Yamada, A., and Konagi, M. 1996, Jpn. J. Appl. Phys., 35, L1089.
114. Amman, M., Sleight, J.W., Lombardi, D.R., Welser, R.E., Deshpande, M.R., Reed, M.A., and Guido, L.J. 1996, J. Vac. Sci. Technol. B, 14, 54.

115. Guise, O., Ahner, J., Yates Jr, J.T., and Levy, J. 2004, Appl. Phys. Lett., 85, 2352.
116. Zeng, Z.M., Tian, X.B., Kwok, T.K., Tang, B.Y., Fung, M.K., and Chu, O.K. 2000, J. Vac. Sci. Technol. A, 18, 2164.
117. Zhang, S., Zeng, X.T., Xie, H., and Hing, P. 2000, Surf. Coat. Technol., 123, 256.
118. Broers, A.N. 1964, in Proceedings of the First International Conference on Electron and Ion Beam Technology, Wiley, New York, 181.
119. Molzen, W.W., Broers, A.N., Cuomo, J.J., Harper, J.M.E., and Laibowitz, R.B. 1979, J. Vac. Sci. Technol., 16, 269.
120. Ueta, A., Avramescu, A., Uesugi, K., Suemune, I., Machida, H., and Shimoyama, N. 1998, Jpn. J. Appl. Phys., 37, 272.
121. Djenizian, T., Santinacci, L., and Schmuki, P. 2001, Appl. Phys. Lett., 78, 2840.
122. Miura, N., Yamada, A., and Konagai, M. 1997, Jpn. J. Appl. Phys., 36, L1275.
123. Sieber, I., Hildebrand, H., Djenizian, T., and Schmuki, P. 2003, Electrochem. and Solid-State Lett., 6, C1.
124. Djenizian, T., Macak, J., and Schmuki, P. 2002, in Nano- and Micro-Electromechanical Systems (NEMS and MEMS) and Molecular Machines, Mat. Res. Soc. Symp. Proc., Boston, 79–83.
125. Djenizian, T., Santinacci, L., and Schmuki, P. 2001, J. Electrochem. Soc., 148, 197.
126. Scharifker, B. Hills, G. 1983, Electrochim. Acta, 28, 879.
127. Scharifker, B. Mostany, J. 1984, J. Electroanal. Chem., 177, 13.
128. Gunawardena, G., Hills, G., Montenegro, I., and Scharifker, B. 1982, J. Electroanal. Chem., 138, 225.
129. Scherb, G. Kolb, D.M. 1995, J. Electroanal. Chem., 396, 151.
130. Vereecken, P.M., Strubbe, K., and Gomes, W.P. 1997, J. Electroanal. Chem., 433, 19.
131. Stiger, R.M., Gorer, S., Craft, B., and Penner, R.M. 1999, Langmuir, 15, 790.
132. Oskam, G., van Heerden, D., and Searson, P.C. 1998, Appl. Phys. Lett., 73, 3241.
133. Pasa, A.A. Schwarzacher, W. 1999, Phys. Status Solidi A, 173, 73.
134. Rashkova, B., Guel, B., Pötzschke, R.T., Staikov, G., and Lorenz, W.J. 1998, Electrochim. Acta, 43, 3021.
135. Djenizian, T., Santinacci, L., and Schmuki, P. 2004, J. Electrochem. Soc., 151, G175-G180.
136. Djenizian, T., Santinacci, L., Hildebrand, H., and Schmuki, P. 2003, Surf. Sci., 524, 40.
137. Stewart, D.K., Doyle, A.F., and Casey, J.D.J. 1995, SPIE, 2437, 276.
138. Xu, X. Melngailis, J. 1993, J. Vac. Sci. Technol. B, 11, 2436.
139. Frey, L. Lehrer, C. 2003, Praktische Metallographie, 40, 184.
140. Hooghan, K.N., Wills, K.S., Rodriguez, P.A., and O′Connell, S. 1999, in ASM International, Materials Park, Ohio, 247.
141. Van Doorselaer, K., Van den Reeck, M., Van Den Bempt, L., Young, R., and Whitney, J. 1993, in Proc. 19th International Symposium for Testing and Failure Analysis, ASM International, Materials Park, Ohio, 405.
142. Verkleij, D. 1998, Microelectron. Reliab., 38, 869.
143. Nikawa, K. 1994, IEICE Trans. Fund. Electr., E77, 174.
144. Hahn, L.L., Abramo, M.T., and Coutu, P.T. 1991, in Proc. 17th International Symposium for Testing and Failure Analysis, ASM International, Materials Park, Ohio, 1.
145. Walker, J.F., Reiner, J.C., and Solenthaler, C. 1995, Inst. Phys. Conf. Ser., 146, 629.
146. Ishitani, T. Yaguchi, T. 1996, Microsc. Res. Techn., 35, 320.
147. Rai, R., Subramanian, S., Rose, S., Conner, J., Schani, P., and Moss, J. 2000, in Proc. 26th International Symposium for Testing and Failure Analysis, 415.
148. Stevie, F.A., Irwin, R.B., Shofner, T.L., Brown, S.R., Drown, J.L., and Giannuzzi, L.A. 1998, in, American Institute of Physics, Woodburry, NY, 868.
149. Vasile, M.J., Xie, J., and Nassar, R. 1999, J. Vac. Sci. Technol. B, 17, 3085.
150. Ishitani, T., Ohnishi, T., Madokoro, Y., and Kawanami, Y. 1991, J. Vac. Sci. Technol. B, 9, 2633.

151. Vasile, M.J., Grigg, D., Griffith, J.E., Fitzgerald, E., and Russel, P.E. 1991, J. Vac. Sci. Technol. B, 9, 3569.
152. Olbrich, A., Ebersberger, B., Boit, C., Niedermann, P., Hänni, W., Vancea, J., and Hoffmann, H. 1999, J. Vac. Sci. Technol. B, 17, 1570.
153. Hill, A.R. 1968, Nature, 218, 292.
154. Guharay, S.K., E., S., and Orloff, J. 1999, J. Vac. Sci. Technol. B, 17, 2779.
155. Orloff, J. Swanson, L.W. 1975, J. Vac. Sci. Technol. B, 12, 1209.
156. Kohn, V.E. Ring, G.R. 1975, Appl. Phys. Lett., 27, 479.
157. Orloff, J. 1993, Rev. Sci. Instrum., 64, 1105.
158. Frey, L. Lehrer, C. 2003, Applied Physics A, 76, 1017.
159. Wang, K., Chelnokov, A., Rowson, S., Garoche, P., and Lourtioz, J.M. 2000, J. Phys. D: Appl. Phys., 33, L119.
160. Gruning, U., Lehmann, V., Ottow, S., and Bush, K. 1996, Appl. Phys. Lett., 68, 747.
161. Schmidt, B., Bischoff, L., and Teichert, J. 1997, Sens. Actuators A, 61, 369.
162. Brugger, J., Beljakovic, G., Despont, M., de Rooiji, N.F., and Vettiger, P. 1997, Microelectron. Eng., 35, 401.
163. Chen, W., Chen, P., Madhukar, R., Viswanathan, R., and So, J. 1993, Mater. Res. Soc. Proc., 279, 599.
164. Cummings, K.D., Harriott, L.R., Chi, G.C., and Ostermayer, F.W. 1986, Proc. SPIE Int. Soc. Opt. Eng., 93.
165. Arimoto, H., Kosugi, M., Kitada, H., and Miyauchi, E. 1989, Microelectron. Eng., 9, 321.
166. Rennon, S., Bach, L., König, H., Reithmaier, J.P., Forchel, A., Gentner, J.L., and Goldstein, L. 2001, 57–58, 891.
167. D′Arrigo, G. Spinella, C. 2001, Mater. Sci. Semicond. Proc., 4, 93.
168. Spinella, C. 1998, Mater. Sci. Semicond. Proc., 1, 55.
169. Garozzo, G., La Magna, A., Coffa, S., D′Arrigo, G., Parasole, N., Renna, M., and Spinella, C. 2002, Comp. Mater. Sci., 24, 246.
170. Schmuki, P., Erickson, L.E., and Lockwood, D.J. 1998, Phys. Rev. Lett., 80, 4060.
171. Schmuki, P. Erickson, L.E. 2000, Phys. Rev. Lett., 85, 2985.
172. Spiegel, A., Erickson, L.E., and Schmuki, P. 2000, J. Electrochem. Soc., 147, 2993.
173. Spiegel, A., Staemmler, L., Dobeli, M., and Schmuki, P. 2002, J. Electrochem. Soc., 149, C432.
174. Berger, M.J. and Seltzer, S.M. 1964, NASA-SP-3012.
175. Fitting, H.J. 1974, Phys. Stat. Sol. A, 26, 525.
176. Gibbons, J.F., Johnson, W.S., and Mylroie, S.W. 1975, Projected Range Statistics, Stroudsburg.

Wet Chemical Approaches for Chemical Functionalization of Semiconductor Nanostructures

Rabah Boukherroub and Sabine Szunerits

1 Introduction

As there are already several reports and reviews dealing with chemical modification of crystalline and nanocrystalline surfaces in the literature [1–9], this book chapter is organized in several parts dealing with the progress made in nanocrystalline silicon and germanium surface modification with a special focus on: (1) methods for the preparation of hydrogen-terminated porous silicon, different strategies developed for organic layers formation, their characterization, and some examples of their applications; (2) techniques for porous germanium preparation, characterization, and surface modification; and (3) conclusions and perspectives.

Chemical derivatization of hydrogen-terminated semiconductor-nanostructured surfaces holds considerable promise from both fundamental and applied research aspects. Surface chemistry of semiconductor nanostructures emerged just after the discovery of the room-temperature bright photoluminescence (PL) [10] from porous silicon, as a crucial point of the material. This is due to the chemical instability of the native hydride surface termination of porous silicon, prepared via electrochemical or chemical dissolution of crystalline silicon in HF-based solutions. The silicon–hydrogen (Si–H) and germanium–hydrogen (Ge–H) bonds terminating porous silicon (PSi) and germanium (PGe), respectively, are chemically reactive and react readily in ambient air to form an oxide submonolayer. This uncontrolled oxidation reaction introduces surface active defects responsible for the degradation of the electronic and optical properties of the sample. This limitation restricts the use of PSi and PGe in the fabrication of commercial devices. Although, the hydride termination of PSi can be easily converted to silicon oxide in a controlled way by different means and therefore explore the well-known silanization reaction, there are some drawbacks associated with such a chemical transformation. The amorphous

R. Boukherroub (✉)
Biointerfaces Group, Interdisciplinary Research Institute (IRI), FRE 2963, IRI-IEMN, Avenue Poincaré-BP 60069, 59652 Villeneuve d'Ascq, France
e-mail: rabah.boukherroub@iemn.univ-lille1.fr

P. Schmuki, S. Virtanen (eds.), *Electrochemistry at the Nanoscale,* Nanostructure Science and Technology, DOI 10.1007/978-0-387-73582-5_5,

character of the silicon oxide layer makes very difficult to control the density of the Si–OH groups generated on the surface that are necessary for chemical linkages and the silanization reaction itself is complicated by the difficult-to-control polymerization occurring during the reaction. This leads to nonreproducible results and may cause pore narrowing and blocking. Furthermore, the Si–O bond is prone to hydrolysis under physiological or wet chemical conditions and thus limits the lifetime and stability of the device. In the case of germanium, the instable nature of the germanium oxide in aqueous solutions in well documented.

On the other hand, replacing Si–H and Ge–H bonds by silicon–carbon (Si–C) and germanium–carbon (Ge–C) bonds offers a high stability to the interface. The Si–C and Ge–C bonds are both thermodynamically and kinetically stable due to the high bond strength and low polarity of the bonds. Advantages associated with these transformations include the existence of a wide range of chemical functionalities compatible with the Si–H and Ge–H bonds terminating the PSi and PGe surfaces, the ease of carrying out the chemical reactions and, finally, the very well-established organometallic chemistry in solution. Furthermore, chemical grafting of organic layers directly on the semiconductor surface (without an interfacial silicon oxide layer) brings the functionality close to the semiconductor surface and is thus expected to lead to more sensitive devices.

From the fundamental viewpoint, designing simple approaches are desirable for the immobilization of chemical and biological species on the semiconductor surface without the need to build up thick organic layers or postmodification strategies to limit nonspecific adsorption of the target. Furthermore, understanding the reaction mechanism on the surface is of considerable interest to establish a general trend between chemical reactivity on the surface- and solution-phase organometallic chemistry. The presence of a small band gap in the semiconductor nanostructures will influence the reaction pathway and product distribution.

Using the well-established microfabrication techniques will allow to prepare and to integrate various microcomponents into devices such as biological microelectromechanical systems (bioMEMS) with potential applications in microscale, high-throughput biosensing and medical devices. Furthermore, chemical grafting of organic species directly on hydrogen-terminated nanostructured surfaces (with no intervening oxide layer) will allow to fully take advantage of the electronic properties of the semiconductor to develop sensitive devices for label-free detection of biomolecular interactions. Finally, scaling down to the molecular level will open new opportunities for a new generation of devices.

2 Porous Silicon

PSi thin-films formation was first observed by Uhlir in 1956 during electropolishing of crystalline silicon (*c*-Si) [11]. He found that under appropriate conditions of applied current and solution composition, the silicon substrate did not dissolve uniformly but instead a brownish layer composed of fine holes was produced. Similar films were, however, observed in HF/HNO_3 solutions without applying any external electrical bias to the silicon substrate [12]. The reports were followed by detailed studies of the resulting films regarding the formation mechanism and their properties without any evidence of the porous nature of the layers [13, 14]. The porous nature of the films was first reported by Watanabe and co-workers and found that PSi layers can be oxidized more easily than single crystal silicon [15, 16]. Since then the oxidized PSi layers have generated a significant interest and were successfully employed for dielectric isolation processes in Si integrated circuits [17, 18].

The report of Canham on the tunable efficient, room temperature light output from PSi has marked a breakthrough point in PSi research [10]. The study was followed by an independent report by Lehman and Gösele [19] claiming that PSi layers exhibit an increased band gap compared to bulk silicon, and two-dimensional quantum confinement was used to explain this property and the PSi formation mechanism. Further studies have shown that not only the PL of the PSi can be tuned from red to green [20] but also efficient visible electroluminescence (EL) had been achieved [21]. Then followed a significant volume of work devoted to structural, optical, and electronic aspects of the material to understand the origin of the PL and to develop applications in solid-state electroluminescent devices [22]. Considerable efforts have been made to explain the origin of the PL and the physical aspects of the material. It is now believed that the PL results from the quantum confinement of carriers with silicon nanocrystals even though contributions from the surface species are not excluded.

2.1 Preparation of Porous Silicon

PSi can be prepared by chemical or electrochemical etching processes of *c*-Si in HF-based solutions and consists of nano- or microcrystalline domains with defined pore morphology. The diameter, geometric shape, direction of the pores, the porosity, and the thickness of the porous layer depend on surface orientation, doping level and type, temperature, the composition of the etching solution, the current density, and anodization duration [22–24]. Ethanol is frequently added in the electrolyte solutions to reduce the surface tension and wet the interior parts of the hydrophobic porous layer, and to minimize hydrogen bubble formation during anodization. PSi films fabricated in dilute aqueous or ethanolic HF solutions are generally mesoporous, while PSi layers prepared in concentrated aqueous HF solutions are microporous [25]. PSi samples composed of macropores with diameters in the range of 25–100 nm are obtained by anodizing heavily doped p-type silicon ($R = 10^{-3}$ Ohm cm) in 25% ethanolic HF solution [26] or by etching p-type silicon in dry solvents such as acetonitrile or DMF in a moisture-free environment [27]. PSi films with cylindrically shaped pores and pore diameters that can be tuned from 5 nm to 1200 nm were obtained by electrochemical dissolution of heavily doped (10^{-3} Ohm-cm) p-type Si(100) in ethanolic HF at ambient temperatures [28]. Because silicon dissolution requires holes (h^+) supply during anodization, illumination is required to generate luminescent PSi from lightly n-type Si substrates [29]. In the dark, however, PSi formation was only observed at high applied voltages. In both

Table 1 Pore size classification

Type of porous silicon	Corresponding size (nm)
Microporous	≤2
Mesoporous	2–50
Macroporous	>50

cases macroporous layers formation was observed. Detailed structural studies of the widely varying morphologies resulting from photoelectrochemical etching of such substrates have been undertaken in the literature [30, 31]. For heavily doped ($\sim 10^{19}$ cm^{-3}) n-type Si substrates, PSi formation was observed even in the dark.

PSi films also can be obtained by "stain etching", that is, without applying any external electrical bias or light illumination [32, 33]. In this technique, the silicon substrate is simply immersed in the appropriate etching solution. Because the method does not require any electrical bias, it is very useful for porosification of Si-based structures on electrically insulating substrates. However, stain-etched samples display lower PL efficiencies and homogeneity than those prepared by anodization, and the technique suffers from lack of reproducibility [34, 35].

Porous materials are classified according to IUPAC guidelines, which define ranges of pore size exhibiting characteristic adsorption properties (Table 1). Pore size has however only a clear meaning when the geometrical shape of the pores is well defined, which is the case for macroporous and mesoporous Si. However, the term "microporous" has been too loosely applied in the PSi literature and bears no resemblance to the strict criteria now universally accepted by porous carbons and glasses communities.

PSi is a complex network of zero-dimensional nanocrystallites and 1-D crystalline nanowires. Figure 1a displays a top-view scanning electron microscopy (SEM) image of a PSi surface obtained from highly doped p-type Si(100) substrate that has been etched at a current density of 400 mA/cm^2 for 10 s in 3:1 (v:v) HF/EtOH at room temperature. At a magnification of $\times 500\,000$, the image clearly shows average pore sizes on the order of 10 nm. A cross-sectional SEM view (Fig. 1b) displays the cylindrical morphology of the pores and the well-defined parallel arrangement of the pores. Under the conditions that were used, the typical thickness of the porous layer was 5 μm.

2.2 Reactive Surface Species on PSi Surfaces

2.2.1 Hydrogen Termination

As described above, PSi samples are prepared in HF-based (commonly HF/EtOH) solutions under various conditions (chemical, electrochemical, and photoelectrochemical) from different silicon wafers of different doping type and level, and

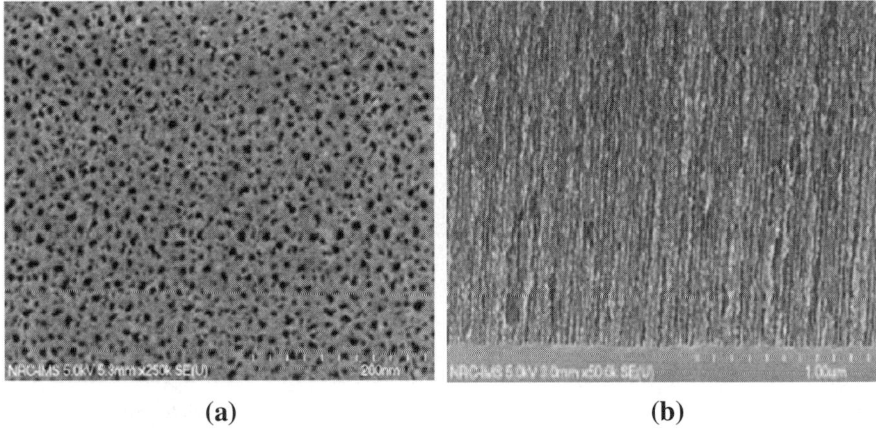

Fig. 1 SEM images of a porous silicon layer obtained by anodization of p-type Si wafers (<100>, B-doped, 0.0018–0.0022 Ωcm of 5 μm) at 400 mA/cm^2 for 10 s in HF/EtOH (3:1 v/v). (**a**) plan view showing the approximate pore size of *ca.* 10 nm; (**b**) cross-sectional view allowing a determination of the porous layer thickness of 5 μm and displaying the well-defined cylindrical pore morphology of the porous layer

crystallographic orientations. Anodization under galvanostatic conditions is generally the preferred method for reproducibly attaining wide ranges of porosity and thickness. The silicon dissolution leaves a hydrogen termination. The resulting PSi layers are transparent in the IR region, and spectra can be conveniently acquired using diffuse reflectance (DRIFT) or transmission FTIR spectroscopy. A typical transmission FTIR spectrum of a freshly prepared PSi sample is shown in Fig. 2. It exhibits a typical tripartite band for Si–H$_x$ ($x = 1$–3) stretching modes $\nu_{\text{Si-H}x}$ ($\nu_{\text{Si-H1}}$ at ~2089, $\nu_{\text{Si-H2}}$ at ~2115 cm^{-1}, and $\nu_{\text{Si-H3}}$ at ~2139 cm^{-1}), Si–H$_2$ scissor mode $\delta_{\text{Si-H2}}$ at ~912 cm^{-1}, and $\delta_{\text{Si-H}x}$ at ~669 cm^{-1} and ~ 629 cm^{-1}. A small peak at around 1037 cm^{-1} (Si–O–Si stretching mode), which is present in all PSi samples, results most likely from a small oxidation of the reactive surface.

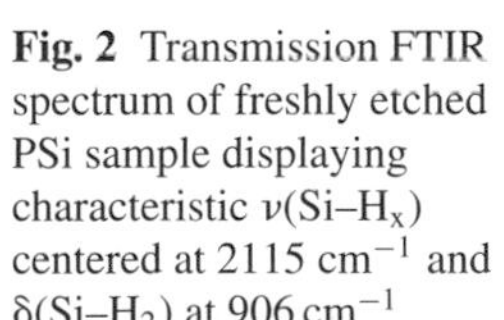
Fig. 2 Transmission FTIR spectrum of freshly etched PSi sample displaying characteristic ν(Si–H$_x$) centered at 2115 cm^{-1} and δ(Si–H$_2$) at 906 cm^{-1}

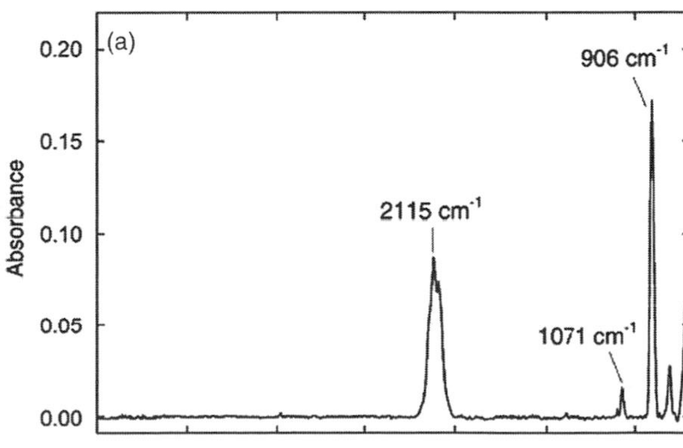

2.2.2 Deuterium Termination

Deuterated PSi surfaces are prepared by common electrochemical etching of crystalline silicon in DF-ethanol-D_6 or 48% DF/D_2O:EtOD (1:1) [36–39] electrolytes. The FTIR spectrum of deuterium-terminated PSi surface displays stretching modes at 1518, 1529, and 1554 cm^{-1} assigned to Si–D, Si–D_2, and Si–D_3, respectively. This is in a good agreement with the theoretically predicted values in which the ratio of the Si–H and Si–D stretching frequencies must be equal to 1.376 [40]. Upon oxidation of the surface, the Si–D stretching modes shift from ~1529 cm^{-1} to ~1630 cm^{-1} [36].

2.2.3 Halide Termination

Halide-terminated PSi surfaces are very reactive and tend to hydrolyze in ambient conditions. Thus, halide-terminated surfaces must be handled in an inert atmosphere. Halide termination is generally achieved by chemical treatment of hydrogen-terminated PSi surfaces with halide gas (X_2, X = Cl, Br, I) or alkyl halides (RX) under inert atmosphere or vacuum. Exposure of a freshly etched hydride-terminated PSi surface to molecular iodine (<1 torr of I_2(g) for 1 min) results in efficient Si–I bond formation (Scheme 1) [41, 42]. The presence of Si–I bonds on the PSi surface was revealed by XPS (I $3d_{5/2}$ at ~619 eV and a chemical shift of Si $2p$ from 99 eV to 102 eV). The reaction proceeds by initial attack of I_2 at the weak Si–Si bonds as evidenced by the unchanged of Si–H and Si–H_2 infrared stretching modes on I_2 exposure. This result contrasts the observed reaction pathway on crystalline silicon where the reaction takes place by Si–H substitution by the halide [43].

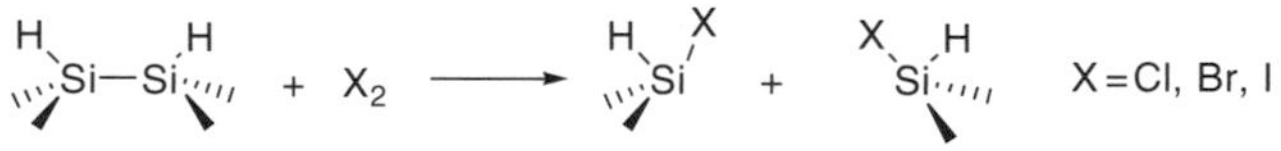

H–Si—Si–H + X_2 ⟶ H–Si–X + X–Si–H   X=Cl, Br, I

Scheme 1

In situ iodination of hydrogenated PSi surfaces was carried out using iodoform (CHI_3) under visible light illumination using 15 W ordinary white lamp for 6 h or under thermal activation at 45 °C for 8 h [44]. The role of the visible light or heat is to generate iodine radicals from CHI_3. In contrast to the iodination of PSi–H with I_2(g) [41], the reaction of CHI_3 with the hydrogenated PSi surface takes place with Si–H consumption, which is in agreement with a free-radical mechanism (Scheme 2).

Chlorine termination of PSi was obtained by ultraviolet (UV) irradiation of the hydrogen-terminated PSi surface in the presence of Cl_2 gas and the resulting Cl-capped PSi surface was used for chemical grafting of ferrocene groups [45]. Treatment of hydrogen-terminated PSi surfaces with chlorinated solvents: CCl_4 vapor for 6 h [46], boiling in CCl_4 for 30 min [47] or trichloroethylene for various periods (0–10 min) at room temperature [48] has been found to promote hydrogen removal, but no evidence for Si–Cl bonds formation. However, based on the fast oxidation of

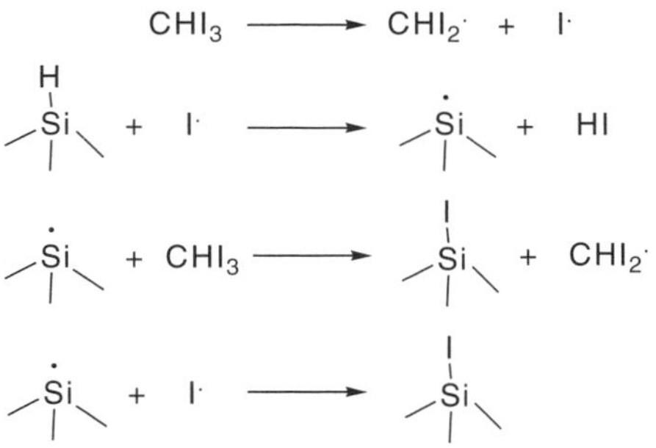

$$CHI_3 \longrightarrow CHI_2^{\cdot} + I^{\cdot}$$

$$\equiv Si\text{-}H + I^{\cdot} \longrightarrow \equiv Si^{\cdot} + HI$$

$$\equiv Si^{\cdot} + CHI_3 \longrightarrow \equiv Si\text{-}I + CHI_2^{\cdot}$$

$$\equiv Si^{\cdot} + I^{\cdot} \longrightarrow \equiv Si\text{-}I$$

Scheme 2

the resulting surfaces, one can hypothesize on Si–Cl intermediates formation during the chemical process. Halide-terminated PSi surfaces are also prepared by reaction of freshly prepared PSi sample with alkyl halides (R-X) under microwave irradiation [49]. Under these conditions, however, both Si-X and Si-R are obtained on the PSi surface and the probable reaction pathway is outlined in Scheme 3.

$$\equiv Si\text{-}H + R\text{-}X \longrightarrow \equiv Si\text{-}X \xrightarrow{H_2O} SiO_2$$

Scheme 3

The Si-X termination was not characterized in the report, but its formation as an intermediate was assumed from the final surface oxidation of the PSi surface after rinsing in ambient. The reaction mechanism is, however, not clear. Comparable results were obtained by Guo et al. by reacting PSi–H surfaces with α-bromo, ω-carboxy alkanes under microwave irradiation [50]. The Si–H_x bonds consumption accompanied by a chemical shift of the Si–H_x stretching modes from 2100 cm^{-1} to 2253 cm^{-1} is a clear evidence for covalent attachment of organic moieties on the surface and an indirect evidence for Si–Br intermediate formation during the chemical process. The final surface is composed of alkyl chains and oxygen back-bonded Si–H_x (O_2)Si–H_2 and (O_3)Si–H species. The surface oxidation may result from hydrolysis of the reactive Si–Br intermediates formed on the surface through homolytic decomposition of the alkyl halide, upon rinsing and handling the surface in ambient.

2.3 Chemical Derivatization of PSi Surfaces Through Si–C

2.3.1 Metal-Induced Hydrosilylation

Late-transition metal complexes based on platinum, palladium, and rhodium were tested and found effective for the hydrosilylation reaction of alkynes and alkenes on hydride-terminated PSi surfaces [51]. Hydrosilylation reaction of 1-dodecyne and phenylacetylene on PSi–H proceeds at room temperature in toluene in the presence of Wilkinson's catalyst, $RhCl(PPh_3)_3$. Attempts for bis-silylation reaction of 1-dodecyne or phenylacetylene mediated by $PdCl_2(PEt_3)_2$ or $Pd(OAc)_2$/1,1,3,3-tetramethyl-butyl isocyanide, in refluxing toluene resulted exclusively in hydrosilylation reaction, as indicated by appearance of ν(C=C) at 1602 and 1597 cm^{-1}, respectively [51]. The hydrosilylation reaction was accompanied by surface oxidation and metal deposition. The latter originated from simple immersion plating readily observed in PSi [52, 53]. A quantitative PL quenching was also observed on the resulting surfaces due most likely to the presence of metal deposits on the surface [54–56].

Platinum-mediated hydrosilylation reaction using alkenes have been investigated by Ghadiri et al. [57]. They found that only Karstedt's catalyst, a highly active colloidal Pt-based species, gave the desired reactivity pattern. Concomitant oxidation of the PSi surface was obtained along the hydrosilylation reaction. The presence of oxygen is crucial for the occurrence of the hydrosilylation. Indeed, in the absence of air (reaction carried out under N_2 atmosphere), the PSi surface remained unchanged with no apparent alkyl incorporation or surface oxidation. It was also demonstrated that oxidation of the silicon atoms to OSi-H species increases the hydrosilylation reaction rate, and under these conditions the presence of oxygen is not necessary. The role of oxygen in these reactions is stabilizing Pt colloids [58]. On the other hand, if $Rh_2(OAc)_4$ is used as a catalyst in the presence of diazo compounds to promote insertion of carbene species into Si–H bonds of hydride-terminated PSi surface, FTIR spectroscopy of the resulting surfaces is consistent with a chemical process taking place with no apparent surface oxidation. The scope and the utility of the technique were demonstrated using different functional diazo starting reagents.

To avoid late transition metals and their potential for metal deposition on the porous layer and surface oxidation, $EtAlCl_2$ was utilized to mediate hydrosilylation of readily available alkynes and alkenes to yield surface-bound vinyl and alkyl groups, respectively (Scheme 4) [59, 60]. The method is tolerant of a wide variety of functional groups, as demonstrated for the formation of hydroxyl-, nitrile-, and ester-terminated surfaces.

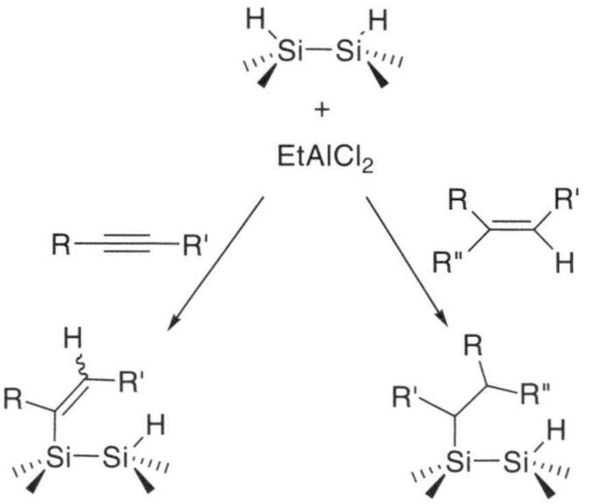

Scheme 4

For example, addition of 10 μl (10 μM) of a commercial 1.0 M hexanes solution of $EtAlCl_2$ to a freshly prepared PSi surface followed by 3 μl (14 μM) of 1-dodecyne under nitrogen resulted in dodecenyl groups grafting on the surface. FTIR analysis of the modified PSi surface shows features due to ν(C–H) stretches of the decyl chain between 2960 cm^{-1} and 2850 cm^{-1} and δ (C-H) methylene and methyl bending modes at 1466 and 1387 cm^{-1} (Fig. 3a). The decrease of the ν(Si–H) intensity at ~2100 cm^{-1} and the presence of a stretching band at 1595 cm^{-1} ν(Si–C=C) is a direct evidence for a hydrosilylation reaction.

To clearly demonstrate that the stretch observed at 1595 cm^{-1} is indeed an olefinic stretch, the dodecenyl-terminated PSi surface was subjected to hydroboration (Scheme 5). Reaction of the surface with 0.8 M BH_3.THF in THF under nitrogen atmosphere followed by copious rinsing with THF in air led to the disappearance of the stretch at 1595 cm^{-1} and the concomitant appearance of a new stretch at 1334 cm^{-1} due to B–O stretching (Fig. 3b).

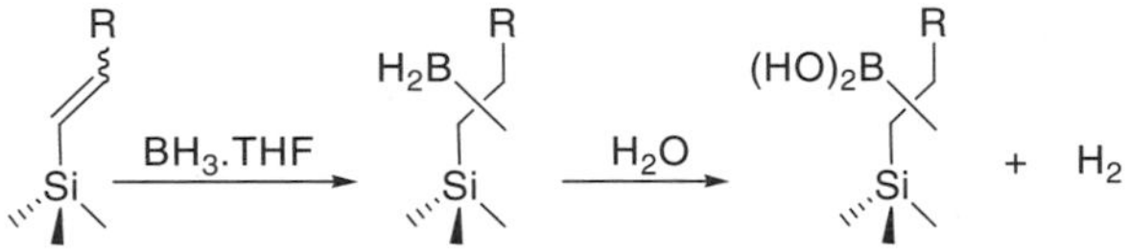

R, Si —(BH_3.THF)→ H_2B, R, Si —(H_2O)→ $(HO)_2B$, R, Si + H_2

Scheme 5

Because the Lewis-acid-mediated hydrosilylation takes place without any apparent surface oxidation, integration of ν(Si–H) before and after reaction gives the percent of the Si–H bonds consumed during the chemical process or efficiency (%E):

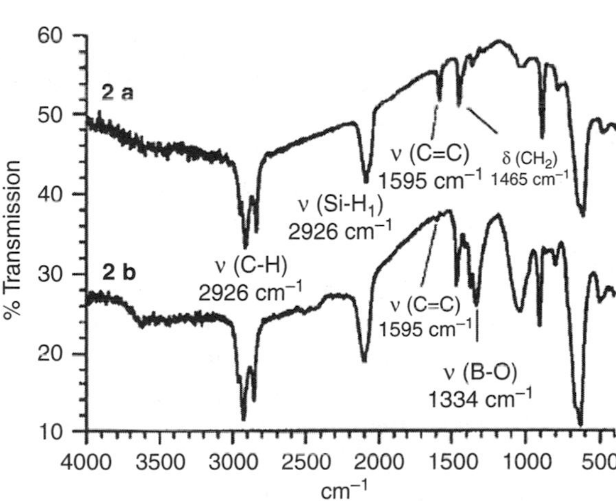

Fig. 3 FTIR spectra of PSi–H surfaces modified through (**a**) hydrosilylation of 1-dodecyne mediated by $EtAlCl_2$ and (**b**) hydroboration of the olefin of the dodecenyl group with 0.8 M BH_3.THF. Reprinted with permission from J.M. Buriak, M.J. Allen, J. Am. Chem. Soc., Vol. 120 (1998), pp. 1339–1340. Copyright (1998) American Chemical Society

Table 2 Relative efficiencies of the Lewis-acid-mediated hydrosilylation reaction

Entry	Hydrosilylated substrate	Average efficiency ***E*** (%)
1	1-pentyne	19
2	1-dodecyne	17
3	2-hexyne	14
4	1-pentene	28
5	1-dodecene	28
6	*cis*-2-pentene	20
7	*trans*-2-hexene	11
8	2,3-dimethyl-2-pentene	11

$$\%E = (A_0 - A_1)/A_0$$

where A_0 and A_1 are the baseline to peak areas of the freshly prepared and the hydrosilylated samples, respectively. The average efficiency is sensitive to the regiochemistry and stereochemistry of the starting molecule and the best results were obtained with 1-alkenes (28%) (Table 2). This indicates that about 70% of the surface Si–H bonds remain intact. Furthermore, there is a clear difference between *cis*- and *trans*-alkenes. A lower efficiency was obtained with the *trans*-alkenes. Internal alkenes incorporate less readily as compared with terminal alkenes and alkynes.

The hydrophobic alkyl and alkenyl-terminated PSi surfaces display a high stability in harsh environments: HF/EtOH rinsing, boiling in aerated water for 2 h, and boiling in aerated KOH (pH 10) solutions. Furthermore, chemography [61] experiments based on silver halide reduction by emitted silane gas due to hydrolysis of Si–SiH_3 bond, releasing SiH_4, were carried out on the PSi surfaces [60]. Reduction of silver halide on photographic plates generates images in which the optical density will depend on the rate of silane release. As a comparison, hydrogen- and dodecenyl-terminated PSi surfaces were exposed to 100% humidity. After 1 h, the hydride-terminated PSi sample heavily reduced the photographic plate, leading to a dark spot while the dodecenyl-terminated PSi sample yielded a very faint trace, indicating that the rate of silane emission from this sample is very slow. After a 4-h exposure, the dodecenyl-terminated surface showed a light, circular pattern. The results point out that the chemical modification consumed a large fraction of the surface hydrides –SiH_3 and thus reduces the rate of surface hydrolysis and oxidation.

The effects of different functionalities, incorporated through Lewis-acid-mediated hydrosilylation, on the PL characteristics of PSi was investigated in detail [62]. It was found that hydrosilylation of aromatic alkynes resulted in quantitative quenching of the PL (>95% quenching). Conversely, alkyl or alkenyl termination, obtained through insertion of an alkene or alkyne into a Si–H bond, induced ~80% quenching of the initial PL intensity. Furthermore, the PL intensity of functionalized PSi surfaces, notably surfaces **1** and **2**, formed from hydrosilylation of 1-dodecene and 1-dodecyne, respectively, increased upon either soaking in HF solutions (1:1 49% HF/EtOH) at room temperature for 30 min or boiling in aerated basic solutions (3:1 KOH/EtOH, pH 10) for 2 h, as displayed in Table 3.

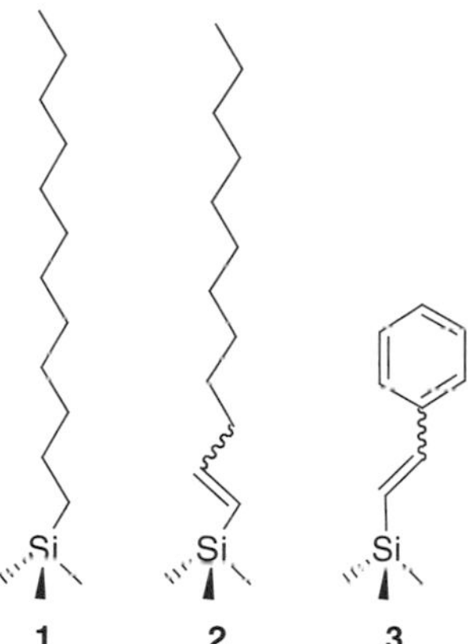

These treatments, however, did not affect the PL intensity of the PSi surface modified with phenylacetylene (surface **3**). It is important to notice that $EtAlCl_2$ induced almost quantitative quenching of the PSi PL and no PL recovery was obtained after exposure to HF/EtOH solution. Immersion of the surface in KOH solution led to complete dissolution of the porous layer.

2.3.2 Photochemical Hydrosilylation

Chemical functionalization of PSi–II through white-light-mediated hydrosilylation of alkenes and alkynes was demonstrated by Stewart and Buriak [63, 64]. Illumination with an ELH bulb (GE slide projector bulb) of moderate intensity (22–44 $mW\,cm^{-2}$), filtered through a 400–600 nm window to eliminate all UV and IR components, of a "photoluminescent" hydride-terminated PSi surface wetted with an alkyne or alkene under controlled atmosphere leads to the formation of organic layers covalently attached to the surface through Si–C bonds (Scheme 6).

Table 3 Relative PL quenching at 670 nm of PSi surfaces functionalized with as compared to the freshly etched porous silicon (100%)

Surface	PL quenching (%)	PL quenching following HF treatment (%)	PL quenching following basic treatment (%)
1	84	65	74
2	82	56	24
3	95	97	96
Neat $EtAlCl_2$	91	94	0

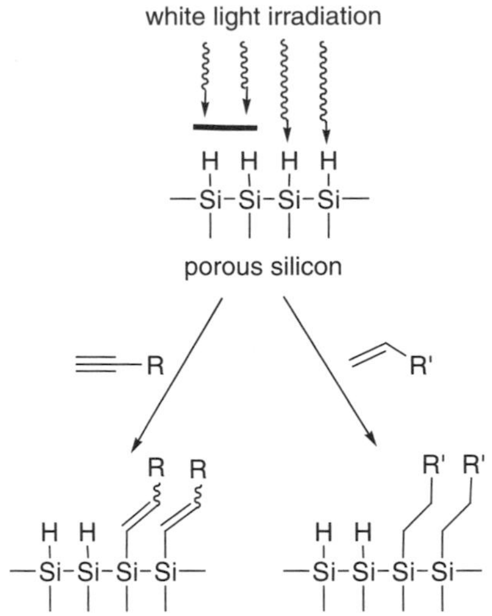

white light irradiation

porous silicon

Scheme 6

While the reaction is not limited by the doping type and morphology of the substrate, it requires, however, the PSi samples to be "photoluminescent". The white-light-induced hydrosilylation reaction yields predominately *trans*-alkenyl product when alkynes are used instead of *cis* stereochemistry seen for the Lewis-acid-induced hydrosilylation of alkynes [59, 60]. The presence of a strong *trans*-olefinic out-of-plane deformation mode γ (=CH) at $\sim$980 cm^{-1} in the FTIR spectrum of the PSi surface modified with alkynes suggests that the chemical process yields mainly *trans* adducts. Reaction efficiencies for the white-light-induced hydrosilylation are in the range of 1–16% and depend typically on the side group of the unsaturated bond, namely, the steric factors and conjugation of the alkynes. Simple alkenes gave the best results (13–16%). In the case of 5-hexynenitrile and tris(ethylene glycol) methyl vinyl ether, surface oxidation was the dominant pathway. For comparison, the Lewis-acid-mediated hydrosilylation of 1-dodecene proceeded with 28% consumption of Si–H bonds, while only 13% of the Si–H bonds were replaced by Si–C bonds for the white light hydrosilylation reaction.

PL is largely preserved after derivatization of PSi surfaces. For example, reaction with 1-dodecene retains almost all the PL intensity ($\sim$97%) whereas dodecenyl termination obtained through 1-dodecyne hydrosilylation maintains $\sim$60% of the original PL intensity, relative to freshly prepared PSi samples. On the other hand, alkenyl termination induces $\sim$ 6% quenching, and hydrosilylation of conjugated alkynes (phenylacetylene and its derivatives) yields complete PL quenching (Fig. 4).

The modified PSi surfaces showed higher chemical stability than freshly prepared surfaces in boiling aerated, aqueous KOH solution. For example, the decyl-terminated PSi surface exposed to boiling alkaline solution of pH 14 showed a corrosion resistance for 5 min, while hydride-terminated PSi was completely

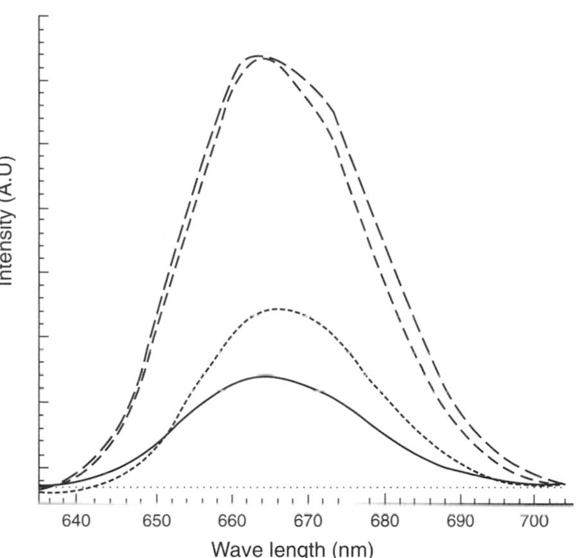

Fig. 4 Photoluminescence spectra of various terminations on porous silicon in order of decreasing intensity: Si-H (–), dodecyl (- -), phenethyl (···), octenyl (-), styrenyl (- - -). Reprinted with permission from M.P. Stewart, J.M. Buriak, J. Am. Chem. Soc., Vol. 123 (2001), pp. 7821–7830. Copyright (2001) American Chemical Society

corroded in seconds under the same conditions. The advantage of using white light is the possibility of photopatterning the surface through a mask (Fig. 5). For instance, a white light pattern was focused on a PSi surface wetted with neat alkene or alkyne and irradiated for 15 min or longer. Figure 5(a) and (b) display optical images of the PSi surfaces that have been photopatterned with 1-dodecyne and 1-dodecene,

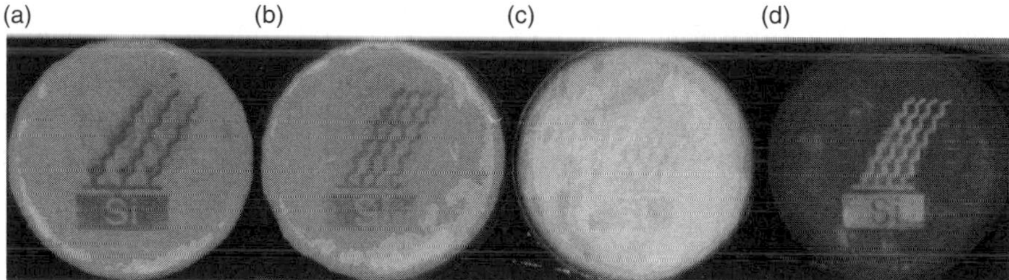

Fig. 5 Photographs of por-Si samples (12 mm diameter) prepared through light-promoted hydrosilylation of 1-dodecyne (surface 2) and 1-dodecene (surface 1) through masking procedures. (**a**) The functionalized section of surface 2 appears as the darkened, (red-shifted) patterned area when illuminated with a 365 nm hand-held UV lamp. The other areas of the wafer are unfunctionalized (native Si-H termination). (**b**) Surface 1 (dark patterned area) upon illumination with 365 nm light. (**c**) Sample from (**b**) after boiling in aerated, aqueous KOH solution (pH 12) for 15 seconds. The unfunctionalized porous silicon (*grey area*) has dissolved, while the hydrosilylated surface (surface 1) remains intact. (**d**) Illumination of the surface from (**c**) with a 365 nm hand-held UV lamp. The PL of the hydrosilylated area (surface 1) remains intact while most of the unfunctionalized PL is destroyed. From M.P. Stewart, J.M. Buriak, Photopatterned Hydrosilylation on Porous Silicon, Angew. Chem. Int. Ed. (1998), VL. 37, pp. 3257–3260. Copyright Wiley-VCH Verlag GmbH & Co. KgaA. Reproduced with permission.

respectively, under PL conditions. The functionalized regions appear red-shifted relative to nonfunctionalized (hydrogen-terminated) areas. Chemical treatment of the photopatterned sample in boiling solution of KOH (pH 10, 15 s) leads to a complete dissolution of the nonfunctionalized porous layer while keeping the derivatized regions intact. Excitation with a UV handheld UV lamp shows PL emission in the functionalized areas (golden area in Fig. 5d), while almost no PL is visible in the dissolved hydride-terminated areas.

Because the hydrosilylation reaction requires relatively low-energy irradiation ($\lambda > 400$ nm) and the porous sample to be photoluminescent, it is believed that excitons generated in situ are responsible for the hydrosilylation reaction, as opposed to Si–H homolysis proposed for UV irradiation on crystalline silicon [65]. The reaction mechanism for the white-light-induced hydrosilylation is outlined in Fig. 6. It suggests the formation of a complex between surface-localized hole and an adsorbed alkene or alkyne. Nucleophilic attack of the π-electron-rich triple or double bond on an electrophilic silicon center leads to the formation of Si–C bond and a carbocation stabilized by a β-silyl group [66, 67]. The acidic carbocation intermediate abstracts a hydride from an adjacent Si–H bond to yield the neutral organic termination.

The proposed reaction mechanism was corroborated by the experiments performed in the presence of PL quenching agents such as ferrocene, 9, 10-diphenylanthracene, and decamethylruthenocene. In fact, efficient quenching of the light emission from PSi prevented the hydrosilylation reaction. The observed reaction pathway has no parallel in the solution phase.

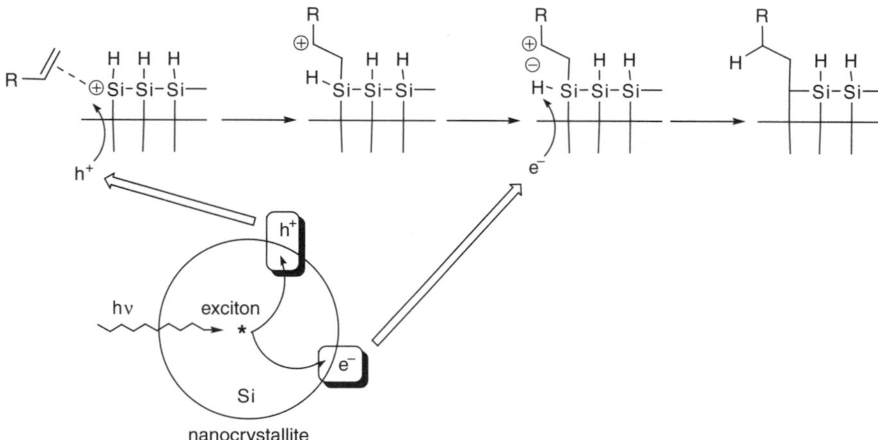

Fig. 6 Proposed mechanism for the exciton-mediated hydrosilylation reaction. An unbound exciton produced by light absorption leads to a surface-localized supra-band gap positive charge. This surface charge can then interact with an alkene and form a silylated β-carbocation upon Si–C bond formation. This carbocation can then abstract a hydride (formally $H^{\cdot+}$ electron from exciton) from an adjacent Si–H bond, yielding the neutral organic termination

2.3.3 Thermal Hydrosilylation

Organic derivatization of PSi–H surfaces by direct reaction with alkenes and alkynes under thermal conditions was first reported by Bateman et al. [68]. In a typical experiment, the PSi-H surfaces were immersed in 1 M solution of unsaturated hydrocarbons in toluene and heated at reflux (110–180°C) for 18–20 h. Surface oxidation was found to compete with alkylation during the chemical process. Based on the FTIR analysis, about approximately 30% of Si–H intensity loss was attributable to oxidation. Electrochemically-active PSi layers were also prepared via hydrosilylation of vinyl ferrocene and the electronic properties of the functionalized PSi surfaces were evaluated using cyclic voltammetry (CV) measurements. The chemically modified PSi showed PL quenching with no PL wavelength shift being observed. The alkyl-terminated PSi surfaces show a significant improvement in their chemical stability in boiling KOH solution (pH 12) for 1 h while unmodified PSi surfaces completely degraded and dissolve within minutes under the same experimental conditions.

The thermal reaction of PSi–H with functional alkenes was further investigated in details by Boukherroub et al. [55, 56, 69–74]. Reaction of freshly etched PSi with neat or a solution of 1-alkene in hexane conducted at a moderately elevated temperature ($<120\,^{\circ}C$) leads to the formation of organic monolayers covalently grafted to the PSi surface through Si–C bonds (Scheme 7):

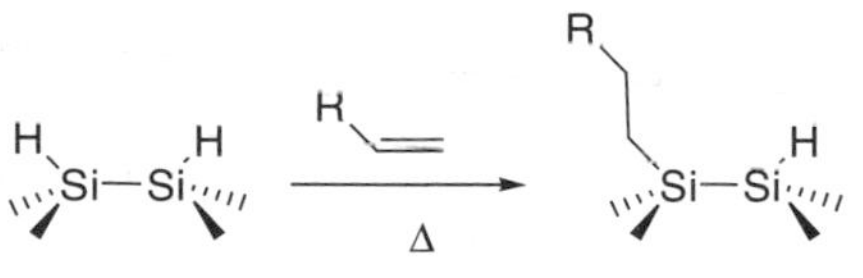

Scheme 7

The reaction takes place with no or little surface oxidation and about 30–50% of the Si–H_x bonds are consumed during the chemical process. The thermal reaction tolerates a wide variety of functional groups to be incorporated in one step on the PSi substrate. The surfaces produced by the thermal method are shown in Fig. 7:

The chemically functionalized PSi surfaces were characterized using Raman, diffuse reflectance FTIR (DRIFT), transmission FTIR, XPS, Auger, and near-edge X-ray absorption fine structure (NEXAFS). Raman spectroscopy shows that the porous nanostructure and the average nanocrystallites diameter are not affected by the thermal treatment. The Raman spectrum of a freshly etched sample is shown in Fig. 8A. It consists of an intense peak at 516 cm^{-1} and weaker features at 630 and 960 cm^{-1} due to scattering from the porous layer. The Raman frequency of crystalline silicon is 520 cm^{-1}. The peak centered at 2115 cm^{-1} comprising three bands at 2089, 2113, and 2139 cm^{-1} (inset) are attributed to Si–H vibrations. The decyl-terminated PSi surface displays identical peaks in feature position and line shape with 30% lower intensity to that of the initial hydride-terminated PSi surface (Fig. 8B).

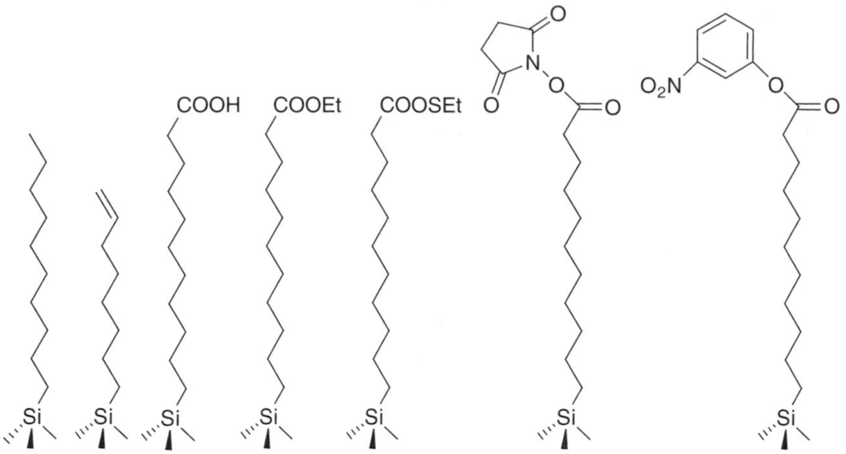

Fig. 7 Examples of PSi surfaces prepared by thermal hydrosilylation of simple and functional alkenes and dienes

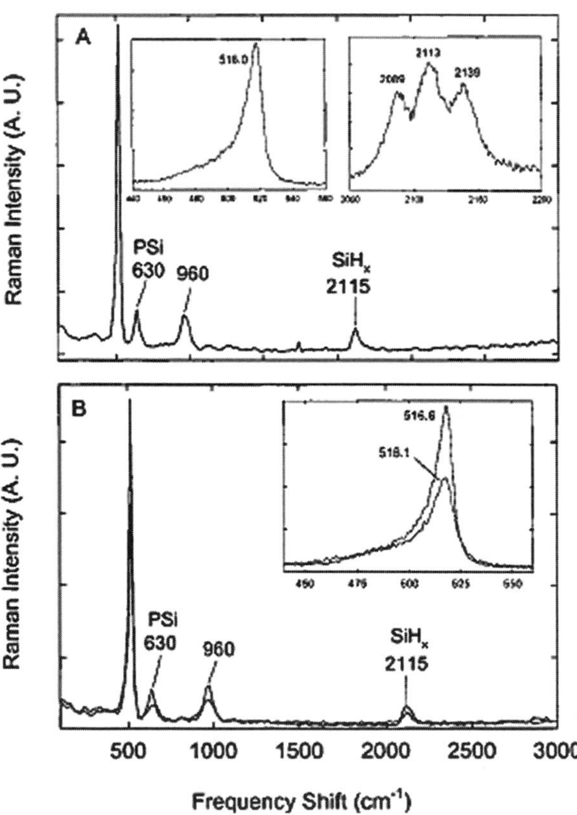

Fig. 8 (**A**) Raman spectrum of freshly prepared PSi recorded at room temperature with a resolution of 8 cm^{-1} and a step size of 20 cm^{-1} and (**B**) Raman spectra of freshly prepared PSi at room temperature before and after derivatization with 1-decene. The insets show details of the spectrum recorded with a resolution of 3 cm^{-1} and step size of 1 cm^{-1}. Reprinted with permission from R. Boukherroub, S. Morin, D.D.M. Wayner, F. Bensebaa, G.I. Sproule, J.-M. Baribeau, D.J. Lockwood, Chem. Mater. Vol. 13 (2001), pp. 2002–2011. Copyright (2001) American Chemical Society

FTIR analysis of PSi surfaces modified with simple alkenes display typical peaks at 2857–2960 cm^{-1} and at 1470 cm^{-1} associated with C–H stretching and methylene bending modes of the alkyl chain, respectively. Additional peaks at ~1715 or 1740 cm^{-1} due to $\nu_{C=O}$ stretching mode were observed in the FTIR spectra of hydrogen-terminated PSi surfaces derivatized with terminal alkenes bearing an acid [55, 56, 74] or an ester [72] terminal group. XPS spectra of the derivatized PSi surfaces show typically peaks at ~99, 151, and 285 eV due to Si 2p, Si 2s, and C 1s, respectively for the samples reacted with simple 1-alkenes. Additional peak at ~532 eV due to O 1s appears in the porous samples functionalized with undecylenic acid and different alkenes bearing a terminal active ester group. The hydrosilylation reaction occurs with little or no surface oxidation, except for the thermal reaction of PSi–H with 3-nitrophenyl undecylenate [72]. In this case, a significant surface oxidation concurrent with the hydrosilylation reaction was observed.

Auger depth profiles of a PSi surface modified with 1-decene displays a constant atomic concentration of carbon throughout the porous layer, indicating that the surface is uniformly modified. The presence of Si–C bond between the PSi and the capping molecules was identified in the C, O, and Si K-edge near-edge X-ray absorption fine structures (NEXAFS) spectra, measured in both the surface-sensitive total electron yield (TEY) and the interface and bulk-sensitive fluorescence yield (FLY) [73].

Brillouin scattering spectroscopy was performed on the derivatized PSi samples to probe the acoustic waves of the chemically modified PSi samples [75–78]. Brillouin spectra of 3.0-μm thick PSi films derivatized with 1-decene, undecylenic acid, and ethyl undecylenate revealed the presence of three acoustic phonon peaks associated with the surface Rayleigh, bulk transverse, and longitudinal acoustic modes, respectively. Relative to the longitudinal acoustic mode frequency of hydride-terminated PSi sample, chemical derivatization resulted in an increase of the acoustic-mode frequency and a shift to higher frequencies. These variations are less pronounced for the decyl-terminated PSi surface. The magnitude of the frequency shifts is assigned to a variation in the average elastic constants and densities due to the difference in bonding, chain length, molecular weight, and/or presence of a dipole moment in the capping organic layer.

The thermal reaction of PSi–H surfaces with 1-alkenes consumes Si–H bonds, as evidenced by FTIR results. This is in agreement with the addition of the Si–H bond across the carbon–carbon double bond. Based on these observations, three different reaction pathways are proposed for the thermal hydrosilylation, as outlined in Fig. 9 [71]. The first reaction mechanism is a free-radical chain reaction, which is initiated by adventitious radical formation (Fig. 9a). This is similar to the mechanism proposed by Chidsey et al. [79] for alkene addition to hydrogen-terminated Si(111)–H single crystal surfaces. The second mechanism involves residual fluoride anions on the PSi surface or in the pores catalysis (Fig. 9a). Nucleophilic attack of the surface silicon atom by F^- anion results in the formation of a pentavalent silicon intermediate, which could then transfer a hydride to the double bond to form a carbanion ($RCH_2CH_2^-$). Subsequent nucleophilic attack of the carbanion on the polarized silicon center leads to Si–C bond formation. The last mechanism

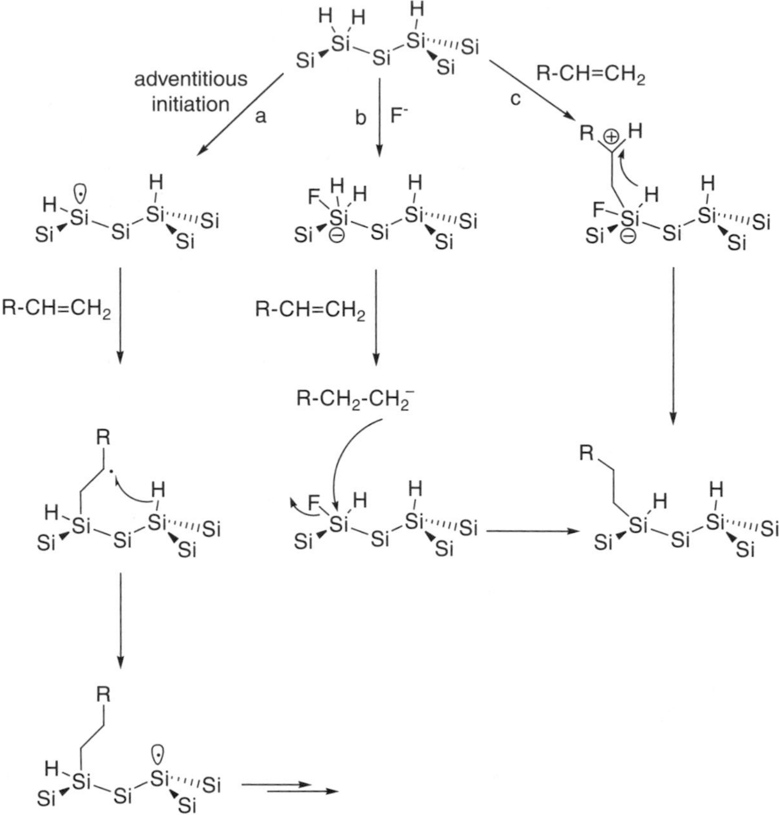

Fig. 9 Possible reaction pathways for the thermal hydrosilylation of alkenes. Path (**a**) is a surface radical-based mechanism. Path (**b**) involves a fluoride-assisted mechanism, resulting from residual fluoride derived from the HF etch. Path (**c**) is based upon direct nucleophilic attack of silicon center by the alkene

is based on the nucleophilic attack of the π-electron-rich double bond on a surface silicon atom to generate a pentavalent Si intermediate, followed by a [1,3]-hydride shift (Fig. 9c). Based on deuterium labeling of the reagents and the surface termination, FTIR spectroscopy and theoretical investigations, Horrocks et al. proposed a radical chain mechanism for the thermal hydrosilylation of PSi with alkenes [37, 39].

The PSi surfaces modified with simple 1-alkenes are chemically robust in harsh environments and can stand the following sequential treatments: sonication in CH_2Cl_2, boiling in $CHCl_3$, boiling in water, and immersion in 1.2 N HCl at 75 °C without apparent chemical degradation of the monolayer or surface oxidation, as evidenced by FTIR spectroscopy [71]. The chemical stability is not affected by long-term immersion in 48% HF or KOH solutions (pH 13) at room temperature. The hydrophobic character of the alkyl chain and the high surface coverage prevents the permeation and diffusion of molecules inside the porous layer. Furthermore, the chemical resistance of the PSi surface modified with undecylenic acid and 1-decene were evaluated by chemography [74]. The derivatized surfaces were exposed to air at 100% humidity and the silane release through hydrolysis reaction of the Si–Si backbonds was monitored [60, 61]. The treatment indicates the extent to which the chemical derivatization protects the surface from oxidation. After 4-h exposure to 100% humidity, the acid-terminated PSi surface was corroded, but to a lower degree than the hydrogen-terminated PSi sample. The decyl-terminated surface was found to be more stable than the acid-terminated one under the same experimental environment [80]. This may be due to the high surface coverage of the decyl-terminated surface and the hydrophobic nature of the methyl group terminating the monolayer as compared to the hydrophilic COOH group of the acid-terminated PSi surface. On the other hand, a partially oxidized PSi surface that has been functionalized with ethyl undecylenate showed an unprecedented stability when exposed to 100% humidity [81].

The effect of the thermal derivatization on the optical properties of the PSi substrate was studied in detail [71]. For example, the PL intensity of the decyl-terminated PSi surface is only 40% of the as-prepared surface. This is less pronounced than 25% of the PL intensity retained by the Lewis-acid-induced hydrosilylation of 1-dodecene [60]. Treatment of the dodecyl-terminated PSi surface with HF resulted in partial recovery to ∼50% of the original PL intensity. In the case of the thermally derivatized PSi surfaces, this treatment had no effect on the PL intensity, which is in agreement with an effective passivation of the surface by the organic coating layer. Aging the sample in air for several months does not induce any change in the PL intensity or in the PL peak position, while exposure to 100% humidity air at 70 °C for 6 weeks caused a PL intensity increase by a factor 3. Aging the freshly etched sample under the same conditions resulted in PL intensity increase by a factor 50. This behavior was assigned to a gradual oxidation of the PSi skeleton, as evidenced by XPS and FTIR spectroscopies [82]. The results were corroborated by Raman spectroscopy, as evidenced by the formation of a transparent silicon oxide layer on the stream-treated PSi–H surface.

By combining controlled surface oxidation and thermal functionalization, it was possible to prepare highly luminescent PSi surfaces with a high chemical and PL stabilities [81, 83]. Figure 10a displays a PL spectrum of freshly anodized PSi sample. The peak is centered at 1.8 eV, characteristic of 70% porosity. After anodic oxidation [47, 84, 85] in 1 M H_2SO_4 at 3 mA/cm^2 for 5 min, an increase of the PL intensity by a factor ∼100 was observed (Fig. 10b). Reaction of the partially oxidized PSi surface with 1-decene at 120 °C for 24 h led to a decrease of the PL intensity by ∼25% (Fig. 10c). The decrease in the PL intensity upon chemical derivatization was also observed for PSi that has not been post-anodized, which implies the introduction of nonradiative recombination centers by the chemical treatment.

Thermal hydrosilylation of alkenes in the gas phase was also examined on hydrogen-terminated PSi surfaces [86]. Reaction of hydrogen-terminated PSi surface with unsaturated hydrocarbons at 350 °C led to the incorporation of organic species on the PSi surface. The derivatized surfaces were characterized using

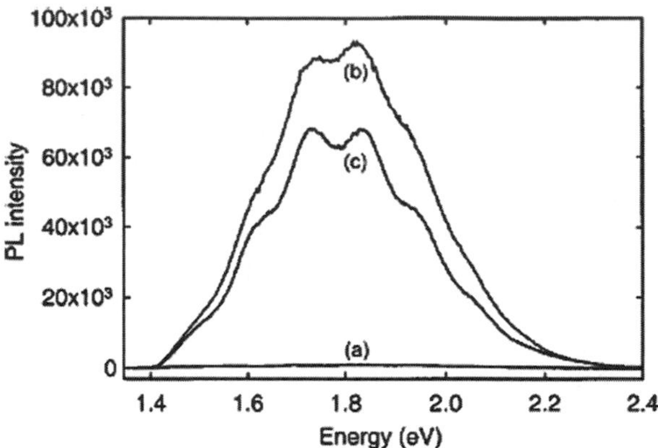

Fig. 10 PL spectrum of the PSi surface etched at 5 mA/cm^2 in HF/EtOH = 1/1 for 8 min (**a**) before electrochemical anodization, and anodized in 1 M H_2SO_4 at 3 mA/cm^2 for 5 min (**b**) before derivatization, and (**c**) after derivatization with 1-decene. Reused with permssion from R. Boukherroub , D.D.M. Wayner, D.J. Lockwood, Applied Physics Letters, 81, 601 (2002). Copyright 2002, American Institute of Physics

transmission FTIR. Comparable results were obtained with ethylene, propene, and cyclohexene. A free-radical mechanism was proposed as a consequence of a homolytic cleavage of Si–H bonds (hydrogen desorption) at 350 °C to generate surface-reactive dangling bonds. The authors did not comment, however, the probable formation of multilayers expected for ethylene under free-radical reaction conditions.

2.3.4 Microwave-Irradiation-Induced Hydrosilylation

Microwave (MW) irradiation was successfully used for the hydrosilylation of simple and functional alkenes with a high chemical yield and a substantial increase in the reaction rate [87]. The specific absorption of MW energy by silicon is believed to generate confined heating in the near vicinity of the surface of the material and thus leading to an efficient hydrosilylation reaction. The MW effect on the hydrosilylation reaction is emphasized by carrying out the reaction under classical heating in an oil bath. At the same temperature and for a given time, the reaction of hydrogen-terminated PSi surface with 1-dodecene at 180 °C leads to very little incorporation of organic species on the surface (Fig. 11d and e). However, a reaction efficiency of 38% was obtained for the hydrosilylation reaction of 1-dodecene at 170 °C for 30 min (Fig. 11c). This is much higher than the average reaction efficiency of 28% reported for the Lewis-acid-induced hydrosilylation of 1-dodecene [60].

The technique allows the incorporation of reactive functional groups on the surface. For example, hydrosilylation reaction of ethyl undecylenate with PSi–H under MW irradiation at 180 °C for 10 min results in the formation of an organic layer, covalently attached to the surface, bearing an ester ($\nu_{C=O} = 1740$ cm^{-1}) terminal group (Fig. 12a). In a similar way, acid end functional groups ($\nu_{C=O} = 1715$ cm^{-1};

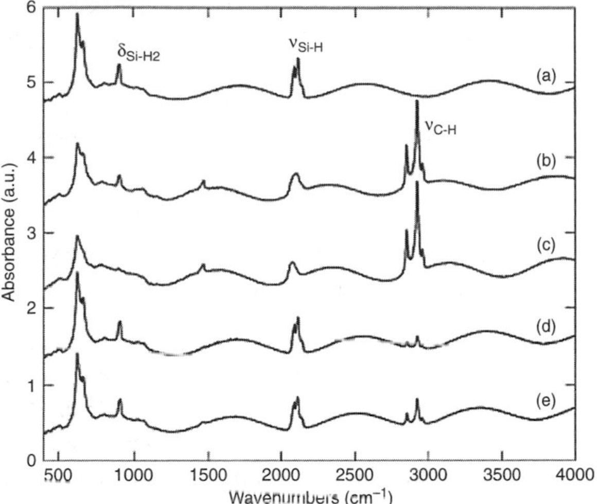

Fig. 11 Transmission-mode FTIR spectra of freshly prepared porous silicon before (**a**), and after chemical functionalization with 1-dodecene under microwave irradiation for 15 min at 170 °C (**b**) and 30 min at 170 °C (**c**), and after thermal derivatization with 1-dodecene at 180 °C for 15 min (**d**) and 30 min (**e**)

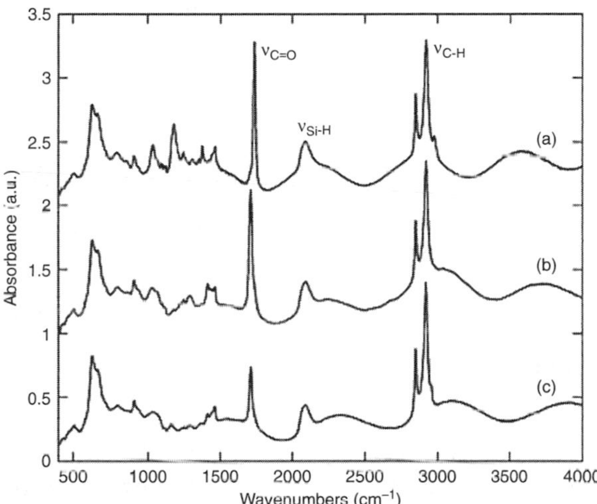

Fig. 12 Transmission-mode FTIR spectra of porous silicon surfaces chemically derivatized with ethyl undecylenate under microwave irradiation for 10 min at 180 °C (**a**), surface (**a**) treated with 2.4 M HCl at 100 °C for 9 h (**b**), and porous silicon surface functionalized with a mixture of 1-dodecene/undecylenic acid (1/1) for 50 min at 180 °C

$\nu_{OH} = 3080\ \text{cm}^{-1}$) can be incorporated to the hydrogen-terminated PSi surface in one step using MW activation (Fig. 12c). FTIR spectroscopy proves that the reaction takes place exclusively on the carbon–carbon double bond and neither surface oxidation nor grafting through the carboxyl groups occurs. The result is comparable to that obtained by thermal-induced hydrosilylation of undecylenic acid [74] with the advantage of a significant increase of the rate of the hydrosilylation reaction for the MW-induced derivatization.

The modified PSi surfaces display high chemical stability in boiling 2.4 M HCl and in HF and KOH aqueous solutions at room temperature. Figure 12b exhibits the

FTIR spectrum of a derivatized PSi surface with ethyl undecylenate that has been treated with 2.4 M HCl at 100 °C for 9 h. It shows only the ester hydrolysis (shift of the $\nu_{C=O}$ from 1740 cm^{-1} to 1715 cm^{-1}), which is in agreement with the formation of very robust organic layers on the PSi surface. The absence of $\nu_{Si\text{–}O\text{–}Si}$ at around 1050 cm^{-1} is a good indication of the ideal passivation of the PSi surface by the organic monolayer.

2.3.5 Hydrosilylation of Alkenes and Alkynes Initiated by Hydride Abstraction

A room-temperature approach for PSi surface functionalization, involving hydride abstraction and subsequent Si–C bond formation by hydrosilylation of terminal alkenes and alkynes was developed by Schmeltzer et al. [88]. Exposure of a hydride-terminated PSi surface to Ph_3CBF_4 in the presence of alkene or alkyne for 3 h at room temperature allows preparing a wide variety of alkyl or alkenyl-terminated PSi surfaces under mild conditions (Scheme 8).

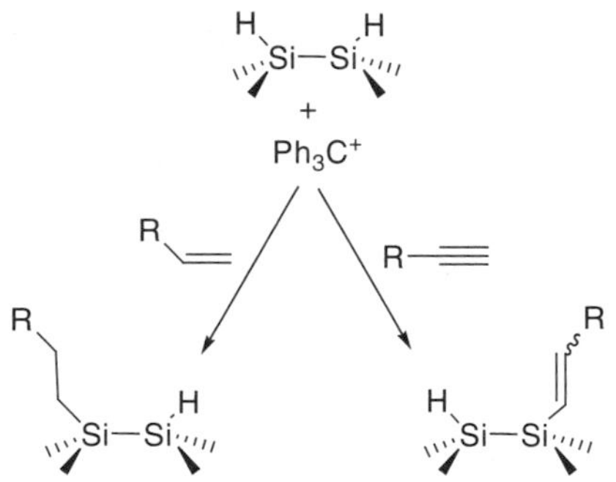

Scheme 8

The presence of alkenyl groups on the PSi surface upon reaction with 1-dodecyne in the presence of the carbocation salt was evidenced by the presence of $\nu_{C\text{–}H}$ stretching modes in the region of 2850–2960 cm^{-1} and $\nu_{C=C}$ vibration at 1600 cm^{-1} in the FTIR spectrum. A band at 979 cm^{-1}, assigned to extraplanar C–H bending of a *trans*-disubstituted alkene was also apparent in the spectrum, suggesting that at least a fraction of the functionalized surface has *trans*-stereochemistry. Further chemical transformation of the double bond by hydroboration reaction is a direct chemical proof for dodecenyl termination of the PSi surface. The reaction takes place with Si–H consumption. The reaction efficiency of ~27% and 19% for the alkenes and alkynes studied, respectively, was determined by the integration of the $\nu_{Si\text{–}H}$ before and after hydrosilylation in the FTIR spectra. The results are comparable to that obtained for Lewis-acid-mediated hydrosilylation [59, 60].

The functionalized PSi surfaces are more stable to oxidation upon immersion in aerated boiling water and ethanolic solutions. The increase in the stability is believed to arise from the hydrophobic character of the derivatized surfaces that retards diffusion of water and nucleophiles inside the porous layer.

By analogy to the proposed mechanism for the solution-phase reaction of silylenium cation with 1,1-diphenylethene to form β-silyl carbocation that are stable at room temperature in solution [66, 89], the authors proposed that the first step corresponds to hydride abstraction by the carbocation. The resulting surface silyl cation then reacts with an alkene or alkyne to form β-silyl carbocation. The propagation step occurs through abstraction of adjacent hydride, leading to new positively charged silicon site on the surface (Scheme 9).

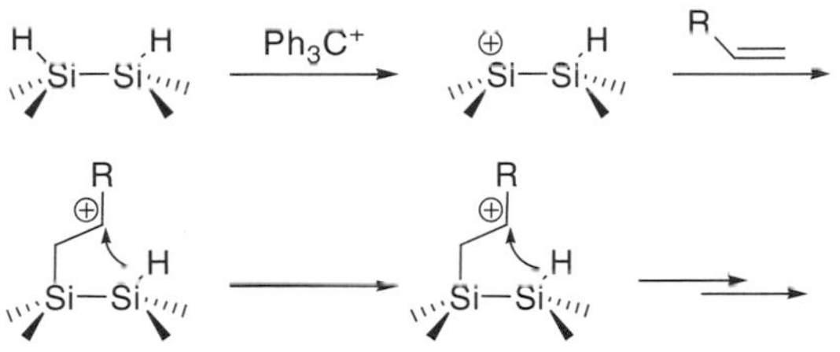

Scheme 9

2.3.6 Reaction with Grignard and Alkyl Lithium Reagents

The analogy drawn between hydrogen-terminated semiconductor reactivity and organosilanes has led to the exploration of various schemes for PSi surface chemical modification. Direct electrochemical formation of Si–C on PSi–H was first reported by Chazalviel et al. [90, 91]. Surface methylation was obtained by anodization of hydride-terminated PSi surface in an electrolyte consisting $CH_3Li.LiI$ or CH_3MgI. The latter reagent is more appropriate, because it provides better electrolyte properties and does not require any addition of supporting salt. In a typical experiment, p-type silicon substrate was positively biased in the presence of the methylating agent for 10 min at 0.5–1 mA/cm^2. The reaction efficiency was over 80% with little surface oxidation. The methyl group was investigated for surface stabilization, because of its small size and thus represents an ideal candidate to achieve high substitution of Si–H bonds. The stability of the modified PSi surfaces is an order of magnitude higher than that of freshly prepared PSi upon ageing in air at 100 °C.

The dominant Faradaic process proposed for the Grignard reagents is the reagent electrolysis as evidenced by the large current flow and magnesium deposition at the counter electrode. Furthermore, the charge passed through during a typical treatment (10 mA/cm^2 $\times$ 10 min) is 6 C/cm^2 is much larger than the Faradaic charge required for oxidizing the surface Si–H bonds (10^{18}) present in a typical sample. This implies that the modification process represents a minority reaction pathway, and the dominant process at the anode being the decomposition of the electrolyte:

$$\mathrm{CH_3MgX + h^+ \longrightarrow CH_3^{\cdot} + MgX^+}$$

Another plausible route proposed by the authors is the generation of I˙ radical intermediates during the electrolyte decomposition:

$$\mathrm{CH_3MgI + h^+ \longrightarrow I^{\cdot} + CH_3Mg^+}$$

The Si–H surface bonds may react with methyl or iodine radicals to generate surface silyl radicals according to:

$$\mathrm{\equiv SiH + CH_3^{\cdot} \longrightarrow \equiv Si^{\cdot} + CH_4}$$

$$\mathrm{\equiv SiH + I^{\cdot} \longrightarrow \equiv Si^{\cdot} + HI}$$

The silyl radical can either react with another I˙ radical to generate iodine-terminated surface or directly react with the Grignard reagent to produce methyl termination. Nucleophilic substitution of Si–I bonds by CH_3– groups yields a similar surface composition:

$$\mathrm{\equiv SiI + CH_3MgI \longrightarrow \equiv SiCH_3 + MgI_2}$$

$$\mathrm{\equiv Si^{\cdot} + CH_3MgI \longrightarrow \equiv SiCH_3 + MgI^+ + e^-}$$

The direct formation of organic layers on PSi–H surfaces through Si–C bonds using a variety of organolithium [92–94] and Grignard [95] reagents at room temperature without intervening chlorination step or photo- or electrochemical activation was demonstrated. The reaction mechanism proposed for Si–C bond formation involves nucleophilic attack of the Si–Si bond by the carbanion (R^-) as outlined in Scheme 10. The cleavage of the Si–Si bond leads to the formation of Si–R and Si–Li surface species.

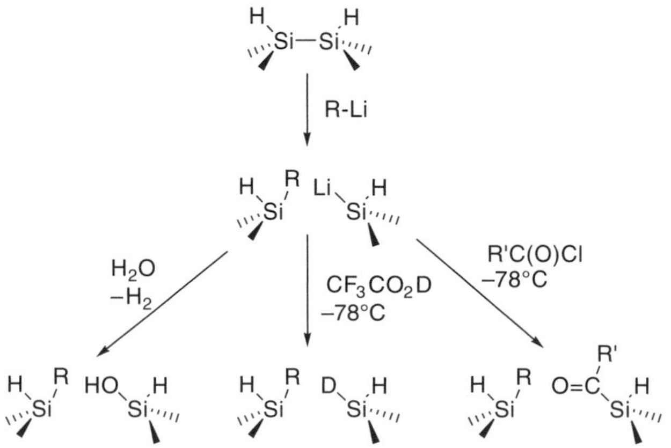

Scheme 10

The highly reactive Si–Li or Si–MgX group is rapidly hydrolyzed, which accounts for the large amount of oxide observed upon exposure of the surface to air. The generation of reactive silyl anions provides a convenient means for introducing different electrophiles on the PSi surface, offering the possibility to form mixed layers (Scheme 10).

The PL properties of the modified PSi surfaces have been investigated in some detail. Surface methylation was found to have little effect on the PL intensity of p-type PSi surfaces [90]. In some cases, a slight increase of the PL intensity was observed with a small blue shift. The blue shift was more obvious upon ageing in air due to surface oxidation.

While PSi surfaces treated with CH_3Li and PhLi showed almost a complete PL quenching after exposure to air, rinsing the surfaces with ethanolic HF solution led to partial recovery of the PL intensity. On the other hand, an irreversible quenching of the PL was observed for PSi surfaces treated with a conjugated aryllithium (PhC≡CLi) exposed to air. Subsequent HF rinsing of the surfaces resulted in a slight (5%) recovery of the PL. Moreover, the PSi surface that has been protonated at low temperature after reaction with PhLi showed a PL quenching, but to lesser extent [92, 93]. However, PSi samples derivatized with decyl and 4-fluorophenylmagnesium bromides continued to exhibit photoluminescent behavior with only small changes in the resulting PL intensity [95].

2.3.7 Reaction with Alkyl Halides under Microwave Irradiation

Direct reaction of PSi–H surfaces with alkyl halides using microwave irradiation (MO) as a source of energy was exploited for alkyl moieties incorporation [49]. Reaction of freshly prepared PSi sample with bromodecane at 120 °C for 30 min under MO led to the covalent attachment of decyl groups on the surface. The chemical grafting of the alkyl groups was accompanied by surface oxidation. IR analysis confirms the covalent grafting of decyl groups on the surface as indicated by the presence of vibrations due to C–H stretching at ~2900 cm^{-1} (Fig. 13b). Peaks associated with chemical oxidation of the PSi surface are also observed at

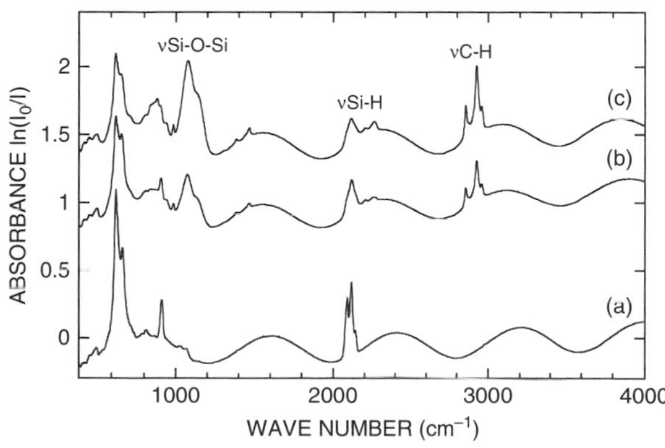

Fig. 13 Transmission-mode FTIR spectra of freshly prepared porous silicon surfaces before (**a**) and after chemical derivatization with bromodecane under microwave irradiation for 30 min at 120 °C (**b**) and 30 min at 150 °C (**c**)

1100 cm^{-1} ($\nu_{Si-O-Si}$), and at 2200 and 2250 cm^{-1} (ν_{Si-H} bonds bearing oxygen in their backbonds with different oxidation states). Increasing the temperature to 150 °C produces a higher concentration of grafted alkyl chains and a significant surface oxidation (Fig. 13c). Alkyl iodides gave similar results under the same experimental conditions. A similar behavior has been observed during the thermal treatment of p-type hydrogen-terminated crystalline silicon Si(111) with tetradecyl bromide [96].

Furthermore, organic layers with terminal acid functional groups are introduced on PSi–H surface by reacting the surface with α-bromo, ω-carboxy alkanes under MO [50]. The resulting PSi surfaces were characterized using FTIR, XPS, and field emission scanning electron microscopy (FESEM). The presence of the COOH end terminal was confirmed by the $\nu_{C=O}$ stretching at 1722 cm^{-1} in the FTIR spectrum and a peak at ~289 eV in the high-resolution XPS spectrum of C 1s. The chemical process occurs with Si–H$_x$ bonds consumption and significant surface oxidation (presence of an intense peak at 103 eV in the high-resolution XPS spectrum of Si 2p).

The reaction of freshly prepared PSi–H with alkyl halides under MO takes place with Si–H consumption and surface oxidation. Because PSi substrate absorbs effectively MW energy, it is believed that a confined heating is produced in the near vicinity of the surface, which can dissociate surface Si–H bonds to yield silicon and hydrogen radicals. The silicon radical propagate the chain reaction as proposed in Scheme 11. The surface Si radical reacts with alkyl halide to form a silicon–halide bond and an alkyl radical. The latter reacts with a Si radical to produce Si–R termination on the PSi surface. Because of the high polarity of the silicon–halide bond, simple exposure to ambient air or solvent rinsing will lead to surface oxidation. However, one cannot exclude the homolytic decomposition of the alkyl halide to yield both alkyl and halide radicals. These radicals are able to react with the Si–H bonds terminating the PSi surface by hydrogen abstraction followed by chemical bond formation of Si–R and Si–X.

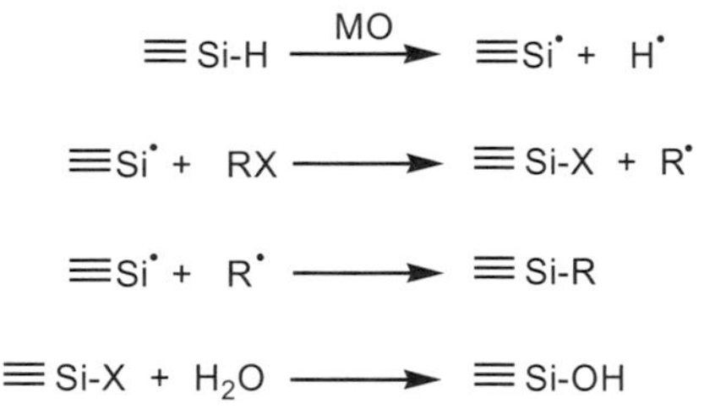

$$\equiv \text{Si-H} \xrightarrow{\text{MO}} \equiv \text{Si}^{\bullet} + \text{H}^{\bullet}$$

$$\equiv \text{Si}^{\bullet} + \text{RX} \longrightarrow \equiv \text{Si-X} + \text{R}^{\bullet}$$

$$\equiv \text{Si}^{\bullet} + \text{R}^{\bullet} \longrightarrow \equiv \text{Si-R}$$

$$\equiv \text{Si-X} + \text{H}_2\text{O} \longrightarrow \equiv \text{Si-OH}$$

Scheme 11

The reaction of free-silyl radicals with organic halides in solution is well documented in the literature [97]. Silanes are used as effective reagents to reduce alkyl halides to the corresponding alkanes in a free-radical process. In other words, reduction of alkyl halides with silanes leads to the formation of silyl halides in a free-radical process [97]. If a parallel exists between molecular solution and the surface-related chemistry, one would expect to obtain a fully halogenated PSi surface when the PSi–H surface is reacted with alkyl halides. However, the presence of both alkyl chains and oxidized silicon–silicon bonds (resulting most likely from hydrolysis of Si–X bonds) on the PSi surface excludes the reaction pathway proposed for free-radical reduction of alkyl halides with silanes. This is in agreement with the conclusions drawn in a previous report [96].

2.3.8 Electrochemical Functionalization

Reduction of Organo Halides

Electrochemical reduction of alkyl iodides, alkyl bromides, and benzyl bromides yields organic layers covalently bonded to p- or n-type PSi surfaces through Si–C bonds with high surface coverage [98]. The reductions are performed in 0.2–0.4 M of the organo halides in dry, deoxygenated acetonitrile or mixtures of acetonitrile and tetrahydrofuran containing 0.2 M $LiBF_4$ by application of a cathodic current of 10 mA/cm^2 for short (<2 min) reaction times (Scheme 12).

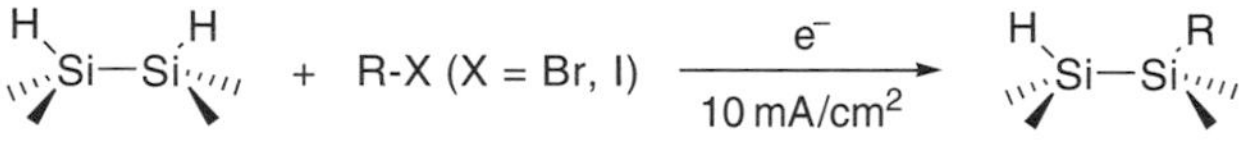

$$\text{H(Si–Si)H} + \text{R-X (X = Br, I)} \xrightarrow[10\ \text{mA/cm}^2]{e^-} \text{H(Si–Si)R}$$

Scheme 12

The electrochemical method leads to a 20–40% decrease in the Si–H stretching band in the FTIR spectrum. Si–C stretching mode at 766 cm^{-1} was observed in the FTIR spectrum of the methyl-terminated PSi surface obtained by reduction of methyl iodide. Isotopic labeling (^{13}C and ^{2}H) studies were further used to identify unambiguously species characteristic of the Si–C stretching vibrational modes [99]. The surfaces produced by the electrochemical method are shown in Fig. 14:

The proposed mechanism suggests the reduction of alkyl or benzyl halides to yield alkyl or benzyl radical species that react with the surface Si–H bonds to generate silicon dangling bonds (or silicon radicals) as outlined in Eqs. (1) and (2).

$$\text{R-X} + e^- \rightleftharpoons [\text{R-X}]^{\cdot -} \longrightarrow \text{R}^{\cdot} + \text{X}^- \quad (1)$$

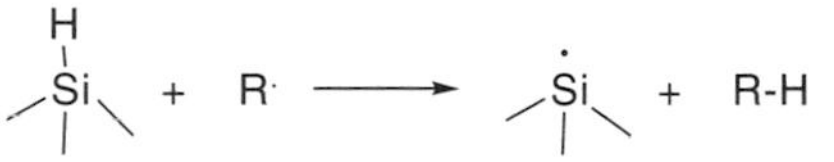

$$\equiv\text{Si–H} + \text{R}^{\cdot} \longrightarrow \equiv\text{Si}^{\cdot} + \text{R-H} \quad (2)$$

The functionalization step may take place by the simple reaction of the surface silicon radical with an alkyl or benzyl radical [Eq. (3)] or by reduction of the silicon radical to silyl anion followed by nucleophilic attack on the organo halide [Eqs. (4) and (5)].

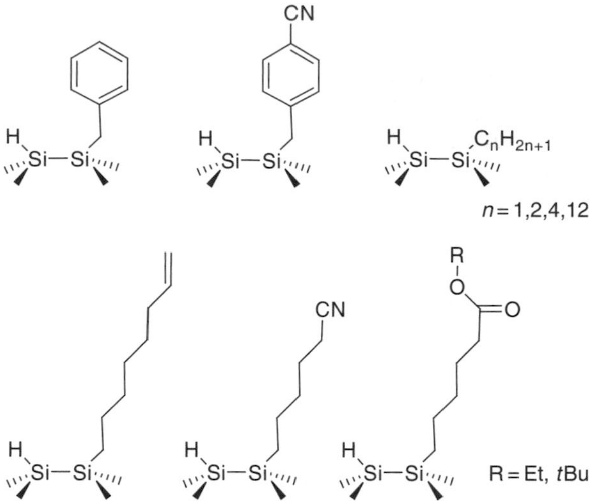

Fig. 14 Examples of PSi surfaces prepared by cathodic reduction of alkyl and benzyl halides

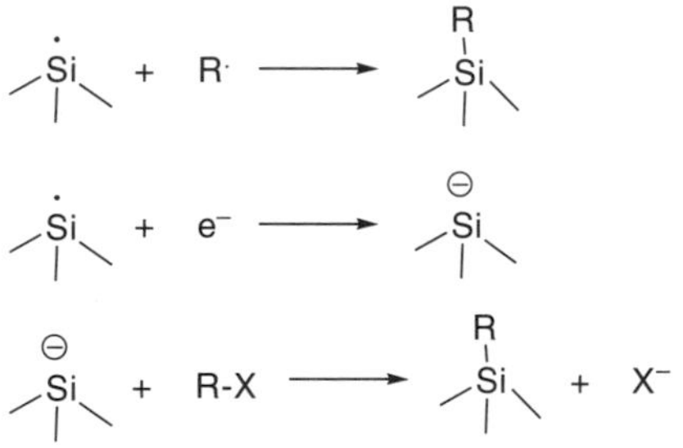

$$\equiv\text{Si}^{\bullet} + \text{R}^{\bullet} \longrightarrow \equiv\text{Si–R} \quad (3)$$

$$\equiv\text{Si}^{\bullet} + e^{-} \longrightarrow \equiv\text{Si}^{\ominus} \quad (4)$$

$$\equiv\text{Si}^{\ominus} + \text{R-X} \longrightarrow \equiv\text{Si–R} + \text{X}^{-} \quad (5)$$

Another plausible mechanism implies the in situ reduction of the alkyl or benzyl radicals to carbanions, which can react with the weak Si–Si bonds to generate Si–R species and silicon anions [Eqs. (7) and (8)], similar to Grignard [95] and alkyllithium reagents [92].

$$\text{R}^{\cdot} + e^{-} \rightleftharpoons \text{R}^{-} \quad (7)$$

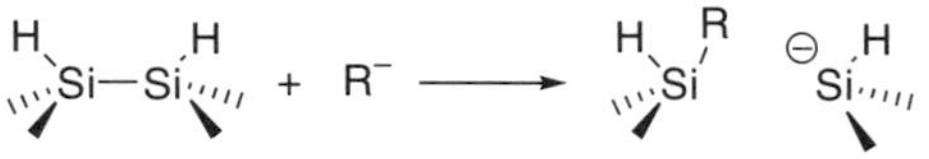

$$\text{H–Si–Si–H} + \text{R}^{-} \longrightarrow \text{H–Si–R} \;\; {}^{\ominus}\text{Si–H} \quad (8)$$

The resulting surfaces are stable in organic solvents and ethanolic HF. Furthermore, the chemical stability of PSi surfaces that have been functionalized in a two-step process, consisting of attachment of the functional group of interest followed by surface methylation (reduction of iodomethane) was studied in detail [100]. Reacting a previously derivatized PSi surface with a smaller reagent to "cap off" the remaining Si–H bonds on the surface leads to Si–H stretching band intensity decrease of ~80%. The resulting surfaces display a greater stability compared to hydride-terminated PSi surfaces or the surfaces modified in one step, when subjected to dimethyl sulfoxide (DMSO), aqueous Cu^{2+}, or 10% ethanol in a solution of phosphate buffered (pH = 7.4). While DMSO oxidizes PSi–H by oxygen transfer [101], reducible transition metal ions are known to oxidize PSi by electron or H atom transfer, generating metallic deposits within the pores [52, 53, 102, 103]. Similar improvements in stability are observed toward aqueous media, such as sodium acetate or $Na_2CO_3/NaHCO_3$ in ethanol–water.

The PL of PSi samples that have been derivatized upon reduction of alkyl iodides was entirely quenched. This was assigned to the presence of residual iodo species on the surface (as detected by energy dispersive X-ray analysis, EDX) [41]. Attempts to remove these species or to recover the PL were unsuccessful. Functionalization with alkyl and benzyl bromides (in $LiBF_4$ electrolyte) on the other hand led to partial loss (~70%) of the initial PL intensity and subsequent rinsing of the PSi samples with ethanolic HF resulted in the PL recovery to 70–90% of the original intensity.

Reaction with Alkynes

Alkynes can be grafted to hydride-terminated PSi surfaces under negative (anodic electrografting, AEG) or cathodic bias (cathodic electrografting, CEG) [104]. FTIR spectroscopy suggests that CEG binds directly alkynes to the PSi surface, whereas AEG produces surface alkylation. CEG of pent-1-yne on the PSi surface reveals the broadening of $Si–H_x$ stretches and appearance of $C–H_x$ vibrations. The absence of a $\nu_{(\equiv CH)}$ vibration mode at ~3300 cm^{-1} and the presence of a sharp peak at 2179 cm^{-1}, which is identical to the silylated molecular analog 1-trimethylsilylpent-1-yne is a good indication for pentynyl termination (Fig. 15a). Furthermore, the disappearance of the $\nu_{(C\equiv C)}$ stretching and appearance of a new vibration band at

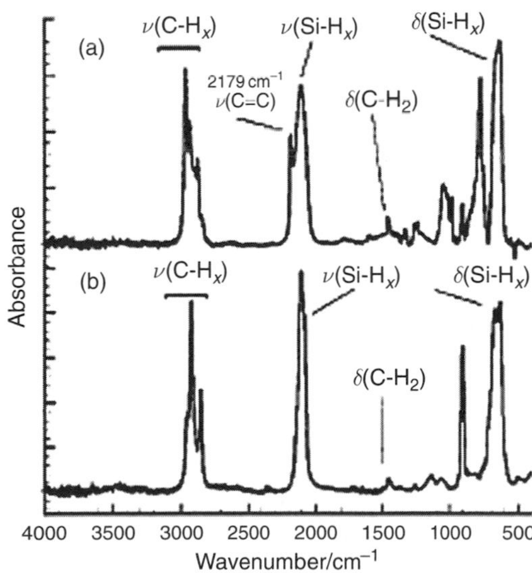

Fig. 15 Transmission FTIR of derivatized porous silicon: (**a**) CEG of pent-1-yne, (**b**) AEG of dodec-1-yne. From E.G. Robins, M.P. Stewart, J.M. Buriak, Chem. Commun. (1999), pp. 2479–2480. Reproduced by permission of The Royal Society of Chemistry

1580 cm^{-1} in the FTIR spectrum upon hydroboration of the pentynyl-terminated PSi surface with a THF solution of disiamylborane is a direct proof for alkyne termination. Positive bias of PSi surface in the presence of alkynes leads also to Si–C bond formation and C≡C reduction. AEG of 1-dodecyne on hydrogenated PSi surface yields a surface comprised of unsaturated bonds. FTIR analysis shows only new features due to aliphatic C–H_x vibrations (Fig. 15b). The lack of C≡C and C=C stretching modes in the FTIR spectrum and the saturation of the C≡C bonds may arise from bis-silylation reaction or oligomeric material formation.

The probable reaction mechanism for the CEG of alkynes on hydrogen-terminated PSi surfaces suggests the formation of silyl anion intermediate by reduction of surface Si–H bond in a space charge layer to yield $H^{\cdot}$ or $\frac{1}{2}H_2$. The formation of silyl anion intermediate was previously proposed for the electrochemical silylation of phenylacetylene with chlorotrimethylsilane [105, 106]. The protonation of the silyl anion by abstraction of H^+ from the alkyne leads to the generation of a carbanion. Subsequent nucleophilic attack of the carbanion on the Si–Si backbonds leads to covalent grafting of alkynyl moieties on the PSi surface accompanied with silyl anion formation (Scheme 13). The reaction pathway was suggested for other carbanions [92, 93, 95, 98]. It was also demonstrated that other weakly acidic moieties, such as alkanethiols can be grafted *via* the CEG reaction.

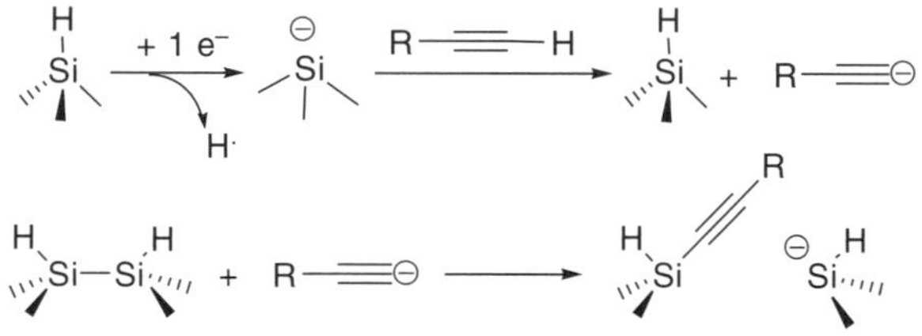

Scheme 13

In the case of the AEG reaction, the incorporation of alkyl groups on the PSi surface is believed to occur through nucleophilic attack by alkyne of positively charged surface silicon sites, followed by hydride transfer. This can be the initiation step for a successive hydrosilylation reaction. The surface-initiated cationic hydrosilylation mechanism is responsible for the Si–C bond formation (Scheme 14).

The derivatized-PSi surfaces display higher chemical resistance in boiling $CHCl_3$, NaOH solution (pH 10), and extended treatment with ethanolic HF compared to unfunctionalized surfaces.

The chemical process retains between *ca.* 5–15% of the initial PL intensity, depending on the surface type. However, a complete PL quenching was observed for arynyl-terminated PSi surfaces.

In Situ Reaction with Alkynes

To date, the covalent derivatization of hydrogen-terminated PSi surfaces was achieved ex situ by exposure of the freshly prepared surface to deoxygenated

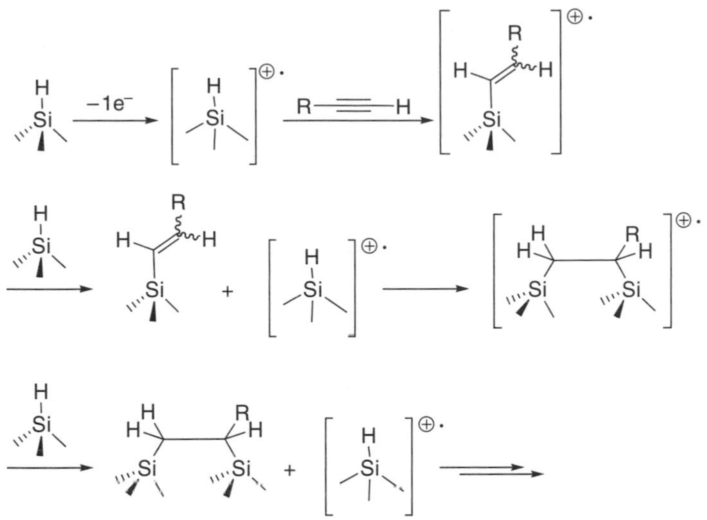

Scheme 14

solution of the organic molecule under different conditions [4]. Mattei and Valentini have developed a new technique for the in situ preparation–functionalization of PS layers [107]. The method involves functionalizing the PSi during its electrochemical formation process. The modification was achieved via dissolving unsaturated organic molecules (alkenes and alkynes) directly in the HF-based etching solution. FTIR analysis of PSi surface prepared by electrochemical anodization of p^+-type crystalline silicon in HF/EtOH (1/2) containing 1-heptyne revealed the presence of features attributable to Si-bonded organic moieties. The procedure is simple, fast, and quite effective with functionalization efficiency of 50–60%. In a control experiment, it was found that no significant functionalization was achieved upon exposure of hydrogen-terminated PSi surface to 0.7 M 1-heptyne HF ethanoic solution, for a time as long as 1000 s.

The influence of the organic molecules present in the etching solution during anodization on the PS morphology was investigated using scanning electron microscopy [108]. It was found that both the thickness and porosity of the PSi layer were reduced, compared to PSi layers etched under the same conditions in the absence of organic molecules, suggesting that the electrochemical functionalization and etching are parallel and competing processes. The formation of nonpolar and chemically stable Si–C bonds during the electrochemical process leads to termination of the etching process on the modified sites. The behavior is dependent on the organic molecule concentration. Increasing the concentration of the organic reactant leads to a decrease of the PS layer thickness and porosity.

The work was further extended to functional molecules such as 6-heytynoic acid and the covalent linkage of the organic species to the PSi surface was demonstrated by their ability to take part in subsequent chemical reactions [108]. For the in situ functionalization of the PSi 1 M of either 1-heptyne or heptynoic acid was added to the electrolyte prior to etching. Even though the reaction mechanism is unknown, the overall expected chemical reactions can be schematically represented as in Scheme 15:

$$\text{PSi-H} \;+\; \text{HC}{\equiv}\text{C(CH}_2)_4\text{-CH}_3 \xrightarrow{e^-} \text{PSi-CH=CH-(CH}_2)_4\text{-CH}_3$$

$$\text{PSi-H} \;+\; \text{HC}{\equiv}\text{C(CH}_2)_4\text{-COOH} \xrightarrow{e^-} \text{PSi-CH=CH-(CH}_2)_4\text{-COOH}$$

Scheme 15

It is believed that the in situ electrochemical modification involves at least one electrochemical step since dipping hydrogen-terminated PSi surface into an organic solution containing electrolytes without the application of a potential bias did not result in any surface modification, as evidenced by FTIR. The presence of the C=C double bond stretching at 1628 cm^{-1} and the absence of C≡C triple bond in the FTIR spectrum of the derivatized PSi surface supports the assumption that the attachment proceeds via the triple bond (hydrosilylation process). A further evidence of the presence of the C=C double bond was its acidic hydration. FTIR shows the disappearance of the C=C double bond at 1628 cm^{-1} and appearance of new peaks at 1120 and 3550 cm^{-1} due to C–O and OH bonds, respectively, when a heptenyl-terminated PSi surface was subjected to acidic hydration.

$$\text{PSi-CH=CH-(CH}_2)_4\text{-CH}_3 + \text{H}_2\text{O} \xrightarrow{H^+} \text{PSi-CH}_2\text{-}\underset{\text{OH}}{\underset{|}{\text{CH}}}\text{-(CH}_2)_4\text{-CH}_3 \;\text{ or }\; \text{PSi-}\underset{\text{OH}}{\underset{|}{\text{CH}}}\text{-CH}_2\text{-(CH}_2)_4\text{-CH}_3$$

Another interesting feature of the acid-terminated PSi surfaces is the availability of the acid function for further chemical transformations. Reaction with methanol at a temperature $>100\,^{\circ}\text{C}$ in the presence of sulfuric acid led to surface esterification (shift of the C=O stretching from 1721 cm^{-1} to 1744 cm^{-1}) without any hydration of the C=C double bond.

$$\text{PSi-CH=CH-(CH}_2)_4\text{-COOH} + \text{CH}_3\text{-OH} \xrightarrow[\Delta]{H^+} \text{PSi-CH=CH-(CH}_2)_4\text{-COOCH}_3$$

The in situ functionalization has a pronounced effect on the PSi PL. The process leads to a decrease of the PL intensity without affecting the PL wavelength.

Reduction of Alkylammonium and Alkylphosphonium Reagents

Alkyl groups derived from electrochemical reduction of alkylammonium, alkyl pyridinium, and alkylphosphonium salts can be covalently bonded to PSi surfaces through Si–C bonds [109]. The electrochemical functionalization was carried out by passing a constant cathodic current (–10 mA/cm^2) to the hydride-terminated PSi surface in a 0.1 M ammonium (or phosphonium/pyridinium) salt solution in dichloromethane for a period of several minutes. FTIR spectra of the resulting PSi surfaces display alkyl vibrations in the region 2800–3000 cm^{-1}. In the case of a mixed phosphonium salt (triphenylmethylphosphonium salt), only the methyl termination was obtained. There was no evidence for phenyl group transfer to the PSi surface. The reaction efficiency was estimated to be in the range of 7–17%, comparable to surface coverages obtained by the white light exciton-mediated alkene hydrosilylation [64], but lower to those found for Lewis-acid-mediated hydrosilylation of alkenes [60]. Prolonged reaction times do not induce any significant increase of the surface coverage, but results in higher level of surface oxidation.

The authors have proposed three different reaction mechanism pathways to explain the surface composition of the derivatized PSi surfaces: (1) reduction of the tetra-alkyl ammonium NR_4 cation into neutral NR_3 moieties and alkyl radicals R. The reactive alkyl radicals can initiate radical-based surface alkylation:

$$\overset{\oplus}{NR_4} + e^- \longrightarrow NR_3 + R^{\cdot}$$

$$\text{H–Si–Si–H} + R^{\cdot} \longrightarrow \dot{\text{Si}}\text{–Si–H} + RH$$

$$\dot{\text{Si}}\text{–Si–H} + R^{\cdot} \longrightarrow \text{R–Si–Si–H}$$

An alternative route is the in situ reduction of the alkyl radicals to carbanions (R^-), which can attack the relatively weak Si–Si back bonds to yield surface alkylation and silyl anions, similar to the reaction of Grignard [95] and alkyl lithium [92] reagents. Addition of 0.1 M HCl in ether blocks both the surface alkylation and oxidation, most likely due to protonation of the silyl anions or carbanions:

$$R^{\cdot} + e^- \longrightarrow R^-$$

$$\text{H–Si–Si–H} + R^- \longrightarrow \text{H(R)Si} \quad \overset{\ominus}{\text{Si}}\text{–H}$$

Another plausible mechanism is the in situ formation of silyl anions, which are susceptible for nucleophilic attack on the trialkylammonium cations, similar to the side reaction of the Hoffman elimination. The involvement of silyl anions is more probable, because the electrografting reaction does not proceed in the presence of alkenes (which rules out the radical-based mechanism) and acids (proton source).

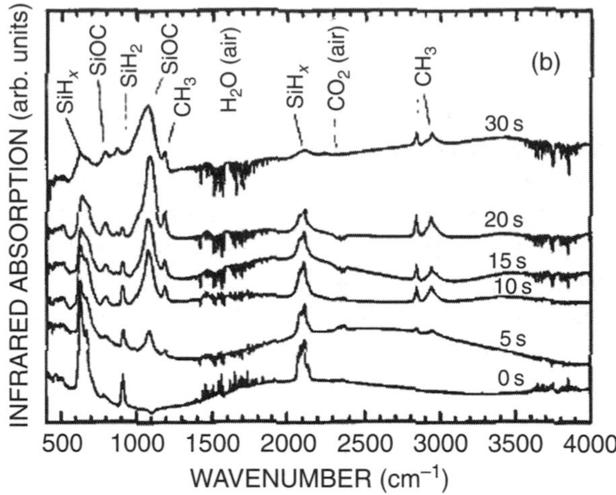

Fig. 16 Infrared absorption spectra of the PSi layer before and after anodic derivatization in anhydrous methanol electrolyte, at a current density of 10 mA/cm^2, and for various durations. The spectra are referenced to that of the bare Si plate. From C. Vieillard, M. Warnthes, F. Ozanam, J.N. Chazalviel, ECS Conf. Proc. Vol. 95–25 (1995), pp. 250–258. Reproduced by permission of The Electrochemical Society

2.4 Chemical Derivatization of PSi Surfaces Through Si–O–C

2.4.1 Electrochemical Methoxylation

Electrochemical anodization of p-type hydride-terminated PSi surfaces in a (methanol + 0.1 M lithium perchlorate) electrolyte led to surface methoxylation [1, 110, 111]. A typical derivatization was obtained by applying a current of 10 mA/cm^2 during 15–20 s. A prolonged treatment up to a coulombic charge in excess of ~300 mC/cm^2 (for a 2-μm porous layer) results in the porous layer dissolution. FTIR analysis of the PSi surface after anodization in methanol electrolyte confirms the grafting of OCH_3 units on the surface: peaks at 2945 cm^{-1} and 2840 cm^{-1} (ν CH_3), 1190 cm^{-1} (ρ CH_3), 1085 cm^{-1} (ν_{as} Si-OC), and 800 cm^{-1} (ν_s Si–OC) (Fig. 16). In a control experiment, it was shown that prolonged exposure of the PSi surface to saturated vapor pressure or liquid methanol without any bias for a period up to 1 day at room temperature does not result in any detectable methoxy groups on the surface by FTIR.

The proposed reaction mechanism suggests an electrochemical initiation step through hole capture by a surface Si–H bond, followed by the reaction of methanol (electron injection into the conduction band) with the activated surface, according to:

$$\equiv\text{Si–H} + h^{+} \longrightarrow \equiv\text{Si–H}^{+\bullet}$$

$$\equiv\text{Si–H}^{+\bullet} + CH_3OH \longrightarrow \equiv\text{Si–OCH}_3 + 2\,H^{+} + e^{-}$$

The dissolution was, on the other hand, ascribed to a chemical process. The silicon atoms bonded to methoxy groups become more prone to dissolution, due to induced polarization and increased reactivity of their backbonds:

$$\equiv\text{Si-SiOCH}_3 + CH_3OH \longrightarrow \equiv\text{SiH} + =\text{Si(OCH}_3)_2$$

The reaction can be repeated until the final dissolution of the grafted silicon atom as $Si(OCH_3)_4$. The existence of such a pathway was established by performing the reaction in deuterated methanol (CH_3OD) instead of methanol. The presence of Si–D_x groups in the IR spectrum of the modified PSi surfaces is a direct proof for the existence of the chemical process, since the direct isotopic exchange between CH_3OD and Si–H species is very unlikely in view of the poor lability of Si–H group.

The methoxy-terminated PSi surfaces exhibit better stability and improved optical characteristics against aging than hydride-terminated samples. The PL of the methoxy-terminated PSi surfaces is slightly blue-shifted upon aging. However, a loss of the PL, accompanied by surface oxidation occurs on a timescale of weeks. This is most likely due to the incomplete substitution of Si–H groups by Si–OCH_3 species.

2.4.2 Photoderivatization with Carboxylic Acids

Lee et al. [112] showed that PSi esterification can be achieved by anodization (a bias of +0.3 V was applied to the Si working electrode relative to the Pt counter electrode) of n-type hydrogen-terminated PSi in neat HCOOH containing 1 M HCOONa under white light illumination. No ester formation was evident when the PSi sample was treated under identical conditions without exposure to light or illuminated at open-circuit potential. In a similar way, acetate and trifluoroacetate groups were grafted on the PSi surface by operating in CH_3COOH/0.5 M CH_3COONa and CF_3COOH/1 M CF_3COONa, respectively [113]. Figure 17 displays the evolution of the FTIR spectrum of the PSi surface during the course of the reaction. In the region below 2000 cm^{-1}, new peaks at ~1070, 1157, and 1710 cm^{-1} assigned to ν(Si–O), ν_{as} (Si–O–C), and ν(C=O) stretching modes, respectively, appear and grow over time. In the hydride region, the intensity of the $\nu(Si–H_1)$ and $\nu(Si–H_2)$

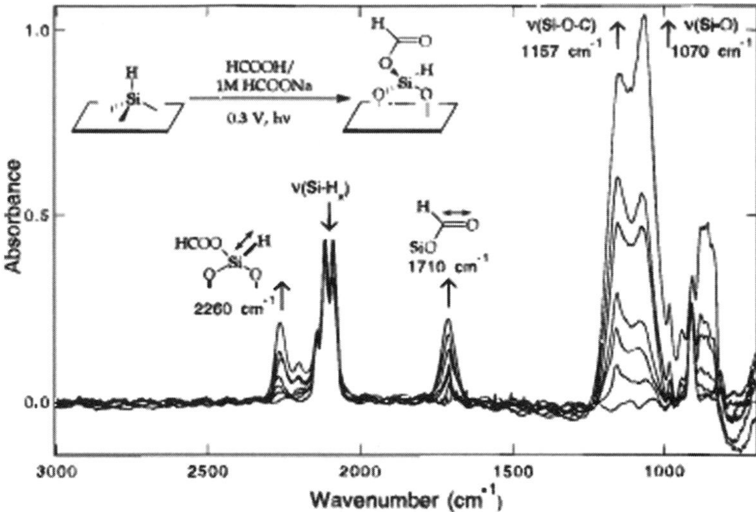

Fig. 17 Infrared absorption spectra of porous silicon taken following photo electrochemical reaction in HCOOH/1 M HCOONa for 0, 1, 3, 5, 10, 18, and 30 min. Reprinted with permission from E.J. Lee, T.W. Bitner, J.S. Ha, M.J. Shane, M.J. Sailor, J. Am. Chem. Soc. Vol. 118 (1996) pp. 5375–5382. Copyright (1996) American Chemical Society

absorptions decrease, with monohydride experiencing the greatest loss. A new peak at 2260 cm^{-1} to ν(OSi–H) stretch was observed and grows over time. The results are in accordance with ester species covalently bonded to the PSi surface through a single Si–OC bonding (binding through a bidentate mode is expected to display three absorptions between 1600 cm^{-1} and 1400 cm^{-1}). The ν(OSi–H), ν(C=O), ν_{as} (Si–O–C), and ν(Si–O) grow at approximately the same rate, with ν(OSi–H) growing faster than ν_{as} (Si–O–C) at longer times. This is in agreement with conversion of surface ester groups to oxide over time.

The absence of Si–H loss, together with the results of the anodization reaction in DCOOD/1 M DCOONa suggest that the reaction mechanism involves Si–Si bond-breaking and silyl radical formation. The silicon surface is activated toward photochemical reaction by applying a positive potential across n-type Si. This bends the conduction and the valence bands of the silicon such that, upon illumination, holes are driven to the surface. The electron-deficient Si surface is susceptible for nucleophilic attack, breaking Si–Si bonds and generating a silicon-ester termination. The reaction produces also a surface silicon radical, which abstracts $H^{\cdot}$ from the solution to form Si–H bond (Scheme 16).

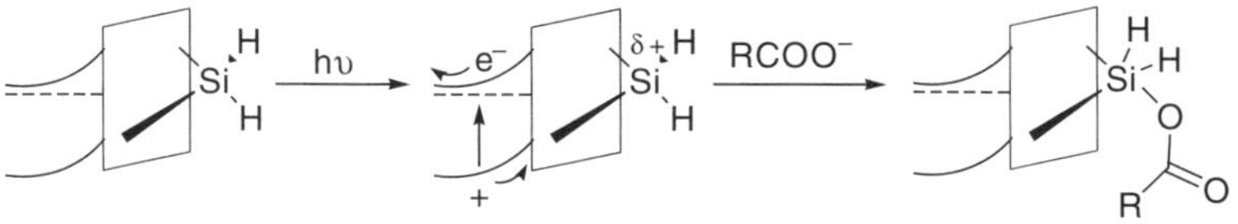

Scheme 16

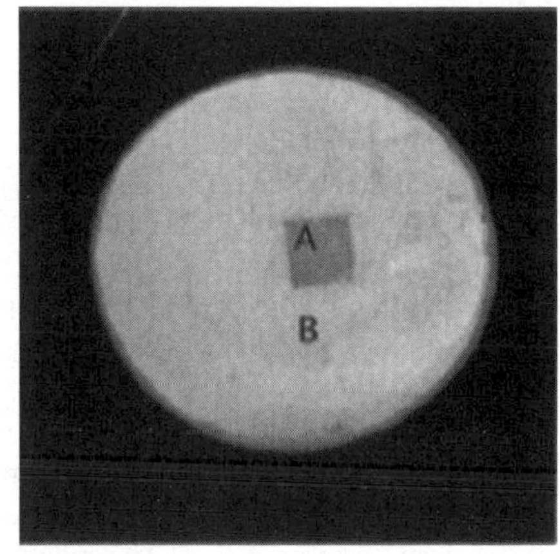

Fig. 18 Photograph of the PL of a PSi surface patterned with trifluoroacetic ester in a 1.5 × 1.5 mm^2. The region modified with the ester (labeled **A**) is less luminescent compared to the rest of the sample (labeled **B**). The PSi was excited at 365 nm. Reprinted with permission from E.J. Lee, T.W. Bitner, J.S. Ha, M.J. Shane, M.J. Sailor, J. Am. Chem. Soc. Vol. 118 (1996) pp. 5375–5382. Copyright (1996) American Chemical Society

The sensitivity of the reaction to light intensity allows the Si to be photopatterned with ester by illumination the surface through a mask during derivatization. Figure 18 displays a photograph of the PL of a PSi surface patterned with trifluoroacetic ester (region A on the figure). The region modified with the ester (A) has a lower PL quantum yield, because the ester acts as nonradiative recombination center. The PL intensity is even more attenuated by increasing the concentration of the grafted ester.

Another interesting aspect regarding the ester termination is its ability to be displaced by organomethoxysilanes [$R(CH_3)_2SiOCH_3$]. Reaction of a trifluoro ester-terminated PSi surface with a 1% solution of $CH_3(CH_2)_7(CH_3)_2SiOCH_3$ in toluene results in the replacement of the ester species with organosilyl groups. The reaction is specific to the ester termination as demonstrated by the chemical reaction of the patterned PSi surface with an organomethoxysilane. The attachment of the silane takes place exclusively on the esterified region.

2.4.3 Thermal Reaction of H-Terminated PSi with Alcohols

Thermal reaction of hydride-terminated PSi surface with methanol was first investigated by Glass et al. [114]. They found that the initial reaction of methanol with the porous layer occurs only at temperatures ≥600 K. At temperatures above 600 K, hydrogen desorption takes place to generate open sites for methanol reaction. In addition to oxygen (Si–O–Si) and carbon (Si–C) incorporation, Si–OCH_3, Si–CH_3, and Si–H surface species were formed when excess of CH_3OH(g) was used. The observed Si–O–Si and Si–CH_3 originate most likely from the decomposition of surface Si–O–CH_3 species at high temperatures. Increasing the annealing temperature for the reaction of 9 Torr of CH_3OH with hydride-terminated PSi surface generated more intense Si–O–Si, Si–C, and Si–CH_3 features. On the other hand, immersion of hydride-terminated PSi surfaces in boiling methanol for 15 min was found to result in a slight surface methoxylation accompanied by a significant oxidation [47]. The relatively high level of oxidation was assigned to the reaction of the surface with residual water present in the solvent. Kim et al. [115] have investigated the thermal reaction of aromatic and long alkyl chain alcohols with hydride-terminated PSi surfaces. Direct reaction of the hydrogen-terminated PSi surface in neat alcohols or a solution in anhydrous dioxane (0.1 M) at temperatures between 20 °C and 90 °C under N_2 atmosphere for 0.5–24 h leads to the formation of organic layers covalently attached to the PSi surface through Si–O–C bonds [115]: The method was successfully used to graft a range of alcohols onto PSi, including phenol, 3-phenylpropanol, 10-undecenol, 11-bromoundecanol, ethyl glycolate, and ethyl 6-hydroxyhexanoate. FTIR analysis was consistent with the formation of O–SiH_x species and a relative decrease of the Si–H_3 species at 2139 cm^{-1} relative to the changes for Si–H (2089 cm^{-1}) and Si–H_2 species (2115 cm^{-1}) upon reaction with alcohol. This observation suggests that the SiH_3 groups are either more reactive or removed from the PSi surface during reaction. The mechanistic aspects of the reaction were further explored by reacting PSi–H with CH_3CH_2OD. DRIFTS analysis of PSi surface that was exposed to the deuterated alcohol for 1 h at 45 °C displayed peaks at 1517–1636 cm^{-1} that are due to Si–D bonds. The presence of Si–D vibrations implies a reaction between the PSi–H surface and alcohols occurring with cleavage of Si–Si bonds:

$$\text{(H)Si–Si(H)} + \text{RO-D} \longrightarrow \text{(H)(OR)Si} \quad \text{(D)(H)Si}$$

The reaction mechanism of alcohols with hydrogen-terminated PSi surfaces was further investigated in detail using infrared spectroscopy and deuterium labeling of the reagents and the surface termination [39]. The reaction kinetics was consistent with either a dissociative adsorption [115] or an electrochemical corrosion type of mechanism. The loss of Si–H species on the surface was attributed to the reaction of PSi surface with trace water in the reaction mixture.

Functional monolayers containing ferrocenylphosphines were covalently attached to H-terminated PSi surfaces through Si–O–C bonds by refluxing the PSi surface in a dry acetonitrile solution of the ferrocenyl derivative (0.02–0.07 M) for 3 h under nitrogen [116]. The absence of O–H stretch and the significant decrease of the Si–H_x intensity in the FTIR spectrum of the PSi surface derivatized with ferrocenylmethanol is an evidence for chemisorption through Si–O–C linkage.

The effect of the derivatization on the photoluminescent properties of the PSi was examined. In general, the thermal treatment with alcohols produced no change in the PL intensity or frequency, regardless of the alcohol [115]. For the PSi samples modified in boiling methanol, there was almost no change in the PL intensity, although a small blue shift and a noticeable increase in the decay time were observed. However, it is hard to assign precisely the efficiency of the process, since the treatment was accompanied by a significant surface oxidation [47].

2.4.4 Thermal Reaction of Halogenated PSi with Alcohols

Lee et al. have developed a two-step reaction to graft alcohols on PSi surfaces [117]. The PSi surface was first brominated by exposure to Br_2 vapor for 5 min, and then reacted with a variety of alcohols for $\sim$10 min to produce organic layers covalently bonded through Si–OC bonds. The authors proposed the following reaction mechanism: exposure of hydride-terminated PSi surface to halogen vapors leads to the cleavage of the relatively weak Si–Si bonds to form Si–X bonds [41, 42, 117]; the halogenated surfaces are very reactive and undergo facile nucleophilic attack with alcohols to yield organic layers covalently attached to the surface through Si–OC bonds.

$$\mathrm{H{-}Si{-}Si{-}H} + Br_2 \longrightarrow \mathrm{H{-}Si{-}Br} + \mathrm{Br{-}Si{-}H}$$

$$\mathrm{H{-}Si{-}Br} + \mathrm{ROH} \longrightarrow \mathrm{H{-}Si{-}OR} + \mathrm{HBr}$$

The derivatized PSi surfaces retain 10–40% of the initial luminescence.

High-quality organic layers of long-chain alcohols can be formed on H-terminated PSi surfaces under mild conditions, in a one-step reaction using iodoform as in situ iodinating agent [44]. Typically, the alcohol (0.25 M of 1-octadecanol or 1 M of 1-dodecanol) and traces of CHI_3 (2.5 mM) were dissolved in deoxygenated toluene and reacted with PSi–H under visible light (15 W ordinary white lamp) or thermal conditions (45 °C) for 24 h.

Iodine termination involves thermal or visible light-induced decomposition of CHI_3 into $CHI_2^{\cdot}$ and $I^{\cdot}$, followed by the subsequent abstraction of surface H-atoms by $I^{\cdot}$, generating surface Si radicals. The surface Si radicals abstracts $I^{\cdot}$ from CHI_3 or combines with another $I^{\cdot}$ to form Si–I termination. FTIR analysis showed a complete disappearance of Si–H stretching upon iodination step. This is in contrast with previous reports that suggested that the iodination of a H-terminated silicon surface with I_2 vapor proceeds through Si–Si cleavage [41, 42]. The iodinated PSi surfaces undergo facile nucleophilic attack with alcohols to yield organic layers covalently attached to the surface through Si–O–C bonds.

The PL of the iodinated PSi surfaces was totally quenched, while about 80% of the PL intensity was recovered for the in situ iodinated/alcohol functionalized PSi surfaces. Attempts to derivatize PSi surfaces with alcohols via a two-step chlorination/substitution method resulted in a complete quenching of the PL during the chlorination step. Further reaction of the chlorinated PSi surfaces with alcohols, water, alkyl lithium reagents failed to recover the PSi PL.

2.4.5 Thermal Hydrosilylation of Aldehydes

While the reaction of hydrogen-terminated PSi surfaces with alcohols leads to Si–Si bond breaking, thermal hydrosilylation of aldehydes produces organic layers covalently bonded to the PSi layers through Si–O–C bonds [69–71]. The efficiency of the reaction as determined by the Si–H_x stretch that was consumed during the chemical process is 30–50%. The hydrosilylation reaction takes place without apparent surface oxidation, as evidenced by the Si–O–Si stretching peak, which does not increase (in contrast to the reported thermal hydrosilylation of alkenes and alkynes [68]).

In a similar way, the reaction of a PSi–H surface with decanal under MO at 180 °C for 15 min yields an organic monolayer covalently attached to the surface through Si–O–C bonds [49]. FTIR spectrum of the decyloxy-terminated PSi surface displays an asymmetric C–H stretching mode at 2925 cm^{-1} and a methylene-bending mode at 1470 cm^{-1}, characteristic of the alkyl chain, and a broad peak in the region of 1000–1100 cm^{-1} assigned to Si–O–C stretching modes. A contribution from partial oxidation of the surface during the chemical process will lead to vibration modes of Si–O–Si in the same region. The reaction takes place with Si–H consumption as evidenced by a net decrease of the Si–H intensity after the chemical treatment.

Ketones react as well with hydrogen-terminated PSi surfaces under MO [49]. Indeed, the reaction of PSi–H with 2-decanone at 180 °C for 15 min yields an organic monolayer covalently attached to the PSi surface through Si–O–C bonds. IR analysis showed a similar spectrum to that obtained for decanal, but with a higher degree of surface oxidation.

Raman spectroscopy showed that the PSi structure was not affected by the thermal process [71]. The modified PSi surfaces displayed identical peaks in feature position and line shape (516, 630, and 960 cm^{-1}) with 30% lower intensity to that of the initial hydride-terminated PSi surface. High-resolution XPS spectra of the PSi surfaces modified with aldehydes are characterized by chemically shifted C 1s and Si 2p peaks, ascribed to the respective carbon–oxygen (C–O) and silicon–oxygen (Si–O) linkages. The results were corroborated by near-edge X-ray absorption fine structures (NEXAFS) [73]. The presence of Si–O bond between the PSi and the capping molecules was identified in the C, O, and Si K-edge NEXAFS spectra, measured in both the surface-sensitive TEY and the interface- and bulk-sensitive FLY.

The elastic properties of the chemically modified PSi samples were investigated using Brillouin scattering spectroscopy [75–78]. Brillouin spectra of 3.0-μm thick PSi films derivatized with decyl aldehyde revealed the presence of three acoustic phonon peaks associated with the surface Rayleigh, bulk transverse, and longitudinal acoustic modes, respectively. Relative to the longitudinal acoustic-mode frequency of hydride-terminated PSi sample, chemical derivatization with decanal resulted in a decrease of the acoustic-mode frequency and a shift to lower frequencies.

The thermal reaction of aldehydes takes place with Si–H consumption as indicated by FTIR spectroscopy. This is consistent with the addition of the weak Si–H bond across the unsaturated C=O double bond in contrast to Si–Si bond cleavage observed for the thermal reaction with alcohols [115], alkyllithium [92, 94], and Grignard [95] reagents. Three different reaction pathways for the hydrosilylation of aldehydes were proposed as for 1-alkenes (**Section 2.3.3.**). Two of the three mechanisms hypothesize on the formation of pentavalent surface silicon anion intermediate. Although the ease of expanding coordination at silicon is well known in solution phase [118], there are no reports in the literature for analogous processes on the surface.

The PSi surfaces derivatized with aldehydes are chemically robust and can stand the following sequential treatments: sonication in CH_2Cl_2, boiling in $CHCl_3$, boiling in water, immersion in 1.2 N HCl at 75 °C, and immersion in 48% HF for 65 h at room temperature without apparent chemical degradation of the monolayer as evidenced by FTIR spectroscopy. The hydrophobic character of the alkyl chain and the high surface coverage prevents the permeation and diffusion of molecules inside the porous layer [71].

The chemical treatment did not have a significant effect on the PL intensity and peak position. The PSi sample modified with octyl aldehyde exhibits an orange-red PL comparable to that of hydride-terminated sample. Samples modified with decanal showed a reduction in the PL intensity of ~60%. The PL was stable during aging the derivatized PSi samples in air for several months. On the other hand, aging for 3 days in 100% humidity air at 70 °C induced an increase of the PL intensity by about 40%. However, aging the freshly etched sample under the same conditions resulted in PL intensity increase by a factor 50. This behavior was assigned to a gradual oxidation of the PSi skeleton, as evidenced by XPS and FTIR spectroscopies [82]. Treatment of the modified PSi surfaces with HF had no effect on the PL intensity, which is in agreement with an effective passivation of the surface by the organic coating layer.

2.4.6 Reaction with Benzoquinone

Reaction of hydride-terminated PSi surface with a 0.02 M toluene (or tetrahydrofuran) solution of 1,4-benzoquinone for 30 min results in surface-bound *p*-hydroxylphenoxy (hydroquinone) species on the surface [119]. The reaction rate is sensitive to the polarity of the solvent and irradiation. FTIR absorbance spectra after derivatization show the appearance of bands characteristic *p*-hydroxylphenoxy termination: $\nu_{(C\text{–}C)}$ at 1518 cm^{-1}, $\nu_{(O\text{–}H)}$ at 3414 cm^{-1}, aromatic $\nu_{(C\text{–}H)}$ at 3052 cm^{-1}, and $\nu_{(C\text{–}O)}$ at 1229 cm^{-1}.

The enhancement of the derivatization rate in polar solvents suggests hydride abstraction from surface Si–H species (Scheme 17) although hydrogen atom

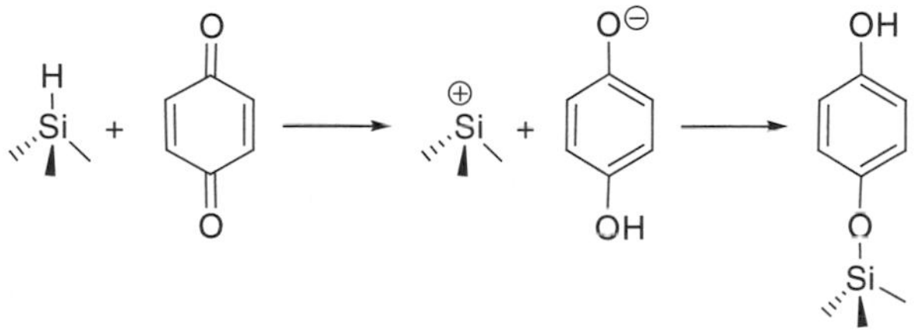

Scheme 17

abstraction cannot be ruled out. Hydride transfer mechanism has already been suggested for the reaction of molecular silanes with quinines [120].

Chemical treatment of the hydroquinone-terminated PSi surface in a solution of $HCl/H_2O/EtOH$ removes >60% of the hydroquinone moieties, accompanied by concomitant surface oxidation. The photoluminescent properties of the PSi surface were slightly affected by the chemical process. The PL intensity of the derivatized PSi surface is reduced to 14% of its original intensity.

2.5 *Miscellaneous*

2.5.1 Formation of Organic Layers through Si–Si Bonds

Li et al. [121] have examined the efficiency of zirconocene and titanocene catalyst systems for direct coupling of molecular silanes ($RSiH_3$) to hydrogen-terminated PSi surfaces through dehydrogenative coupling. It was found that zirconocene catalysts are more effective than the titanium-based catalysts and the level of incorporation of trihydroarylsilanes on the PSi surface was substantially higher than that of aliphatic trihydrosilanes. Several parameters were studied in order to optimize silane incorporation and limit surface oxidation, including adding small amounts of toluene, dropping the catalyst solution directly to the PSi followed by silane, adding the silane first followed by active catalyst solution, and premixing of active catalyst and silane for various lengths of time. FTIR analysis shows that the $\nu_{\text{Si–H}}$ intensity increases during $RSiH_3$ incorporation, since the attachment consumes only one Si–H but leaves two bonded to the original molecule. Integration of the $\nu_{\text{Si–H}}$ stretching region in the FTIR spectra was used to evaluate the incorporation level of different silanes on the PSi surface. The aromatic trihydrosilanes show 14–19% overall increase of the $\nu_{\text{Si–H}}$ intensity, while the aliphatic silanes are substantially lower (3–8%). This was assigned to the lower intrinsic reactivity of aliphatic compared to aromatic silanes, and the high steric hindrance of the surface. Tof-SIMS analyses for PSi samples that has been derivatized via zirconocene-catalyzed dehydrocoupling with phenylsilane showed an order of magnitude higher level of carbon compared to the freshly prepared PSi sample and the modification took place throughout the porous layer.

2.5.2 Plasma Modification

Hydrocarbon-coated PSi surfaces have been prepared using methane decomposition by RF (13.5 MHz) plasma [122]. FTIR spectroscopy shows that the derivatized PSi surface is composed of a strong band at $\sim$2900 cm^{-1}, characteristic of C–H stretching modes and weak bands at $\sim$1600 and $\sim$1710 cm^{-1} due to C=C and C=O stretching modes, respectively. However, the authors did not comment on the origin of these bands. The current-voltage and capacitance-voltage responses of the CH_x-coated PSi surfaces were successfully used for gas sensing [123–125].

2.6 Preparation of Polymer/PSi Hybrids

Irradiation of hydride-terminated PSi surface in the presence of an ethereal solution of diazomethane using the $\lambda = 365$ nm mercury line resulted in the formation of polymeric hydrocarbon species on the surface [126]. FTIR analysis showed the presence of $-CH_2-$ groups rather than terminal $-CH_3$, which is in agreement with an oligomerization reaction taking place most likely via attack of singlet methylene on C–H bonds or via a radical process. The alkylation reaction was accompanied by significant surface oxidation. The oxidized silicon species can be easily removed by rinsing with aqueous HF solutions without affecting the C–H stretching bands observed in the FTIR spectrum.

Another approach consisting on thermal hydrosilylation reaction of vinyl silicone was developed to covalently bind polydimethylsiloxane to hydrogen-terminated PSi surface [127]. The presence of CH_3 vibration and deformation modes at 2961 cm^{-1} and 1440 cm^{-1}, respectively, and Si–O stretching at 1260 cm^{-1} in the transmission FTIR spectrum confirms the grafting of the PDMS polymer on the PSi surface. A hydrosilylation efficiency of 13.4% was estimated from $Si–H_x$ peak integration before and after reaction. The FTIR results were corroborated by contact angle measurements. The PDMS-terminated PSi surface displayed an advancing contact angle of 107°, as compared to 120° for the hydride-terminated PSi surface. Interferometric reflectance spectra indicate that the PSi nanostructure was retained after grafting PDMS. The optical properties of the PSi surface were significantly altered by the functionalization. While the PL peak maximum was not affected by the functionalization, the PL intensity was reduced by 75%. The PDMS termination offers a good stability to the PSi surface in harsh environments, such as boiling in alkaline solutions.

Ring-opening metathesis polymerization (ROMP) was used for *in situ* polymerization and cross-linking of pores in a free-standing PSi multilayered dielectric stacks [128]. The PSi film was prepared by periodic anodic etching of B-doped, p^{++} type, Si(100)-oriented silicon wafers of resistivity <1.0 mΩ cm [129, 130]. A polymer composite was prepared in two steps by loading the ROMP catalyst (tricyclohexylphosphine[1,3-bis(2,4,6-trimethylphenyl)-4,5-dihydroimidazol-2-ylidene][benzylidine]ruthenium(IV) dichloride, at room temperature, and then 3 M solution of norbornene in dichloroethane. Heating at 100 °C for 30 min led to the formation of a composite PSi/polymer material in which the polymer is covalently attached to the surface through Si–C bonds. FTIR spectrum of the resulted polymer composite showed vibrations due to C–H and C=C bonds at 2780–3090 cm^{-1} and 1550–1810 cm^{-1}, respectively. Optical reflectance confirmed that the polymer infiltrated the pores of the porous layer and red shift of 90 nm was observed upon introduction of the polymer. The chemical and mechanical stability of the chemically cross-linked PSi matrix are significantly improved compared to PSi alone. The PSi/polymer composite prepared from hydride-terminated PSi multilayer can stand treatments in HF or alkaline (3 M KOH) for several days without any apparent degradation, while the PSi/poly(norbornene) composite material obtained from oxidized PSi surface readily dissolves in HF-based solutions or removed by soaking in THF. This clearly suggests that the polymerization in a freshly etched PSi silicon template leads to covalent attachment *via* hydrosilylation reaction.

2.7 Covalent Immobilization of Biomolecules

Direct synthesis of oligonucleotides using automated solid-phase DNA synthesis on chemically functionalized PSi surfaces was investigated by Lie et al. [131]. Hydrogen-terminated PSi surface was reacted with protected difunctional alkene: undecenol-DMT by reflux in toluene under nitrogen atmosphere. Deprotection of the DMT group led to the formation of an organic monolayer bearing primary alcohol functional group capable of reacting with nucleobase phosphoramidites. The amount of DNA synthesized on the PSi surface was quantified by attaching ^{32}P-labeled 5′-phosphate groups to complementary DNA sequences or by retaining the 5′-dimethoxytrityl protecting group at the end of synthesis and determining spectrophotometrically the DMT^+ cation released upon deprotection in a small volume of trichloroacetic acid/dichloromethane. Surface coverages of synthesized oligonucleotides of $\sim 10^8$ mol cm^{-2} are obtained on PSi compared to *ca.* 5×10^{-12} mol cm^{-2} on Si(111) [132]. A hybridization efficiency of 40 ± 5% was determined by spectrophotometry. The lack of improvement of the hybridization efficiency by addition of EtOH suggests that wetting of the PSi layer was not the reason for incomplete hybridization, but is a consequence of the packing of the DNA molecules on the surface itself.

Single-strand DNA was immobilized on NHS-terminated PSi surfaces, and its hybridization reaction was evaluated using FTIR spectroscopy and optical reflectivity [133]. The NHS termination was obtained through photochemical hydrosilylation of N-hydroxysuccinimide ester of undecylenic acid (UANHS). While optical reflectivity spectra of PSi surfaces after s-DNA immobilization and hybridization with its complementary showed shifts of 31 nm and 33 nm, respectively, the authors did not give any detail regarding the nature of the chemical bonding associated with the DNA or surface coverage or hybridization efficiency.

Covalent attachment of biomolecules to active PSi surfaces has been demonstrated by Reynolds et al. using a multistep approach [134, 135]. First, amine termination on hydrogen-terminated PSi was obtained in a two-step process based on Lewis-acid-mediated hydrosilylation of hex-5-ynenitrile followed by $LiAlH_4$ reduction or on photochemical of *N*-BOC-amino-3-butene followed by acid deprotection of the BOC group. The chemical versatility of the $-NH_2$ functional group was further explored to introduce –NHS termination using heterobifunctional protein cross-linking reagents *N*-succinimidyl-3-(2-pyridyldithio)propionate (SPDP) and *N*-(γ-maleimidobutyryloxy)succinimide (GMBS). The presence of an activated ester (NHS) provided by GMBS at the end of the linker provides an electrophilic platform readily available for coupling with nucleophilic species. Dansyl, biotin, streptavidin species [134] and enzymes [135] bearing terminal primary amine functional groups have been successfully immobilized on the PSi surface through amide bond formation.

Bovine serum albumin (BSA) and lysozyme (Lys) were covalently immobilized on acid-terminated PSi surface through amide bond formation [50]. The acid-terminated PSi surfaces were prepared by the reaction of α-bromo, ω-carboxy compounds with hydrogenated PSi substrate MO. The acid termination was converted to an activated NHS ester by a simple chemical route using *N*-hydroxysuccinimide in the presence of *N*-ethyl-*N'*-(3-dimethylaminopropyl) [74] or *N,N'*-dicyclohexyl carbodiimide [50]. The NHS termination can then easily be coupled to any biomolecule bearing a primary amine.

Chemical grafting of anti-fouling components onto PSi surface such as oligo(ethylene glycols) through Si–C bond formation, and active groups to allow the covalent immobilization of biorecognition elements has been reported by Schwartz et al. [136]. Biofouling is crucial for preventing and suppressing the nonspecific adsorption of biomolecules on the surface during device operation and for designing biocompatible coatings. PEG-terminated organic layer was prepared in a multistep process from an acid-terminated PSi surface. The acid termination was obtained by thermal hydrosilylation of undecylenic acid. PEG incorporation was achieved by coupling of amino-dPEG$_4$-*t*-butyl ester in the presence of *N,N'*-dicyclohexylcarbodiimide (DCC). Finally, a terminal acid functional group was formed through acidic deprotection of the *t*-butyl ester. The resulting surface was exposed to a PBS solution of sucrose and the change in reflectivity peak maximum versus time for different concentrations was recorded.

2.8 Applications

2.8.1 Surface Passivation

Earlier work by Canham et al. demonstrated that low-porosity microporous films display a beneficial form of bioactivity, the in vitro promotion of hydroxyapatite growth [137]. On the other hand, it was found that high-porosity mesoporous films exhibit substantial dissolution under simulated physiological conditions [138]. Whereas for biomedical applications, material corrosion is desirable, for biosensors a stable interface of the porous layer in biological environment is required [28].

PSi samples derivatized through Lewis-acid-mediated hydrosilylation of 1-dodecyne [59] were incubated at 37 °C in simulated body fluid (SBF) for periods ranging from hours to weeks [139]. After 4 weeks (700 h) incubation, layer thinning of the derivatized PSi surface was estimated to be $\leq$25 nm while almost the whole porous layer ($\sim$250 nm) of the underivatized PSi surface was dissolved after only 70 h. The result indicates that the corrosion rate was reduced by a factor of 100 by replacing surface Si–H bonds by more stable Si–C bonds. Consequently, the ability to tune bioreactivity via surface chemistry opens ways for new applications in the field of biosensors and biomedical research.

2.8.2 Electroluminescence Stabilization

The effects of surface termination of PSi samples on the electrochemiluminescence (ECL) were investigated by Buriak et al. [140]. The PSi surfaces were functionalized through cathodic electrografting (CEG), Lewis-acid-mediated hydrosilylation, and anodic electrografting (AEG) and the ECL was induced through formic acid/sodium formate electron injection system [141]. Alkyl-terminated PSi surfaces obtained through anodic electrografting of 1-dodecyne showed the brightest emission, even though ECL emission was about half as compared to that of hydride-terminated PSi surface. On the other hand, the lifetime was extended by a factor 2, and the derivatized surfaces exhibited recharging phenomenon upon application of a cathodic potential. The dodecenyl-terminated PSi surfaces prepared through Lewis-acid-mediated hydrosilylation of 1-dodecyne showed the lowest ECL emission intensity almost unobservable under the experimental conditions used. The lifetime, however, was an order of magnitude longer than that of PSi–H surface. Comparable results were obtained for dodecyl-terminated PSi surfaces obtained through Lewis-acid-mediated hydrosilylation of 1-dodecene. Finally, the styrenyl-terminated PSi surface, produced through hydrosilylation of phenylacetylene, showed absolutely no light emission. Conjugated termination consisting of conjugated aryls is known to quench also quantitatively the PSi PL [62, 92], suggesting that these organic moieties act as nonradiative recombination centers for both PL and ECL emission.

Stabilization of EL properties of PSi-based light-emitting devices by chemical derivatization has been demonstrated by Gelloz and collaborators [142–144]. The PSi diodes are formed from n^+ (111) (0.018 Ωcm) substrates. First, a superficial compact PS layer is formed by anodization in the dark for 30 s at 5 mA/cm^2 in a HF (10 vol.%) solution. Optically active PS is subsequently formed by photo-anodization at 3 mA/cm^2 for 10 min, under illumination with a 500 W tungsten lamp mounted at a distance of 20 cm, in HF (40 vol.%), at 0 °C. The total thickness of the porous layer is about 3 μm, but the optically active PS thickness should be limited to about 1 μm because of a large porosity gradient apparent in SEM pictures. Electrochemical oxidation of PS is then carried out until the intensity of electrochemical EL during the treatment reaches a maximum [145, 146]. The partially oxidized PSi surfaces were derivatized by thermal reaction with 1-alkenes [81, 83]. Figure 19 shows the normalized EL efficiency as a function of time for different devices operated at a constant voltage. The EL efficiency of the as-prepared device (just after electrochemical oxidation) dropped by half of its original intensity in only 40 min. The decrease in efficiency was assigned to surface oxidation occurring during device operation, which introduces surface-active defects acting as nonradiative recombination centers. In contrast, devices functionalized with 1-decene and ethyl undecylenate under thermal conditions showed an outstanding EL stability. The results can be explained by the higher chemical stability of Si–C bonds, as compared to Si–H bonds, which prevents surface oxidation. More interestingly, EL efficiency of the PSi-based diode derivatized with undecylenic acid increases as a function of time then saturates, and remains stable for hours under continuous operation. Although thermal modification of PSi-based devices with 1-decene and ethyl undecylenate

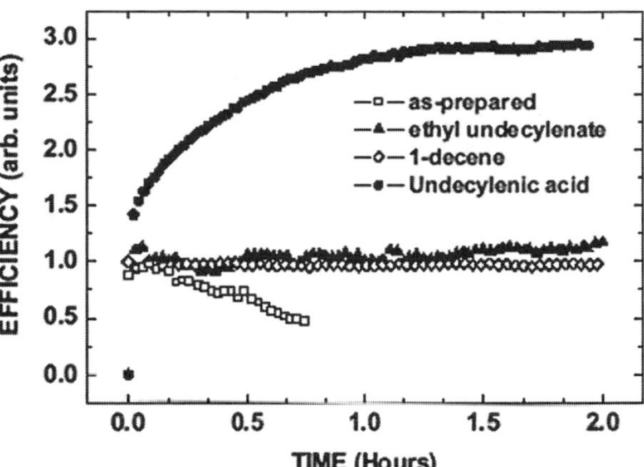

Fig. 19 Normalized EL efficiency as a function of time. From B. Gelloz, H. Sano, R. Boukherroub, D. D. M. Wayner, D. J. Lockwood, N. Koshida, Stable electroluminescence from passivated nano-crystalline porous silicon using undecylenic acid, Physica Status Solidi C (2005) Vol. 2, pp. 3273–3277. Copyright Wiley VCH Verlag GmbH&Co. KgaA. Reproduced with permission

leads to reduced EL efficiency and EL output intensity, the device modified with undecylenic acid shows both EL efficiency and output intensity comparable to the reference device. EL can be observed for hours under DC operation.

The spectral characteristics of the chemically functionalized PSi-based LED are similar to that of as-prepared ones. The EL peak position at 650 nm, shape, and the width at half-maximum (~240 nm) remain unchanged after thermal derivatization. Furthermore, the EL peak and shape of the device modified with undecylenic acid are not affected upon operation for several hours (Fig. 20). The PL characteristics were not affected by the prolonged operation as well.

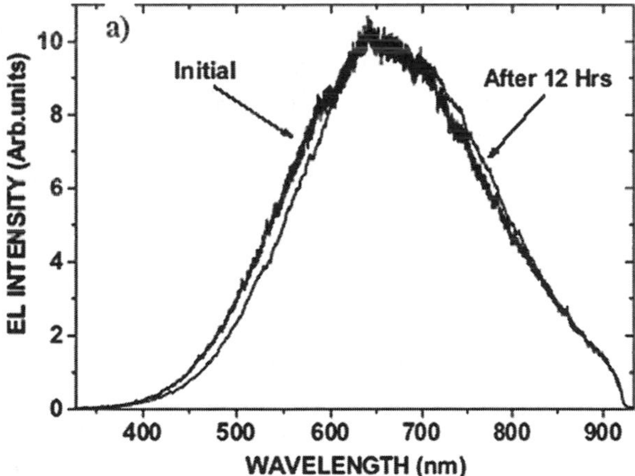

Fig. 20 Normalized EL spectra (**a**) of a device treated with undecylenic acid, before and after 12 h of operation. From B. Gelloz, H. Sano, R. Boukherroub, D. D. M. Wayner, D. J. Lockwood, N. Koshida, Stable electroluminescence from passivated nano-crystalline porous silicon using undecylenic acid, Physica Status Solidi C (2005) Vol. 2, pp. 3273–3277. Copyright Wiley VCH Verlag GmbH&Co. KgaA. Reproduced with permission

2.8.3 Desorption/Ionization on Silicon (DIOS)

Siuzdak and Buriak have shown that PSi samples can be used as sample plates for desorption/ionization mass spectrometry [147]. The role of the porous layer is to assist desorption/ionization of the sample molecules instead of the matrix compounds used in classical matrix-assisted laser desorption/ionization (MALDI). Because organic matrix is not required in DIOS, mass spectra with significantly low background at low mass range can be obtained. This property allows reliable analysis of low molecular weight molecules such as ligands and drugs. Even though the influence of different parameters such as PSi preparation, pore size and depth, sample composition, and optical characteristics on the DIOS performance have been investigated in the literature [148–151], the mechanism by which the ionization occurs is still not understood. Several compounds with molecular weights ranging from 150 Da to 12,000 Da, including carbohydrates, peptides, glycolipids, natural products, and small molecules, were analyzed and their molecular ion was observed with little or no fragmentation. It was also demonstrated that phenylethyl termination obtained through Lewis-acid-mediated hydrosilylation of styrene provides better signals [147]. Chemical derivatization of PSi samples with ethyl undecylenate or 10-undecenoic acid under thermal conditions improved the DIOS performance providing better ionization efficiency and signal stability with lower laser energy [151]. It was suggested that the organic moieties on the chemically derivatized PSi surfaces aid proton transfer reaction in DIOS without increasing the background noise.

PSi surfaces with tailored chemical properties are of particular interest to selectively capture small or biological molecules from complex mixtures and perform DIOS analysis. Moreover, covalent attachment of analytes on PSi surfaces enables spatially addressable and multistep synthesis on the chip. For example, thermal functionalization of hydrogen-terminated PSi surface using *N*-(4-vinylphenyl)maleimide at room temperature leads to the incorporation of Diels–Alder moieties on the PSi surface. The termination allows small molecules to be both covalently attached to the PSi surface and detected by mass spectrometry [152].

2.8.4 Sensing

The photoluminescent properties of the modified PSi surface can be used to tune the sensor properties by efficient quenching of the PL of the material in the presence of different gazes (Fig. 21). PSi surface derivatized with hydroquinone moieties imparts a greater sensitivity to water vapor and reduces the sensitivity to benzene vapor, compared to hydrogen-terminated PSi [119]. In fact, water vapor quenches 86% of the PL from hydroquinone-terminated PSi, whereas it only quenches 3% of the hydride-terminated PSi. On the other hand, benzene vapor reduces the PL intensity of the hydroquinone modified PSi surface by only 26%, while it quenches 70% of the PSi–H surface.

The effect of chemical surface treatment on the sensitivity, specificity, and stability of mesoporous silicon thin film vapor sensors was studied in detail [153]. The sensor operates by measurement of the shift of the wavelength of the Fabry-Pérot

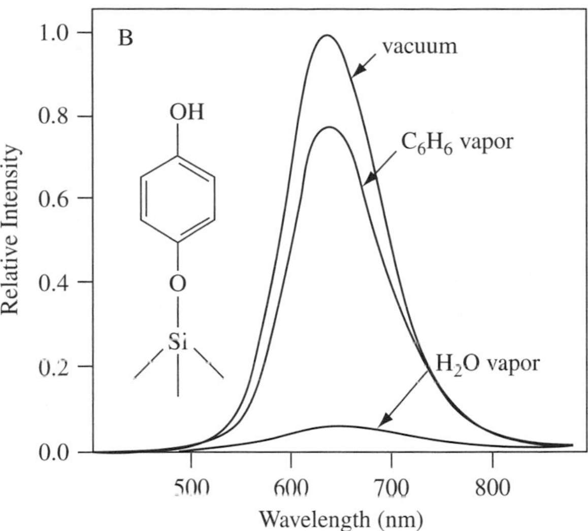

Fig. 21 Steady-state PL spectra of hydroquinone-terminated PSi after exposure to benzene (90 Torr) and water (21 Torr) vapors relative to vacuum. Reprinted with permission from T.F. Harper, M.J. Sailor, J. Am. Chem. Soc. Vol. 119 (1997) pp. 6943–6944. Copyright (1997) American Chemical Society

fringes induced by the change in the refractive index of the porous layer upon exposure to organic vapors. A methyl-terminated PSi, prepared by electrochemical reduction of CH_3I [98], was found more stable to vapors that are corrosive to the hydride-terminated PSi surface. Furthermore, both hydrogenated and methylated PSi surfaces are more sensitive to hydrophobic analyte relative to oxidized PSi samples.

The ability to couple biological molecules to PSi is particularly attractive due to the possibility of label-free detection. The chemical functionalization of hydrogen-terminated PSi surfaces by means of Si–C bonds offers a better stability of the interface in aqueous and biological environments. DNA probe was immobilized on NHS-terminated PSi surface obtained in a two-step process and its hybridization was detected using reflectivity [154]. First, an acid termination on PSi surface, obtained by thermal hydrosilylation of undecylenic acid, was converted to NHS termination [74]. In the second step, a DNA probe bearing a terminal primary amine was covalently immobilized to the PSi surface through amide bond formation. Reflectivity spectra of the resulting surface after exposure to complementary DNA showed a blue-shift of 8 nm, as compared to 1 nm detected after interaction with the noncomplementary DNA.

The PSi luminescence was exploited as a direct transduction mode for the detection of enzymatic activity [135]. β-Glucuronidase enzyme was covalently immobilized on a GMBS-functionalized PSi substrate prepared in five steps from hydrogen-terminated surface [134]. The glucuronidase-functionalized PSi samples were mounted in a Teflon flow cell fitted with a quartz window and then exposed to the substrate *p*-nitrophenyl-β-D-glucuronide in order to assay the enzyme activity. The enzyme activity was monitored in real time by recording the PL in situ at 560 nm for substrate concentrations varying from 25 μM to 250 μM. A decrease of the PL intensity was correlated with the increase of the *p*-nitrophenol produced by the immobilized enzyme upon exposure to the substrate. The PL quenching process is reversible and believed to occur via a charge-transfer mechanism. On the other hand, it was demonstrated that in the absence of covalent binding between the enzyme and the PSi surface, no activity can be detected.

The same group has investigated organophosphorus acid anhydrolase (OPAA) enzyme immobilization and activity toward the hydrolysis of acetylcholinesterase inhibitors [155]. The OPAA-functionalized PSi device was exposed to *p*-nitrophenyl-soman substrate and the PL intensity was monitored at the same time. Figure 22 displays two sets of experiments for substrate concentrations of 50 μM and 25 μM. A significant decrease in the PL intensity associated with the injection of substrate was observed. There is an inverse linear relationship between the amount of the initial substrate concentration and the decrease in PL. The transduction is fast (less than 2 min), sensitive (10% decrease for 25 μM), and reversible. The device is stable over time and retains activity for at least 6 months.

Chemical modification of PSi surfaces was successfully used to control the loading and release of a hydrophobic drug [156]. The loading of the steroid dexamethasone was followed by FTIR spectroscopy, while the release rates were measured using optical reflectivity. It was demonstrated that dodecyl-terminated PSi surfaces,

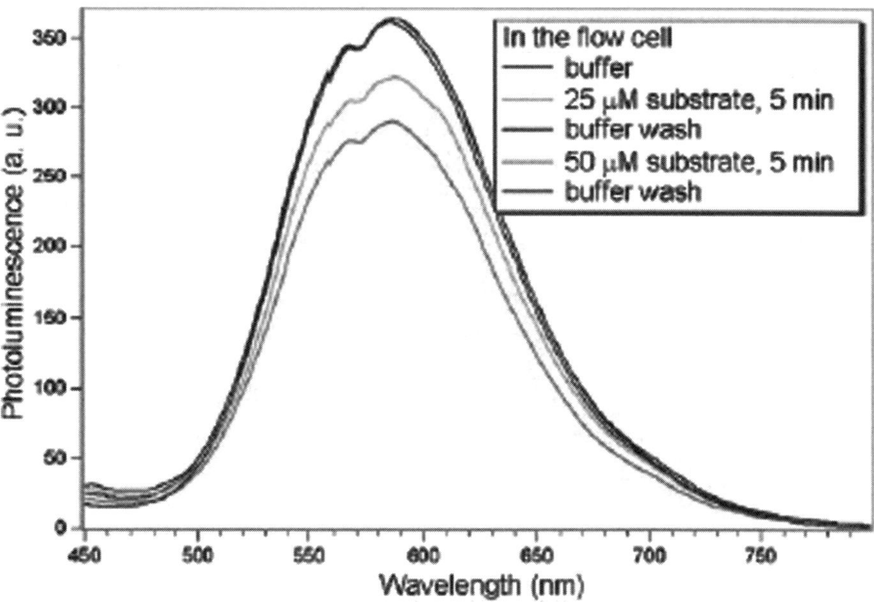

Fig. 22 Emission spectra ($\lambda_{ex} = 290$ nm) of OPAA-functionalized PSi before exposure to p-nitrophenyl-soman, during exposure to 25 mM and 50 mM substrate concentrations, and following PBS buffer washes. From S.E. Letant, S. R. Kane, B. R. Hart, M. Z. Hadi, T.-C. Cheng, V. K. Rastogi, J. G. Reynolds, Chem. Commun. (2005) pp. 851–853. Reproduced by permission of The Royal Society of Chemistry

obtained by thermal hydrosilylation of 1-dodecene, are more stable at physiological pH and exhibit a slower release rates (reduced by a factor ~20) of the drug relative to freshly prepared PSi. On the one hand, the dexamethasone molecule infiltrates the hydride-terminated PSi samples easily, but on the other hand, it does not penetrate in the porous layer of the dodecyl-terminated samples. This was assigned to PSi size reduction upon chemical modification with 1-dodecene. To achieve a higher loading rate in the chemically modified PSi films, the pores were expanded before hydrosilylation using a chemical treatment with an aqueous solution containing HF and DMSO. The mechanism of drug release is believed to involve a combination of leaching and matrix dissolution.

3 Porous-Based Germanium Materials

There is a huge work devoted to the chemical functionalization of PSi surfaces with the final goal to either stabilize the optical properties of the material or to interface the semiconductor surface with a biological matter. However, there are only few reports regarding the preparation of PGe and PGe-based materials ($PSi_{1-x}Ge_x$), and the chemical derivatization of these materials was studied to a lesser extent. There is only one study on the chemical derivatization of hydrogen-terminated PGe surface reported in the literature. The investigation of PGe-based materials followed the discovery of visible PL of PSi [10] at room temperature with the objective to prepare bright photoluminescent materials for optoelectronics and photonics.

The biocompatibility of PGe substrates with various porosities was explored in the literature [157]. It was demonstrated that PGe films allow passive deposition of biological salts, whereas crystalline Ge does not. The finding opens new opportunities to envisage applications in biotechnology.

3.1 Formation

3.1.1 Porous Germanium

Because of the similarities between silicon and germanium lattices, a parallel between silicon etching conditions was made in order to explore the PGe formation. Photoluminescent PGe was prepared by stain and electrochemical etching of germanium surfaces [158]. The stain etching consisted on chemical treatment of Ge single-crystal surfaces with 48% HF or 48% HF/30% H_2O_2 (1:50), while the electrochemical dissolution was performed in 5% KOH under illumination. The resulting surfaces were characterized using Raman spectroscopy, FTIR spectroscopy, XPS, secondary ion mass spectrometry (SIMS), and PL measurements. The IR and Raman spectra of stain-etched Ge show the presence of Ge–O_2 stretching vibrations. XPS and SIMS analysis revealed the presence of GeO_2, GeO, and OH groups on the surface. Under visible laser excitation, the stain-etched PGe displays a PL band at 525 nm (Fig. 23), while a band centered at 400 nm was observed under UV

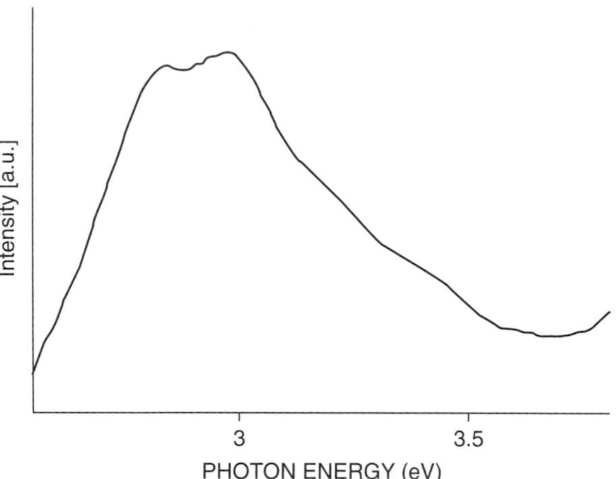

Fig. 23 Photoluminescence excitation spectrum for the 525 nm band of stain-etched germanium. Reproduced with permission from Thin Solid Films, Vol. 255, M. Sendova-Vassileva, N. Tzenov, D. Dimova-Malinovska, M. Rosenbauer, M. Stutzmann, K. V. Josepovits, Structural and luminescence studies of stain-etched and electrochemically etched germanium, pp. 282–285. Copyright Elsevier (1995)

excitation. Time-resolved spectra (Fig. 24) at 100 K for the 420-nm (UV-excited) band shows that the luminescence lifetime is in the order of 1 ms. The luminescence of all the samples decays under laser irradiation. The origin of the PL was assigned to the presence of defects in the oxide GeO_2.

Hydrogen-terminated PGe was prepared by electrochemical anodization of p-type c-Ge(100) in aqueous HF solutions under illumination [159]. IR analysis of the surface showed the presence of distinct vibrational modes around 580, 830, and 2300 cm^{-1} due to rocking, bending, and stretching modes of hydrogen bonded to

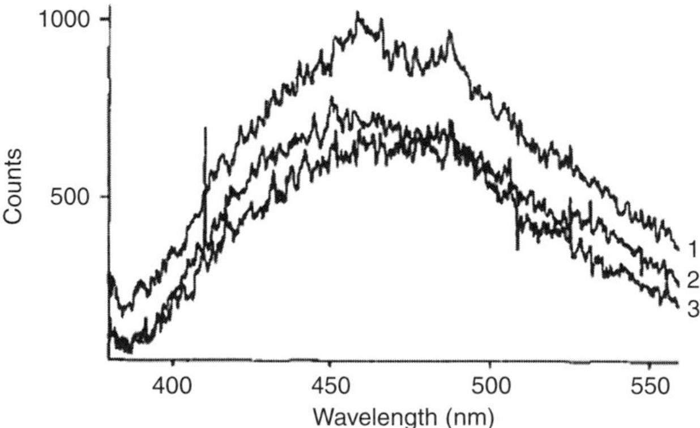

Fig. 24 Time-resolved luminescence spectra of stain-etched germanium with time delays of (1) 60 μs, (2) 5 ms, and (3) 12 ms. The spectra were accumulated for 1 ms starting at the indicated delay time after the laser pulse. Reproduced with permission from Thin Solid Films, Vol. 255, M. Sendova-Vassileva, N. Tzenov, D. Dimova-Malinovska, M. Rosenbauer, M. Stutzmann, K.V. Josepovits, Structural and luminescence studies of stain-etched and electrochemically etched germanium, pp. 282–285. Copyright Elsevier (1995)

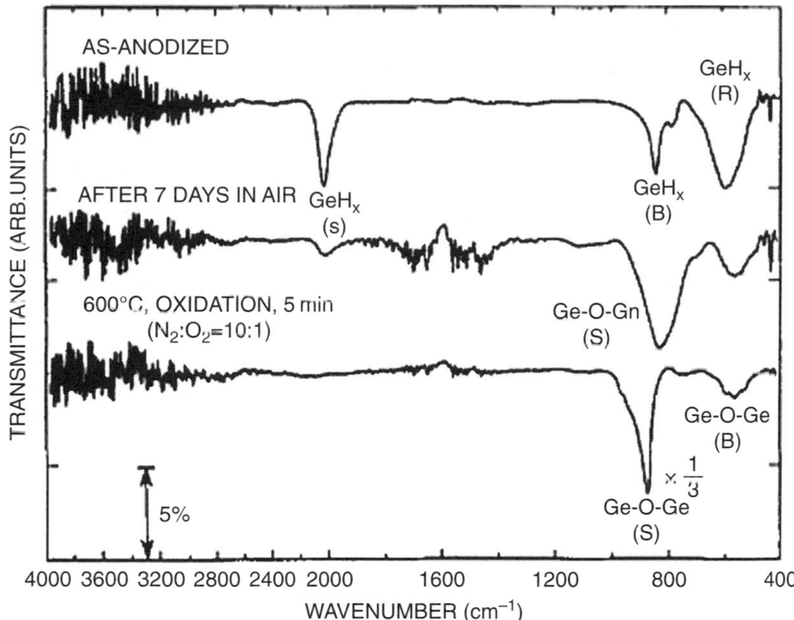

Fig. 25 IR absorption spectra for an as-anodized PGe layer, after exposure to air for 7 days and after 600 °C oxidation for 5 min. Reproduced with permission from Thin Solid Films, Vol. 255, S. Miyazaki, K. Sakamoto, K. Shiba and M. Hirose, Photoluminescence from anodized and thermally oxidized porous germanium, pp. 99–102. Copyright Elsevier (1995)

the Ge, Ge–H$_x$ (Fig. 25). No apparent absorption originating from germanium oxide was observed after anodization. Exposure of the freshly prepared PGe surface to ambient shows a decrease in the IR intensity of the Ge–H$_x$ peaks and appearance of Ge–O–Ge stretching band. However, no frequency shift of the Ge–H$_x$ stretching vibrational modes is evident in the IR spectrum after 7-day exposure to ambient air [160]. It may be inferred that (O_xGeH_{4-x}, x = 1–3) species resulting from surface oxidation are less stable than their silicon counterparts. A clear shift of the Si–H$_x$ stretching bands from 2100 cm^{-1} to 2250 cm^{-1} was seen when freshly prepared PSi was oxidized in air or electrochemically [83]. Thermal treatment of the PGe surface at 600 °C in 10% O_2 diluted in N_2 for 5 min led to the complete oxidation of the surface (disappearance of the Ge–H$_x$ stretching vibrational modes).

A broad PL band at 1.17 eV was observed for the as-anodized PGe at room temperature. Thermal oxidation of the PGe surface induces blue shift and intensity enhancement of the PL. A new PL peak appears at 2.15 eV whose spectral shape remains unchanged with progressive oxidation (Fig. 26). This suggests that the luminescence from oxidized PGe occurs through localized states at or near the Ge/GeO_2 interface. PGe surfaces prepared using similar anodization process to that for PSi display weaker PL by a factor of about 10 and rougher surfaces [161]. The PL behavior and the temperature dependence of the PL properties of spark-processed

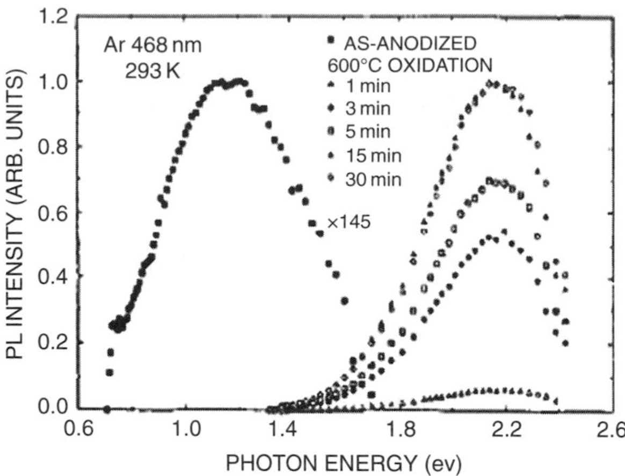

Fig. 26 Room temperature PL spectra for as-anodized and thermally oxidized PGe layers. Reproduced with permission from Thin Solid Films, Vol. 255, S. Miyazaki, K. Sakamoto, K. Shiba and M. Hirose, Photoluminescence from anodized and thermally oxidized porous germanium, pp. 99–102. Copyright Elsevier (1995)

and anodically etched (porous) germanium have been studied [162]. The authors found that the emitted light wavelengths are relatively similar. However, the origin of the PL was attributed to molecular effects in spark-processed Ge, whereas quantum effects seem to be prevalent for PGe. PGe films with a high concentration of Ge nanocrystals were synthesized by a combination of stain etching with subsequent annealing in hydrogen. Detail analysis of the structural properties of PGe surfaces were investigated using Raman scattering and X-ray techniques [163–165]. Further investigation of the chemical composition of stain-etched Ge samples in HF/H_2O_2 at 1:50 volume ratio (samples 1) (4) and in $HF{:}H_3PO_4{:}H_2O_2$ at 34:17:1 volume ratio (samples 2) [163] using energy-dispersive X-ray (EDX), Raman and FTIR spectroscopy, and the near-edge X-ray absorption structure (XANES) indicate that the samples are composed of nonstochiometric Ge oxides [166]. The samples 1 are composed of GeO_x and are free from Ge nanocrystals (from the Raman spectra), while the samples 2 contain Ge nanocrystals with an average size of 2–3 nm, surface covered with oxygen. The samples 1 display visible PL at 2.3 eV, similar to that of c-GeO_2, while the samples 2 showed a weak PL peak at 2.3 eV when excited with the same wavelength (442 nm). It was found that the PL intensity increased by an order after annealing of the samples 2 in air, and the PL was almost completely quenched after annealing in H_2, even though the PL peak wavelength did not change. Moreover, it was found that annealing in H_2 of samples 2 caused an increase in the Ge average diameter size (8–9 nm), as concluded from the Raman spectra [163]. The results suggest that the observed PL from the stain-etched samples originates from GeO_x, rather than quantum confinement. Ion beam was used to create several micron deep sponge-like structures of amorphous Ge [167]. Irradiation of germanium with high energetic heavy ions led to volume expansion of the bombarded regions, forming a buried layer with a porous structure.

Hydrogen-terminated PGe can be prepared by a bipolar electrochemical etching technique. It consists on anodization followed by cathodization of the germanium surface [168]. Upon anodization it is believed that the surface is composed of Ge–OH or Ge–Cl groups (lack of Ge–H_x vibrational modes in the transmission FTIR spectrum). Without anodization, only hydrogen evolution occurs during the cathodic process. The subsequent cathodic step is critical for the formation of the porous layer and hydride termination (Scheme 18).

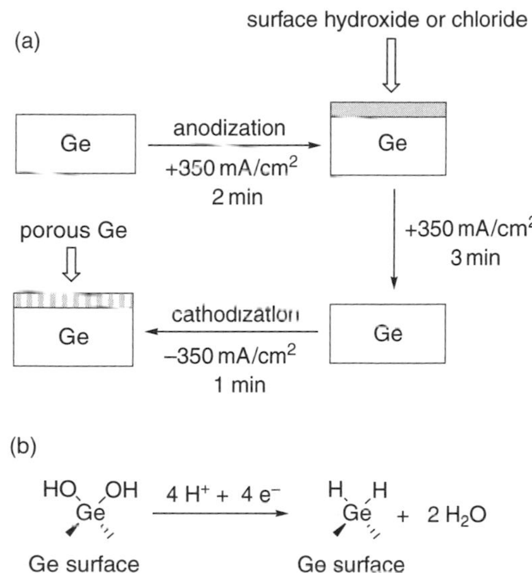

Scheme 18 (**a**) Schematic representation of the bipolar electrochemical etching (BEE) of Ge(100) wafers. (**b**) Surface electrochemical conversion of oxidized Ge(100) surfaces to hydride-terminated Ge(100) under cathodic bias in an acidic medium (from ref. [10])

The resulting samples display a homogeneous red PL upon excitation at 77 K with 365 nm irradiation. The PL intensity is, however, much weaker than that observed from PSi. The PGe samples etched only anodically emit yellow-white PL under 254 nm illumination at 77 K. The observed PL is believed to arise from germanium oxide layers rather than from germanium nanoparticles, since there are no observable Ge–H_x bonds by FTIR or a porous layer by SEM, after anodization. This statement was corroborated by the disappearance of the observed PL upon washing the sample with 25% aqueous HF. Visible room-temperature PL (broad PL peak at 2.3 eV) from oxidized germanium has already been reported in the literature [169]. The red PL from the Ge–H_x terminated PGe surface is assumed to result from Ge nanoparticles, since it is not affected by HF washing.

Recently, PGe thin films were grown by high-density, inductively coupled plasma chemical vapor deposition (ICPCVD) [170]. The process leading to the porous layer formation is illustrated in Schemes 19 and 20:

The technique is based on the chemical vapor decomposition of GeH_4 catalyzed by gold nanoparticles. The process was used for the synthesis of nanowires and known as VLS (vapor–liquid–solid) growth process [171, 172]. Germanium (from the decomposition of GeH_4) and Au form a liquid alloy when the temperature is higher than the eutectic point. The liquid surface has a large accommodation coefficient and is therefore a preferred deposition site for incoming Ge vapor. After the

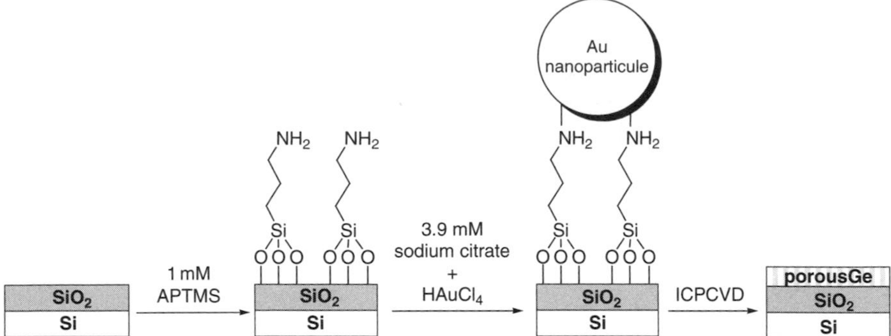

Scheme 19 Schematic illustration of the process by which the citrate reduced Au nanoparticles were deposited on the SiO_2 substrate (from ref. [11])

Scheme 20 Schematic illustration depicting the growth mechanism of the porous thin film: (**a**) gold nanoparticles dispersed on silicon oxide, (**b**) germanium vapor dissolves preferentially into the gold nanoparticles; the solid precipitates and raises the catalysts, (**c**) gold nanoparticles are removed during germanium growth, (**d**) the porous structure develops as a result of incomplete surface coverage and nanotip aggregation

liquid alloy becomes supersaturated with Ge, Ge nanowire growth occurs by precipitation at the solid–liquid interface. The gold nanoparticles were removed during germanium growth, and the porous structure develops as a result of incomplete surface coverage and nanotip aggregation.

3.1.2 Macroporous Germanium

Fabrication of macroporous (pore size $>$ 50 nm) periodic structures of germanium was mainly motivated by the expected potential application of such structures in photonics, chemical sensors, and membranes. Solid silica colloidal crystals (opals) were employed as templates to achieve such a goal [173, 174]. A network of air spheres in germanium was prepared by a three-step approach involving: (1) infiltration and hydrolysis of tetramethoxygermane (germanium precursor) in a crystalline silica template to give rise to GeO_2 formation, (2) reduction of GeO_2 in H_2 atmosphere at 550 °C, and (3) selective removal of the silica spheres of the matrix to yield Ge inverse opals [173].

Figure 27a displays a SEM image of GeO_2 infiltrated opal. Oxide crystallites with an average cluster size of around 100 nm are homogeneously distributed throughout the template void lattice. The reduction process of GeO_2 to yield Ge opal resulted in a huge increase of the cluster size and a poor connectivity of the

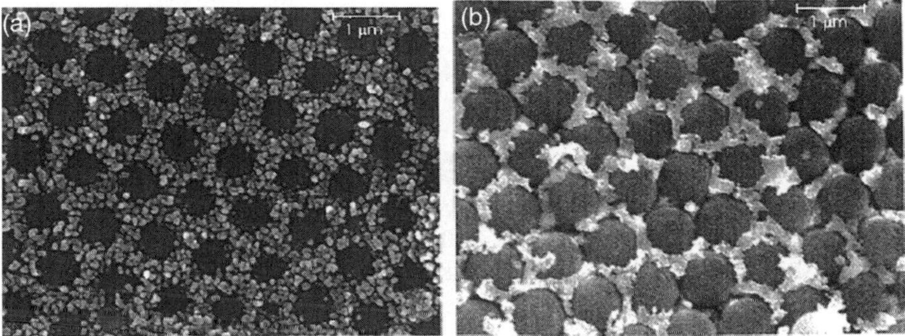

Fig. 27 Views of a GeO_2 (**a**) and Ge (**b**) infiltrated opal made of 850-nm diameter spheres. Reprinted with permission from H. Miguez, F. Meseguer, C. Lopez, M. Holgado, G. Andreasen, A. Mifsud, V. Fornes, Langmuir Vol. 16 (2000) p. 4405, Copyright (2000) American Chemical Society

clusters. A Ge lattice with a high connectivity was obtained after five cycles of the GeO_2 formation/reduction (Fig. 27b). Germanium inverse opals were also made by a chemical vapor deposition technique (CVD) [173]. The germanium was deposited from digermane (Ge_2H_6) precursor. By varying the deposition conditions, the degree of germanium infiltration can be controlled, which enables to tailor the optical properties of the material. The resulting materials display photonic band gap behavior in the near-infrared region. More recently, an electrochemical approach was described for macroporous Ge formation [175]. Electrochemical deposition of germanium in a template formed from dried suspension of silica spheres gives a macroporous germanium–air sphere matrix after selective dissolution of the silica template. The plating bath consisted on $GeCl_4$ in ethylene or propylene glycol. Figure 28 displays a SEM image of the electrodeposited sample after silica matrix removal in 5% HF. It clearly shows a three-dimensional network of air spheres in germanium. The

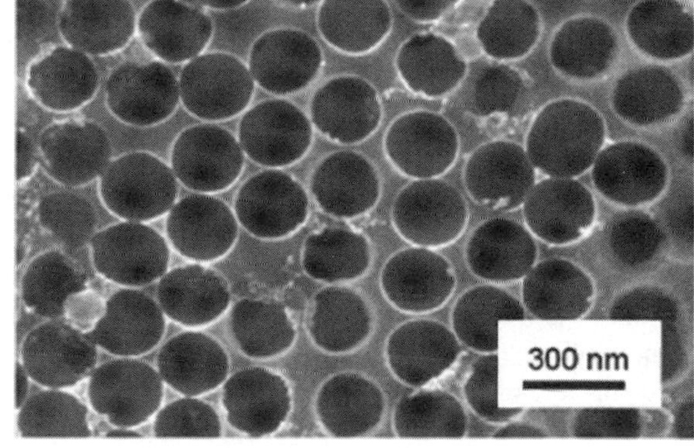

Fig. 28 SEM image of a macroporous amorphous germanium layer electrodeposited from a propylene glycol solution after removal of the template. From L. K. van Vugt, A. F. van Driel, R. W. Tjerkstra, L. Bechger, W. L. Vos, D. Vanmaekelbergh, J. J. Kelly, Chem. Commun. (2002) pp. 2054–2055. Reproduced by permission of The Royal Society of Chemistry

as-electrodeposited Ge is amorphous but can be crystallized by high-temperature annealing in a nitrogen atmosphere.

3.1.3 PSi–Germanium

Silicon–germanium ($Si_{1-x}Ge_x$) alloys are interesting semiconductor materials since their bandgaps can be tailored between those of silicon and germanium single crystals by varying the alloy composition. Following the report on the visible PL from PSi at room temperature [10], many attempts have been made to prepare PSi–germanium (PSiGe) with visible luminescence. Porous layers of $Si_{1-x}Ge_x$ are prepared by electrochemical anodization [176–181] or stain etching [182, 183]. It was found that the shape of the porous structure and pore morphology of PSiGe are similar to those of PSi [178, 179]. The microstructure of anodized $Si_{1-x}Ge_x$ films was characterized using Raman spectroscopy and electron microscopy. It was demonstrated that as the film porosity increases, the size of the nanocrystals decreases, and the film composition modifies in favor of Ge [180, 181]. It was also shown from these studies that silicon was preferentially etched in $Si_{1-x}Ge_x$ films during electrochemical anodization. The spectral shape of PL emission from PSiGe is very similar to that of PSi, except that it slightly red-shifted. The luminescence intensity decreases, however, with increasing the Ge fraction [181]. PSiGe samples showed a much faster recombination time than PSi. Time-resolved PL spectra reveal lifetimes in the order of 10^{-2} ns range and directly related to the concentration of germanium [184–186]. The origin of visible PL at room temperature from as-anodized PSiGe films is assigned to a quantum confinement effect (like for PSi) and the evolution of the SiGe Si like-band structure [185].

3.1.4 Chemical Functionalization of Hydride-Terminated Porous Germanium

Most of the studies regarding the PGe matrix were linked with the optical properties and their stabilization upon thermal oxidation. There is only one report on the chemical reactivity of the hydrogen-terminated PGe [168]. The Ge–H_x terminating the PGe surface prepared by bipolar electrochemical etching undergo hydrogermylation reaction with alkenes and alkynes to give surfaces terminated with alkyl and alkenyl groups, respectively. FTIR analysis of freshly prepared PGe surface by electrochemical etching in ethanoic HCl solution reveals the presence of broad absorption peak around 2015 cm^{-1}, assigned to Ge–H_x stretching modes (Fig. 29a).

The surface is expected to be terminated by mono-, di-, and trihydrides, like PSi [187]. In order to prove that the termination of the PGe surface by Ge–H_x bonds, a deuterated PGe surface was prepared by same etching procedure in deuterated ethanol and DCl. The Ge–D_x stretching band appear at 1450 cm^{-1} as expected, with a very weak absorption peak of Ge–H_x at 2020 cm^{-1} (Fig. 29b). Refluxing freshly prepared hydrogen-terminated PGe surface in a 20% 1-dodecene solution in mesitylene (v/v) for 2 h resulted in incorporation of dodecyl moieties on the PGe surface. Infrared analysis shows the presence of peaks due to C–H_x stretching modes (2850–2960 cm^{-1}) and peaks due to the stretching of the remaining Ge–H_x

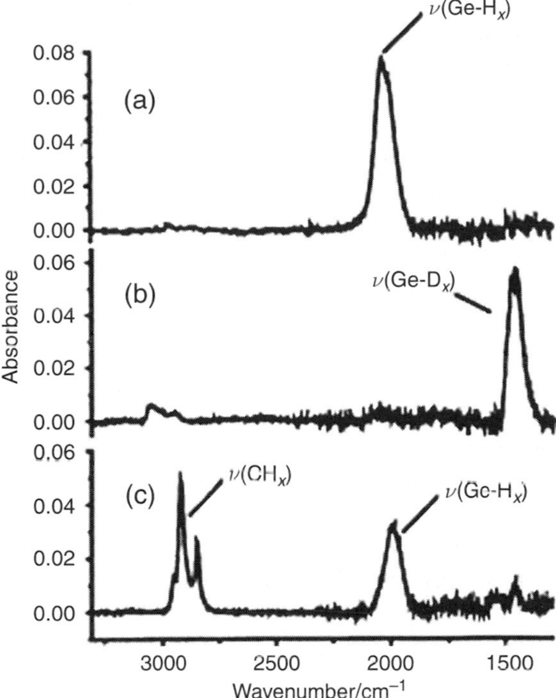

Fig. 29 FTIR spectrum of PG: (**a**) Etched by BEE using ethanoic HCl, (**b**) using deuterated ethanol and DCl, and (**c**) thermally hydrogermylated using 1-dodecene. From H. Choi, J.M. Buriak, Chem. Commun. (2000) pp. 1669–1670. Reproduced by permission of The Royal Society of Chemistry

bonds at 2020 cm^{-1} (Fig. 29c). The result is consistent with a hydrogermylation process consuming Ge–H$_x$ bonds of the PGe surface. Application of other reaction conditions known for the formation of Ge–C bonds on flat Ge, including Lewis-acid-mediated hydrogermylation, were unsuccessful [188].

4 Conclusions and Perspectives

This book chapter has summarized work carried out in the area of organic modification of hydrogen-terminated PSi and germanium surfaces. It is important to notice that most of the work was undertaken on PSi substrates, because silicon is the basic material in microelectronics, and is cheap compared to germanium. Moreover, the report does not cover the work related to surface modification of oxidized PSi, because the chemistry has some drawbacks regarding its reproducibility and the stability of the organic layers obtained through silanization.

While the initial step was focused on the elaboration and designing of new chemical routes for anchoring organic functional groups on the surface in a controlled way, a huge amount of work was lately devoted to the understanding of the reaction mechanism involved in the functionalization process. Even though there is a parallel between organometallic surface chemistry of PSi and germanium, and molecular chemistry, the thermal and white-light-induced hydrosilylation of alkenes and alkynes has no equivalent in organosilicon molecular chemistry. This may be attributed to the influence of the small band gap of silicon-nanostructured substrate compared to silane molecules. The potential of the resulting organic layers covalently bonded to the PSi surface to stabilize its optical properties was examined in detail. It was demonstrated that the chemical process can effectively passivate the PSi surface. Finally, PSi was investigated as a new platform for label-free sensing of biomolecular interactions and to record enzymatic activity in real time.

There are several advantages associated with chemical modification of high surface area hydrogen-terminated semiconductor nanostructured substrates compared to other surfaces: (1) preservation of the nanostructure, (2) compatibility of the Si–H and Ge–H bonds terminating the surface with the main organic and organometallic reactions and a wide variety of organic functional groups, (3) formation of highly stable organic layers covalently bonded to the surface through nonpolar Si–C and Ge–C bonds, (4) use of the well-established methods for the integration of chemical and biochemical functionality into microelectronic platforms, and (5) take advantage of intrinsic properties of the substrate for label-free detection molecular events occurring on the surface with high sensitivity. However, there are some limitations to overcome before reaching stable, specific, and reusable devices based on organic monolayer/semiconductor hybrids and using the electronic properties of the material for label-free detection of biomolecular events on the surface. The main drawback is the number of Si–H and Ge–H bonds remaining unsubstituted after the chemical process. Because of the steric hindrance on the surface, the maximum surface coverage reachable is 40%. The remaining bonds on the surface are not completely preserved against oxidation, which may introduce electronic active-surface defects or nonradiative recombination centers on the surface. Another limiting factor is the slow diffusion of large biomolecules inside mesoporous layers, which may affect the device response and sensitivity. Finally, the improvement of the monolayers quality to resist biofouling is a real challenge to design selective and sensitive devices for monitoring biomolecular interactions.

The field of chemical functionalization and assembly of organic layers on hydrogen-terminated PSi and germanium surfaces will remain very active for the coming years. This will be mainly driven by potential applications of such hybrid structures in different fields ranging from surface passivation and stabilization to label-free biosensors, biological microelectromechanical systems (bioMEMS), and biomedical area.

Acknowledgements The authors are grateful for the permissions from Elsevier, ACS, and AIP for providing useful material in the chapter. The Centre National de la Recherche Scientifique (CNRS) and the Nord-Pas-de Calais region are gratefully acknowledged for financial support.

References

1. Chazalviel, J. N., Ozanam, F., *Mater. Res. Soc. Symp. Proc.* **1999**, 536, 155–166.
2. Buriak, J. M., *Chem. Commun.* **1999**, 1051–1060.
3. Stewart, M. P., Buriak, J. M., *Adv. Mater.* **2000**, 12, 859–869.

4. Buriak, J. M., *Chem. Rev.* **2002**, 102, 1272–1306.
5. Wayner, D. D. M., Wolkow, R. A., *J. Chem. Soc., Perkin Trans.* **2002**, 2, 23–34.
6. Boukherroub, R., *Curr. Opin. Solid State Mater. Sci.* **2005**, 9, 66–72.
7. Sailor, M. J., Link, J. R., *Chem. Commun.* **2005**, 1375–1383.
8. Shirahata, N., Hozumi, A., Yonezawa, T., *Chem. Rec.* **2005**, 5, 145–159.
9. Buriak, J. M., *Phil. Trans. R. Soc. A* **2006**, 364, 217–225.
10. Canham, L. T., *Appl. Phys. Lett.* **1990**, 57, 1046.
11. Uhlir, A., *Bell Syst. Tech. J.* **1956**, 35, 333–347.
12. Fuller, C. S., Ditzenberger, J. A., *J. Appl. Phys.* **1957**, 27, 544.
13. Turner, D. R., *J. Electrochem. Soc.* **1958**, 105, 402.
14. Archer, R. J., *J. Phys. Chem. Solids* **1960**, 14, 104.
15. Watanabe, Y., Sakai, T., *Rev. Electron. Commun. Labs.* **1971**, 19, 899.
16. Watanabe, Y., Arita, Y., Yokoyama, T., Igarashi, Y., *J. Electrochem. Soc.* **1975**, 122, 1351.
17. Arita, Y., Kato, K., Sudo, T., *IEEE Trans. Electron Devices* **1977**, 24, 756.
18. Unagami, T., Kato, K., *Jpn. J. Appl. Phys.* **1977**, 16, 1635.
19. Lehmann, V., Gösele, U., *Appl. Phys. Lett.* **1991**, 58, 856.
20. Canham, L. T., *Phys. World* **1992**, 5, 41.
21. Halimaoui, A., Oules, C., Bromchil, G., Bsiesy, A., Gaspard, F., Herino, R., Ligeon, M., Muller, F., *Appl. Phys. Lett.* **1991**, 59, 304.
22. Cullis, A. G., Canham, L. T., Calcott, P. D. J., *J. Appl. Phys.* **1997**, 82, 909–965.
23. Jung, K. H., Shih, S., Kuong, D. L., *J. Electrochem. Soc.* **1993**, 140, 3046.
24. Halimaoui, A., In *Porous Silicon Science and Technology*, J.-C. Vial and J. Derrien Eds. Springer, Berlin, 1995, p. 33.
25. Canham, L. T., Groszek, A. J., *J. Appl. Phys.* **1992**, 72, 1558.
26. Hérino, R., in "Properties of porous silicon" Datareview Ser. N°18; Canham: London, 1997; pp. 89–96.
27. Rieger, M. M., Kohl, P. A., *J. Electrochem. Soc.* **1995**, 142, 1490.
28. Janshoff, A., Dancil, K.-P. S., Steinem, C., Greiner, D. P., Lin, V. S.-Y., Gurtner, C., Motesharei, K., Sailor, M. J., Ghadiri, M. R., *J. Am. Chem. Soc.* **1998**, 120, 12108–12116.
29. Koshida, N., Koyama, H., *Jpn. J. Appl. Phys.* **1991**, 30, L1221.
30. Lévy-Clément, C., Lagoubi, A., Ballutaud, D., Ozanam, F., Chazalviel, J. N., Neumann-Spallart, M., *Appl. Surf. Sci.* **1993**, 65/66, 408.
31. Lévy-Clément, C., Lagoubi, A., Tomkiewicz, M., *J. Electrochem. Soc.* **1994**, 141, 958.
32. Fathauer, R. W., George, T., Ksendzov, A., Vasquez, R. P., *Appl. Phys. Lett.* **1992**, 60, 995.
33. Shih, S., Jung, K. H., Hsieh, T. Y., Sarathy, J., Campbell, J. C., Kwong, D. L., *Appl. Phys. Lett.* **1992**, 60, 1863.
34. Kelly, M. T., Chun, J. K. M., Bocarsly, A. B., *Appl. Phys. Lett.* **1994**, 64, 1693.
35. Chandler-Henderson, R. R., Coffer, J. L., Filessesler, L. A., *J. Electrochem. Soc.* **1994**, 141, L166.
36. Belogorokhov, A. I., Gavrilov, S. A., Kashkarov, P. K., Belogorokhov, I. A., *Phys. Stat. Sol. (a)* **2005**, 202, 1581–1585.
37. de smet, L. C. P. M., Zuilhof, H., Sudhölter, E. J. R., Lie, L. H., Houlton, A., Horrocks, B. R., *J. Phys. Chem. B* **2005**, 109, 12020–12031.
38. Matsumoto, T., Masumoto, Y., Nakashima, S., Koshida, N., *Thin Solid Films* **1997**, 297, 31–34.
39. Bateman, J. E., Eagling, R. D., Horrocks, B. R., Houlton, A., *J. Phys. Chem. B* **2000**, 104, 5557–5565.
40. Ipatova, I. P., Chekalova-Luzina, O. P., Hess, K., *J. Appl. Phys.* **1998**, 83, 814.
41. Lauerhaas, J. M., Sailor, M. J., *Science* **1993**, 261, 1567–1568.
42. Lauerhaas, J. M., Sailor, M. J., *Mater. Res. Soc. Symp. Proc.* **1993**, 298, 259–263.
43. Lopinski, G. P., Eves, B. J., Hul'ko, O., Mark, C., Patitsas, S. N., Boukherroub, R., Ward, T. R., *Phys. Rev. B* **2005**, 71, 125308.

44. Joy, V. T., Daniel, M., *CHEMPHYSCHEM* **2002**, 973–975.
45. Gun′ko, Y. K., Perova, T. S., Balakrishnan, S., Potapova, D. A., Moore, R. A., Astrova, E. V., *Phys. Stat. Sol. (a)* **2003**, 197, 492–496.
46. Lavine, J. M., Sawan, S. P., Shieh, Y. T., Bellezza, A. J., *Appl. Phys. Lett.* **1993**, 62, 1099–1101.
47. Hory, M. A., Hérino, R., Ligeon, M., Muller, F., Gaspard, F., Mihalcescu, I., Vial, J. C., *Thin Solid Films* **1995**, 255, 200–203.
48. Seo, Y. H., Lee, H.-J., Jeon, H. I., Oh, D. H., Nahm, K. S., Lee, Y. H., Suh, E.-K., Lee, H. J., Kwang, Y. G., *Appl. Phys. Lett.* **1993**, 62, 1812–1814.
49. Boukherroub, R., Petit, A., Loupy, A., Chazalviel, J.-N., Ozanam, F., *ECS Conf. Proc.* **2004**, 2004–19, 13–22.
50. Guo, D.-J., Xiao, S.-J., Xia, B., Wei, S., Pei, J., Pan, Y., You, X.-Z., Gu, Z.-Z., Lu, Z., *J. Phys. Chem. B* **2005**, 109, 20620–20628.
51. Holland, J. M., Stewart, M. P., Allen, M. J., Buriak, J. M., *J. Solid State Chem.* **1999**, 147, 251–258.
52. Coulthard, I., Jiang, D.-T., Lorimer, J. W., Sham, T. K., Feng, X.-H., *Langmuir* **1993**, 9, 3441–3445.
53. Tsuboi, T., Sakka, T., Ogata, Y. H., *J. Appl. Phys.* **1998**, 83, 4501–4506.
54. Hilliard, J. E., Nayfeh, H. M., Nayfeh, M. H., *J. Appl. Phys.* **1995**, 77, 4130.
55. Boukherroub, R., Wayner, D. D. M., Lockwood, D. J., Zargarian, D., *Physica Status Solidi (a)* **2003**, 197, 476–481.
56. Boukherroub, R., Zargarian, D., Reber, R., Lockwood, D. J., Carty, A. J., Wayner, D. D. M., *Appl. Surf. Sci.* **2003**, 217, 125–133.
57. Saghatelian, A., Buriak, J. M., Lin, V. S.-Y., Ghadiri, M. R., *Tetrahedron* **2001**, 57, 5131–5136.
58. Lewis, L. N., *J. Am. Chem. Soc.* **1990**, 112, 5998–6004.
59. Buriak, J. M., Allen, M. J., *J. Am. Chem. Soc.* **1998**, 120, 1339–1340.
60. Buriak, J. M., Stewart, M. P., Geders, T. W., Allen, M. J., Choi, H. C., Smith, J., Raftery, D., Canham, L. T., *J. Am. Chem. Soc.* **1999**, 121, 11491–11502.
61. Canham, L. T., Saunders, S. J., Heeley, P. B., Keir, A. M., Cox, T. I., *Adv. Mater.* **1994**, 6, 865.
62. Buriak, J. M., Allen, M. J., *J. Lumin.* **1999**, 80, 29–35.
63. Stewart, M. P., Buriak, J. M., *Angew. Chem. Int. Ed.* **1998**, 37, 3257–3260.
64. Stewart, M. P., Buriak, J. M., *J. Am. Chem. Soc.* **2001**, 123, 7821–7830.
65. Cicero, R. L., Linford, M. R., Chidsey, C. E. D., *Langmuir* **2000**, 16, 5688.
66. Lambert, J. B., Zhao, Y., Wu, H., *J. Org. Chem.* **1999**, 64, 2729.
67. Lambert, J. B., *Tetrahedron* **1990**, 46, 2677.
68. Bateman, J. E., Eagling, R. D., Worrall, D. R., Horrocks, B. R., Houlton, A., *Angew. Chem. Int. Ed.* **1998**, 37, 2683–2685.
69. Boukherroub, R., Morin, S., Wayner, D. D. M., Lockwood, D. J., *Phys. Stat. Sol. (a)* **2000**, 182, 117–121.
70. Boukherroub, R., Morin, S., Wayner, D. D. M., Lockwood, D. J., *Solid State Commun.* **2001**, 118, 319–323.
71. Boukherroub, R., Morin, S., Wayner, D. D. M., Bensebaa, F., Sproule, G. I., Baribeau, J.-M., Lockwood, D. J., *Chem. Mater.* **2001**, 13, 2002–2011.
72. Wojtyk, J. T. C., Morin, K. A., Boukherroub, R., Wayner, D. D. M., *Langmuir* **2002**, 18, 6081–6087.
73. Hu, Y. F., Boukherroub, R., Sham, T. K., *J. Electr. Spectrosc. Related Phenom.* **2004**, 135, 143–147.
74. Boukherroub, R., Wojtyk, J. T. C., Wayner, D. D. M., Lockwood, D. J., *J. Electrochem. Soc.* **2002**, 149, H59–H63.
75. Fan, H. J., Kuok, M. H., Ng, S. C., Boukherroub, R., Lockwood, D. J., *Appl. Phys. Lett.* **2001**, 79, 4521–4523.

76. Fan, H. J., Kuok, M. H., Ng, S. C., Boukherroub, R., Baribeau, J.-M., Fraser, J. W., Lockwood, D. J., *Phys. Rev. B* **2002**, 65, 165330.
77. Fan, H. J., Kuok, M. H., Ng, S. C., Boukherroub, R., Lockwood, D. J., *Semicond. Sci. Technol.* **2002**, 17, 692–695.
78. Fan, H. J., Kuok, M. H., Ng, S. C., Lim, H. S., Liu, N. N., Boukherroub, R., Lockwood, D. J., *J. Appl. Phys.* **2003**, 94, 1243–1247.
79. Linford, M. R., Fenter, P., Eisenberger, P. M., Chidsey, C. E. D., *J. Am. Chem. Soc.* **1995**, 117, 3145–3155.
80. Boukherroub, R., Lockwood, D. J., Wayner, D. D. M., Canham, L. T., *Mater. Res. Soc. Symp. Proc.* **2001**, 638, F11.
81. Boukherroub, R., Wayner, D. D. M., Sproule, G. I., Lockwood, D. J., Canham, L. T., *Nano Lett.* **2001**, 1, 713–717.
82. Maruyama, T., Ohtani, S., *Appl. Phys. Lett.* **1995**, 65, 1346–1348.
83. Boukherroub, R., Wayner, D. D. M., Lockwood, D. J., *Appl. Phys. Lett.* **2002**, 81, 601.
84. Vial, J. C., Bsiesy, A., Gaspard, F., Hérino, R., Ligeon, M., Muller, F., Romenstain, R., Macfarlane, R. M., *Phys. Rev. B* **1992**, 45, 14171.
85. Muller, F., Hérino, R., Ligeon, M., Gaspard, F., Romenstain, R., Vial, J. C., Bsiesy, A., *J. Lumin.* **1993**, 57, 283.
86. Geobaldo, F., Rivolo, P., Ugliengo, P., Garrone, E., *Sens. Actuat. B* **2004**, 100, 29–32.
87. Boukherroub, R., Petit, A., Loupy, A., Chazalviel, J. N., Ozanam, F., *J. Phys. Chem. B* **2003**, 107, 13459–13462.
88. Schmeltzer, J. M., Porter, L. A., Stewart, M. P., Buriak, J. M., *Langmuir* **2002**, 18, 2971–2974.
89. Lambert, J. B., Zhao, Y., *J. Am. Chem. Soc.* **1996**, 118, 7867.
90. Dubois, T., Ozanam, F., Chazalviel, J.-N., *ECS Conf. Proc.* **1997**, 97–7, 296–310.
91. Ozanam, F., Vieillard, C., Warntjes, M., Dubois, T., Pauly, M., Chazalviel, J. N., *Can. J. Chem. Eng.* **1998**, 76, 1020–1026.
92. Song, J. H., Sailor, M. J., *J. Am. Chem. Soc.* **1998**, 120, 2376–2381.
93. Song, J. H., Sailor, M. J., *Inorg. Chem.* **1999**, 38, 1498–1503.
94. Kim, N. Y., Laibinis, P. E., *J. Am. Chem. Soc.* **1999**, 121, 7162–7163.
95. Kim, N. Y., Laibinis, P. E., *J. Am. Chem. Soc.* **1998**, 120, 4516–4517.
96. Fellah, S., Boukherroub, R., Ozanam, F., Chazalviel, J.-N., *Langmuir* **2004**, 20, 6359.
97. Chatgilialoglu, C., *Chem. Rev.* **1995**, 95, 1229.
98. Gurtner, C., Wun, A. W., Sailor, M. J., *Angew. Chem. Int. Ed.* **1999**, 38, 1966–1968.
99. Canaria, C. A., Lees, I. N., Wun, A. W., Miskelly, G. M., Sailor, M. J., *Inorg. Chem. Commun.* **2002**, 5, 560–564.
100. Lees, I. N., Lin, L., Canaria, C. A., Gurtner, C., Sailor, M. J., Miskelly, G. M., *Langmuir* **2003**, 19, 9812–9817.
101. Song, J. H., Sailor, M. J., *Inorg. Chem.* **1998**, 37, 3355–3360.
102. Andsager, D., Hillard, J., Hetrick, J. M., AbuHassan, L. H., Plisch, M., Nayfeh, M. H., *J. Appl. Phys.* **1993**, 74, 4783–4785.
103. Andsager, D., Hillard, J., Nayfeh, M. H., *Appl. Phys. Lett.* **1994**, 64, 1141–1143.
104. Robins, E. G., Stewart, M. P., Buriak, J. M., *Chem. Commun.* **1999**, 2479–2480.
105. Jouikov, V., Salaheev, G., *Electrochim. Acta* **1996**, 41, 2623–2629.
106. Jouikov, V. V., *Russ. Chem. Rev.* **1997**, 66, 509.
107. Mattei, G., Valentini, V., *J. Am. Chem. Soc.* **2003**, 125, 9608–9609.
108. Blackwood, D. J., Bin Mohamed Akber, M. F., *J. Electrochem. Soc.* **2006**, 153, G976–G980.
109. Wang, D., Buriak, J. M., *Surf. Sci.* **2005**, 590, 154–161.
110. Warntjes, M., Vieillard, C., Ozanam, F., Chazalviel, J. N., *J. Electrochem. Soc.* **1995**, 142, 4138–4142.
111. Vieillard, C., Warntjes, M., Ozanam, F., Chazalviel, J. N., *ECS Conf. Proc.* **1995**, 95–25, 250–258.
112. Lee, E. J., Ha, J. S., Sailor, M. J., *J. Am. Chem. Soc.* **1995**, 117, 8295–8296.

113. Lee, E. J., Bitner, T. W., Ha, J. S., Shane, M. J., Sailor, M. J., *J. Am. Chem. Soc.* **1996**, 118, 5375–5382.
114. Glass, J. A., Wovchko, E. A., Yates, J. T., *Surf. Sci.* **1995**, 338, 125–137.
115. Kim, N. Y., Laibinis, P. E., *J. Am. Chem. Soc.* **1997**, 119, 2297–2298.
116. Eagling, R. D., Bateman, J. E., Goodwin, N. J., Henderson, W., Horrocks, B. R., Houlton, A., *J. Chem. Soc. Dalton Trans.* **1998**, 1273–1275.
117. Lee, E. J., Ha, J. S., Sailor, M. J., *Mater. Res. Soc. Symp. Proc.* **1995**, 358, 387.
118. Corriu, R. J. R., Guerin, C., Moreau, J. J. E., *Top. Stereochem.* **1984**, 15, 43.
119. Harper, T. F., Sailor, M. J., *J. Am. Chem. Soc.* **1997**, 119, 6943–6944.
120. Becker, H.-D., In *The Chemistry of the Quinonoid Compounds,* S. Patai, Ed., John Wiley & Sons: London, 1974, Vol. I, pp. 335–423.
121. Li, Y.-H., Buriak, J. M., *Inorg. Chem.* **2006**, 45, 1096–1102.
122. Belhousse, S., Cheraga, H., Gabouze, N., Outamzabet, R., *Sens. Actuat. B* **2004**, 100, 250–255.
123. Cheraga, H., Belhousse, S., Gabouze, N., *Appl. Surf. Sci.* **2004**, 238, 495–500.
124. Belhousse, S., Gabouze, N., Cheraga, H., Henda, K., *Thin Solid Films* **2005**, 482, 253–257.
125. Gabouze, N., Belhousse, S., Cheraga, H., *Phys. Stat. Sol. (C)* **2005**, 2, 3449–3452.
126. Lie, L. H., Patole, S. N., Hart, E. R., Houlton, A., Horrocks, B. R., *J. Phys. Chem. B* **2002**, 116, 113–120.
127. Xia, B., Xiao, S.-J., Wang, J., Guo, D.-J., *Thin Solid Films* **2005**, 474, 306–309.
128. Yoon, M. S., Ahn, K. H., Cheung, R. W., Sohn, H., Link, J. R., Cunin, F., Sailor, M. J., *Chem. Comm.* **2003**, 680–681.
129. Cunin, F., Schmedake, T. A., Link, J. R., Li, Y. Y., Koh, J., Bhatia, S. N., Sailor, M. J., *Nat. Mater.* **2002**, 1, 39.
130. Schmedake, T. A., Cunin, F., Link, J. R., Sailor, M. J., *Adv. Mater.* **2002**, 14, 1270.
131. Lie, L. H., Patole, S. N., Pike, A. R., Ryder, L. C., Connolly, B. A., Ward, A. D., Tuite, E. M., Houlton, A., Horrocks, B. R., *Faraday Discuss.* **2004**, 125, 235–249.
132. Pike, A. R., Lie, L. H., Eagling, R. D., Ryder, L. C., Patole, S. N., Connolly, B. A., Horrocks, B. R., Houlton, A., *Angew. Chem. Int. Ed.* **2002**, 41, 615.
133. De Stefano, L., Rotiroti, L., Rea, I., Moretti, L., Di Francia, G., Massera, E., Lamberti, A., Arcari, P., Sanges, C., Rendina, I., *J. Opt. A: Pure Appl. Opt.* **2006**, 8, S540–S544.
134. Hart, B. R., Létant, S. E., Kane, S. R., Hadi, M. Z., Shields, S. J., Reynolds, J. G., *Chem. Comm.* **2003**, 322–323.
135. Létant, S. E., Hart, B. R., Kane, S. R., Hadi, M. Z., Shields, S. J., Reynolds, J. G., *Adv. Mater.* **2004**, 16, 689–693.
136. Schwartz, M. P., Cunin, F., Cheung, R. W., Sailor, M. J., *Phys. Stat. Sol. (a)* **2005**, 202, 1380–1384.
137. Canham, L. T., *Adv. Mater.* **1995**, 7, 1033.
138. Anderson, S. H. C., Elliott, H., Wallis, D. J., Canham, L. T., Powell, J. J., *Phys. Stat. Sol. (a)* **2003**, 197, 331–335.
139. Canham, L. T., Reeves, C. L., Newey, J. P., Houlton, M. R., Cox, T. I., Buriak, J. M., Stewart, M. P., *Adv. Mater.* **1999**, 11, 1505–1507.
140. Choi, H. C., Buriak, J. M., *Chem. Mater.* **2000**, 12, 2151–2156.
141. Green, W. H., Lee, E. J., Lauerhaas, J. M., Bitner, T. W., Sailor, M. J., *Appl. Phys. Lett.* **1995**, 67, 1468.
142. Koshida, N., Kadokura, J., Gelloz, B., Boukherroub, R., Wayner, D. D. M., Lockwood, D. J., *ECS Conf. Proc.* **2002**, 2002-9, 195.
143. Gelloz, B., Sano, H., Boukherroub, R., Wayner, D. D. M., Lockwood, D. J., Koshida, N., *Appl. Phys. Lett.* **2003**, 83, 2342–2344.
144. Gelloz, B., Sano, H., Boukherroub, R., Wayner, D. D. M., Lockwood, D. J., Koshida, N., *Physica Status Solidi C* **2005**, 2, 3273–3277.
145. Gelloz, B., Nakagawa, T., Koshida, N., *Appl. Phys. Lett.* **1998**, 73, 2021.
146. Gelloz, B., Koshida, N., *J. Appl. Phys.* **2000**, 88, 4319.

147. Wei, J., Buriak, J. M., Siuzdak, G., *Nature* **1999**, 399, 243.
148. Shen, Z., Thomas, J. J., Averbuj, C., Broo, K. M., Engerlhard, J. E., Finn, M. G., Siuzdak, G., *Anal. Chem,* **2001**, 73, 612–619.
149. Kruse, R. A., Li, X., Bohn, P. W., Sweedler, J. V., *Anal. Chem,* **2001**, 73, 3639–3645.
150. Alimpiev, S., Nikiforov, S., Karavanskii, V. A., Minton, T., Sunner, J., *J. Chem. Phys.* **2001**, 115, 1891–1901.
151. Tuomikoski, S., Huikko, K., Grigoras, K., Östman, P., Kostiainen, R., Baumann, M., Abian, J., Kotiaho, T., Franssila, S., *Lab Chip* **2002**, 2, 247–253.
152. Meng, J.-C., Averbuj, C., Lewis, W. G., Siuzdak, G., Finn, M. G., *Angew. Chem. Int. Ed.* **2004**, 116, 1275–1280.
153. Gao, T., Gao, J., Sailor, M. J., *Langmuir* **2002**, 18, 9953–9957.
154. De Stefano, L., Moretti, L., Lamberti, A., Longo, O., Rocchia, M., Rossi, A. M., Arcari, P., Rendina, I., *IEEE Trans. Trans. Nanotechnol.* **2004**, 3, 49–54.
155. Létant, S. E., Kane, S. R., Hart, B. R., Hadi, M. Z., Cheng, T.-C., Rastogi, V. K., Reynolds, J. G., *Chem. Commun.* **2005**, 851–853.
156. Anglin, E. J., Schwartz, M. P., Ng, V. P., Perelman, L. A., Sailor, M. J., *Langmuir* **2004**, 20, 11264–11269.
157. Bayliss, S., Buckberry, L., Harris, P., Rousseau, C., *Thin Solid Films* **1997**, 297, 308–310.
158. Sendova-Vassileva, M., Tzenov, N., Dimova-Malinovska, D., Rosenbauer, M., Stutzmann, M., Josepovits, K. V., *Thin Solid Films* **1995**, 255, 282–285.
159. Miyazaki, S., Sakamoto, K., Shiba, K., Hirose, M., *Thin Solid Films* **1995**, 255, 99–102.
160. Maroun, F., Ozanam, F., Chazalviel, J.-N., *J. Phys. Chem. B* **1999**, 103, 5280.
161. Bayliss, S., Zhang, Q., Harris, P., *Appl. Surf. Sci.* **1996**, 102, 390–394.
162. Chang, S.-S., Hummel, R. E., *J. Lumin.* **2000**, 86, 33–38.
163. Karavanskii, V. A., Lomov, A. A., Sutyrin, A. G., Bushuev, V. A., Loikho, N. N., Melnik, N. N., Zavaritskaya, T. N., Bayliss, S., *Phys. Stat. Sol. (a)* **2003**, 197, 144–149.
164. Karavanskii, V. A., Lomov, A. A., Sutyrin, A. G., Bushuev, V. A., Loikho, N. N., Melnik, N. N., Zavaritskaya, T. N., Bayliss, S., *Thin solid-films* **2003**, 437, 290–296.
165. Lomov, A. A., Bushuev, V. A., Karavanskii, V. A., Bayliss, S., *Crystallogr. Rep.* **2003**, 48, 326–334.
166. Kartopu, G., Bayliss, S. C., Karavanskii, V. A., Curry, R. J., Turan, R., Sapelkin, A. V., *J. Lumin.* **2003**, 101, 275–283.
167. Huber, H., Assmann, W., Karamian, S. A., Mücklich, A., Prusseit, W., Gazis, E., Grötzschel, R., Kokkoris, M., Kossionidis, E., Mieskes, H. D., Vlastou, R., *Nuclear Instr. Methods Phys. Res. B* **1997**, 122, 542–546.
168. Choi, H., Buriak, J. M., *Chem. Comm.* **2000**, 1669–1670.
169. Chen, J. H., Pang, D., Wickboldt, P., Cheong, H. M., Paul, W., *J. Non-Cryst. Solids* **1996**, 198–200, 128–131.
170. Shieh, J., Chen, H. L., Ko, T. S., Cheng, H. C., Chu, T. C., *Adv. Mater.* **2004**, 16, 1121–1124.
171. Hu, J., Odom, T. W., Lieber, C. M., *Acc. Chem. Res.* **1999**, 32, 435–445.
172. Rao, C. N. R., Deepak, F. L., Gundiah, G., Govindaraj, A., *Progress in Solid State Chem.* **2003**, 31, 5–147.
173. Míguez, H., Meseguer, F., López, C., Holgado, M., Andreasen, G., Mifsud, A., Fornés, V., *Langmuir* **2000**, 16, 4405.
174. Míguez, H., Chomski, E., García-Santamaría, F., Ibisate, M., John, S., López, C., Meseguer, F., Mondia, J. P., Ozin, G. A., Toader, O., van Driel, H. M., *Adv. Mater.* **2001**, 13, 1634.
175. van Vugt, L. K., van Driel, A. F., Tjerkstra, R. W., Bechger, L., Vos, W. L., Vanmaekelbergh, D., Kelly, J. J., *Chem. Comm.* **2002**, 2054–2055.
176. Gardelis, S., Rimmer, J. S., Dawson, P., Hamilton, B., Kubiak, R. A., Whall, T. E., Parker, E. H. C., *Appl. Phys. Lett.* **1991**, 59, 2118.
177. Kolic, K., Borne, E., Garcia Perez, M. A., Sibai, A., Gauthier, R., Laugier, A., *Thin Solid Films* **1995**, 255, 279–281.

178. Schoisswohl, M., Cantin, J. L., Chamarro, M., von Bardeleben, H. J., Morgenstern, T., Bugiel, E., Kissinger, W., Andreu, R. C., *Thin Solid Films* **1996**, 276, 92–95.
179. Buttard, D., Schoisswohl, M., Cantin, J. L., von Bardeleben, H. J., *Thin solid-films* **1997**, 297, 233–236.
180. Kartopu, G., Bayliss, S. C., Ekinci, Y., Parker, E. H. C., Naylor, T., *Phys. Stat. Solidi A* **2003**, 197, 263.
181. Kartopu, G., Ekinci, Y., *Thin Solid Films* **2005**, 473, 213–217.
182. Ksendzov, A., Fathauer, R. W., George, T., Pike, W. T., Vasquez, R. P., Taylor, A. P., *Appl. Phys. Lett.* **1993**, 63, 200.
183. Vyatkin, A. F., Linnross, J., Lalic, N., Rosler, M., *Phys. Low-Dimens. Struct.* **1997**, 5/6, 89.
184. Schoisswohl, M., Cantin, J. L., Chamarro, M., von Bardeleben, H. J., Morgenstern, T., Bugiel, E., Kissinger, W., Andreu, R. C., *Phys. Rev. B* **1995**, 52, 11898.
185. Unal, B., Parkinson, M., Bayliss, S. C., Naylor, T., Schröder, D., *J. Porous Mater.* **2000**, 7, 143–146.
186. Bsiesy, A., Vial, J. C., Gaspard, F., Herino, R., Ligeon, M., Muller, F., Romenstain, R., Wasiela, A., Halimaoui, A., Bomchil, G., *Surf. Sci.* **1991**, 254, 195.
187. Gupta, P., Dillon, A. C., Bracker, A. S., George, S. M., *Surf. Sci.* **1991**, 245, 360.
188. Choi, K., Buriak, J. M., *Langmuir* **2000**, 16, 7737–7741.

The Electrochemistry of Porous Semiconductors

John J. Kelly and A.F. van Driel

1 Introduction

The porous semiconductor electrode provides an interesting example for the theme of this volume: electrochemistry at the nanoscale. In porous etching, the anodic reaction can be considered to occur at an array of "nanoelectrodes", the pore tips, while the remainder of the porous matrix remains electrochemically inactive. In this case, conditions are clearly different from those at a macroscopic surface. Porous electrodes can also exhibit another aspect, one in which charge transfer is not restricted to the pore fronts; instead, the whole internal surface of the matrix acts as an electrode with a very large area but with a reduced "thickness", corresponding to the dimensions of the pore wall. Such small dimensions, which can even lead to size quantization, play a critical role in the electrochemistry. In this chapter, we consider the factors that decide whether the electrochemical reaction occurs exclusively at the pore fronts or at the whole internal surface of the porous layer. We review the electrochemistry of the two cases and related chemical and physical properties. In addition, we compare some results of porous-etched single crystals with those of nanoporous electrodes made by deposition from colloidal suspension.

Porous solids are classified according to an IUPAC convention [1] which distinguishes three cases: "macroporous" refers to systems with pore diameters larger than 50 nm while "microporous" describes the range below 2 nm. The term "mesoporous" is used for the intermediate range. Since physical and chemical properties of these materials do not recognize sharp IUPAC definitions, we shall use a mixed notation: micro/mesoporous refers to microporous systems and mesoporous systems with dimensions bordering on microporous. Similarly, we use meso/macroporous to denote macroporous systems and larger-dimension mesoporous systems. One can also classify porous materials on the basis of the morphology. Pores may be

J.J. Kelly (✉)
Condensed Matter and Interfaces, Debye Institute, Utrecht University, P.O. Box 80000, 3508 TA, Utrecht, The Netherlands
e-mail: J.J.Kelly@phys.uu.nl

P. Schmuki, S. Virtanen (eds.), *Electrochemistry at the Nanoscale,* Nanostructure Science and Technology, DOI 10.1007/978-0-387-73582-5_6,

isotropic, that is, with a random orientation, or anisotropic when they follow electric current lines or preferred crystallographic orientations.

Since in the case of micro/mesoporous semiconductors the dimensions of the pore walls can be similar to the radius of the Bohr exciton, effects due to quantum confinement can be expected [2]. Porous silicon is the most spectacular of such systems. While bulk single-crystal silicon shows only a very weak photoluminescence in the near-infrared, porous silicon with pore-wall dimensions less than 10 nm gives a very strong emission in the visible spectral range. The emission shifts to the blue as the porous structures become smaller, that is, with stronger confinement. We can use this phenomenon as a diagnostic tool to give information about the exact location of electrochemical reactions.

The review, which deals with electrochemistry in aqueous solutions, is organized as follows. Chapter 2 deals with the charging of porous semiconductors, and the results are compared with those of conventional (nonporous) electrodes. Chapter 3 focuses on electrochemical reactions. There is a subdivision on the basis of the main type of charge carrier involved (majority, minority, or both). In each case, the process of porous etching is first considered (electrochemistry at the pore fronts), and this is followed by oxidation and reduction reactions (electrochemistry at the pore walls). In Chapter 4 we draw some general conclusions.

2 Charging of Porous Electrodes

In classical semiconductor electrochemistry, the reference point for the electrode is the flat-band potential, U_{fb} (Fig. 1(a)) [3]. At this potential, the surface and bulk concentrations of charge carriers are the same, and there is no electric field present in the semiconductor. If the applied potential U deviates from U_{fb}, then a space charge is created at the surface, as in a Schottky diode (Fig. 1(b,, c, and d). The concentration of majority carriers at the surface may increase (accumulation) or decrease (depletion). Under strong depletion, that is, for a large deviation from the flat-band condition, one may, in principle, observe inversion; the concentration of minority carriers at the surface is larger than that of the majority carriers. However, in aqueous systems, reactions of the minority carriers at the solid–solution interface generally prevents inversion; instead deep depletion is observed (see Section 2.2). In this section, we consider the charging of porous electrodes under conditions in which majority carriers are expected to accumulate (Section 2.1) or to be depleted (Section 2.2) at the surface of a macroscopic porous electrode. The morphology of the porous matrix plays an important role in determining the charging properties.

2.1 Accumulation

In the case of an n-type semiconductor, accumulation is achieved by making the potential of the electrode negative with respect to the flat-band potential ($U < U_{\mathrm{fb}}$),

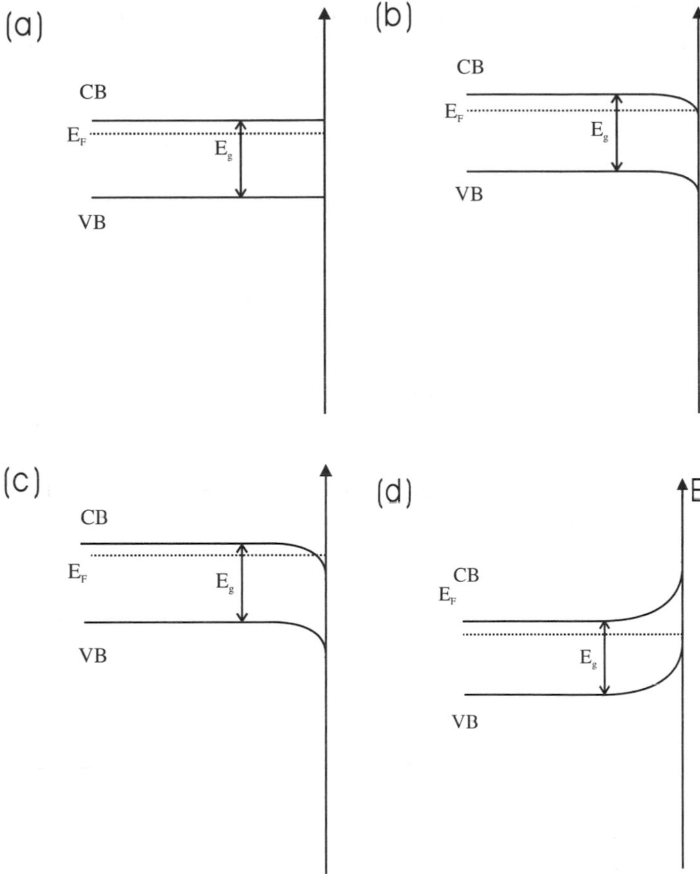

Fig. 1 Schematic band-energy diagram for an n-type semiconductor under flat-band conditions ($U = U_{fb}$) (**a**), accumulation conditions ($U < U_{fb}$) (**b**), under strong negative polarization (i.e., degeneracy, $U << U_{fb}$) (**c**), and depletion conditions ($U > U_{fb}$) (**d**). The conduction band (CB) and valence band (VB) are separated by the bandgap (E_g)

that is, by raising the Fermi level (see Fig. 1(b)). The majority carrier concentration at the surface becomes larger than that in the bulk. Eventually, the Fermi level passes through the conduction-band edge, and the surface becomes degenerate (quasi-metallic) (Fig. 1(c)). In this case, the Fermi level becomes pinned (the band edges at the interface become unpinned (Fig. 1(c)). The applied potential is now dropped across the Helmholtz layer in solution, as at a metal electrode. A corresponding situation holds for a p-type electrode at potentials positive with respect to the flat-band potential ($U > U_{fb}$) [3].

In contrast to the case of depletion (see Section 2.2), the space-charge layer capacitance for accumulation is large (in the case of degeneracy, comparable to the capacitance of the Helmholtz layer at a metal electrode). Consequently, the space-charge layer has a thickness similar to that of the Helmholtz layer. In the case of a porous electrode, one would generally expect the accumulation layer to follow the contours of the internal surface area of the matrix, even for a microporous electrode. The interfacial capacitance should then be proportional to the *total* surface area. This was shown by Peter et al. [4] for n-type silicon during anodic porous etching in 40% HF/ethanol (1/1). The interfacial capacitance was determined by performing a potential sweep to negative potentials with a triangular voltage waveform at various stages in the etching process (Fig. 2(a)). The capacitance was shown to be linearly dependent on the anodic charge passed during etching (Fig. 2(b)) and independent of the current density used for forming the porous layer. By assuming a particular particle geometry (spherical, columnar), the authors were able to calculate typical particle dimensions from the surface area and the volume of silicon in the porous layer (inset Fig. 2(b)). During subsequent chemical dissolution of the porous matrix in the same electrolyte solution, the capacitance decreased with etching time, indicating a decrease in the particle dimensions. This pore broadening due to chemical etching gave rise to an intense photoluminescence, characteristic of porous silicon.

An equivalent charging experiment using cyclic voltammetry is not possible with p-type porous silicon in aqueous solution because of anodic oxidation of the semiconductor. In the n-type case, charging is not hindered by the Faraday reaction because hydrogen evolution has a large overpotential.

The charging curves in the work of Peter et al. [4] do not show the structure which might be expected from quantization effects (filling of discrete shells of quantized structures, electron–electron repulsion, etc.). This is probably due to the large degree of polydispersity in porous silicon. (Luminescence results described in the next section give strong indirect evidence for size quantization.) Striking quantization effects have been observed in porous solids deposited from colloidal suspensions of monodisperse nanoparticles or quantum dots, Fig. 3(a). An electrochemically gated transistor geometry was used to study porous layers formed in this way [5–7]. The insulating substrate was provided with two parallel gold tracks that acted as a source and drain. The electrode potentials of source and drain could be controlled by a potentiostat. A small potential difference between source and drain allowed measurement of the conductance of the layer. Under applied potential, the quantum dot layer equilibrates with the source/drain electrode by adjustment of the electron density in the layer; the charge injected is compensated by incorporation of additional positive ions into the pores of the film. By measuring the current (Fig. 3(a)) or the differential capacitance [5, 7] as a function of potential, the authors could follow the charging of the quantum dots. For both systems studied (ZnO particles with diameters in the 3–6 nm range [5] and CdSe particles with similar dimensions [6, 7]), a stepped charging is observed, which could be attributed to a sequential filling of the S and P orbitals of the quantum dots (Fig. 3(b)). The sequential filling of the discrete levels led to dramatic changes in the conductance of the films (Fig. 3(b)) and in the carrier mobility [5–7]. This provides further evidence for size quantization.

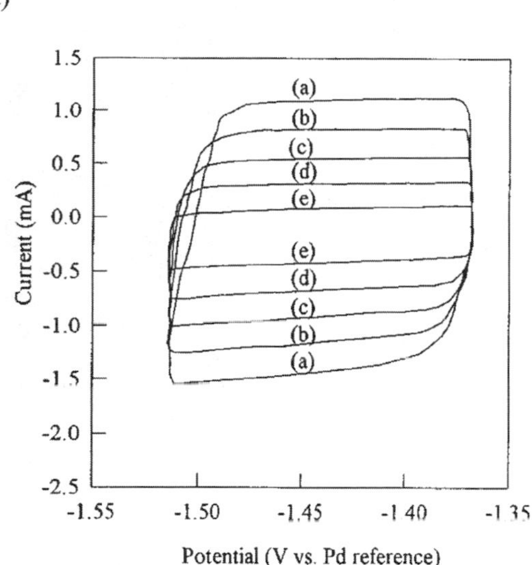

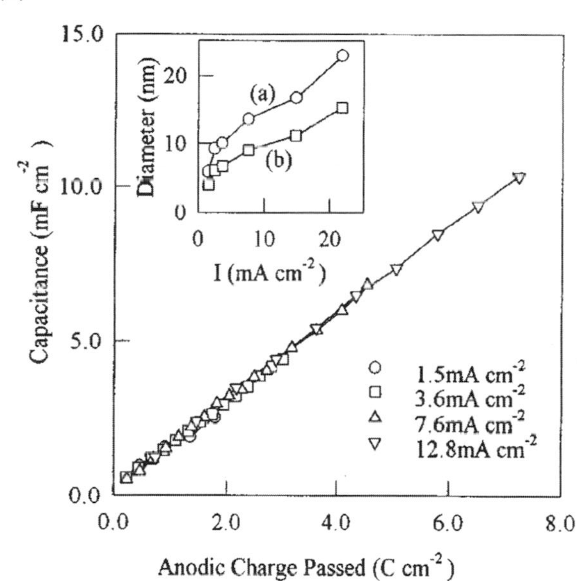

Fig. 2 (**a**) Cyclic voltammograms of n-type porous silicon in 40% HF. The sample was prepared by anodization at 14.8 mA cm^{-2} for 4 min to form a 6.8-μm thick porous layer. The scan rates were 250 (**a**), 200 (**b**), 150 (**c**), 100 (**d**), and 50 (**e**) mVs^{-1}. In (**b**) the interfacial capacitance is shown as a function of the anodic charge passed during formation of the porous layer. The current densities employed during porous etching are indicated in the figure. The inset shows the calculated average diameter as a function of the anodization current density for both a spherical (**a**) and columnar (**b**) geometry. Reprinted with permission from L.M. Peter, D.J. Riley, R.I. Wielgosz, "In situ monitoring of internal surface are during the growth of porous silicon", Applied Physics Letters, Vol. 66, p. 2355 (1995). Copyright (1995), American Institute of Physics

2.2 Depletion

Depletion occurs for an n-type semiconductor polarized positive with respect to the flat-band potential ($U > U_{\text{fb}}$, see Fig. 3(d)) and for a p-type semiconductor negatively polarized ($U < U_{\text{fb}}$) [3]. The thickness of the space-charge layer d_{sc} of a flat Schottky diode under depletion conditions is given by [8]:

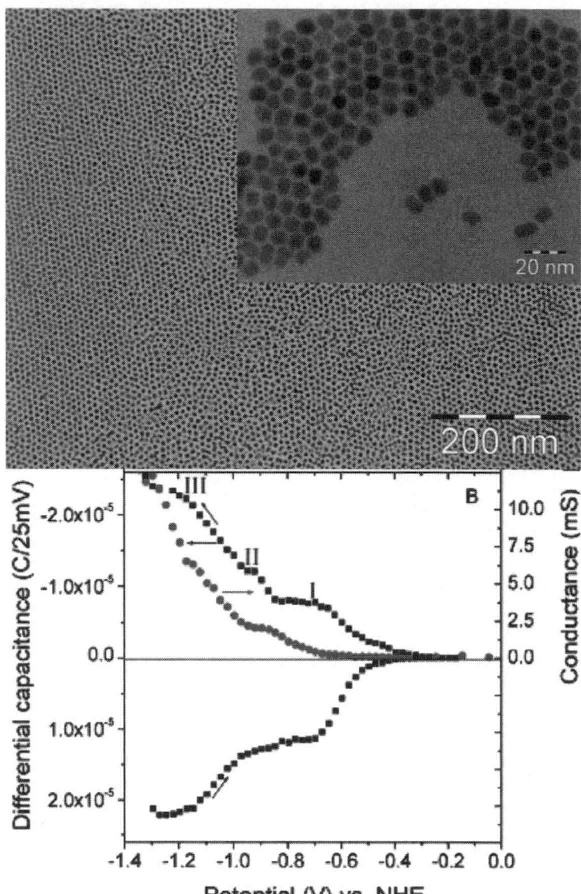

Fig. 3 (*top*) TEM image of 6.2-nm CdSe nanoparticles. The inset shows a magnified image of a monolayer of 8.4-nm CdSe nanoparticles. (*bottom*) Differential capacitance (solid squares) and long-range electronic conduction (open triangles) of 6.4-nm CdSe nanoparticle assemblies under steady-state conditions. Wave I in the charging curve occurs before the onset of electronic conduction; it is attributed to electron injection into localized states in the bandgap. Waves (II) and (III) are due to electron injection into quantized states of the nanoparticles. Reprinted with permission from A.J. Houtepen, D. Vanmaekelbergh, J. Phys. Chem. B, Vol. 109, p. 19634 (2005). Copyright (2005). American Chemical Society

$$d_{sc} = \sqrt{\frac{2\varepsilon\varepsilon_0}{eN_D}U_{sc}} \tag{1}$$

where ε is the dielectric constant, ε_0 the permittivity of free space, U_{sc} the band bending (U-U_{fb}), e the electronic charge, and N_D the donor density. For a typical band bending of 1 eV, d_{sc} can vary between 11 and 332 nm for dopant densities in the range 10^{16}–10^{19} cm^{-3} ($\varepsilon = 10$). Although the space-charge layer will be somewhat different at the surface of a cylindrical pore or a hemispherical pore front, it is clear that the range of values of d_{sc} covers the range of pore and pore-wall dimensions on going from low mesoporous to macroporous. Consequently, the potential distribution is more complicated than in the case of degeneracy, resulting from strong accumulation. We can distinguish a number of cases which depend on the relative thickness of the space-charge layer (d_{sc}) and the pore-wall thickness (W).

(i) Micro/mesoporous electrodes ($d_{sc} > W$)

In this case, the typical dimensions of the porous layer are considerably smaller than d_{sc}, and the whole porous structure will be depleted, even for moderate values of U_{sc} [9]. The depletion layer is located at the interface between the porous layer and the (nonporous) substrate.

The absence of a space charge in the porous matrix is responsible for a somewhat unexpected dependence of the photoluminescence of micro/mesoporous silicon on applied potential. While the presence of a depletion layer in a conventional semiconductor electrode suppresses light emission (the electric field separates the electrons and holes before they can recombine), electrons and holes generated in a "field-free" porous structure are not spatially separated. Radiative recombination within quantized structures gives photoluminescence in the n-type silicon which persists to potentials strongly positive with respect to the U_{fb} value of the nonporous semiconductor. We return to this result in Section 3.3.

(ii) Meso/macroporous ($d_{sc} < \frac{1}{2}W$)

When the pore-wall dimensions are larger than 2 d_{sc}, the space-charge layer can be accommodated within the porous structure, following the contours of the pores (Fig. 4(a)). The depletion-layer capacitance depends in this case on the porosity; it can be orders of magnitude larger than that of the corresponding geometric surface [9–14].

(iii) Meso/macroporous electrodes ($d_{sc} > \frac{1}{2}W$)

Since d_{sc} is directly proportional to the square root of the band bending ($U - U_{fb}$), the space-charge layer can, at higher potential, become thicker than the pore walls ($d_{sc} > \frac{1}{2}W$). In this case, the porous structure becomes fully depleted (as in case (i)) and the depletion-layer capacitance drops drastically; the depletion layer is again located at the porous/nonporous interface. A fairly abrupt transition to a fully depleted structure has been observed in different systems, for example, n-type GaP (see Fig. 5) [10, 11] and n-type SiC [12, 13].

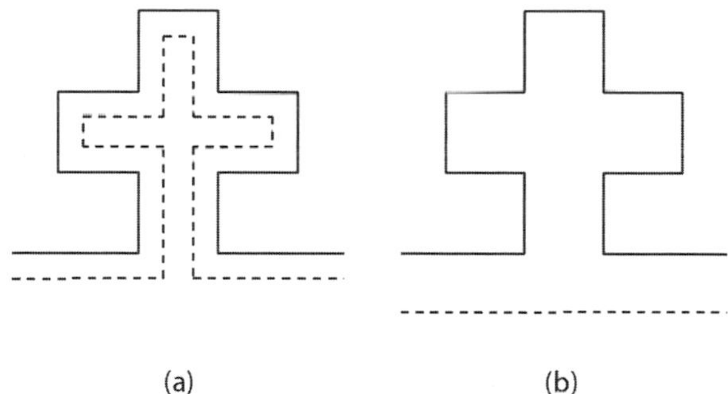

Fig. 4 Schematic representation of a porous structure under depletion conditions. In case (**a**) the depletion-layer width d_{sc} is smaller than half the wall thickness ($\frac{1}{2}$ W). The inner edge of the depletion layer follows the contours of the porous layer and the space-charge layer capacitance is very large. In case (**b**) the porous layer is completely depleted ($d_{sc} > \frac{1}{2}$ W). The depletion-layer edge is not located within the porous layer but at the interface between the porous layer and the substrate. In this case, the space-charge layer capacitance is similar to that of the flat electrode

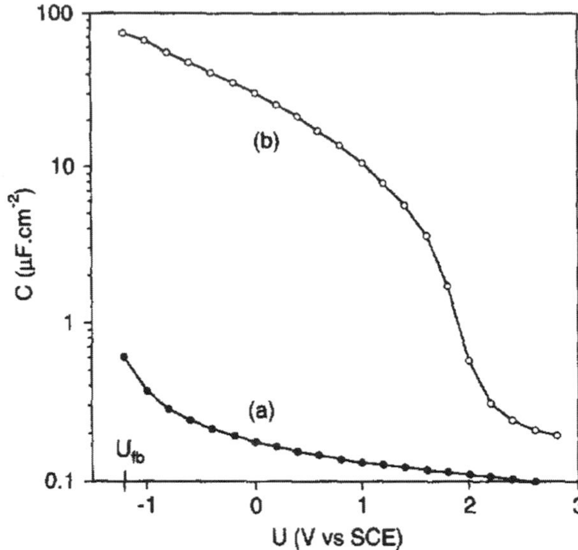

Fig. 5 The potential dependence of the space-charge layer capacitance of n-type GaP. Curve (**a**) refers to the nonporous electrode and curve (**b**) to the porous electrode. The etching charge passed in this case was about 5 C cm^{-2}. In case (**a**) the sharp drop in capacitance at higher potential corresponds to the transition from (**a**) to (**b**) in Fig. 4.The flat-band potential is denoted U_{fb}. Reprinted from Electrochimica Acta, J.J. Kelly and D. Vanmaekelbergh, "Charge carrier dynamics in nanoporous photoelectrodes", Vol. 43, p. 2773, Copyright (1998), with permission from Elsevier

3 Electrochemical Reactions

Reactions at semiconductor electrodes can be classified on the basis of the type of charge carrier involved in the interfacial process. Majority carrier reactions are expected when the surface concentration of holes in a p-type material or electrons in an n-type material is high, that is, in or close to accumulation conditions [3]. Minority carrier reactions can be initiated by injection from a species in solution. Alternatively, electron–hole pairs can be generated with supra-bandgap light. If the charge carriers are effectively separated by the electric field, that is, under depletion conditions, minority carrier reactions can be observed at the interface with solution. Finally, there is a third class of reaction in which recombination of electrons and holes determines the kinetics of the interfacial process; clearly, in this case, the surface concentration of both types of charge carrier is important.

In the context of this chapter, an important electrochemical reaction is the porous etching of the semiconductor. To dissolve a semiconductor valence band holes must be localized in surface bonds at the solution interface. Consequently, anodic etching of p-type electrodes having holes as majority carriers is expected to occur in the dark with an onset close to flat-band potential. To dissolve an n-type semiconductor, minority carriers must be provided. Generally, dissolution of the semiconductor is uniform with the whole surface etching at the same rate. Two factors distinguish porous etching: (i) pores are initiated by localized attack at certain points on the surface and (ii) these pores are able to propagate. A third important process is pore branching which allows a three-dimensional matrix to be formed. There is a huge literature on the porous etching of semiconductors, and a wide range of models have been proposed to describe the various systems. Despite the rich variety of attempts to describe completely the mechanism of porous etching, no consensus has been achieved. The various models focus on specific aspects of porous etching, using their own language and their own level of description. These models have been reviewed by Chazalviel and co-workers [15], who classify them in terms of chemical models [16, 17], physical models [18, 19], and simulation approaches [20, 21]. In this chapter, we limit the discussion to certain "physical" models which reflect the theme of the volume: electrochemistry on the nanoscale. For an overview of porous etching of semiconductors, we refer the reader to review papers [15, 22, 23] and books [24–26].

In Section 3.1, we show how minority carriers can be generated and consider the role of minority carriers in pore formation and redox reactions at porous electrodes. Corresponding majority carrier reactions are considered in Section 3.2. In Section 3.3, reactions are described in which *both* electrons and holes are involved; such processes at micro/mesoporous electrodes give rise to interesting optical effects.

3.1 Minority-Carrier Reactions

There are three main approaches to generating minority carriers in a semiconductor electrode: by "injection" from a redox species in solution, by illumination with supra-bandgap light, and by band-to-band tunneling. Because the third approach is the most important mechanism for porous etching of a wide range of n-type semiconductors, we shall begin by discussing this case.

3.1.1 Electron Tunneling

It is generally accepted that the propagation of pores during anodic etching of n-type semiconductors in the dark is due to enhanced dissolution as a result of electron tunneling from the top of the valence band to the conduction band. This is made possible by the much higher electric field at the tip of the pores. This idea was first proposed by Theunissen [18] and extended by Beale [19] and others [27–29]. This model is able to account for much of the phenomena observed during porous etching of n-type semiconductors in the dark.

The electric field at the pore tip is described by an abrupt one-sided Schottky barrier [8]. During etching, a strong reverse bias is applied, inducing deep depletion at the surface of the semiconductor (Fig. 6). The thickness of the depletion (or space charge) region at a flat n-type semiconductor/electrolyte interface is given by Eq. (1). The electric field in the space-charge region, as shown in Fig. 6, has a maximum E_{max} at the semiconductor surface:

$$E_{\mathrm{max}} = \frac{eN_{\mathrm{D}}d_{\mathrm{sc}}}{\varepsilon\varepsilon_0} \tag{2}$$

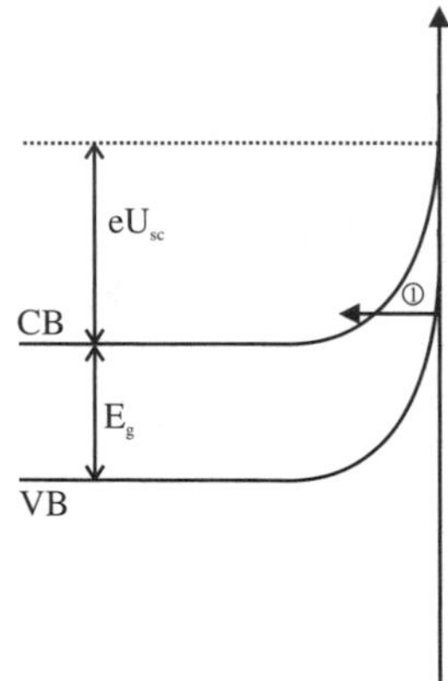

Fig. 6 Energy-band diagram for electron tunneling (step 1) under strongly anodic conditions (deep depletion). At the semiconductor–electrolyte interface, the electron tunnels from the valence band (VB) to the conduction band (CB). eU_{sc} is the potential difference with respect to flat-band conditions

where d_{sc} is the thickness of the space-charge layer. For GaP with a dopant density of 10^{17} cm^{-3} and an applied potential of 5 V with respect to the flat-band potential, the maximum field is 0.4×10^{6} V cm^{-1}. As pointed out by several authors [19, 28, 29], the curvature of the pore tip leads to a local intensification of the electric field. As a consequence of the strong field, "breakdown" occurs selectively at the tip of the pore. Electrons tunnel from the top of the valence band to the conduction band (Fig. 6), thus providing holes at the surface needed for the dissolution of the solid. In the case of GaP in acidic solution, six holes are required for each formula unit of the semiconductor and the trivalent Ga and P products dissolve [10, 11, 25]

$$\mathrm{GaP} + 3\mathrm{H_2O} + 6\,\mathrm{h}^+ \rightarrow \mathrm{Ga(III)} + \mathrm{H_3PO_3} + 3\mathrm{H}^+ \tag{3}$$

Tunneling of electrons through the barrier and avalanche multiplication are well-known phenomena in strongly reverse-biased p–n junctions [8, 30]. The role of the dopant density in breakdown in p–n junctions has been studied extensively [30]. The breakdown voltage increases with decreasing carrier concentration. The same trend is observed for etching of GaP [31] and Si. Furthermore, the dopant density in a p–n junction must be high ($>5 \times 10^{17}$ cm^{-3}), in order to achieve the field required for tunneling [8]. This explains the observation that for porous etching (in the dark), only materials with a relatively high carrier concentration ($>10^{17}$ cm^{-3}) are used. Dark etching of lower-doped materials is not possible or results in nonuniform layers. As for low-doped p–n junctions, avalanche multiplication can occur in low-doped semiconductors, leading to strongly localized etch pits [19, 27, 31].

Two general trends could be observed in the potential dependence of etching of n-type single crystals (Si [27], GaP [31], and InP [32]): first, the pore diameter and the pore spacing both increase with increasing potential. This can be understood on the basis of the breakdown model. At low potential, the threshold field for breakdown is only reached if the radius of curvature is small, that is, only for small pores. Thus, a lower applied potential leads to the development of small pores. Second, the depletion layers around the pores cannot overlap. This is well demonstrated in GaP, where initially all the pores grow isotropically from a point at the surface, while the pores subsequently "push" each other in a direction perpendicular to the surface [33]. A top view of a porous GaP layer (Fig. 7(a)) clearly shows that the pores are randomly distributed but do not overlap. This is related to the thickness of the depletion layer. At higher potential, the space-charge layer is larger (Eq. 1), which results in an increase of the spacing between the pores. A pore growing faster than the surrounding pores obtains more space and therefore a larger radius. A larger pore radius, however, means that the local electric field is weaker and thus the current density is reduced and thereby also the growth rate. This self-correcting mechanism explains the broadly observed flatness of the porous/nonporous interface.

The parallel between a p–n junction and an n-type semiconductor/solution junction under strong reverse bias is further demonstrated in the case of GaP by the observation of hot-carrier light emission [34]. Electroluminescence detected during porous etching of the semiconductor shows the same broad spectral features

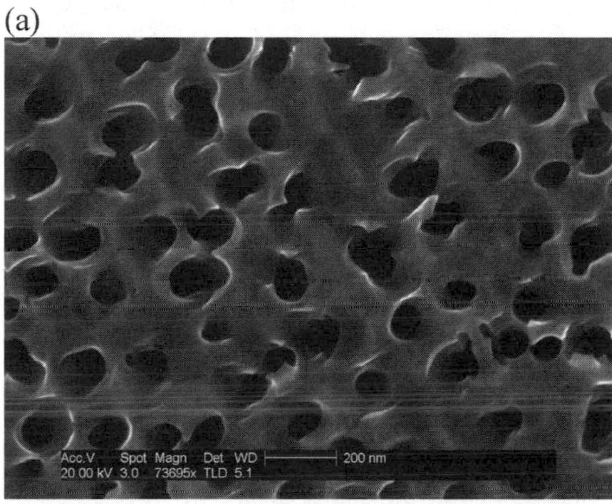

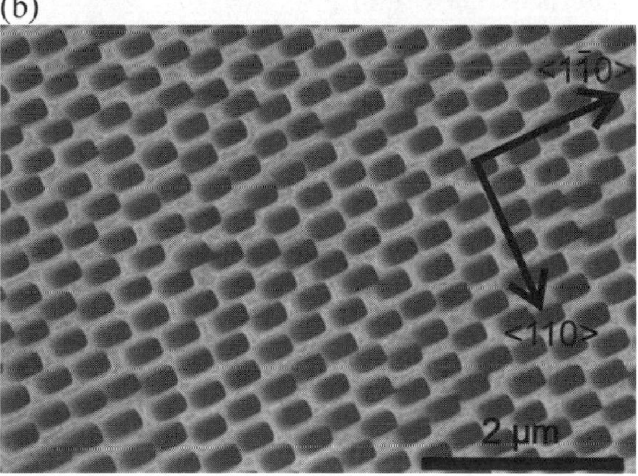

Fig. 7 Scanning electron microscopy picture of the top view of porous GaP. Sample (**a**) was etched at 10 V versus SCE in H_2SO_4 solution, and the pores are randomly oriented. Sample (**b**) was etched at 15 V versus SCE in HBr solution, and the pores have a rectangular shape and show self-organization. (**a**): Reprinted with permission from A.F. van Driel, B.P.J. Bret, D. Vanmaekelbergh, and J.J. Kelly, Applied Physics Letters, Vol. 84, p. 3852 (2004). Copyright (2004), American Institute of Physics. (**b**): Reproduced by permission of ECS – The Electrochemical Society, from J. Wloka, K. Mueller, and P. Schmuki, Electrochem. Solid State Lett. Vol. 8, p. B72 (2005)

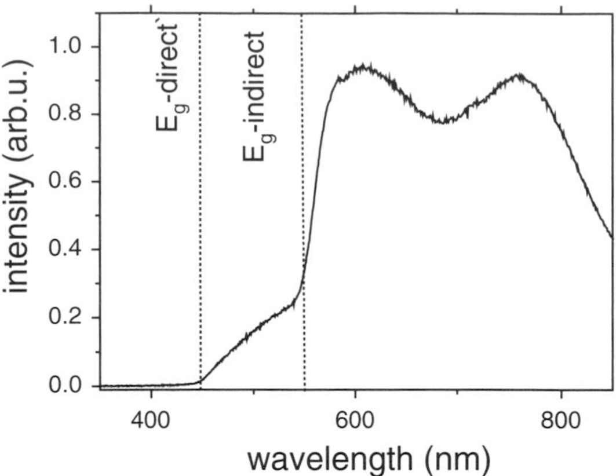

Fig. 8 Electroluminescence spectrum of n-type GaP during etching at 7 V in H_2SO_4 solution. The spectrum is very broad and contains a supra-bandgap contribution. The luminescence was attributed to a hot-carrier process. The direct and the indirect bandgaps are shown. Reprinted from Surface Science, A.F. van Driel, B.P.J. Bret, D. Vanmaekelbergh, and J.J. Kelly, "Hot carrier luminescence during porous etching of GaP under high electric field conditions", Vol. 529, p. 197, Copyright (2003), with permission from Elsevier

as those caused by reverse current flow in the p–n junction [30, 35], including a supra-bandgap contribution (Fig. 8). The electrochemical luminescence, which is generated at the tip of the pores, is related to the strong electric field in the depletion layer. After electron tunneling through the barrier, the electron is accelerated in the electric field. The electron can gain enough kinetic energy to excite a valence-band electron into the conduction band. The luminescence is attributed to the radiative recombination of these carriers. In contrast to the spatially uniform luminescence observed during porous etching of GaP, emission related to avalanche breakdown is strongly localized in bright spots [30]. The source of the luminescence, that is, the active etch front, moves during etching with a constant rate through the crystal. This allows one to measure in situ the wavelength-dependent transmittance of light of the porous layer [36].

In contrast to the random porous structure formed in GaP during etching in H_2SO_4 solution, strongly anisotropic, crystallographically oriented pores are formed during etching in HBr solution [37]. This was accompanied by strong electroluminescence [38], comparable to that observed during etching in H_2SO_4 solution [34]. In HBr solution, oscillations in the anodic current were accompanied by oscillations in light-emission intensity [38]. Current-line-oriented pores can also be grown in n-type InP [39–41]. By modulating the potential or the current density, various groups have grown modulated porous multilayers [33, 42].

3.1.2 Photogeneration

Illumination with supra-bandgap light can generate electrons and holes in a semiconductor electrode. Under depletion conditions, the charge carriers are separated; minority carriers are collected at the solid/solution interface where they can participate in oxidation or reductions reactions, depending on the semiconductor type. On the other hand, under accumulation conditions or if the electric field of the depletion layer is weak, then the electrons and holes are not spatially separated, but recombine. In a direct-bandgap semiconductor, this can lead to light emission (photoluminescence). In this section, both aspects of photogeneration, reaction and recombination, are dealt with. "Reaction" includes photoanodic porous etching of n-type Si and passivation of the porous semiconductor. We also consider the effect of porous morphology on the quantum efficiency of photoanodic processes and on minority-carrier transport in porous semiconductors. Under "recombination", we show how the porous structure can change the kinetics of electron–hole recombination and thus influence the physical and chemical properties.

(i) Reaction

Under illumination, n-type semiconductors can be anodically dissolved at potentials much less negative than those required for breakdown (see Section 3.1.1.). If the electric field of the depletion layer is sufficiently strong to separate effectively photogenerated electrons and holes, then the holes reach the surface and cause oxidation and dissolution of the solid. In most cases, etching of the surface is uniform. Silicon in HF solution is exceptional in a number of respects. At lower-light intensity, the anodic photocurrent–potential curve (Fig. 9(b), lower curve) resembles that of other n-type semiconductors. At negative potential, no photocurrent is observed because of electron–hole recombination. Photocurrent onset results from the increasing electric field of the depletion layer. The current finally levels off. The limiting photocurrent, however, does not show a simple linear dependence on photon density. At very low light intensity, a quantum efficiency of 4 is observed, that is, one photon is responsible for the passage of four charge carriers in the external circuit. To explain this, we must assume that the photogenerated holes create oxidation intermediates that lead to injection of three electrons into the conduction band. As the light intensity is increased, the quantum efficiency drops to 2 and hydrogen

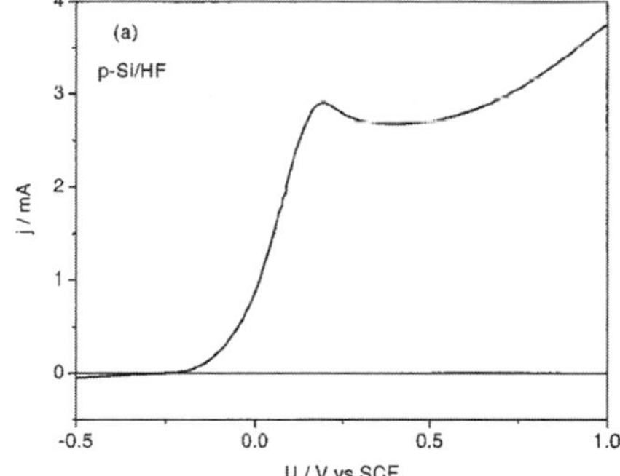

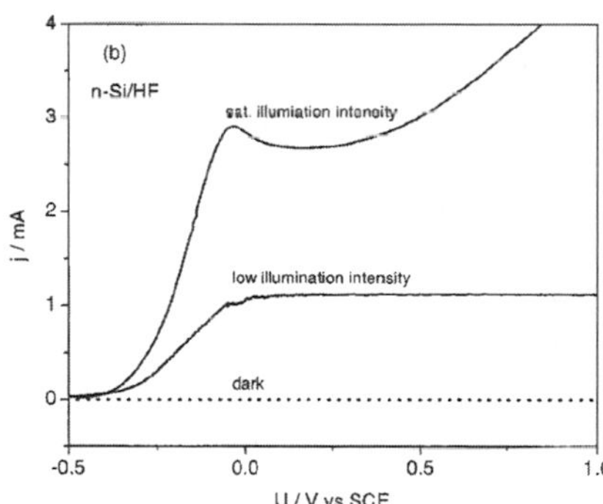

Fig. 9 Current potential curves of (100) silicon in 1% HF solution at room temperature. Curve (**a**) is for p-type Si in the dark and (**b**) for n-type Si in the dark, at low light intensity and under saturated illumination conditions. A standard three-electrode electrochemical cell was used, the potential of the Si working electrode was varied with respect to that of a SCE reference with a scan rate of 20 mV/s. Reproduced with permission from Wiley, from Reference [25]

is evolved. The latter results from a chemical reaction between an intermediate and H^+ or HF. The total reaction can be represented in a simplified form by [16, 43]:

$$Si + 6HF + h^+ \rightarrow SiF_6^{2-} + H_2 + 4H^+ + e^- \quad (4)$$

Silicon dissolves as a hexafluoride species, and the silicon surface is hydrogen terminated under these conditions. At high light intensity, the photocurrent–potential curve becomes independent of the photon flux (Fig. 9(b), upper curve). A peak in the curve indicates the formation of oxide on the electrode. This is a four-hole reaction.

$$Si + 2H_2O + 4\,h^+ \rightarrow SiO_2 + 4H^+ \quad (5)$$

The oxide is soluble in HF solution, and dissolution is mass-transport controlled. At potentials approaching the peak potential, hydride termination is replaced by hydroxide or oxide termination. Beyond the peak potential, illuminated silicon is electropolished.

The most striking feature of the electrochemistry of n-type silicon illuminated in HF solution is the porosity produced by anodic etching. This result, first reported by Uhlir in 1956 [44], is not found with most other semiconductors. At low photocurrent densities, a random array of micro/mesopores is formed (this material is luminescent in the visible), while at higher current densities corresponding to potentials approaching the electropolishing range, the pores become wider (meso/macro) and tend to be anisotropic. Anisotropic-pore growth is very likely due to field-enhanced collection of minority carriers at the pore tips.

In a series of papers starting in 1990, Lehmann and co-workers [45–47] described very elegant work on macropore formation in silicon. Highly anisotropic pores could be formed by photoanodic etching of n-type (100) silicon whose surface had first been provided with etch pits, for example, by dark etching at high potential. The wafers were illuminated from the backside. Holes diffuse from the backside selectively to tip of the pores and dissolve the semiconductor locally. In this way, a porous structure can develop that corresponds to the predefined pore pattern. In order to cross the complete semiconductor width, the hole-diffusion length must be large. The hole-diffusion length increases with decreasing dopant density, and therefore low-doped materials must be used. Pore widths ranged from 0.5 to 100 μm, while a pore depth of 200 μm could be readily achieved (Fig. 10(a) and (b)) [48]. By starting with an ordered array of pits, produced by anisotropic etching through a photolithographically masked pattern, a perfectly ordered two-dimensional array of pores could be produced. Such lattices are interesting for photonic applications. The orientation of the primary pores is crystallographically determined; they grow in the $\langle 100 \rangle$ direction. Lehmann described macropore formation in n-type silicon as a self-adjusting mechanism characterized by a specific current density at the pore tip. At this current density, dissolution changes from a charge-transfer-limited to a mass-transport-limited regime. Passivation of the pore walls is attributed to a depletion of the minority carriers.

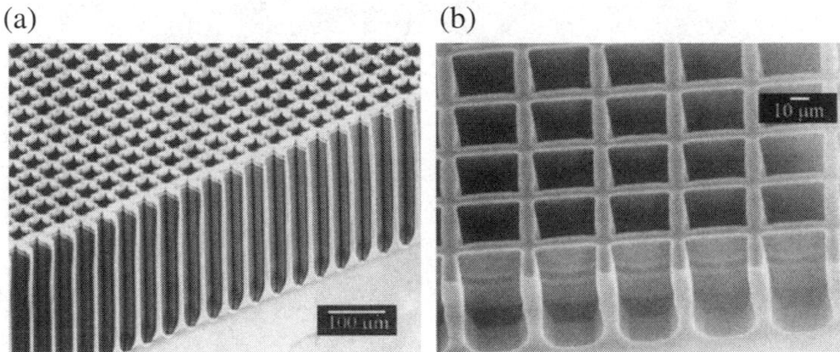

Fig. 10 (**a**) and (**b**) Scanning electron microscopy pictures of macropores formed in silicon. Etching was performed in 3% HF solution, starting with a predefined pore-pattern. The etching conditions allowed control of the pore wall thickness. Reprinted from Materials Science and Engineering B, P. Kleinmann, J. Linnros, and S. Petersson, "Formation of wide and deep pores in silicon by electrochemical etching", Vol. 69, p. 29, Copyright (2000), with permission from Elsevier

As pointed out by Lehmann, there are several limitations to the pore arrays that can be fabricated with n-type silicon [49]. Pores with a diameter down to 0.3 μm have been achieved. The lower limit is established by the diameter at which the intensification of the electric field is large enough to cause breakdown at the tip of the pore (see Section 3.1.1). There does not seem to be an upper limit for the pore radius. Pore widths up to 100 μm have been reported [48]. In principle, the pore spacing can be chosen arbitrarily. However, when the pore spacing is larger than twice the depletion layer thickness ($W > 2d_{sc}$), the porous layer is not completely depleted. Holes can now diffuse between the pores and dissolve the pore walls. Usually low-doped materials are used for macropore formation in Si. This means that the depletion layer is wide (typically 1.5 μm for 10^{15} cm^{-3} doped n-type Si at 2 V) and for porous layers with a spacing smaller than $2d_{sc}$, the porous structure is depleted and thus protected against dissolution.

Föll and co-workers [50] extended earlier work of Lehmann by studying the dependence of macropore formation on the substrate orientation. Samples were cut from a large silicon crystal in various orientations between $\{100\}$ and $\{111\}$. The growth direction of the photoanodically formed macropores was found to be $\langle 100\rangle$ and $\langle 103\rangle$ with a switch over at a critical angle of 43°. Surprisingly, pores obtained by breakdown at high field without illumination always grow in the $\langle 100\rangle$ direction. The reason for such differences is not clear.

The shape of anisotropic macropores in n-type silicon can be tuned by means of the photocurrent density or the potential. Gösele and co-workers [51–53] have succeeded in making three-dimensional structures with a high degree of perfection by combined modulation of the applied potential and the light intensity (Fig. 11(a) and (b)). A subsequent anisotropic etching step could be used to extend the range of structures and to adjust the photonic properties of the system.

In HF-free acidic solution, porous n-type silicon passivates under illumination at positive potentials (depletion conditions) [54]. The oxide formed (see Eq. 5) is

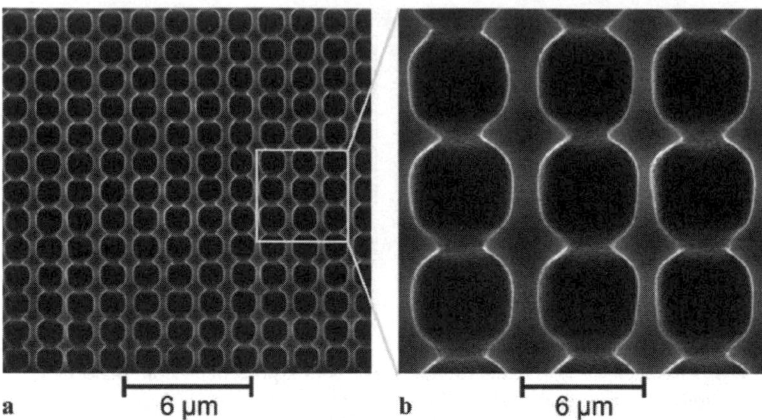

Fig. 11 Scanning electron microscopy pictures of strongly modulated porous n-type silicon samples. In (**a**) an overview is shown and in (**b**) the close-up. During porous etching of the Si crystal, the applied potential was modulated, resulting in a strongly modulated structure. Reproduced with permission by Springer, from S. Matthias, F. Muller, J. Schilling, and U. Gosele, Appl. Physics A, Vol. 80, p. 1391 (2005)

not soluble. In contrast to electrochemical oxide reactions in HF solution, oxidation occurs at the whole internal surface of the porous layer; this can be deduced from the charge involved in the oxidation reaction. Very likely, the pore fronts are passivated first, and this is followed by passivation of the pore walls. Photoanodic oxidation can be caused even by near-infrared light corresponding to the bandgap energy of the silicon substrate. Surprisingly, infrared-induced photoanodic oxidation of porous n-type silicon is accompanied by emission of visible light (electroluminescence). Infrared light cannot excite the wider bandgap porous semiconductor directly. Instead, holes generated under depletion conditions in the substrate must be injected into quantized structures of the porous matrix. This luminescence, which is closely connected to anodic oxidation, is also observed with porous p-type silicon. The mechanism will be discussed in Section 3.3.1.

The quantum efficiency of a photoelectrochemical reaction, that is, the number of charge carriers measured in the external circuit per absorbed photon, depends on (i) the efficiency with which the minority carriers are collected at the surface and (ii) the competition between surface recombination and the charge transfer reaction [12]. The collection efficiency depends on the penetration depth of the light $1/\alpha$ (where α is the absorption coefficient), the minority carrier diffusion length L_p (in the case of holes), and the thickness of the space-charge layer d_{sc}. For the simple case in which surface recombination can be disregarded, that is, all carriers reaching the surface are transferred to solution, the Gärtner model [9] for an illuminated Schottky diode can be used to calculate the quantum efficiency:

$$Q = j_p/\phi_o = 1 - \frac{e^{-\alpha d_{sc}}}{1 + \alpha L_p} \tag{6}$$

where j_p is the flux of holes to the surface and ϕ_o is the absorbed photon flux. The term on the right-hand side is the fraction of carriers lost to recombination in the semiconductor. For a direct-bandgap semiconductor, the absorption coefficient for energies above the bandgap is large and consequently the penetration depth of the light is small. In this case, minority carriers can be collected efficiently by migration ($1/\alpha < d_{sc}$) or by diffusion and migration ($1/\alpha < L_p$). In an indirect semiconductor, however, α above the bandgap is small and the penetration depth of the light may be very large ($1/\alpha > (d_{sc} + L_p)$). For example, in GaP (with a bandgap corresponding to 554 nm) illuminated with 514 nm light only about 1% of the photogenerated carriers reach the surface for a strong band bending of 1 eV. Consequently, the quantum efficiency is very low.

Porosity can affect the properties of photoelectrodes in two important ways [10–12, 25]. When the dimensions of the porous structures are comparable to the illumination wavelength, then the incident light is scattered and internally reflected within the porous layer. Equation (6) is no longer valid. The light is, in fact, absorbed much more effectively than in a nonporous medium, giving rise to a smaller effective penetration depth. In addition, light absorbed in a porous electrode generates minority carriers close to the solid/solution interface. The criterion for minority-carrier collection by the electrolyte is no longer $1/\alpha \leq (L_p + d_{sc})$ but $W/2 \leq (L_p + d_{sc})$. If this requirement is fulfilled, the minority carriers escape "bulk" recombination and reach the interface. The efficiency of reaction of minority carriers is then determined by competition between their transfer to solution and surface recombination. If the kinetics of the former process are efficient, then a high quantum yield can be obtained even for a light penetration depth which, in a nonporous electrode, would give a negligible yield.

There have been various reports of enhanced quantum efficiency due to porosity in photoelectrodes. Results obtained with n-type GaP are among the most spectacular. Curve (a) of Fig. 12 shows the photocurrent spectral response of a nonporous electrode. The quantum efficiency for photons with energy just above the fundamental absorption edge (2.24 eV) is low because, as already pointed out, the penetration depth $1/\alpha$ of the light is much larger than the retrieval length $d_{sc} + L_p$ of the minority carriers. As the photon energy is increased, $1/\alpha$ decreases (α increases), leading to an increase in efficiency. A strong increase is observed at 2.76 eV, which corresponds to the first direct optical transition in GaP. On the other hand, the quantum efficiency for the macroporous electrode (curve (b) in Fig. 12) reaches a value of 1 for photons of energy close to the fundamental absorption. Since the dimensions of the structural units are in the 100-nm range and the minority-carrier diffusion length is 50 nm, the criterion ensuring high efficiency $W/2 \leq (d_{sc} + L_p)$ is met for the potential range in which surface recombination is absent ($U_{sc} > 0.5$ V).

In addition to the efficient transfer of minority carriers across the semiconductor/solution interface, the majority carriers must be transported through the porous layer to the nonporous substrate and the back contact. Many recent experimental results show that electron transport through a mesoporous semiconducting network is a slow process due to carrier trapping. For instance, the average time that electrons need to travel through the system before collection, that is, the transit time, is

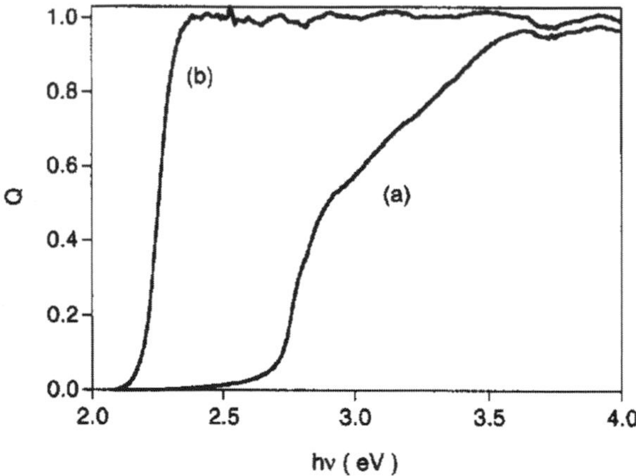

Fig. 12 The dependence of the photocurrent quantum efficiency (Q) on the photon energy (hν) for a polished n-type GaP electrode (**a**), and the same electrode after porous etching (**b**) [16 C cm^{-2}, 10 V vs. SCE]. The photocurrent, measured in 0.5 M H_2SO_4 solution at 1 V versus SCE, was due to anodic dissolution of the semiconductor. Reprinted from Electrochimica Acta, J.J. Kelly and D. Vanmaekelbergh, "Charge carrier dynamics in nanoporous photoelectrodes", Vol. 43, p. 2773, Copyright (1998), with permission from Elsevier

in the millisecond to second range [25]. As a result, photogenerated electrons can be lost before collection, by transfer to the oxidized species in the solution, a process characterized by a time constant τ_{rec}. Electron back transfer forms an important recombination process in the dye-sensitized photoelectrochemical solar cell based on the nanoparticle TiO_2 electrode [25].

(ii) Recombination
In a nonporous direct bandgap semiconductor, photogenerated electron–hole pairs recombining radiatively will give rise to photoluminescence. According to the Gärtner model (see Eq. 6), the PL intensity, I_{PL}, is given by

$$I_{PL} = \frac{\kappa \phi_0}{1 + \alpha L_p} e^{-\alpha d_{sc}} \tag{7}$$

where κ is the ratio of the rate of radiative recombination to the total recombination rate, that is, the photoluminescence quantum efficiency. The dependence of the depletion layer thickness d_{sc} on potential is given by Eq. (1). Figure 13 shows schematically the effect of applied potential on the emission intensity for an n-type semiconductor. Under depletion conditions at positive potential, the photogenerated electron–hole pairs are separated by the electric field preventing radiative recombination; instead, the flow of photocurrent (PC in Fig. 13) is favored. At potentials approaching U_{fb}, the surface concentration of majority carriers (holes in this case) increases and electron–hole recombination is favored. The photocurrent decreases with the onset of photoluminescence. Equation (6) predicts a limiting value of I_{PL} at $d_{sc} = 0$ ($I_{lim} = \kappa\phi_0/(1+\alpha L_p)$).

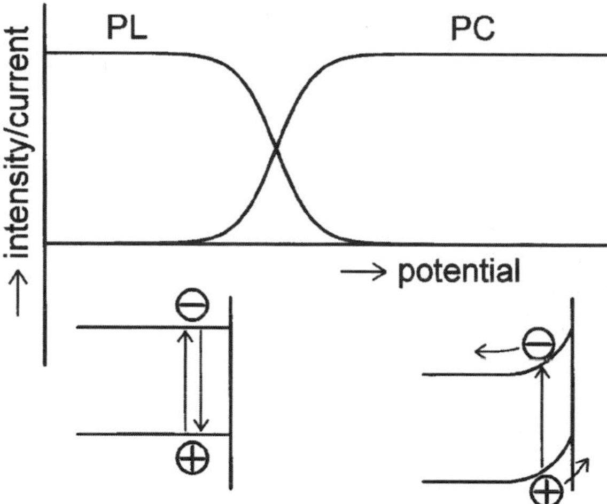

Fig. 13 Schematic representation of the potential dependence of the photocurrent (PC) and the photoluminescence intensity (PL) of an n-type semiconductor electrode in an indifferent electrolyte solution. Energy-band diagrams are shown for the illuminated semiconductor under flat-band (*left*) and depletion (*right*) conditions. Reprinted from Electrochimica Acta, J.J. Kelly, E.S. Kooij, and E.A. Meulenkamp, Vol. 45, p. 561, Copyright (1999), with permission from Elsevier

The potential dependence of the strong PL of micro/mesoporous silicon observed in indifferent electrolyte solution is, on the other hand, markedly different from that of a bulk electrode [55–57]. Under conditions corresponding to depletion in a non-porous electrode, potential-independent emission (PL) is observed (Fig. 14). Charge carriers generated within nanocrystallites are not spatially separated ($d_{sc} > W$) but

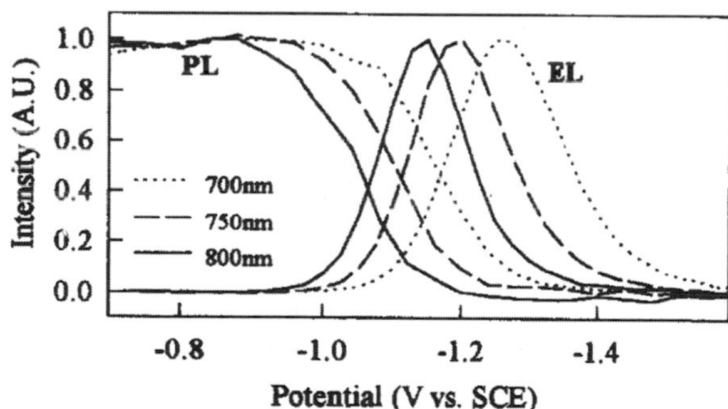

Fig. 14 Potential dependence of the electroluminescence (EL) and the photoluminescence intensity (PL) of a porous n-type Si electrode in an indifferent electrolyte solution (PL) and in $Na_2S_2O_8/H_2SO_4$ (EL) aqueous solution. Results are shown for three emission wavelengths. Note that the potential scale in this figure is reversed. Reprinted from J. Electroanalytical Chemistry, E.A. Meulenkmap, L.M. Peter, D.J. Riley, and R.I. Wielgosz, "On the mechanisms of the voltage tuning of photoluminescence and electroluminescence in porous silicon", Vol. 392, p. 97, Copyright (1995), with permission from Elsevier

recombine. The photoluminescence is quenched at potentials negative with respect to U_{fb}, a process that has been attributed to Auger recombination [57] or to hydrogen-mediated recombination [58]. The potential at which the PL intensity decreases depends on the wavelength of the emitted light: emission at lower energy (i.e., from larger crystallites) is quenched at more positive potentials than the high-energy emission (from the smallest crystallites). This has been attributed to a higher barrier for transfer of an electron from bulk silicon to small, more confined (wider bandgap) structures than to larger structures.

Interesting optical effects are observed during charging of the quantum-dot solids described in Section 2.1. The visible absorption spectrum of CdSe quantum dots in suspension shows characteristic sharp features that can be attributed to discrete excitonic transitions. These features are retained in the quantum-dot solid. As one would expect, such optical transitions are quenched when electrons are injected into the conduction levels at negative potentials. By studying the potential dependence of the absorption, Houtepen and Vanmaekelbergh [7] could conclude that the energy of the conduction S electrons is determined by the quantum confinement energy and by Coulomb repulsion between the S electron and all other electrons in the assembly. Wang et al. [59] observed photoluminescence from CdSe quantum-dot solids in the neutral, and the one- and two-electron charged states, even though the film is transparent at the emission wavelength. This striking result leads the authors to conclude that the threshold for amplified stimulated emission is strongly reduced, a result with important consequences for lasing. These two examples demonstrate the exciting perspectives of electrochemical gating in quantum-dot solid media.

3.1.3 Minority-Carrier Injection

Electrons can be extracted from the valence band of a semiconductor (i.e., holes can be "injected") if the acceptor levels of an oxidizing agent in solution correspond to the valence-band levels of the solid. This is the case for certain "simple" redox species such as Ce^{4+}, $IrCl_6{}^{2-}$, and $Fe(CN)_6{}^{3-}$ at Si [60]. For $Fe(CN)_6{}^{3-}$ in neutral aqueous solution, the reaction can be represented by

$$\mathrm{Fe(CN)_6^{3-}} \rightarrow \mathrm{Fe(CN)_6^{4-}} + \mathrm{h^+(VB)} \tag{8}$$

In an n-type silicon electrode, holes held at the surface by the electric field of a depletion layer cause oxidation and passivation of the semiconductor. The oxide is not soluble in this case. Under accumulation conditions, the injected holes can recombine with conduction band electrons but, since radiative recombination in the indirect bandgap semiconductor is very inefficient, no light emission is observed.

Surprisingly, if holes are injected into n-type (or p-type) micro/mesoporous silicon at open-circuit potential, luminescence typical of the porous semiconductor is observed [60]. This emission, termed chemoluminescence, is related to oxidation of silicon (see Section 3.1.2), and its mechanism will be considered in Section 3.2.1. The fact that visible luminescence is observed shows that holes must be injected directly into quantized structures of the porous layer.

The presence of HF in a solution containing an oxidizing agent prevents passivation (see Eq. 4). Under such conditions, micro/mesoporous silicon can be formed at open-circuit potential. This process has been termed "stain etching", and the most widely used etchant contains HNO_3 as oxidizing agent [61, 62]. The mechanism involves two electrochemical reactions: reduction of the oxidizing agent to give holes (Eq. 8) which are used for oxidation of the semiconductor. Since porous etching occurs under open-circuit conditions, the rates of the two reactions must be equal. Stain etching has the advantage of being relatively simple (a counter electrode and voltage source are not required). However, the range of morphologies and thicknesses is rather limited. When "noble" metal ions are used as oxidizing agents, metal films with very distinct morphologies can be formed [63]. Galvanic exchange of gold on silicon gave, after surface modification of the metal with *n*-dodecanethiol, superhydrophobic surfaces with contact angles up to 165° [64].

Electrochemical oxidation of a reducing agent via the conduction band of a p-type semiconductor provides minority carriers (electrons) in the band. Such a reaction occurs with the methyl viologen radical cation $MV^{+\bullet}$ at silicon [65, 66]

$$MV^{+\bullet} \rightarrow MV^{2+} + e^{-}(CB) \tag{9}$$

The observation of visible luminescence during this reaction at porous p-type silicon under accumulation conditions indicates that electrons are injected into the porous matrix where they subsequently recombine with the majority carriers.

3.2 Majority-Carrier Reactions

3.2.1 p-Type Semiconductors

For most p-type semiconductors in indifferent electrolyte solution, oxidation of the solid is the main reaction involving majority carriers, that is, valence-band holes. (For wide-bandgap covalent semiconductors such as GaN and SiC, oxidation of water to oxygen may compete with oxidation of the semiconductor.) In the context of this chapter, the most interesting reaction is clearly the porous etching of silicon in HF solution. A typical current–potential curve is shown in Fig. 9(a). Initially the anodic oxidation current increases exponentially with increasing potential. Two holes are involved in the reaction, and a hydrogen molecule is formed via a chemical reaction of protons with reaction intermediates [67]. This process is clearly analogous to that observed with illuminated n-type Si at lower light intensity (see Eq. 4). Silicon dissolves as a hexafluoride complex.

$$Si + 6HF + 2h^{+} \rightarrow SiF_6^{2-} + 4H^{+} + H_2 \tag{10}$$

In this potential range, luminescent micro/mesoporous silicon is formed. At more positive potential, a maximum is observed in the current as for n-type Si at high light intensity, and the oxidation mechanism changes. An oxide layer is formed

(Eq. 5). Since the oxide is soluble in HF solution and the rate of dissolution is mass-transport controlled, electropolishing of the silicon surface is observed in the dark.

Macropores can also be grown in p-type Si [68]. This approach has the advantage that illumination is not required. The result is unexpected. To ensure sufficient holes at the surface, a forward bias is required. The pore walls are therefore not protected against dissolution ($d_{sc} < W$), as is the case for n-type Si. The mechanism is not well understood. The presence of surfactants [69] or the high resistivity of the substrate [70, 71] was proposed as key factors in the protection of the pore walls against dissolution. Both models are disputed [72, 73]. The complicating factor in the discussion is that etching is performed close to the flat-band potential; the exact electronic state is not known. Lehmann [72] proposed a mechanism in which it was assumed that the surface is slightly depleted of holes. Therefore, at the pore walls and pore tips, there is a diffusion current toward the surface. However, since the electrode is under slight reverse conditions, there is also a field current in the opposite direction. Under equilibrium conditions, both components compensate. When a positive potential is applied, the field current decreases, but the diffusion current remains constant. The diffusion current now determines etching and is larger at the tip of the pores. Geometrical factors are responsible for this. Because of this, a porous structure can develop. A stability analysis of macropore formation in p-type Si has been presented in Wehrspohn et al. [74].

In contrast to etching in HF solution for which the reaction occurs only at the pore fronts, in HF-free solution, anodic oxidation leads to passivation of the whole surface [54, 75–78]. This can be deduced from the charge required to achieve passivation, which is much higher for a porous electrode. Further evidence that the microporous layer is oxidized comes again from the observation of strong visible luminescence during the anodic reaction, typical of light emission from porous silicon (see Section 3.3.3).

Reducing agents whose donor levels correspond to the valence band of a semiconductor can compete for the surface holes and thus suppress or prevent anodic dissolution or passivation. Electrons may also be transferred from the reducing agent directly to electron-deficient surface bonds (the intermediates of the anodic dissolution process), thereby repairing the bonds and "stabilizing" the semiconductor [54]. It has been shown that the ferrocyanide ion can be oxidized at p-type porous silicon and, in this way, partially stabilize the semiconductor. As would be expected, this process quenches the luminescence. There is evidence to suggest that stabilization occurs via intermediates of the oxidation reaction.

3.2.2 n-Type Semiconductors

For an n-type semiconductor in an aqueous indifferent electrolyte solution, reduction of protons or of water to give hydrogen is the main reaction of the majority carriers. If the acceptor levels of a redox couple in solution overlap with the conduction-band levels of the solid, then the oxidized form can be reduced under accumulation conditions. In a porous semiconductor, this is likely to occur at the whole internal surface of the electrode (see Section 2.1), if mass transport of electroactive species is not a problem. The rate of reduction depends on the electron concentration in the porous layer and thus increases as the electrode potential is made more negative. Two reactions of this type are interesting: electrodeposition of metal and reduction of strong two-electron oxidizing agents. The latter will be considered in Section 3.3.1.

It is possible to cover the whole interior surface of a porous semiconductor with a metal by cathodic reduction of metal ions via the conduction band. An example is gold electrodeposition from gold cyanide solution in macroporous n-type GaP [14]. A metal that forms a blocking contact with the semiconductor will give a Schottky diode with exceptional properties. The contact area of the diode is much larger than the geometric area. Minority carriers photogenerated in such a device can be collected with high efficiency when the dimensions of the pore walls are comparable to or smaller than the minority-carrier diffusion length (see Section 3.1.2).

3.3 Combined Majority/Minority-Carrier Processes

Injection of minority carriers into a semiconductor under accumulation conditions generally leads to recombination with the majority carriers. In that sense, the reactions described in Section 3.1.3 constitute an example of combined electron–hole processes. In this section, we consider some more complex hole- and electron-injecting reactions which, at porous electrodes, give interesting optical results (Sections 3.3.1 and 3.3.2, respectively).

3.3.1 Hole Injection

Reduction of a certain class of two-electron oxidizing agents, for example, the peroxydisulphate anion ($S_2O_8^{2-}$), known to give rise to "photocurrent doubling" at p-type semiconductors, can occur at n-type electrodes in the dark. The interesting feature of such reactions is that the first conduction band step

$$S_2O_8^{2-} + e\ \ (CB) \rightarrow SO_4^{2-} + SO_4^{-\bullet} \tag{11}$$

gives a strongly oxidizing species (in this case the radical anion $SO_4{}^{-\bullet}$) [55, 79, 80] which can inject a hole into the valence band of most semiconductors:

$$SO_4^{-\bullet} \rightarrow SO_4^{2-} + h^+(VB) \tag{12}$$

The injected holes can recombine with the majority electrons to provide electroluminescence. The observation of strong luminescence in the visible range, from micro/mesoporous n-type silicon, is clear evidence that holes are injected directly into quantized structures in the porous matrix.

The potential dependence of the spectral distribution of the electroluminescence (and of the photoluminescence) provides further evidence for quantization (see

Fig. 14) [56–58, 81]. Onset of luminescence is at a potential negative with respect to the onset potential for reduction of $S_2O_8^{2-}$. The emission passes through a maximum and is quenched at negative potential. Quenching is attributed to the same mechanisms as proposed for PL quenching. An interesting aspect of the emission from this system is its "voltage tunability"; the emission maximum shifts to shorter wavelength as the potential is made more negative. It has been suggested that this effect is also due to quantum confinement. Because of band-edge mismatch, supply of majority carriers to porous structures depends on their size. A more negative potential is required to supply the conduction band of smaller structures with electrons; these electrons are required for the first step of the $S_2O_8^{2-}$ reduction to trigger the subsequent hole-injection reaction. A similar voltage tuning is observed for PL (Fig. 14).

Peter and Wielgosz [82] produced electroluminescence in porous p-type silicon in peroxydisulphate solution by illuminating the substrate from the back side. The electrode was cathodically (reverse) biased. Electrons generated by light in the substrate accumulate in the porous layer, where they trigger the peroxydisulphate reduction. Voltage tuning of the emission was observed, similar to that found for porous n-type silicon in the dark.

3.3.2 Electron Injection

Under anodic polarization in HF-free neutral or acidic solution, p-type micro/mesoporous silicon is oxidized and passivates (Fig. 15(a)) [75–79] (see Section 3.2.1). During this process, visible light is emitted (Fig. 15(b)). This was, in fact, the first report of electroluminescence from porous silicon. The maximum in the emission peak shifts to shorter wavelength as oxidation progresses.

That a valence-band process involving holes is essential for inducing light emission could be confirmed by experiments in which n-type silicon, anodically (reverse) biased, was illuminated with infrared light that was absorbed by the substrate but not by the porous layer [54]. The photogenerated electrons and holes are spatially separated, and the electrons are collected at the back contact, giving photocurrent in the external circuit. Under influence of the electric field of the depletion layer at the interface of porous and nonporous regions, the holes accumulate at the substrate/porous silicon interface. On injection into the porous matrix, the holes cause oxidation and, ultimately, passivation of the solid.

For luminescence to occur, conduction-band electrons are also required. Under the conditions of this experiment (reverse bias), the majority carriers are depleted in the porous layer. As described in Section 3.1.2, there is strong evidence to show that intermediates of the anodic oxidation reaction of silicon have energy levels quite high in the bandgap of the semiconductor. Electrons thermally excited from such states into the conduction can recombine radiatively with valence-band holes. A similar anodic electroluminescence was observed for p-type, single-crystal InP in HCl solution [83]. The silicon process can be represented by the schematic reactions.

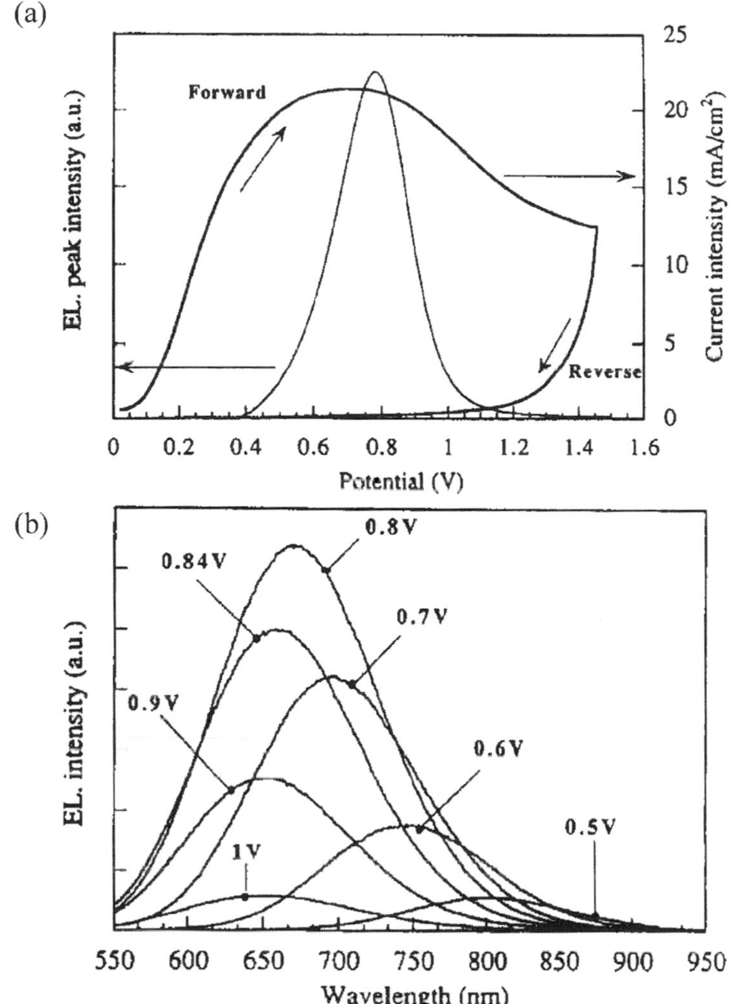

Fig. 15 (**a**) Anodic current density (*right* axis) and electroluminescence (EL) peak intensity (*left* axis) of p-type porous Si in 1 M H_2SO_4 solution, as a function of applied potential versus SCE. Both forward and reverse scans are shown; no EL was observed during the reverse scan. (**b**) The EL spectral evolution during a scan of the potential from 0 V to 1.2 V. Reprinted from Thin Solid Films, S. Billat, F. Gaspard, R. Herino, M. Ligeon, F. Muller, F. Romestain, and J.C. Vial, "Electroluminescence of heavily doped p-type porous silicon under electrochemical oxidation in the potentiostatic regime", Vol. 263, p. 239, Copyright (1995), with permission from Elsevier

$$\mathrm{H{-}Si{-}Si} + h^{+}_{VB} \longrightarrow \mathrm{H{-}Si}^{\oplus} \quad \overset{\bullet}{\mathrm{Si}} \qquad (13)$$

$$\mathrm{H{-}Si}^{\oplus} \quad \overset{\bullet}{\mathrm{Si}} + H_2O \longrightarrow \mathrm{H(OH)Si} \quad \overset{\bullet}{\mathrm{Si}} + H^{+} \qquad (14)$$

$$\mathrm{H(OH)Si} \quad \overset{\bullet}{\mathrm{Si}} \longrightarrow \mathrm{H{-}Si{-}O{-}Si} + H^{+} + e^{-}_{CB} \qquad (15)$$

The blue shift in the visible emission observed during oxidation of both p-type silicon in the dark and n-type silicon under infrared illumination can be explained by size-selective hole injection. The porous layer contains nanostructures (crystallites) with a broad distribution of sizes. The energy barrier for hole transfer from the substrate to quantized structures is smallest for the largest crystallites; the effective bandgap is close to that of bulk silicon (the mismatch of the valence-band edges is small). Such regions of the porous layer will be oxidized first giving longer wavelength emission. As oxidation progresses, these regions become passivated and holes are injected into smaller crystallites, more confined structures. This results in a shift of the emission to shorter wavelengths. Eventually, the luminescence is quenched when the porous matrix is completely passivated.

It is, perhaps, not surprising that the presence of the reducing agent $Fe(CN)_6^{4-}$ in solution quenches the luminescence. As reported in Section 3.2.1, $Fe(CN)_6^{4-}$ can suppress oxidation of silicon by electron injection, very likely into intermediates of the oxidation reactions, for example:

$$\mathrm{H{-}Si}^{\oplus} \quad \overset{\bullet}{\mathrm{Si}} + Fe(CN)_6^{4-} \longrightarrow \mathrm{H{-}Si{-}Si} + Fe(CN)_6^{3-} \qquad (16)$$

thereby "repairing" the electron-deficient bond (this reaction competes with injection of electrons into the conduction band, which is essential for light emission). A similar role is played by the Fe^{2+} ion in suppressing the anodic electroluminescence of p-type InP.

Instead of being supplied from the substrate, as described above, holes can also be injected directly into the porous matrix from an oxidizing agent in solution at open-circuit potential (Sections 3.1.3 and 3.2.1). This, of course, causes passivation in fluoride-free solution. Here again, luminescence is observed, but at open-circuit potential. The emission mechanism is, however, very likely the same as that described above for anodic luminescence.

4 Conclusion

The electrochemical reactions that occur at the internal surface of a porous electrode resemble those observed at macroscopic electrodes. A number of examples are described in this review. The potential dependence of the reaction rate may differ somewhat in the two cases. In confined systems, a larger "overpotential" may be required to drive the reaction. This is caused by the energy barrier at the interface between the smaller bandgap substrate and quantized regions in the micro/mesoporous layer and is a consequence of mismatch of valence- and conduction-band edges. The fact that the electrochemical reaction can occur at both the pore fronts (bulk electrode) and within the porous layer may complicate the analysis of the kinetics of electrochemical reactions. The porosity of the electrode has been shown to give rise to striking effects: strongly potential-dependent space-charge capacitances, unexpectedly high quantum efficiencies for photoelectrochemical reactions, quantized conduction due to charge injection, and characteristic light emission, which is not observed for nonporous electrodes.

Different questions are encountered in the electrochemistry of pore formation. It is, perhaps, not surprising that meso- and macropores nucleate at crystallographic defects at the semiconductor surface. The work of the groups of Lehmann and Gösele has shown that etch-pit patterns, produced photolithographically, can serve to fabricate impressive two- and three-dimensional meso/macroporous structures. The mechanism of nucleation of micro/mesopores is still unclear. Morphological stability models proposed for low-Ohmic p-type and (illuminated) n-type Si give insight into the initial stages of surface development. A kinetic model based on chemical and electrochemical steps is also interesting in this respect, but needs to be further elaborated. Propagation of pores requires localized dissolution at the pore fronts; this means that the pore walls must remain immune. When etching is carried out under depletion conditions, as in photoetching of meso/macropores in n-type Si (Section 3.1.2) or etching of n-type semiconductors in the dark under breakdown conditions (Section 3.1.1), then overlap of adjacent space-charge layers and majority-carrier depletion can explain the stability of the pore walls. In the case of compound semiconductors differences in the reactivity of the different crystal planes may be important, with pore propagation in "easy" crystallographic directions giving rise to facetted pore walls. Pore propagation in micro/mesoporous Si (n-type and p-type) is not well understood. Surface hydride very likely plays a role in pore-wall passivation. Quantum-size effects may be important for the electrochemistry at structures with dimensions under 10 nm. One of the most intriguing questions in the field of porous semiconductors is the reason why Si in HF solution is so different in many respects from other semiconductors.

The relevance of porous semiconductors is clearly attested by the successful series of conferences "Porous Semiconductors: Science and Technology" which have been held biannually since 1998. (The last meeting was in Sitges, Spain, in March 2006.) The proceedings of these meetings [85, 86] give an overview of the wide range of basic science and applications involved. Electrochemistry continues to play an essential role in this field.

References

1. T. J. Barton, L. M. Bull, W. G. Klemperer, D. A. Loy, B. McEnaney, M. Misono, P. A. Monson, G. Pez, G. W. Scherer, J. C. Vartuli, and O. M. Yaghi, Chem. Mat., **11**, 2633 (1999).
2. A. G. Cullis, L. T. Canham, and P. D. J. Calcott, J. Appl. Phys., **82**, 909 (1997).
3. S. R. Morrison, Electrochemistry at Semiconductor and Oxidized Metal Electrodes, Plenum Press, New York, (1980).
4. L. M. Peter, D. J. Riley, and R. I. Wielgosz, Appl. Phys. Lett., **66**, 2355 (1995).
5. A. L. Roest, J. J. Kelly, D. Vanmaekelbergh, and E. A. Meulenkamp, Phys. Rev. Lett, **89**, 036801 (2002).
6. D. Yu, C. J. Wang, and P. Guyot-Sionnest, Science, **300**, 1277 (2003).
7. A. J. Houtepen and D. Vanmaekelbergh, J. Phys. Chem. B, **109**, 19634 (2005).
8. S. M. Sze, Semiconductor Devices: Physics and Technology, 2nd ed., Wiley, New York, (1981).
9. J. J. Kelly and D. Vanmaekelbergh, Electrochim. Acta, **43**, 2773 (1998).
10. B. H. Erné, D. Vanmaekelbergh, and J. J. Kelly, Adv. Mater., **7**, 739 (1995).
11. B. H. Erné, D. Vanmaekelbergh, and J. J. Kelly, J. Electrochem. Soc., **143**, 305 (1996).
12. J. van de Lagemaat, M. Plakman, D. Vanmaekelbergh, and J. J. Kelly, Appl. Phys. Lett., **69**, 2246 (1996).
13. A. O. Konstantinov, C. I. Harris, and E. Janzen, Appl. Phys. Lett., **65**, 2699 (1994).
14. D. Vanmaekelbergh, A. Koster, and F. I. Marin, Adv. Mater., **9**, 575 (1997).
15. J. N. Chazalviel, R. B. Wehrspohn, and F. Ozanam, Mater. Sci. Eng. B-Solid State Mater. Adv. Technol., **69**, 1 (2000).
16. E. S. Kooij and D. Vanmaekelbergh, J. Electrochem. Soc., **144**, 1296 (1997).
17. H. Foll, M. Christophersen, J. Carstensen, and G. Hasse, Materials Science & Engineering, **39**, 93 (2002).
18. M. J. J. Theunissen, J. Electrochem. Soc., **119**, 351 (1972).
19. M. I. J. Beale, J. D. Benjamin, M. J. Uren, N. G. Chew, and A. G. Cullis, J. Cryst. Growth, **73**, 622 (1985).
20. R. L. Smith and S. D. Collins, J. Appl. Phys., **71**, R1 (1992).
21. Y. Kang and J. Jorne, J. Electrochem. Soc., **140**, 2258 (1993).
22. H. Foll, S. Langa, J. Carstensen, M. Christophersen, and I. M. Tiginyanu, Adv. Mater., **15**, 183 (2003).
23. X. G. Zhang, J. Electrochem. Soc., **151**, C69 (2004).
24. V. Lehmann, Electrochemistry of Silicon; instrumantation, science, materials and applications, Wiley-VCH, Weinheim, (2002).
25. J. J. Kelly and D. Vanmaekelbergh, Porous etched semiconductors; formation and characterization, Chapter 4 of *The electrochemistry of nanomaterials*, Wiley-VCH, Weinheim (Germany), (2001).

26. X. G. Zhang, Electrochemistry of silicon and its oxide, Kluwer Academic/Plenum Publishers, Dordrecht, (2001).
27. V. Lehmann, R. Stengl, and A. Luigart, Mater. Sci. Eng. B-Solid State Mater. Adv. Technol., **69**, 11 (2000).
28. P. C. Searson, J. M. Macaulay, and F. M. Ross, J. Appl. Phys., **72**, 253 (1992).
29. X. G. Zhang, J. Electrochem. Soc., **138**, 3750 (1991).
30. S. Mahadevan, S. M. Hardas, and G. Suryan, Phys. Status Solidi A, **8**, 335 (1971).
31. J. Gómez Rivas, A. Lagendijk, R. W. Tjerkstra, D. Vanmaekelbergh, and J. J. Kelly, Appl. Phys. Lett., **80**, 4498 (2002).
32. A. Hamamatsu, C. Kaneshiro, H. Fujikura, and H. Hasegawa, J. Electroanal. Chem., **473**, 223 (1999).
33. R. W. Tjerkstra, J. Gómez Rivas, D. Vanmaekelbergh, and J. J. Kelly, Electrochem. Solid State Lett., **5**, G32 (2002).
34. A. F. van Driel, B. P. J. Bret, D. Vanmaekelbergh, and J. J. Kelly, Surf. Sci., **529**, 197 (2003).
35. M. Gershenzon and R. M. Mikulyak, J. Appl. Phys., **32**, 1338 (1961).
36. A. F. van Driel, D. Vanmaekelbergh, and J. J. Kelly, Appl. Phys. Lett., **84**, 3852 (2004).
37. J. Wloka, K. Mueller, and P. Schmuki, Electrochem. Solid State Lett., **8**, B72 (2005).
38. J. Wloka, D. J. Lockwood, and P. Schmuki, Chem. Phys. Lett., **414**, 47 (2005).
39. T. Takizawa, S. Arai, and M. Nakahara, Jpn. J. Appl. Phys. Part 2 – Lett., **33**, L643 (1994).
40. S. Langa, I. M. Tiginyanu, J. Carstensen, M. Christophersen, and H. Foll, Appl. Phys. Lett., **82**, 278 (2003).
41. S. Langa, I. M. Tiginyanu, J. Carstensen, M. Christophersen, and H. Foll, Electrochem. Solid State Lett., **3**, 514 (2000).
42. S. Langa, M. Christophersen, J. Carstensen, I. M. Tiginyanu, and H. Foll, Phys. Status Solidi A-Appl. Res., **197**, 77 (2003).
43. H. J. Lewerenz, J. Stumper, and L. M. Peter, Phys Rev Lett, **61**, 1989 (1988).
44. A. Uhlir, Bell Syst. Technol., **35**, 333 (1956).
45. V. Lehmann, J. Electrochem. Soc., **140**, 2836 (1993).
46. V. Lehmann and H. Foll, J. Electrochem. Soc., **137**, 653 (1990).
47. S. Ottow, V. Lehmann, and H. Foll, J. Electrochem. Soc., **143**, 385 (1996).
48. P. Kleimann, J. Linnros, and S. Petersson, Mater. Sci. Eng. B-Solid State Mater. Adv. Technol., **69**, 29 (2000).
49. V. Lehmann and U. Gruning, Thin Solid Films, **297**, 13 (1997).
50. S. Ronnebeck, J. Carstensen, S. Ottow, and H. Foll, Electrochem. Solid State Lett., **2**, 126 (1999).
51. S. Matthias, F. Muller, C. Jamois, R. B. Wehrspohn, and U. Goesele, Adv. Mater., **16**, 2166 (2004).
52. S. Matthias, F. Muller, and U. Goesele, J. Appl. Phys., **98**, 023524 (2005).
53. S. Matthias, F. Muller, J. Schilling, and U. Goesele, Appl. Phys. a-Mater, **80**, 1391 (2005).
54. E. S. Kooij, A. R. Rama, and J. J. Kelly, Surf. Sci., **370**, 125 (1997).
55. L. T. Canham, W. Y. Leong, M. I. J. Beale, T. I. Cox, and L. Taylor, Appl. Phys. Lett., **61**, 2563 (1992).
56. L. M. Peter, D. J. Riley, R. I. Wielgosz, P. A. Snow, R. V. Penty, I. H. White, and E. A. Meulenkamp, Thin Solid Films, **276**, 123 (1996).
57. L. M. Peter, D. J. Riley, and P. A. Snow, Electrochem. Commun., **2**, 461 (2000).
58. J. J. Kelly, E. S. Kooij, and D. Vanmaekelbergh, Langmuir, **15**, 3666 (1999).
59. C. J. Wang, B. L. Wehrenberg, C. Y. Woo, and P. Guyot-Sionnest, J. Phys. Chem. B, **108**, 9027 (2004).
60. E. S. Kooij, K. Butter, and J. J. Kelly, J. Electrochem. Soc., **145**, 1232 (1998).
61. D. Mills, M. Nahidi, and K. W. Kolasinski, Physica Status Solidi A, **202**, 1422 (2005).
62. E. Vazsonyi, E. Szilagyi, P. Petrik, Z. E. Horvath, T. Lohner, M. Fried, and G. Jalsovszky, Thin Solid Films, **388**, 295 (2001).

63. Y. Y. Song, Z. D. Gao, J. J. Kelly, and X. H. Xia, Electrochem. Solid State Lett., **8**, C148 (2005).
64. C.-H. Wang, Y.-Y. Song, J.-W. Zhao, and X.-H. Xia, Surf. Sci., **600**, L38–L42 (2006).
65. E. S. Kooij, R. W. Despo, F. P. J. Mulders, and J. J. Kelly, J. Electroanal. Chem., **406**, 139 (1996).
66. E. S. Kooij, R. W. Despo, and J. J. Kelly, Appl. Phys. Lett., **66**, 2552 (1995).
67. M. J. Eddowes, J. Electroanal. Chem., **280**, 297 (1990).
68. E. K. Propst and P. A. Kohl, J. Electrochem. Soc., **141**, 1006 (1994).
69. E. A. Ponomarev and C. Levy-Clement, Electrochem. Solid State Lett., **1**, 42 (1998).
70. R. B. Wehrspohn, J. N. Chazalviel, F. Ozanam, and I. Solomon, Thin Solid Films, **297**, 5 (1997).
71. R. B. Wehrspohn, J. N. Chazalviel, and F. Ozanam, J. Electrochem. Soc., **145**, 2958 (1998).
72. V. Lehmann and S. Ronnebeck, J. Electrochem. Soc., **146**, 2968 (1999).
73. K. J. Chao, S. C. Kao, C. M. Yang, M. S. Hseu, and T. G. Tsai, Electrochem. Solid State Lett., **3**, 489 (2000).
74. R. B. Wehrspohn, F. Ozanam, and J. N. Chazalviel, J. Electrochem. Soc., **146**, 3309 (1999).
75. A. Bsiesy, F. Gaspard, R. Herino, M. Ligeon, F. Muller, and J. C. Oberlin, J. Electrochem. Soc., **138**, 3450 (1991).
76. M. Ligeon, F. Muller, R. Herino, F. Gaspard, J. C. Vial, R. Romestain, S. Billat, and A. Bsiesy, J. Appl. Phys., **74**, 1265 (1993).
77. S. Billat, F. Gaspard, R. Herino, M. Ligeon, F. Muller, F. Romestain, and J. C. Vial, Thin Solid Films, **263**, 238 (1995).
78. A. Bsiesy, B. Gelloz, F. Gaspard, and F. Muller, J. Appl. Phys., **79**, 2513 (1996).
79. A. Bsiesy, F. Muller, M. Ligeon, F. Gaspard, R. Herino, R. Romestain, and J. C. Vial, Phys. Rev. Lett, **71**, 637 (1993).
80. P. M. M. C. Bressers, J. W. J. Knapen, E. A. Meulenkamp, and J. J. Kelly, Appl. Phys. Lett., **61**, 108 (1992).
81. E. A. Meulenkamp, L. M. Peter, D. J. Riley, and R. I. Wielgosz, J. Electroanal. Chem., **392**, 97 (1995).
82. L. M. Peter and R. I. Wielgosz, Appl. Phys. Lett., **69**, 806 (1996).
83. G. H. Schoenmakers, R. Waagenaar, and J. J. Kelly, J. Electrochem. Soc., **142**, L60 (1995).
84. J. J. Kelly, E. S. Kooij, and E. A. Meulenkamp, Electrochim. Acta, **45**, 561 (1999).
85. Phys. Stat. Sol., **202**, 8/9 (2005).
86. Phys. Stat. Sol., **204**, 5/6 (2007).

Deposition into Templates

Charles R. Sides and Charles R. Martin

1 Introduction

The concepts of nanoscience and nanomaterials have recently and dramatically affected the world of scientific research. These fields are of great interest on both fundamental and practical fronts, finding applications in everything from semiconductors to biomedical research. Since these fields are incredibly diverse, a general method for nanomaterial synthesis would be an invaluable tool. The Martin laboratory has pioneered *template synthesis*, a versatile nanofabrication method [1]. The general strategy of this approach is to deposit a precursor to the desired material into the micro- or nanopores of a template. These pores act as nanoscopic beakers. The template may be a variety of porous materials, from a commercially available organic filter [2–9] to an anodized alumina membrane [9–17] to an array of nanospherical beads [18–23]. After the precursor is processed (typically by heating or ageing) into the desired product, the template can either be removed or remain intact. If the template is sacrificial, then the template-synthesized nanotubes or nanowires can be freed from the template membrane and collected. In an alternative approach, if these structures are in contact with a substrate, the result is an ensemble of micro- or nanostructures that protrude from a surface like the bristles of a brush. This template-removal step is usually accomplished via a chemical-dissolution method or a plasma-etching process. Often though, the template may be functional. Common post-host functions of templates are to maintain directional order or to increase mechanical stability.

At this point, we will also establish some naming conventions and terminology that we commonly use in this discussion. We use the term "nano" to classify a structure that has at least one dimension on the nanoscale (<100 nm). Depending on a number of situation-specific variables, including chemical interactions and/or deposition method, the resulting nanostructures are either hollow (nanotubes) or solid (nanowires). For example, a solid gold structure that is 10 nm in diameter and

C.R. Martin (✉)
Department of Chemistry, University of Florida, PO Box 117200, Gainesville, FL 32611, USA
e-mail: crmartin@chem.ufl.edu

P. Schmuki, S. Virtanen (eds.), *Electrochemistry at the Nanoscale*, Nanostructure Science and Technology, DOI 10.1007/978-0-387-73582-5_7,

6 microns in length is considered to be a nanowire. *Nanostructure* is a broad term that refers to either class.

We have used this method to prepare both nanotubes and nanowires that are composed of metals [3, 5, 10, 13, 24–27], conductive polymers [28–31], semiconductors [15, 32, 33], carbon [9, 12, 17, 34], and Li-ion battery electrodes [4, 8, 35–38]. For example, conductive polymers nanowires with diameters as small as 3 nm have been prepared using this method [39]. It is difficult to make nanowires with diameters this small by competing fabrication methods, such as lithography [40]. Such polymeric nanostructures of this type are briefly discussed here, but are reviewed in detail elsewhere [29, 41].

The intent of this chapter is to provide an overview of how a variety of template-synthesis techniques have impacted the study of electrochemistry at the nanoscale. We will start with a brief description of the different types of membranes that are commonly used. Next, we will review familiar deposition strategies. This will transition to highlights of the specific electrochemical-based fields that have already benefited when these templates and strategies are coupled.

2 Templates Used

2.1 Track-Etch Membranes

Micro- and nanoporous polymeric filtration membranes prepared via the "track-etch" method are available from commercial sources (e.g., GE Osmonics) in a variety of materials and pore geometries. Polycarbonate is perhaps the most common example of a track-etch filter material. Other options of materials include polyester, Teflon®, and polyethersulfone [42]. Electron micrographs of the surface and cross-sections of some of these templates are shown in Fig. 1.

The term *track-etch* refers to the pore-production process [43]. The pores of the filters are created by exposing the solid-material film to nuclear-fission fragments, which leave randomly dispersed damage tracks in the film. The high energy (on the order of 2 GeV) of the fragments ensures that the tracks span the entire length of the membrane (typically from 6 to 10 μm). The reactive chain-ends created by the damage track are then etched with a basic chemical solution, and they become pores [43]. One ion creates one track, which in turn becomes one pore. During production, the pore density is controlled by the duration of time that the polycarbonate film is exposed to the charged particles. Typical pore densities of a commercial track-etch membrane are $10^4 - 10^8$ pores/cm^2 [42].

Varying parameters of this etching solution such as temperature, strength, and exposure time dictate the pore diameter. Commercial membranes are available with pore diameters ranging from 10 nm to 20 μm. The microporous material etches to form uniform cylindrical pores, but as the pore diameter is reduced to the smaller nanoscopic dimensions the shape of the pore becomes like a cigar, slightly tapered at the ends [7, 44]. Microscopic investigations of template-synthesized nanostruc-

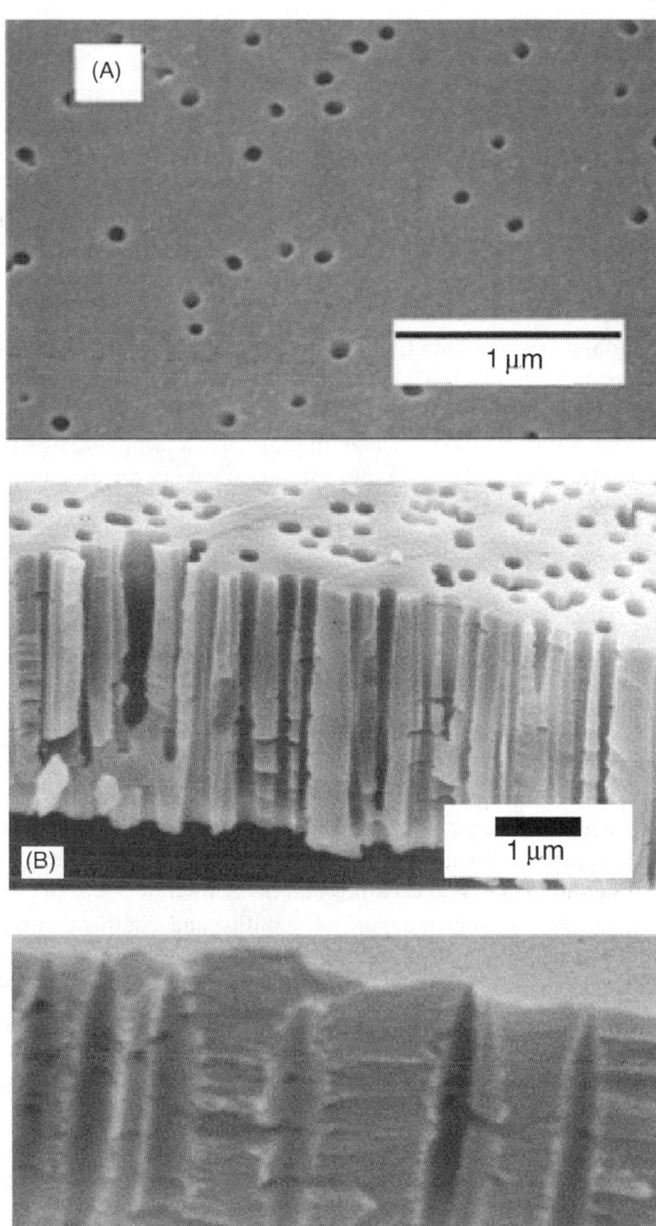

Fig. 1 A few examples of porous structures (templates) produced in thin polymeric films using various methods of irradiation and chemical treatment. (**A**) Surface of polycarbonate. Reproduced from [8]. (**B**) Polypropylene with nearly parallel pores. Reproduced from [46]. (**C**) Polyethylene terephtalate with cigar-like pores. Reproduced from [46]

tures prepared within the pores of such nanoporous membranes have shown that the diameter of the pore in the center of the membrane is larger than the diameter at the membrane surface; that is, cigar-shaped pores are confirmed [4]. It has been suggested that this pore geometry arises because the fission fragment that creates the damage track also generates secondary electrons, which contribute to the damage along the track [45]. The number of secondary electrons generated at the faces of the membrane is lesser than in the central region of the membrane. An alternate suggestion is that the surfactant protective layer adsorbed to the surface of the membrane retards the local etching process [46]. The either suggested mechanism leads to "bottleneck" pores.

An artifact of the random–directional nature of the fission–film interaction is that a tilt may occur in the tracks [43]. When this tilt from the normal is significant, it causes some pores to intersect. Though, this property does not adversely affect this membrane's application as a filter, when employed as a template, this must not be ignored in certain situations. For example, our group's work on modeling the optical properties of metal nanoparticles is significantly affected if the particles are discrete or in contact [13, 47, 48].

2.2 Alumina Membranes

Anodization of aluminum metal in an acidic environment causes the metal to etch in a fashion that leaves a porous structure [49]. These pores are extremely regular, having monodisperse diameters and cylindrical shapes in a hexagonal array. Unlike the track-etch process, this process is systematic and generates an isolated, nonconnected pore structure. The pore densities of these alumina filters can be on the order of 10^{11} pores/cm^2, that is about 1000 times the density available in the track-etch polycarbonate membranes. The porosities can be as high as 50%. In addition, alumina filters have much greater mechanical stability and chemical resistivity than polycarbonate. However, there is an extremely limited selection of commercially available pore sizes, and the smaller pores are branched (see below.) These membranes can be very thick (10–100 μm). It is also notable that this alumina structure is electronically insulating, which may be advantageous in certain situations.

Porous alumina filters are commercially available (Whatman.) This company specializes in the so-called "branched-pore" alumina. The pore structure of this type of alumina membranes is 200 nm in diameter for almost the entire length. However, near the tips of one face, the pores diverge into branches of diameters of either 100 or 20 nm. This is accomplished via a two-step anodization process, where parameters (specifically voltage) are changed during the etching process. Since the applied potential dictates the pore diameter, if the potential is decreased during the etching process then the pores would have two diameters – one larger (200 nm) created at large potentials and one smaller (100 or 20 nm) at smaller potentials. This leads to the branched-pore structure. As these membranes are commercially intended filters, this pore branching allows the membranes to work as sieves (effective diameter down to 20 nm) with a higher volume per pore (higher throughput).

Our lab synthesizes "home-grown" alumina templates [13, 48]. The synthetic process is described in detail elsewhere [48]. Briefly, a section of high-purity Al foil is degreased, electropolished, and then anodized. The anodization is the specific process that forms the pores. Afterward, the porous structured alumina is removed from the bulk aluminum by a voltage-reduction technique. The interfacial oxide layer (barrier layer) can then be removed by exposure to acidic solution. The remaining membrane is equivalent on both sides. Electron micrographs of both a surface and cross-sectional view are shown in Fig. 2. These templates are similar to the commercial filters; however, in our method, there is no branching, which is often preferred

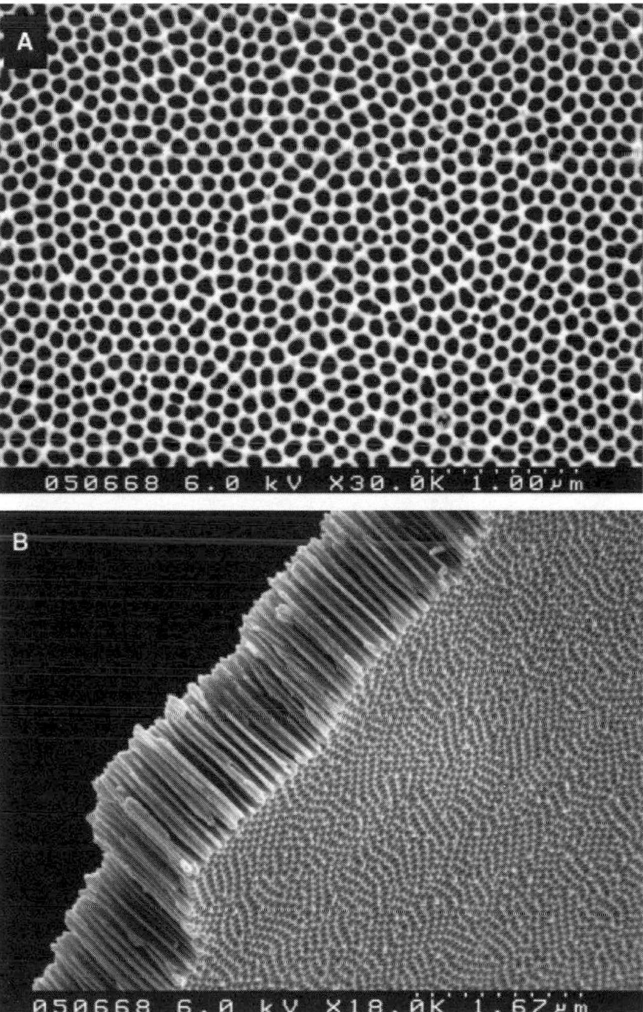

Fig. 2 Scanning electron micrographs of home-grown alumina templates. (**A**) Surface view. (**B**) Cross-section view. Reproduced from [112]

for use as templates. In addition, control over the synthetic conditions allows us the ability to tailor pore diameters and thicknesses to our specific application. The applications of these alumina templates demonstrated in our lab alone are as diverse as separations for enantiomers [16], templates for carbon nanotubes [17], and masks for plasma etching [50].

A slight variation on the synthetic method used for these two membranes involves a pre-anodization patterning of the aluminum foil [51, 52]. Hideki Masuda, of Tokyo Metropolitan University, developed this method (Fig. 3) and has come to master the art of creating highly ordered porous alumina membranes. In his method [52], a SiC mold-master is created via e-beam lithography. This mold is patterned to contain highly ordered convex structures. The mold is then mechanically pressed into a sheet of high-purity aluminum. The Al (previously annealed at 400 °C for 1 h) has sufficient plasticity to deform under this stress, causing indentations (concaves) with the order of the convexes in the SiC mold. This textured Al is then anodized to alumina under similar conditions as the methods discussed early. Figure 4 shows the well-ordered nature of membranes created by this method. The pore diameter is typically 15–40 nm with lengths of about 1 μm. Since the master mold is reusable, this production method provides a low-cost high-throughput template-creation technique.

The SiC master mold can also be constructed with triangle-shaped convexes. This results in triangle-shaped pores in the alumina membrane [53]. Masuda et al. have done extensive work on the membrane synthesis (parameter optimization) and characterization. This ordered alumina has been used to create both (1) duplicate nanohole arrays in different metals [51] and (2) template-synthesized diamond nanostructures [53].

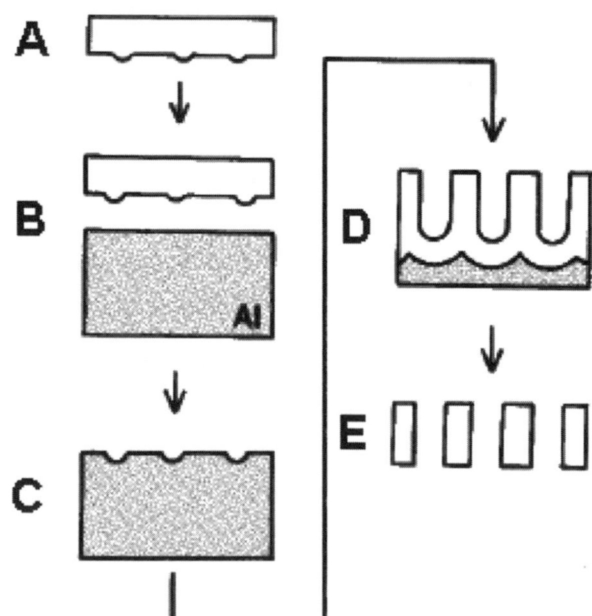

Fig. 3 Process for the ordered-channel array. (**A**) SiC mold with hexagonally ordered array of convexes, (**B**) textured Al, (**C**) molding on the Al, (**D**) anodization and growth of channel architecture (**E**) removal of Al and barrier layer Reproduced from [51]

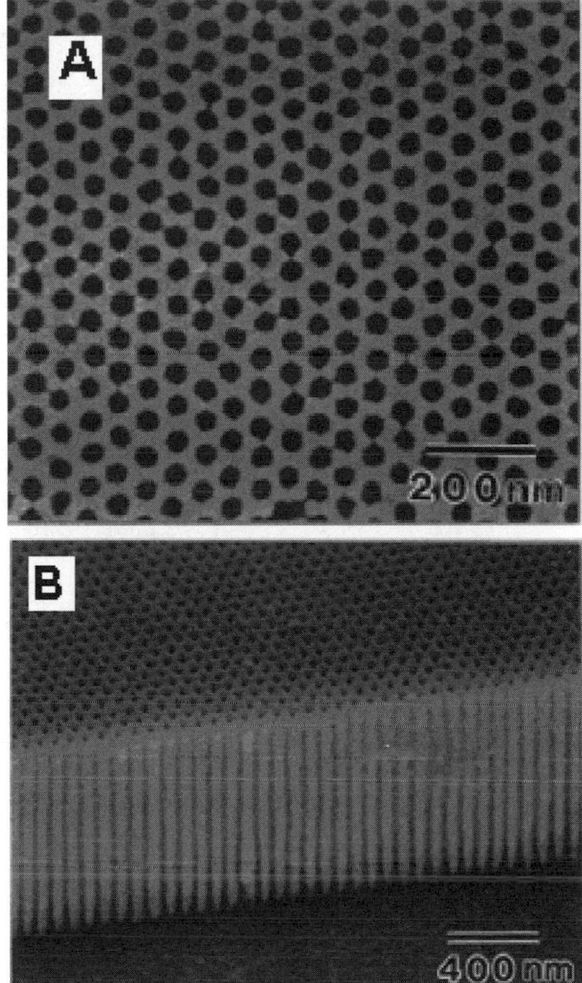

Fig. 4 SEM images of ordered-channel alumina created by Masuda's method (Fig. 3). (**A**) Surface image. (**B**) Cross-sectional image. Reproduced from [52]

2.3 Colloidal Crystals

In the previous examples of templates, the pore structure is cylindrical in general shape. An alternative to this configuration is a nanosphere template (see Ref. [22] for a recent review). In this method, spherical particles with diameters of nano- to microscale dimensions (typically ~100s of nm) are deposited in a close-packed array. This is commonly accomplished by a solvent evaporation technique. If a solvent evaporates at a slow, controlled rate, it imparts an order to the particles. These particles are typically made of polymers (latex, polysterene) or silica. An example of an ordered nanosphere template is shown in Fig. 5. Since they are spherical in

Fig. 5 Micrograph of ordered latex particles that arrange to make a template. The interstitial sites between the spherical particles serve as the pores of the template. Reproduced from [23]

shape, void-volume exists in the interstitial sites, even when close packed. These interstitial sites serve as the porous network and the nanosphere array acts as the template. These close-packed spheres have a theoretical packing efficiency (defined as [volume of space occupied by the spheres/total volume]) depending on specific arrangements of between 68 and 74%. Therefore, the theoretical void volume is approximately 30% in a close-packed array and increases to 48% in a monolayer of spheres.

Perhaps the most important distinction between these nanosphere templates and those previously described is the connectivity of the pore network. While the alumina anodization method creates distinct independent pores, these spheres create an interconnected network, where the pores are exposed to one another via smaller windows. This property can be seen as advantageous when working in the field of photonics, because they have promise as materials with a 3D periodicity on a length scale that approaches that of light [19, 22, 54].

However, this would restrict applications in separation or fundamental transport experiments. These spheres have been shown to exhibit good order out to 1 cm in each direction [20]. Parameters such as nanosphere diameter, density, and packing arrangement are reasonably adjustable. Spherical silica submicron-diameter particles are available on the hundreds of grams scale commercially (Bangs Laboratories, Inc.).

Again by physically restricting particle growth (as opposed to a chemical technique), the deposition strategy becomes general. The requirements of size, shape, and uniformity are transferred from the material to the template. This is demonstrated with examples from the literature of synthesis of a variety of materials, such as metal salts [55, 56], metal particles [57], polymeric materials [58, 59], and evaporated or electroplated metal films [60]. Though the spherical shape is the most common, it is possible that the sphere can be distorted to other shapes, such as an ellipsoid or even a "doughnut" shape [23].

The mesh network that is normally associated as the final product of the nanosphere template can itself be used as a template [61]. In this inverse method, the spherical voids, after nanosphere template removal, are subsequently filled with metals. It has been shown that periodic array of metal nanoparticles can be created via this two-step method [61]. These nanoparticles are thus very similar in dimensions to the original polymer nanospheres.

3 Template-Synthesis Strategies

3.1 Electrochemical Deposition

Electroplating has thrived as an industrial technique for metal deposition for several decades. In general, this electrochemical process has an anode, an electrolyte (the target ion source), and a cathode (target plating site). An applied stimulus (e.g., potential) reduces the target metal ions at the cathode. This process plates the metal as a film. The thickness of this film is governed by quantity of charge (current × time) and the number of electrons required per reduction reaction. Here we discuss how this established process is adapted for the electrochemical deposition of materials (both metals and conductive polymers) into a nanoporous template. Both, anodized aluminum oxide (AAO) and track-etch polycarbonate templates have been used extensively with this deposition technique to create a variety of nanostructures.

3.1.1 Electrodeposition of Metals

Nanowires

Nanometals have interesting optical, electronic, and (for appropriate metals) magnetic properties. The concept of using the pores of a nanoporous membrane as templates for preparing nanoscopic metal wires was first demonstrated by Possin [62]. Nanometal-containing membranes of this type have also been used as selective solar absorbers [63]. Also, magnetic metals have been deposited within the pores of such membranes to make vertical magnetic recording media [64].

The first step of this synthesis is the deposition of a metal conductor (Ag or Au, for example) via either ion sputtering or thermal evaporation unto the backside of the template. This process will coat the face of the template and fill a small portion of the pores. The template, conductive surface intact, serves as the cathode in this configuration for electrochemical deposition. The structures plate from the bottom upward as solid wires. By depositing a small amount of metal, short, squat wires can be obtained; alternatively, by depositing large quantities of metal, long, needlelike wires can be prepared [13, 47, 48]. This ability to control the aspect ratio (length to diameter) of the metal wire is especially important in optical investigations, because the optical properties of nanometals are dependent on this parameter [13, 25, 47, 48]. The length dimension of the electrochemically deposited nanostructures becomes user defined unless the limit occurs when the membrane's opposing face plates over the pores themselves. This method has been used to prepare copper [65], platinum [5], gold [10], silver [24], and nickel [66] wires.

Nanotubes

Though solid nanowires are preferable in several applications, hollow nanotubes allow investigations into other equally important and diverse fields of research. With modest pretreatment, hollow metal nanotubes are possible via this deposition strategy. This is done by derivatizing the pore wall with a "molecular anchor," prior to electrochemical deposition. These anchors ensure that the material forms as a thin "skin" which lines the pore wall (i.e., encourages the metal to plate radially as opposed to axially.) The challenges in synthesizing metal nanotubes, then, are (1) to identify chemistry for forming the metal within the pores of the membrane, (2) to identify a suitable molecular anchor, and (3) to develop chemistry for attaching this anchor to the pore walls in the membrane [10].

For example, organocyanide silane–alumina chemistry (Fig. 6) is well documented. When the pore wall of an alumina template is silanized via this reaction scheme, gold will plate to the wall forming metal nanotubes. Au is deposited from the contacting solution to all exposed surfaces, including both the faces as well as along the walls of the pores in this membrane. In our experiments described by Brumlik and Martin [10], this surface layer was too thin to block the pores in the commercial alumina template membrane but thick enough to cover the surface of the membrane to form an electrode. If the CN-containing silane is not attached to the pore wall prior to Au deposition, solid Au wires are obtained in the pores [10]. The pore wall thickness of this metal nanotube is adjustable (time) and results in decreasing the effective diameter of the tube. This method has demonstrated nanotubes of inner diameter (i.d.) down to 1 nm, which approaches molecular dimensions. Recently, at least one example was published where metal nanotubes (Cu) were synthesized without an anchor [67].

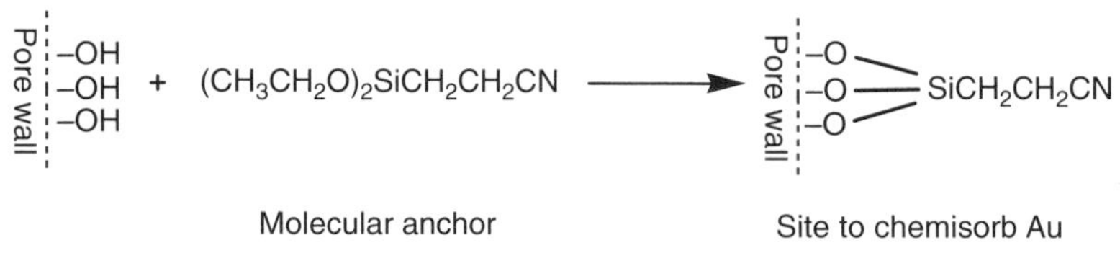

Pore wall –OH –OH –OH + $(CH_3CH_2O)_2SiCH_2CH_2CN$ → Pore wall –O –O –O $SiCH_2CH_2CN$

Molecular anchor   Site to chemisorb Au

Fig. 6 Reaction used to attach a molecular anchor, (2-cyanoethyl)triethoxysilane, to the hydroxyl groups on the pore wall of an alumina template membrane. Reproduced from [10]

3.1.2 Electrodeposition of Polymers

Nanostructures of conductive polymers are also possible by electrodeposition. This has been demonstrated with such conductive polymer systems as polypyrrole, polyaniline, or poly(3-methylthiophene) [68, 69]. These polymers grow on pore walls of the template, so they inherently grow as tubes. The chemical structures of these systems are shown in Fig. 7. The reason for this is well documented [70]. Monomers of these structures are soluble, but the polycationic forms are completely insoluble. There is a solvophobic interaction that pulls polymer to the wall. Also, electrostatic interactions pull the cationic polymer to the anionic sites on the pore wall [70]. This occurs in short time regimes. If this is allowed to propagate the i.d. of the tube decreases, until a solid wire is formed, as in the case of polypyrroie [71].

3.2 Electroless Deposition

Metal nanostructures can also be deposited from solution into a template via a chemical reduction technique [72]. This process is referred to as *electroless deposition*, as opposed to *electrochemical deposition*. An important distinction between these two methods is that this process does not require the template to be electronically conductive. We have developed methods from which gold and other metals are deposited from solution onto both alumina and polymeric templates in the absence of any external applied electric field [26].

The strategy of electroless gold deposition into a polycarbonate template membrane is well documented [3, 9, 26]. Therefore, it serves as our case-study and is described below. Initially, the template membrane is "sensitized" by immersion into a $SnCl_2$ solution which results in deposition of Sn(II) onto all of the membrane's surfaces (both pore walls and membrane faces.) The sensitized membrane is then

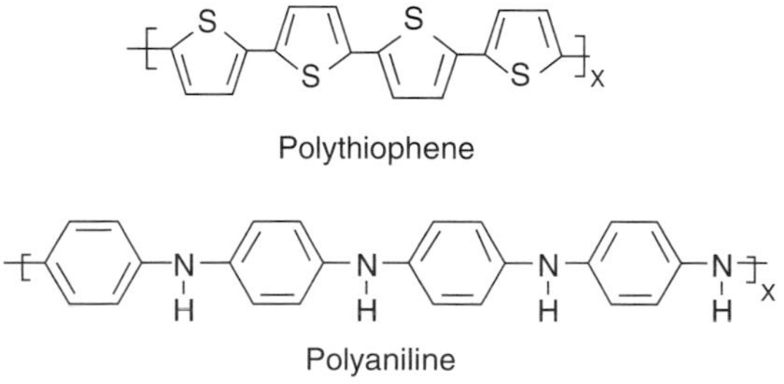

Polypyrrole

Polythiophene

Polyaniline

Fig. 7 Structures of some of the template-synthesized electrochemically conductive polymers [113]

immersed into $AgNO_3$ solution, and a surface redox reaction occurs (Eq. (1)) which yields nanoscopic metallic Ag particles on the membrane surfaces.

$$Sn(II)_{surf} + 2Ag(1)_{aq} \quad \rightarrow \quad Sn(IV)_{surf} + 2Ag(0)_{surf} \tag{1}$$

The Subscripts "surf" and "aq" denote species adsorbed to the membrane surfaces and species dissolved in solution, respectively. The membrane is then immersed into a commercial gold plating solution (Oromerse SO Part B, Technic, Inc.) and a second surface redox reaction occurs to yield Au nanoparticles on the surfaces (Eq. (2)).

$$Au(1)_{aq} + Ag(0)_{surf} \quad \rightarrow \quad Au(0)_{surf} + Ag(1)_{aq} \tag{2}$$

These surface-bound Au nanoparticles are good autocatalysts for the reduction of Au(1) to Au(0) using formaldehyde as the reducing agent [9]. As a result, Au deposition begins at the pore walls, and Au tubes are obtained within the pores. This series of chemical processes are described in Fig. 8, and 9 is a pictorial representation of the process. Unlike the electrochemical deposition method, this chemical-based strategy indiscriminately plates at any membrane surface exposed to the contacting solution phase. Therefore, the structures span the entire length of the template, and both faces are plated.

The key feature of the electroless deposition process is that metal deposition in the pores starts at the pore wall. Therefore, after short deposition times, a hollow metal tube is obtained within each pore, while long deposition times result in solid metal nanowires. Figure 10 shows how adjusting the user-defined parameter of time dictates the i.d. of the nanotubes down to 0 nm (i.e., when the tube becomes a wire). This ability to control the i.d. of a nanotube is a distinct advantage over electrochemical deposition. Of course, the outside diameter of the structures is determined by the diameter of the pores in the template membrane.

The i.d.'s reported in Fig. 10 are estimates made by a gas-transport method using the template-synthesized Au nanotubes [9]. In this measurement, a tube-containing membrane is placed in a gas-permeation cell, and the upper and lower half-cells are evacuated. The upper half-cell is then pressurized, typically to 20 psi with H_2, and the pressure–time transient associated with leakage of H_2 through the nanotubes is measured using a pressure transducer in the lower half-cell. The pressure–time transient is converted to gas flux (Q, mol s^{-1}) which is related to the radius of the nanotubes (r, cm) via

$$Q = (4/3) * (2\pi/MRT)^{1/2} * (nr^3 \Delta P/l) \tag{3}$$

where ΔP is the pressure difference across the membrane (dynes cm^{-2}), M is the molecular weight of the gas, R is the gas constant (erg K^{-1} mol^{-1}), n is the number of nanotubes in the membrane sample, l is the membrane thickness (cm) and T is the temperature (K).

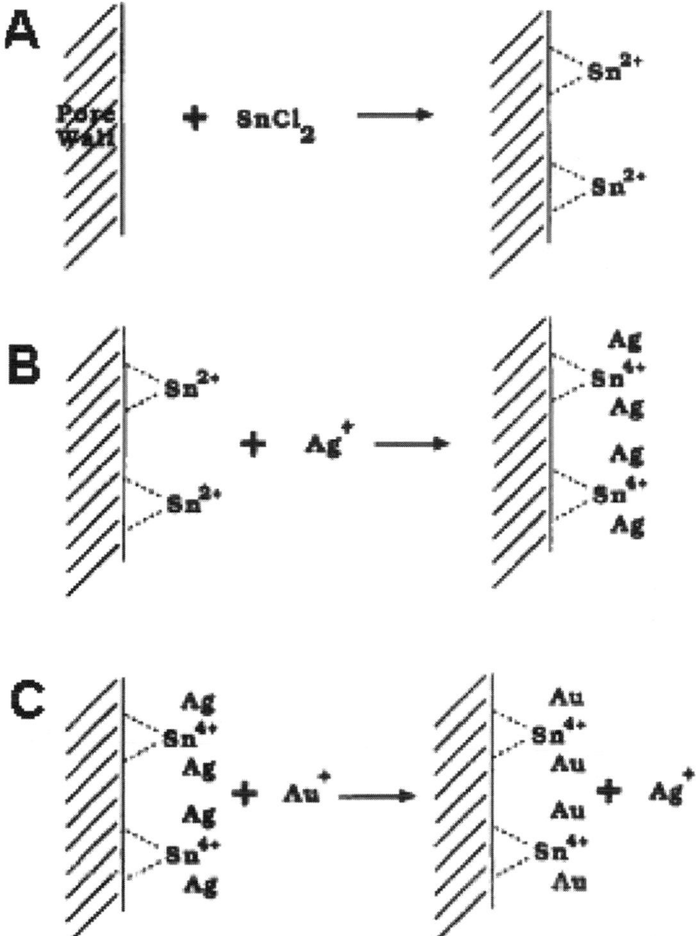

Fig. 8 Schematic diagram of the electroless procedure used to deposit gold in the pores of a polycarbonate template membrane. (**A**) Sn^{2+} sensitizer is applied to the membrane's surface. (**B**) Ag coats the pore wall as discrete nanoparticles. (**C**) Au particles galvanically displace Ag particles. Reproduced from [26]

In Eq. (3), we assume: (1) that we know the number of nanotubes (n) in the membrane sample; (2) that the nanotubes have a constant inside diameter down their entire length; and (3) that the mechanism of gas transport through the membrane is Knudsen diffusion in the nanotubes. We have discussed the validity of each of these assumptions in detail in a recent review [44]. The key points to make are: (1) that nanotubes with the smallest inside diameters have bottlenecks at the membrane surfaces. Hence, the values reported by the gas-flux method must be regarded as effective inside diameters (EIDs); (2) that at long plating times the gas flux measurement reports nanotube EIDs that are of molecular dimensions (<1 nm); and (3) that such

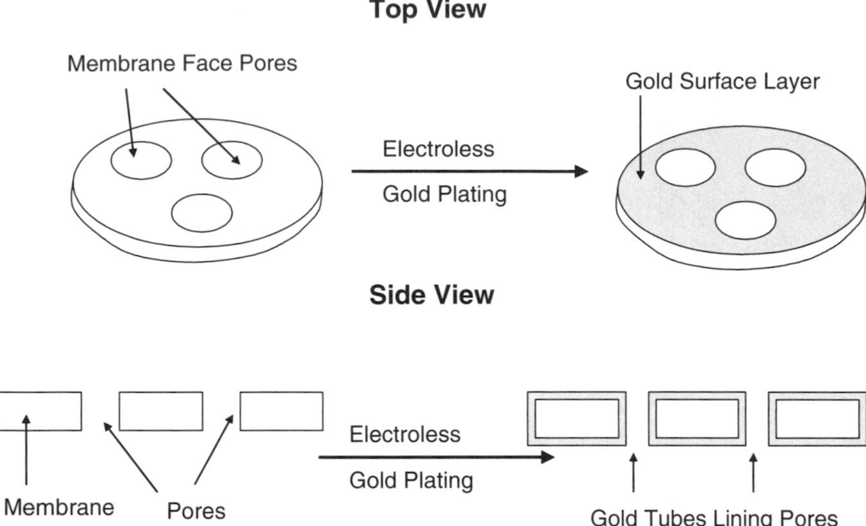

Fig. 9 Conceptual representation of the electroless deposition of Au into the pores of a polycarbonate membrane [114]

Fig. 10 Variation of the nanotube-effective inner diameter with plating time Reproduced from [14]

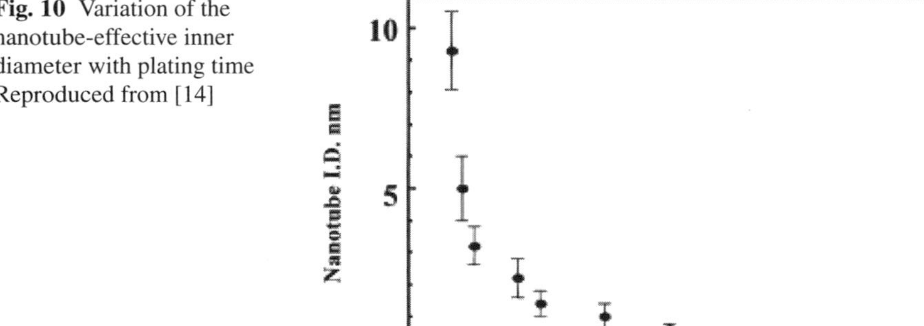

membranes can be used to cleanly separate small molecules on the basis of molecular size, which is the ultimate proof that the nanotubule EID is, indeed, of molecular dimensions [70].

In summation of the comparison of electroless and electrochemical deposition strategies, the *electroless* method yields tubes of adjustable i.d. that traverse the entire length of the template; whereas, the *electrochemical* method yields wires of adjustable length. Of course, there are many experimental and situation-specific parameters that undoubtedly complicate this generalization.

3.3 Sol–gel Deposition

Sol–gel chemistry has recently evolved as a powerful approach for preparing inorganic materials such as glasses and ceramics. This method for the synthesis of inorganic materials has a number of advantages over more conventional synthetic procedures. For example, high-purity materials can be synthesized at a lower temperature. In addition, homogeneous multicomponent systems can be obtained by mixing precursor solutions; this allows for easy chemical doping of the materials prepared [33]. Finally, the rheological properties of the sol and the gel can be utilized in processing the material, for example, by dip coating of thin films, spinning of wires, etc. [73]. Such a versatile deposition technique partners well with the template-synthesis nanofabrication method.

This sol–gel process involves hydrolysis of a solution of the precursor molecule to obtain first a suspension of colloidal particles (the sol) and then a gel composed of aggregated sol particles. The amorphous gel may then be thermally treated to yield a more crystalline product. We have recently conducted various sol-gel syntheses within the pores of the alumina and polycarbonate membranes to create both tubes and wires of a variety of inorganic oxide materials, including semiconductors [15, 32] and Li-ion battery intercalation materials [4, 8, 36]. First, the template membrane is immersed into a sol for a given period of time, and the sol deposits on the pore walls. After hydrolysis, either a tube or wire of the gelled material is formed within the pores. As with other template synthesis techniques, longer immersion times yield wires, while brief immersion times produce tubes.

The formation of tubes after short immersion times indicates that the sol particles adsorb to the template membrane's pore walls. This is expected, because the pore walls are negatively charged, while the semiconductor sol particles used to date are positively charged (a similar situation to what was described for conductive polymers) [32]. It has also been found that the rate of gelation is faster within the pore than in bulk solution. This is most likely due to the enhancement in the local concentration of the sol particles owing to adsorption on the pore walls.

The formation of TiO_2 semiconductor nanostructures is a useful case study [15]. The mechanism of formation of TiO_2 from acidified titanium alkoxide solutions is well documented. In the early stages of the synthesis, sol particles held together by a network of –Ti–O–bonds are obtained. These particles ultimately coalesce to form a three-dimensional infinite network, the gel. The fact that tubes are initially obtained when this process is done in the alumina membrane indicates that the sol particles adsorb to the pore walls. It is well known that at the acidic pH values, the sol particles are weakly positively charged. Tubes are formed because these positively charged particles interact with anionic sites on the alumina pore wall [15]. Several other inorganic oxides, specifically MnO_2, Co_3O_4, ZnO, WO_3, and SiO_2 have been synthesized in a similar fashion.

3.4 Chemical Vapor Deposition

Chemical Vapor Deposition (CVD) is a versatile process in which gas-phase molecules are decomposed to reactive species, leading to film or particle growth. The most prevalent application of CVD in the literature is the deposition of a hydrocarbon precursor to make carbon films [12, 17, 74]. However, CVD processes can be used to deposit a wide range of conducting, semiconducting, and insulating materials. As a result, it has been a mainstay of the industrial processing of solid thin films. A recent thrust of CVD techniques has been the controlled fabrication of nanomaterials in porous hosts, including zeolite nanochannels [75]. Two advantages of the CVD method are the ability to controllably create films of widely varying stoichiometry and to uniformly deposit thin films of materials, even onto nonuniform shapes. This method is certainly similar to the atomic layer deposition (ALD) method discussed later, but it does have its distinctions.

The high pore density and heat tolerance of the alumina membrane makes it a good template host for CVD. When using an alumina template, the template may be dissolved post-CVD, leaving an array of tubes which are open at both ends. The nature of dual accessibility (from front and rear) and thin wall thickness allow for use in concentric nanotube designs [34]. In this embodiment, a carbon nanotube (CNT) array then, itself, acts as a template. While CVD is a good partner with the alumina template, there are examples of this deposition technique being used with nanosphere templates to form structures with interesting geometries and properties [76].

Most instances of CVD use a catalyst, though it is not necessarily a requirement. Highly ordered graphitic nanowires can be obtained when the carbon wires are formed and annealed in the presence of a nickel catalyst [77]. Cobalt and iron catalyst have also been used [78]. A major hurdle in applying CVD techniques to template synthesis has been that deposition rates are often too fast. As a result, the surface of the pores becomes blocked before the chemical vapor can traverse the length of the pore. We have, however, developed two template-based CVD syntheses that circumvent this problem.

The first entails the CVD of carbon within porous alumina membranes, which has been achieved by our group and others [9, 12, 17, 34]. This involves placing an alumina membrane in a high-temperature furnace ($\sim$700 °C) and passing a gas such as ethylene or propylene through the pores of the membrane. Thermal decomposition of the gas occurs throughout the pores, resulting in deposition of carbon films along the length of the pore walls (i.e., carbon tubes are obtained within the pores.) The thickness of the walls of the carbon tubes is dependent on total reaction time and precursor pressure. An inert gas (argon) flows during both the heating and cooling stages of the process. During the heating process, this ensures that deposition begins only after the desired temperature is obtained. Of course, the inert atmosphere is required during cooling the furnace to keep the freshly deposited carbon from oxidizing to CO_2.

The second CVD technique uses a template-synthesized structure as a substrate for deposition. For example, we have used a CVD method to coat an ensemble of Au nanotubes with concentric TiS_2 outer nanotubes [35]. The first step of this process is the electroless plating of Au nanotubes or wires into the pores of a template membrane. The Au surface layer is removed from one face of the plated membrane, and the membrane is dissolved away. The resulting structure is an ensemble of Au tubes or wires protruding from the remaining Au surface layer like the bristles of a brush. This structure is exposed to the specific precursor gases used for the CVD of TiS_2. The Au structures become coated with outer TiS_2 tubes. When the flow-through CVD method is combined with another template-synthesis technique, it is possible to achieve a concentric nanotube membrane. This hyphenated method (ELD-CVD) provides an interesting alternative to segmented nanowires for the synthesis of multicomponent nanostructures.

3.5 Atomic Layer Deposition

Atomic Layer Deposition (ALD) is a "self-limiting" variation to the gas-phase CVD deposition technique. ALD pulses precursor gas that absorbs to the surface of a solid template, followed by reactive-precursor gas that forms a tight monolayer (via a binary sequence of self-limiting chemical reactions) of solid-phase product molecules, sequentially followed by inert purge gas to remove excess precursor molecules [79]. A representation of the deposition of ZnS (from $ZnCl_{(g)}$ and $H_2S_{(g)}$) is shown in Fig. 11. A thin film of material can be deposited by repeating this reaction sequence (approximately a monolayer at a time). By controlling the number of cycles, the deposition-layer thickness can be controlled with a precision unmatched by other strategies. This ability is clearly demonstrated by George et al. [80], with the deposition of either SiO_2 or TiO_2 into the pores of an alumina membrane (Fig. 12).

ALD is performed at temperatures in the range of 100–400 °C, much lower than those used during CVD. This lower operating temperature allows ALD to be used with polymeric templates [81], which are precluded from use with the high temperatures associated with CVD. In addition to these polymeric templates, it has also been applied to alumina templates [82], carbon nanotubes [83], or even patterned self-assembled monolayers [84]. The low deposition temperature has another effect. It is typically not high enough to promote crystallization, so the product is often in the amorphous state. When in the amorphous state, it will form a smooth uniform conformal-deposition layer. If the material deposits in the polycrystalline form, then particle nucleation may increase the layer's surface roughness [85].

This is a powerful technique, but it is predicated on the ability to identify suitable precursor gases for the desired final product. Precursors must chemisorb on the surface of the template or react rapidly with the surface groups and aggressively with each other [86]. To date, most examples in the literature are the deposition of metal oxides (WO_3 [85], RuO_2 [83], V_2O_5 [87], and ZrO_2 [81], in addition to those described earlier). Metal nitrides and metal-film deposition have also been

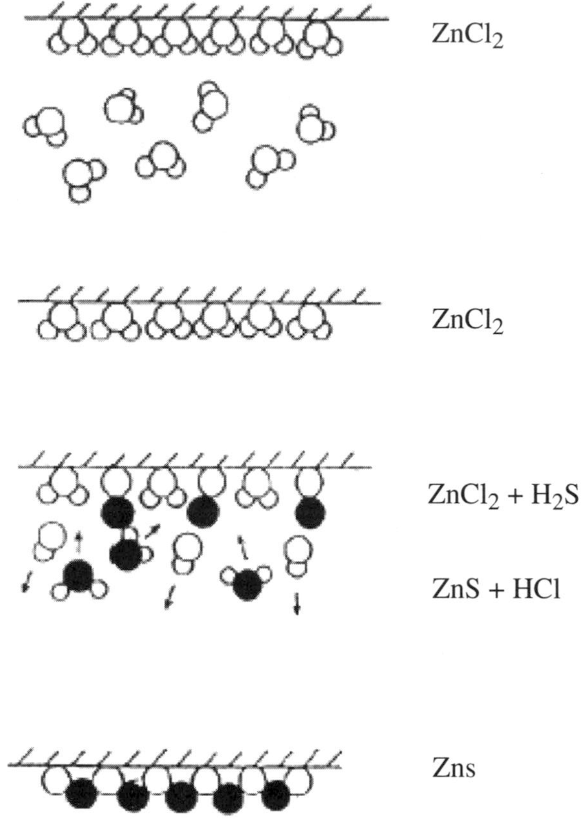

Fig. 11 A schematic representation of the basic principle of the ALD process showing the growth of ZnS film from the gaseous precursors $ZnCl_2$ and H_2S. Reproduced from [86]

demonstrated [86]. The self-limiting nature of this method makes it a lengthy process to deposit films of larger thickness, as each cycle produces a depth less than 1 nm. A complete deposition process can exceed 100 cycles. Therefore, automated gas valves and a well-designed reactor are a virtual necessity to accomplish film growth of sizeable thickness on a reasonable timescale. There are commercial available instruments for this process (Cyclic 4000, Genitech). This system can deposit as quickly as 500 nm h^{-1}, under near-ideal experimental conditions. However, slow-reacting precursors or large ligand groups will decrease this throughput significantly.

ALD is a strategy that produces thin films with ultraprecise control of their thickness. The low-temperature deposition makes it amenable to a variety of templates. It is also easy to envision a process that alternates layers of deposited material (i.e., for concentric tubes). The key challenge of this technique is the identification of suitable precursor gases. Also, thicker films are time consuming to produce.

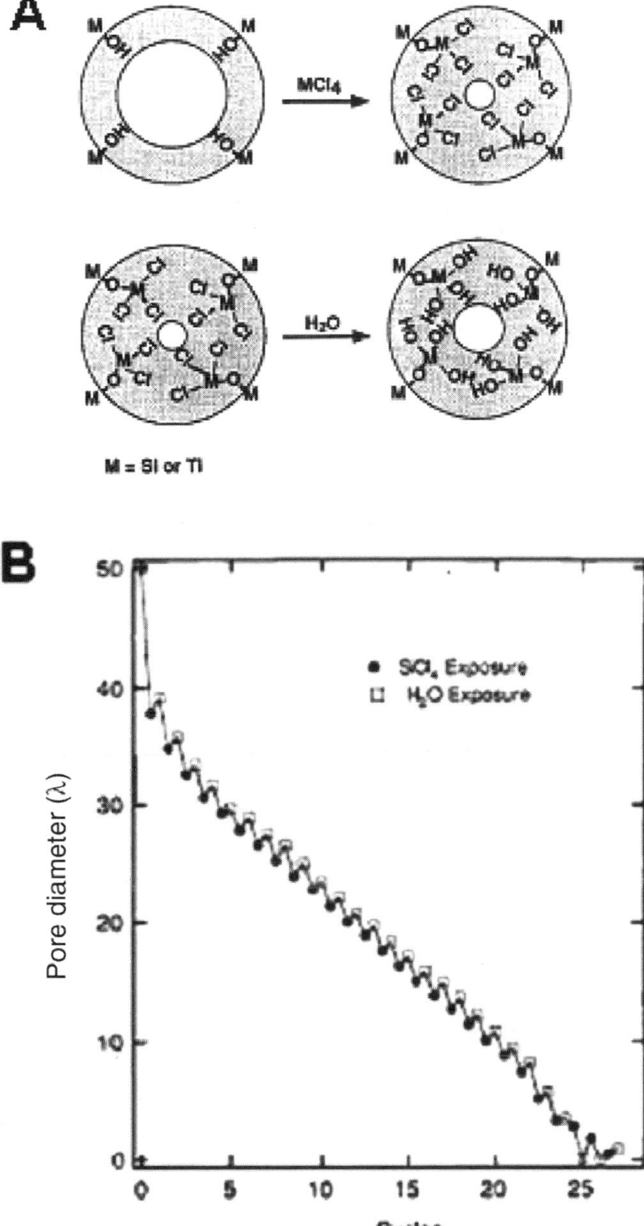

Fig. 12 (**A**) Illustration of the changes in pore diameter after MCl_4 and H_2O alternating exposures during SiO_2 or TiO_2 deposition. M represents either Si or Ti. (**B**) Pore diameter versus cycle number during SiO_2. Note the diameter increases slightly as hydroxyl groups replace the large chloride substituent, but does continually decrease after each complete deposition cycle. Reproduced from [80]

4 Applications in Nanoelectrochemistry

4.1 Gold Nanoelectrodes

One of the first fields where these template-synthesized nanomaterials demonstrated superior functional capabilities is fundamental electrochemical investigations. Nanoelectrodes offer opportunities to perform electrochemical experiments to investigate the kinetics of redox processes that are too fast to measure at conventional macroscopic electrodes [88, 89]. (By macroscopic electrodes we mean disk-shaped electrodes with diameters of the order of 1 mm.) Also, they have the ability to serve as useful electrodes even in highly resistive media [90]. Here we discuss our early work with template-synthesized gold nanowires to measure electrochemical response of redox molecules at trace concentrations [3, 6, 26].

4.1.1 Fabrication

Using the electroless Au deposition procedure, Au nanowires (plating time ~24 h, see Fig. 10 for details) are synthesized within the pores of a polycarbonate track-etch membrane. This process plates Au on all solution-contacting membrane surfaces (i.e., inside the pores and onto both faces). If one of the surface Au films is removed, the disk-shaped ends of the Au nanowires traversing the membrane are exposed. These nanodisks can be used as active elements in a nanoelectrode ensemble (NEE). Electrical contact is made to the remaining surface layer which acts as a common current collector for all the nanoelectrode elements [3, 26]. Even the relatively low pore density of the polycarbonate template is capable of generating millions of elements per cm^2 of current collector area. Micrographs of the NEEs created in differing pore density templates are shown in Fig. 13.

A consistent problem associated with micro- and nano-electrodes is achieving an efficient seal between the conductive element and the host material. If a good seal is not achieved, solution can creep into this junction resulting in significantly higher values of the background or double-layer charging currents [5]. In the case of the NEE, the polycarbonate is stretch oriented during fabrication to improve mechanical properties. Upon heating above the glass-transition temperature (~150 °C), the membrane relaxes, shrinks, and seals the junction between the Au nanowires and the polymer membrane [3, 26]. A schematic of this assembly is shown in Fig. 14.

4.1.2 Current Response of the NEE

Two different electrochemical response-limiting cases can be observed at a NEE, the "total overlap" and "radial" response [3]. Which limiting case is achieved depends strongly upon the distance between the electrode elements and the timescale (e.g., voltammetric scan rate) of the electrochemical experiment. When the electrode elements are in close proximity and the scan rate is relatively slow, the diffusion layers at each electrode element overlap. This overlap results in a single diffusion layer that covers the entire geometric area of the NEE. Linear diffusion occurs to the entire

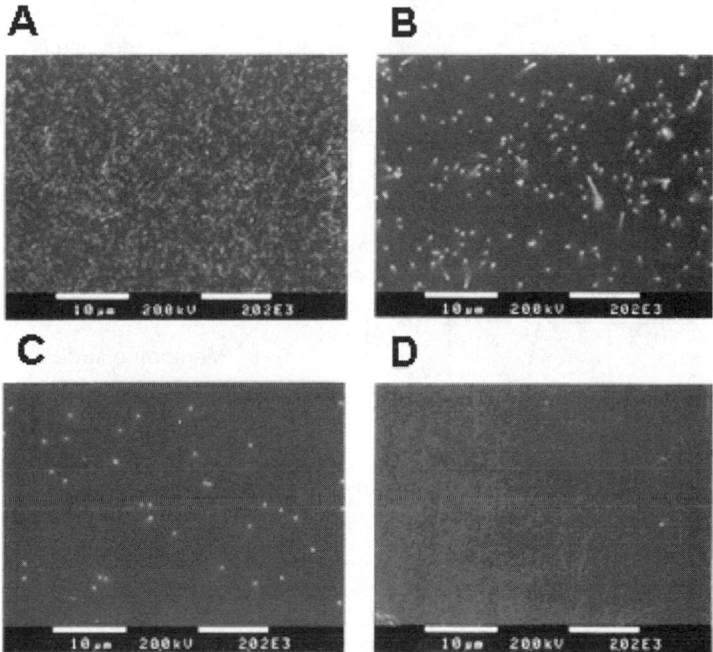

Fig. 13 Micrographs of Au NEE synthesized with polycarbonate templates of decreasing pore density (**A–D**) Each sample was sputtered with a thin (<50 nm) layer of Au-Pd to eliminate surface charging associated with the polymer. Each scale bar equals 10 μm. Reproduced from [3]

NEE surface, and conventional peak-shaped voltammograms are obtained [5, 26]. Also, the total faradaic current is equivalent to that obtained at an electrode of equivalent geometric area whose entire surface area is gold.

If the electrode elements are located far apart and the timescale of the experiment is relatively fast, the diffusion layers at each electrode act individually resulting in a radial-diffusion field at each individual electrode element [3]. The voltammogram has a sigmoidal shape, and the predicted total faradaic current is equivalent to the sum of the current generated at each individual electrode element within the NEE [3, 91]. Depending on template selection (pore density), either response is possible. We have demonstrated that by selecting templates of differing pore densities we can observe both limiting cases. The low pore-density membranes correspond to the radial-diffusion case, but high pore density correspond to the total overlap case. Examples of these two diffusion regimes and their respective anticipated current responses are shown in Fig. 15.

4.1.3 Detection Limits

One application of these NEEs is the ultratrace detection of an electroactive species. We have shown that NEEs with 10-nm diameter disks operating in the total overlap

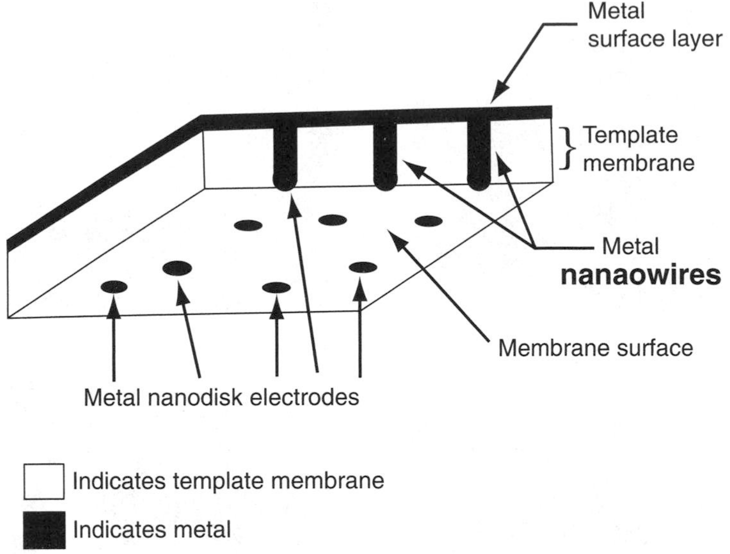

Fig. 14 Schematic of an edge view of a nanoelectrode ensemble. The nanometal wires running through the pores of the template membrane are shown. The *lower* ends of the wires define nanodisks which serve as the electrodes. The opposite *upper* ends of the nanowires are connected to a common metal film which is used to make electrical contact to the nanodisks Reproduced from [115]

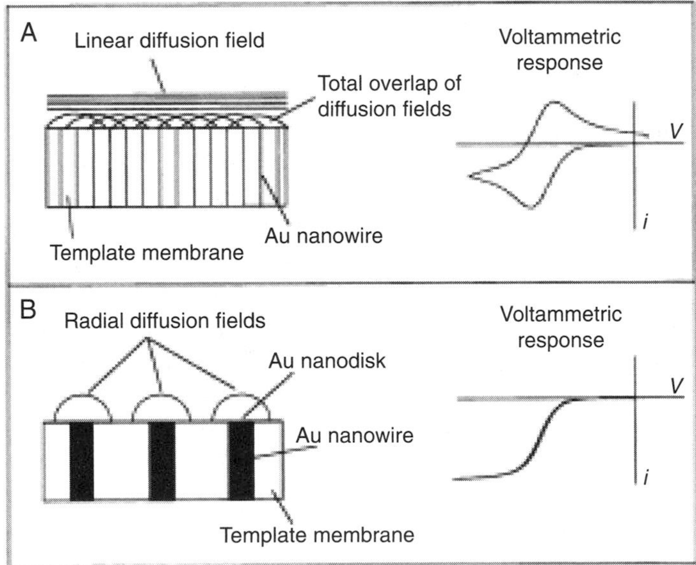

Fig. 15 Schematic of a side view of NEEs and the corresponding diffusion fields for the total overlap (**A**) and radial (**B**) limiting electrochemical response. Reproduced from [115]

mode show electroanalytical detection limits that are 3 orders of magnitude lower than detection limits obtained at macroscopic Au disk electrodes of comparable geometric area [26]. This occurs because in the total overlap mode, the total faradaic signal generated at the NEE is equivalent to that obtained at the conventional macroelectrode of equivalent geometric area. However, the background double-layer charging current is significantly less, because these currents are proportional only to the active Au area. The ratio of active area to geometric area for a 10-nm NEE is approximately 10^{-3} [26]. As a result, the background current is reduced by 3 orders of magnitude.

An example of this enhancement in detection limits at an NEE is shown in a series of voltammograms at a conventional Au macroelectrode at decreasing concentrations of $TMAFc^{+}$ (Fig. 16). As expected, the faradaic signal eventually vanishes into the double-layer charging currents as the concentration of $TMAFc^{+}$ decreases. Contrast this to the responses from the NEE, where the faradaic currents are distinguishable down to concentrations 3 orders of magnitude lower than with the macroelectrode. The detection limit at the macroelectrode was determined to be ~2 mM while the detection limit at the NEE was ~2 nM [26].

4.1.4 Supporting Electrolyte Effect

There is another interesting comment to make about our investigations with these NEEs. This concerns the effect of the supporting electrolyte. The concentration of both the electroactive species (~μM) and the supporting electrolyte (~mM) are low in these experiments [26]. Low concentrations were used, because we have discovered that the reversibility for the voltammetric waves for all couples investigated improves as the concentration of the supporting electrolyte decreases. Figure 17 shows a series of cyclic voltammograms with differing supporting electrolyte concentrations. Note that the peak currents decrease and the ΔEpk values increase as the concentration of supporting electrolyte increases [26].

Template synthesis has been shown to provide a simple means of creating nanoelectrode ensembles. These NEEs have favorable properties for electrochemical investigations. The high double-layer charging currents, typical of cyclic voltammetry, limit its use as an electroanalytical technique. However, the electrochemical response at the NEE demonstrates that in this case cyclic voltammetry was able to achieve the very respectable detection limit of less than 2 nM. This is 3 orders of magnitude lower than the detection limits obtained at conventional macroelectrodes under similar experimental conditions. Therefore, NEEs are capable of serving as useful tools for ultratrace detection of electroactive species.

4.2 Carbon Nanoelectrodes

Carbon nanotubes are of great current interest in both fundamental and applied science. Our group has used template-synthesized carbon nanotubes for fundamental electrochemical investigations, specifically as tools to monitor and control

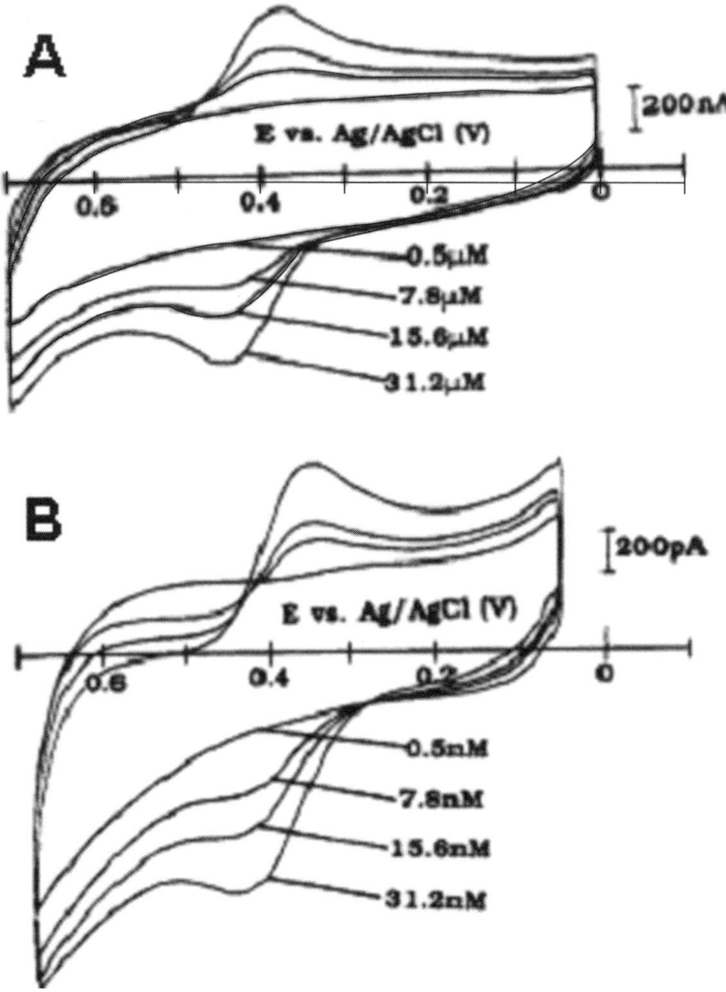

Fig. 16 Cyclic voltammograms at 100 mV s^{-1} in aqueous $TMAFc^+$ at (**A**) a gold macrodisk electrode in 50 mM sodium nitrate and (**B**) a 10-nm NEE in 1 mM sodium nitrate. $TMAFc^+$ concentrations are as indicated; electrode geometric areas in both cases are 0.079 cm^2. Reproduced from [26]

electroosmotic flow (EOF) [17, 74]. EOF refers to the movement of solution past a stationary surface due to an externally applied electric field. It is a consequence of the way ions are distributed near surfaces [92]. Whether its existence serves to aid or to complicate experiments, understanding its role in fluid flow may be critical, especially in microchannel devices and electrophoresis experiments. Here we discuss the work of our group to control both the rate and direction of EOF, independent of the electric field that drives EOF.

Fig. 17 Cyclic voltammograms illustrating the effect of supporting electrolyte concentration at a 10-nm NEE for 5 μM $TMAFc^+$ in aqueous $NaNO_3$ at the indicated concentrations of $NaNO_3$. Scan rate is 100 mV s^{-1}. Reproduced from [26]

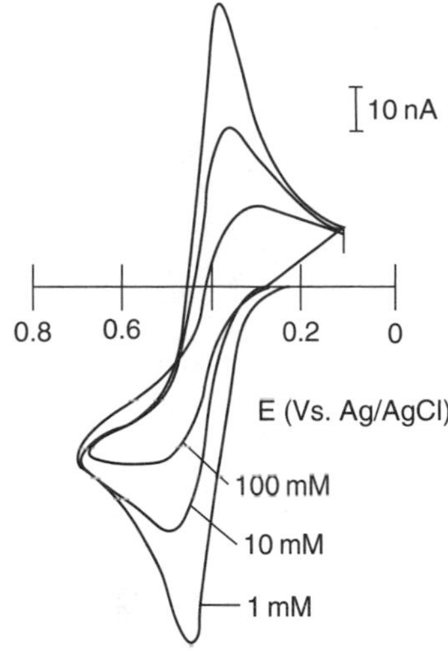

4.2.1 Fabrication

For these investigations, we have created carbon nanotube membranes (CNMs) by the chemical-vapor deposition of carbon into an alumina template. Carbon nanotubes with outside diameters as small as 20 nm have been prepared via this method [75]. Using an alternative template material and method, Joo et al. have prepared carbon nanotubes with outside diameters of 9 nm [93]. Ethylene gas serves as our carbon source (See [17] for method details).

A cross-sectional image of the CVD carbon nanotubes, post dissolution of the alumina template, is shown in Fig. 18A. As with other template-synthesis techniques, the outside diameter of the tubes is defined by the pore diameter of the template membrane as the length is determined by the membrane thickness. Figure 18B shows a TEM montage image of an individual carbon nanotube that was separated from the carbon surface films via ultrasonication. This image shows that the representative tube has a uniform wall thickness down its entire length. By lengthening or shortening the time of the CVD step, the wall thickness and thus the i.d. of the nanotubes can be controlled.

4.2.2 Measuring EOF in CNM

EOF can be driven across the CNMs by allowing the membrane to separate two electrolytic solutions and using an electrode in each solution to pass a constant ionic current through the nanotubes. The conditions of these experiments have been thoroughly described elsewhere [9, 17]. Briefly, the CNM was mounted in a supporting assembly and placed in a simple U-tube permeation cell. The permeation half-cells were filled with phosphate buffer. The feed half-cell contains the small, electrically neutral, chromophoric probe molecule phenol. As per White et al., the rate of EOF

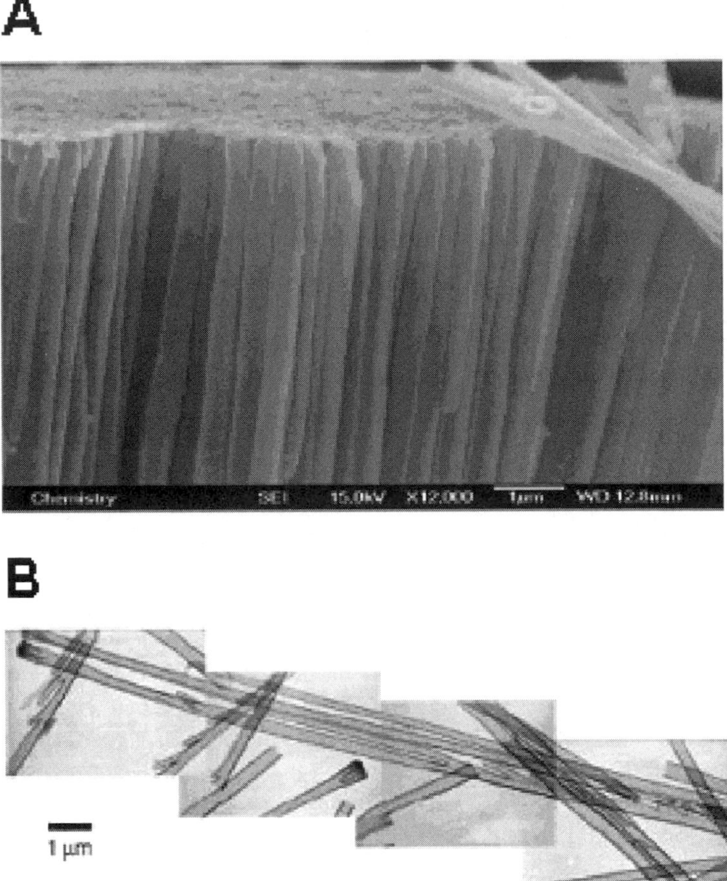

Fig. 18 (**A**) SEM image of a CNM cross-section after removal of the alumina template. (**B**) TEM image montage of individual carbon nanotubes removed from a CNM. Reproduced from [17]

was determined by monitoring the transport of the probe molecule across the membrane and into the permeate half-cell [94]. A constant current, passed through the CNM, was responsible for the EOF in the membrane. The following sign convention was employed: For a positive applied current, the anode was in the feed half-cell and the cathode was in the permeate half-cell. In this case, cations migrate from feed to permeate and anions from permeate to feed.

Figure 19 is a plot of micromoles of phenol transported across the CNM at various values of applied current density (applied current divided by the total pore area of the membrane) [17]. This plot consists of five linear segments, and the slope of each segment provides the flux of phenol, N_J, at that applied current density (Table 1). The first linear segment in Fig. 19 (0–8.5 min) was obtained with no applied current, and the flux in this case is simply the diffusive flux of phenol from

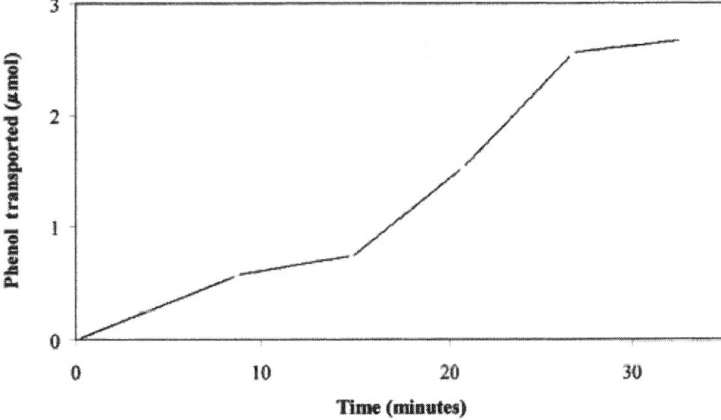

Fig. 19 Micromoles of phenol transported versus time at various values of applied current density (see text for description of currents.) Reproduced from [17]

Table 1 Effect of Applied Current Density on the Flux, Enhancement Factor, and Electroosmotic Velocity in the Carbon Nanotube Membrane. Reproduced from [17]

J_{app}, mA/cm^2	Flux, nmol/(min.cm^2)	E	υ_{eo}, μm/s
−10.1	25.8	0.280	−35.0
−7.2	39.3	0.426	−24.5
0.00	92.1	1.000	0.0
+7.2	187.1	2.030	26.3
+10.1	237.9	2.581	37.6

the feed across the membrane and into the permeate. The second linear segment (9–14.5 min) was obtained with small negative applied current. The flux of phenol across the membrane was observed to decrease, indicating EOF in the opposite direction of the diffusive flux (EOF from permeate solution to feed solution). This is confirmed by data in the third linear segment (15–20.5 min) which were obtained with a small positive current; now the net flux is higher than the diffusive flux, indicating EOF in the direction of diffusive transport. The final two linear segments were obtained at higher positive and negative applied currents, and further enhancement and diminution, respectively, in the flux was observed.

These data were analyzed to determine an enhancement factor of the flux at each applied current [17]. Following Srinivasan and Higuchi [95], the enhancement factor was used to calculate the electroosmostic velocity, υ_{eo} (Table 1). This indicates that the pore walls of the carbon nanotubes have fixed negatively charged sites. This is not surprising in that it is well known that exposure of carbon surfaces to air results in the formation of acidic surface functionalities [96]. This experiment demonstrates our ability to both measure and control the rate of EOF.

4.2.3 Redox Control of EOF

Miller et al. [74] recently also describe the ability to control both the rate and direction of EOF, independent of the electric field that drives EOF. This method entails coating the inner walls of the carbon nanotubes within the CNM with a redox-active polymer film. The redox polymer, poly(vinylferrocene) (PVFc), can be reversibly electrochemically switched between an electrical neutral and a polycationic form. This provides a mechanism for controlling both the magnitude and the sign of the surface charge on the nanotube walls, which in turn allows for control of both the rate and direction of EOF through the CNM. Again EOF is measured by monitoring the flux of phenol.

Equations (4) and (5) show that electroosmotic velocity (υ_{eo}) is related to zeta-potential (ζ), which is in turn related to the magnitude and sign of the surface-charge density (σ)

$$\upsilon_{eo} = -(\varepsilon\zeta J_{app}\rho)/\eta \tag{4}$$

$$\zeta = (\sigma\kappa^{-1})/\varepsilon \tag{5}$$

where κ^{-1} is the Debye length at the solid/electrolyte solution interface. Equation 5 suggests that, if the magnitude and sign of σ can be changed at will, the rate and direction of EOF can be modulated. Our new redox-polymer-based method allows for reversible control over both the sign and magnitude of σ [74].

An ultrathin film of PVFc is polymerized on the exposed surface (inner walls of the CNM). The novel feature of these experiments is that prior to determining υ_{eo} a known potential was applied to the CN/PVFc membrane to totally oxidize, partially oxidize, or totally reduce the PVFc within the nanotubes. After equilibration, the membrane was then removed from potentiostatic control, and υ_{eo} was measured with the PVFc in that selected oxidation state. This redox reaction is shown below in Eq. (6).

$$\text{PVFc} \longleftrightarrow \text{PVFc}^+ + e^- \tag{6}$$

At the potential extreme of +0.7 V the redox molecule is completely in the polycationic state ($PVFc^+$); conversely, at the extreme of −0.2 V it is completely in the neutral state (PVFc). The Nernst equation dictates that the ratio of moles of $PVFc^+$ to the moles of PVFc can be adjusted to any desired value by applying potentials that are intermediate between these two extremes. Therefore, controlling the potential directly controls the excess positive surface charge, and thus the electroosmotic flow rate.

We have proven this point by measuring υ_{eo} after applying to the CN/PVFc membrane various potentials between these extremes. Again, the potential was applied until equilibrium had been achieved, the membrane was released from potentiostatic control, and υ_{eo} was measured. The data obtained were processed as a plot of ζ, which as Eq. (4) shows is directly proportional to υ_{eo}, versus potential applied to the CN/PVFc membrane. The interplay of Eqs. (4) and (5) and results in Fig. 20 A show that we can, in fact, control υ_{eo}. Figure 20B shows that ζ increases from its negative limit when the potential applied to the membrane is more than 0 V and increases continuously until an applied potential of ~+0.5 V is reached. The advantage of this nonfaradic approach is that the channel itself does not need to be electronically conductive. The specific advantages of our approach is that much smaller voltages are needed than in other techniques and the voltage must be applied for only a few seconds to change the oxidation state of the polymer [74].

4.3 Li-ion Battery Nanoelectrodes

During the 1990s the explosion of high-end personal electronics created a demand for a lightweight rechargeable power source. Li-ion batteries have come to dominate that market. Though these batteries have seen great commercial success (a \$3 billion annual market) [97], there is now a vast international research initiative aimed adapting this current technology to power new more demanding pulse-power applications.

These batteries operate by each electrode (anode and cathode) reversibly intercalating Li-ions. During the discharge process, Li-ions deintercalate from the anode, migrate through a Li-ion conducting electrolyte, and then intercalate into the cathode. An example of this reversible intercalation reaction is demonstrated in Eq. (7) using the cathodic electrode material V_2O_5.

$$\mathrm{xLi^- + xe^- + V_2O_5} \underset{\text{Charge}}{\overset{\text{Discharge}}{\rightleftharpoons}} \mathrm{Li}_x\mathrm{V_2O_5} \tag{7}$$

Electroneutrality dictates than an electron must compensate for the movement of each Li^+. This electron travels through the electrical circuit, as the Li-ion travels through the electrolyte. A schematic of this process is shown in Fig. 21. When these Li-ions intercalate into an electrode, they are not able to solid-state diffuse rapidly enough to compensate for the facile nature of the insertion flux (D ~ 10^{-8}–10^{-11} cm^2/s.) This results in concentration–polarization of Li-ions at the surface during intercalation; conversely, the polarization is in the core during the deintercalation process [98]. This is a practical limitation to the stoichiometric quantity of charge (mAh g^{-1}) that is theoretically possible. The discharge-rate (current) determines this rate of insertion flux; therefore, at the high currents of a demanding application, this concentration–polarization problem is exacerbated. Since the diffusion coefficient is an intrinsic property of the electrode material, we work with the strategy of decreasing the size of the electrode particles. This allows the intercalation sites to be closer to the surface, shortening the distance over which the sluggish solid-state diffusion process must propagate [8, 36, 99, 100].

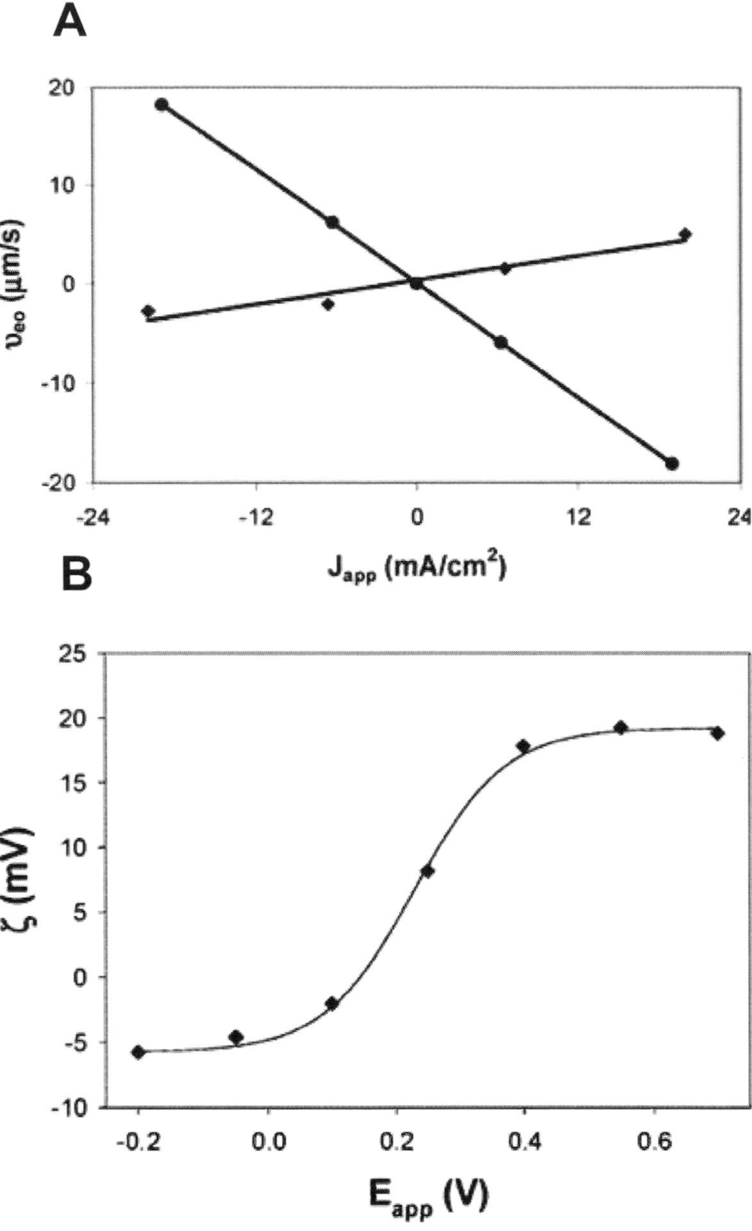

Fig. 20 (**A**) Plots of electro-osmotic flow velocity (υ_{eo}) versus applied current density for totally oxidized (*dot*) and totally reduced (*diamond*) CN/PVFc membranes (**B**) Plot of ζ versus potential applied to the CN/PVFc membrane. Therefore we have tunable control of ζ. Reproduced from [74]

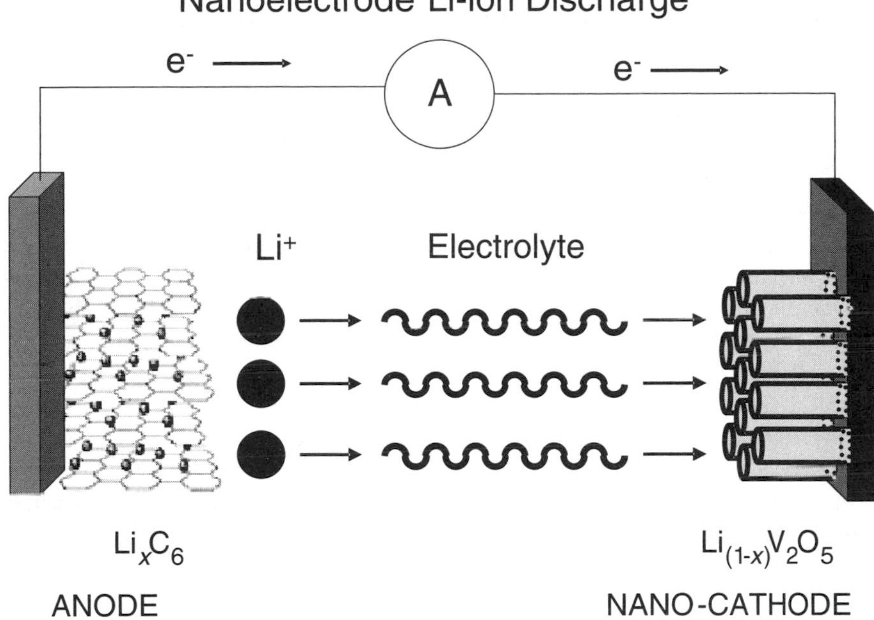

Fig. 21 Schematic of the discharge reaction of a Li-ion battery with a V_2O_5 nanocathode and a carbon anode

This plan of using smaller particles has another attribute. The increased surface area per total-volume fraction serves to decrease the effective current density at any given rate. This works to offset any sluggish electrokinetics of the system. Since the electrodes must exhibit both good electronic and ionic conductivity, a structure comprised of electronically addressable small particles is extremely favorable. This has been demonstrated by template synthesis of both cathodic [8, 37, 99, 100] and anodic materials [36, 101]. In every case, the nanostructured material was able to deliver higher specific capacity (mAh g^{-1}) at any given discharge rate (C, C $= h^{-1}$) than the microstructured control electrode.

4.3.1 Fabrication

A variety of nanostructured Li-ion battery electrodes have been synthesized by the template synthesis method. Each of these procedures has been defined in detail elsewhere [4, 8, 36, 37]. In general, a precursor solution is allowed to impregnate the pores of a polycarbonate track-etch template membrane. A section of this membrane is attached to a current collector. Platinum is a good choice, because of its high electronic conductivity and general inertness (to heat, chemicals, and extreme potentials). This assembly is then gently heated to evaporate the solvent of the precursor. A reactive-ion etching system uses oxygen plasma to preferentially etch the organic template. The result is an electrode that consists of an array of nanostructures that mirror the density and geometry of the pores of the chosen template. These structures extend from the surface of a current collector like the bristles of a brush. Often it is necessary to heat this electrode to high temperatures in order to increase its crystallinity and achieve the correct phase for reversible Li-ion intercalation. Electrodes of V_2O_5 [4, 8, 100], $LiFePO_4$ [38], $LiMn_2O_4$ [37], SnO_2 [36], and TiS_2 [35] have all been synthesized by slight variations to this general method. A micrograph of the V_2O_5 wires is shown in Fig. 22.

Notice during the general synthetic process detailed above there is no addition of electronically conductive carbon or binder. Other labs, including the commercial-unit production industry, synthesize electrodes by combining large (several-micron diameter) active particles with conductive carbon. Slurry is then made with a polymeric binder (PVDf, for example), and this mixture is spread onto a mesh current collector [102]. Because of the parallel electronic-conduction mechanism of our template-synthesized nanostructrued electrodes, this carbon is not required [8]. No binder is necessary because the structures are directly attached to the metal foil that serves as both a physical substrate and a common current collector. The absence of nonactive components in our electrodes also assists in increasing volumetric energy densities.

4.3.2 Measuring Rate Capabilities

A three-electrode cell is used to perform the electrochemical characterization of these electrodes [8]. The charge/discharge reactions for these electrodes were investigated in an electrolytic solution that was 1 M $LiClO_4$ dissolved in a solvent mixture of nonaqueous carbonates. Li metal ribbons served as the counter and reference electrodes. All potentials are quoted here versus the Li/Li^+ reference. Over the potential window of 2.8–3.8 V, 1 mol of V_2O_5 is known to reversibly intercalate 1 mol of Li^+, (i.e., $x = 1$ in Eq. 7) [103]. This corresponds to a maximum specific (per g) charge-storage capacity of 148 mAh g^{-1}.

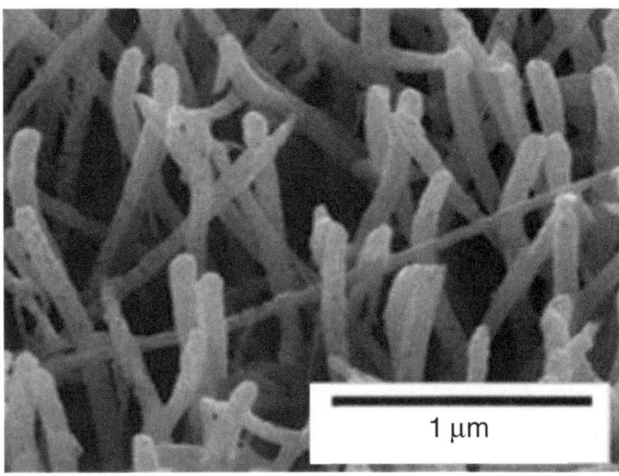

Fig. 22 SEM image of a nanowire V_2O_5 electrode created by sol–gel template synthesis. Reproduced from [8]

The protocol for determining the rate capabilities of an electrode is to charge–discharge the electrode at increasing C rates, while measuring the specific capacity from the potential versus discharge time data. All electrodes will eventually have their capacities decreased because of this concentration polarization phenomenon, but the electrode that has the shortest solid-state diffusion distance is able to delay the onset of this power-limiting factor. This is demonstrated in the data shown in Fig. 23 that compares the specific capacity of a 70-nm diameter nanowire electrode to that of a 0.8-μm diameter microwire electrode at a series of normalized discharge rates.

By using different pore-diameter polycarbonate templates, we were able to compare electrodes composed of identical material, but which differ by nanowire radii (i.e., the solid-state diffusion distance) [8, 37]. We show that at every discharge rate the electrode with the smallest wire diameter delivers a higher portion of its theoretical capacity. This is shown by every point residing above the ratio of 1 (where the capacities would be equal). This original citation has expanded discussion on this effect at decreasing temperatures, as well as an explanation of this array configuration's influence on the electrode's electronic conductivity [8].

4.3.3 Other Electrochemical Studies

For an electrode to function in a rechargeable system, it must be able to maintain its ability to be charged and discharged for many cycles. This cycle–life parameter is measured as the charge–discharge cycle is repeated at the same rate. Our nanostructured electrodes have shown to be able to be cycled for greater than 1400 times at the high rate of 58 °C, without diminishing the quantity of charge delivered per cycle [36].

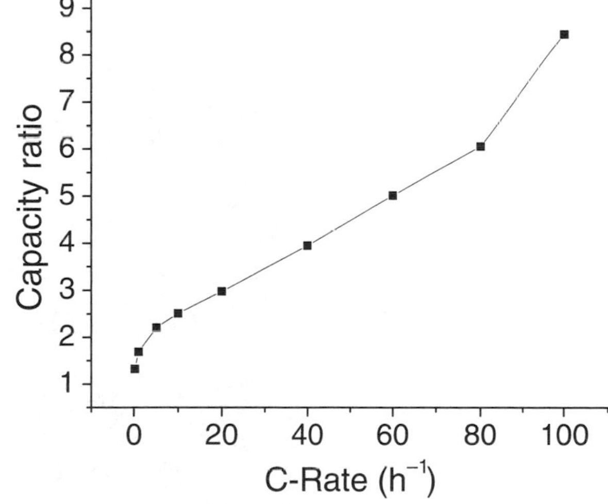

Fig. 23 Discharge C-Rate (normalized current) versus Capacity Ratio. Capacity Ratio is defined as the specific capacity (mAh g^{-1}) of the nanoelectrode divided by the specific capacity (mAh g^{-1}) the microelectrode. Reproduced from. [8]

As demonstrated in the above discussion, electrodes comprised of small particles are advantageous. Void volume is used to keep these particles small. While this has no effect on the gravimetric capacity (mAh g^{-1}), it does on volumetric capacity (mAh L^{-1}). Volummetric restrictions of an application may be just as stringent as gravimetric restrictions. Our lab has shown the ability to "refill" a portion of this void volume with more active material to increase the charge stored per liter [100]. In this investigation, the volumetric charge delivered by the template-synthesized nanostructured electrode at high rates of discharge actually surpassed that delivered by the thin-film control.

4.3.4 Nanosphere-Templated Structures

Other groups (most notably from the University of Minnesota) have worked with a template method that uses nanospheres to create similar short-diffusion-distance materials [21]. An array of monodisperse PMMA spheres of an approximate diameter of 300 nm serve as the template. Approximately ten layers of these spheres construct an array. The void volume of these close-packed spheres hosts a Sn-based precursor solution. After processing the precursor into a state capable of reversible Li-ion intercalation, the PMMA-template spheres were removed via calcinations. The electrode structure shrinks as a result of this heating. Average solid-state diffusion distance for Li-ions is about 170 nm. This electrode is pictured in Fig. 24 and shows good order to an area of 100 μm^2. The resulting electrode was characterized electrochemically [21] and was able to reversibly store and deliver charge under conditions similar to those previously described.

4.4 Ion Channels

Ion channels are protein pores that span cell membranes and which open and close in response to stimuli, such as changes in the transmembrane potential, binding of a

Fig. 24 SEM image of a nanosphere-templated Sn-based anode for a Li-ion battery system. This was heated to 800°C in a N_2/air mixture for 1 h. Reproduced from [21]

ligand, or mechanical stress [104]. When open, ions pass through the pore, and when closed, ion transport is disallowed. Hence, these channels are nanodevices that have a current-rectification function. Elucidating this mechanism on a molecular level is a common goal of many labs.

A step toward accomplishing this goal is to design an abiotic system that has a similar current-voltage response as the ion channel. Siwy et al. have shown that a polymeric membrane that contains a single, conically shaped nanopore also acts as an ion-current rectifier [105, 106]. These conical pores are created from the damage-track polymer film (see the Track-etch section for details). Again, a basic medium is used to etch the track into a pore, but movement of the OH^- ions into the track is hindered by an applied potential [106]. This potential serves to protect the tip of the pore; therefore, the pore etches as a cone as opposed to cylinder. A neutralizing acid medium stops the etching reaction upon film breakthrough (i.e., the moment that a track becomes a pore). This strategy is detailed in Fig. 25. Our group's strategy is to combine our ability to control surface charge (via electroless deposition of gold and thiol chemisorption) with these single conical pores [105].

4.4.1 Electrochemical Measurements

The single conical Au nanotube membrane was mounted between the two halves of a conductivity cell, and a Ag/AgCl electrode was inserted into each half-cell solution (0.1 M KCl). Voltages were then applied to the electrodes in 100-mV steps, and the transmembrane ionic current was measured. At any absolute value of transmembrane potential, the current is higher at negative potentials ("on" state) than at positive potentials ("off" state). These data and rectification phenomenon are shown in Fig. 26 [105].

This effect is probed further in a similar experiment (data also shown in Fig. 26), where KF was substituted as the electrolytic salt. This system did not show the rectification seen with the KCl salt. Chloride ions will absorb to the Au surface, but

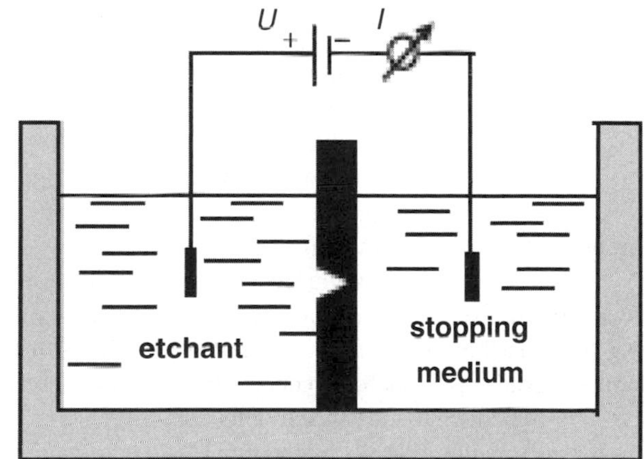

Fig. 25 Scheme of the experimental setup with the conductivity cell. During etching the left compartment is filled with etchant (base) and the right compartment with stopping medium (acid). For current measurements both compartments are filled with KCl solution. Reproduced from [106]

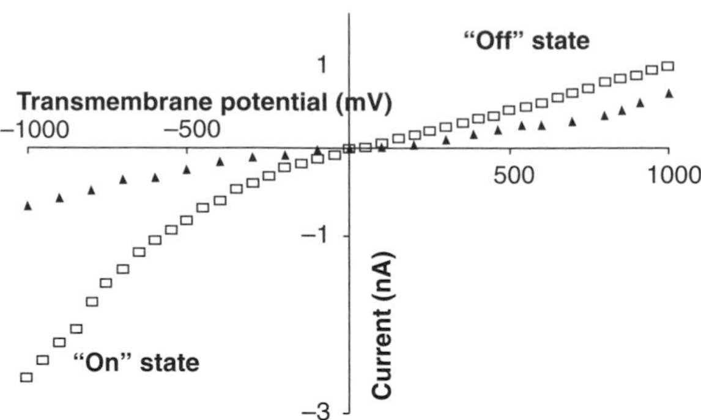

Fig. 26 I–V curves in 0.1 M KCl (*open square*) and 0.1 M KF (*filled triangles*). Reproduced from [105]

the fluoride ions will not. Therefore, surface charge changes this rectification ability. Using a thiol that was chemisorbed to the Au surface (under conditions that would give the walls a negative charge), the KF experiment was performed again, and rectification occurred. In a pH environment that made the surface charge positive, no rectification was observed [105].

Further experiments reveal that if the surface of the nanotube has fixed negatively charged sites, and the radius of the pore approaches the electrical double layer, then the cation is allowed into the pore and the anion is rejected.

This single conical Au-lined pore is an example of a voltage-gated potassium ion channel. It shows current rectification when subjected to a transmembrane potential. This is a successful step toward understanding the mechanism of the ion channels in nature.

4.5 DNA Ion Channels

Development of the ion channels described in the previous section is a major step toward creating an abiotic system that mimics the current–voltage response of its biological analog. It is clear that ion current rectification in biological ion channels is more complicated and involves physical movement of an ionically charged portion of the channel in response to a change in the transmembrane potential [107]. Recently, we have achieved another progression [108], an artificial ion channel that was designed to rectify the ion current flowing through it via "electromechanical" mechanism [109]. This is accomplished by attaching a single-stranded DNA to the mouth of the gold conical pore.

The templates were single-tracked polycarbonate foils that have been etched to form a conical shape (5-μm diameter base to 60-nm diameter mouth). Electroless plating created a thin (~10 nm) layer of gold throughout the length of the pore wall. Using familiar gold-thiol attachment chemistry, single-stranded thiol-terminated DNA was covalently bonded to the surface. This conical structure was left intact with the polycarbonate template. An electrochemical setup similar to that described previously was used to characterize the system (10 mM phosphate buffer at pH = 7, two Ag/AgCI electrodes, and 100 mM KCI in a U-tube permeation cell). A transmembrane potential was applied, while the ionic current was measured [108].

The mouth of this DNA ion channel is larger than that described previously [105], so cation permselectivity is not observed. However, current rectification is seen in this case with attached DNA. That is, they show an on-state at negative transmembrane potentials (anode facing the mouth of nanotube) and an off-state at positive potentials. The extent of this rectification (r_{max} = |Current at −1 V|/|Current at +1 V|) is dependent on the length of the DNA chain. This r_{max} parameter can also be controlled by holding the chain length constant and by varying the diameter of the nanotube mouth [108].

We propose that rectification in these nanotubes entails the electrophoretic insertion of the DNA chains into (off-state, Fig. 27C) and out of (on-state, Fig. 27B)

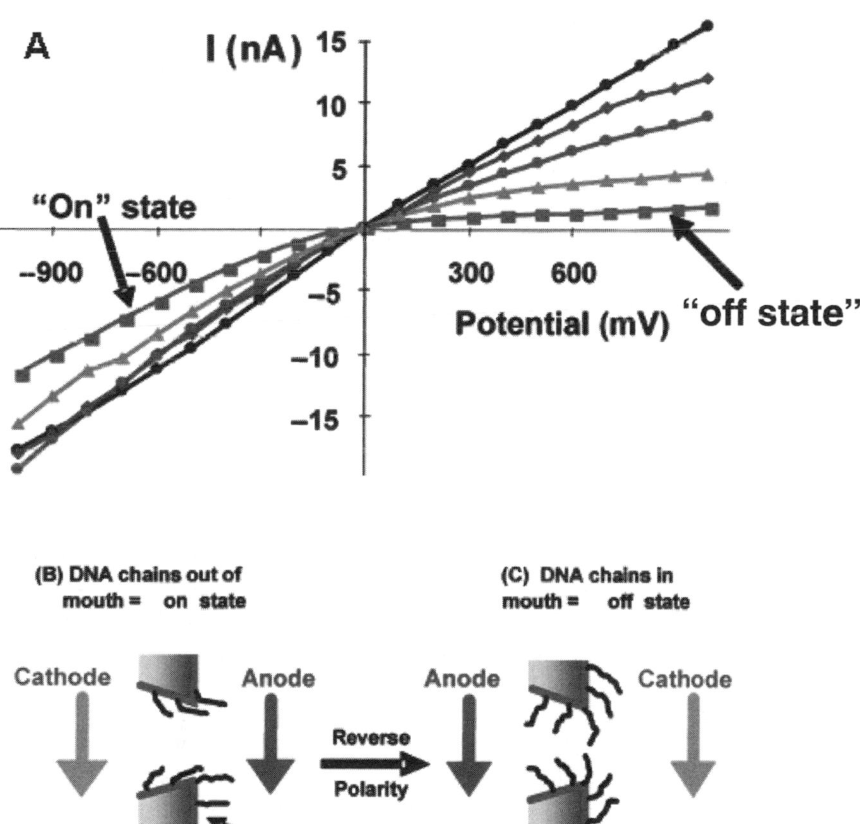

Fig. 27 (A) I–V curves for nanotubes with a mouth diameter of 40 nm containing no DNA (uppermost curve on the right side of the plot), and attached 12-mer, 15-mer, 30-mer, and 45-mer DNAs (curves decreasing in the sequence of increasing DNAs). (**B**) and (**C**) Schematics showing electrode polarity and DNA chain positions for on (**B**) and off (**C**) states. Reproduced from. [108]

the nanotube mouth [108]. The off-state is obtained because, when inserted into the mouth, the chains partially block the pathway for ion transport, yielding a higher ionic resistance for the nanotube. Simulations of the electric field strength in conical nanopores, identical to those used here, show that the field in the electrolyte solution in the mouth of the nanopore is 1×10^6 V m^{-1} when the total voltage drop across the membrane is only 1 V [110]. This focusing of the electric field strength at the nanotube mouth means that there is ample field to extend the DNA chains toward the anode.

This is supported by the I–V curves, because the magnitudes of the on-state currents in Fig. 27A decrease with increasing DNA chain length. This is because even in the on-state, the DNA chains partially occlude the mouth of the nanotube and increase the nanotube resistance. Additional data suggest that the DNA chain can be too long, relative to the mouth diameter (~13 nm), to allow for efficient rectification. Another supporting case can be made in considering the effect of DNA-chain flexibility. This issue was explored by comparing the extent of rectification for a nanotube containing the conventional 30-base DNA and that of an identical tube containing a hairpin 30-base DNA. The hairpin [111] DNA folds back on itself because the bases at one end of the chain are complementary to the bases at the other end, making the chain much less flexible. The nanotube containing this 30-base hairpin DNA is a very poor rectifier, much worse than the same nanotube with the conventional 30-base DNA [108]. Here we demonstrated yet another step toward creating an abiotic ion channel and elucidating the mechanism of the complex biological ion-channel process.

5 Conclusion

The template-synthesis nanofabrication method has found a myriad of applications in laboratories throughout the world. This method is able to create nanostructures by constricting particle growth to the boundaries of the porous template. The genius of this technique is in its generality. We have described templates of differing geometries, sizes, pore densities, and composition. Several strategies to deposit precursors of differing phases and constituents have been discussed. The combination of these diverse templates and deposition routes results in the ability to tailor a variety of nanomaterials to specific controllable dimensions. Each application then can be addressed by a number of possible routes. This chapters details the work done by us and others, which investigates electrochemistry at the nanoscale. Research at the interface of the traditional fields of electrochemistry and material science with the emerging field of nanoscience has proven to be both interesting and advantageous.

So what does the future hold for the template synthesis? Our lab is now expanding into the fields of biological investigations (separations, transport, and sensor design), tailoring the shape of the templates (conical, square, and test tubes), and creating nanoscale energy-storage devices.

Acknowledgements This work would not have been possible without the efforts of a number of hardworking and highly motivated graduate students and postdocs, both past and present. They include Reginald Penner, John C. Hulteen, Vinod P. Menon, Charles J. Brumlik, Charles J. Patrissi, Scott Miller, Naichao Li, Marc Wirtz, Zuza Siwy, Punit Kohli, and C. Chad Harrell. Financial support from the Office of Naval Research and the Department of Energy is also gratefully acknowledged. We also thank The Electron Microscopy Core Laboratory, Biotechnology Program, and the Major Analytical Instrument Center, both at University of Florida.

References

1. C. R. Martin, *Science*. **1994**, *266*, 1961.
2. K. B. Jirage, J. C. Hulteen and C. R. Martin, *Science*. **1997**, *278*, 655.
3. J. C. Hulteen, V. P. Menon and C. R. Martin, *Journal of the Chemical Society, Faraday Transactions*. **1996**, *92*, 4029.
4. C. J. Patrissi and C. R. Martin, *Journal of the Electrochemical Society*. **1999**, *146*, 3176.
5. R. M. Penner and C. R. Martin, *Analytical Chemistry*. **1987**, *59*, 2625.
6. M. Wirtz and C. R. Martin, *Advanced Materials*. **2003**, *15*, 455.
7. S. Yu, N. Li, J. Wharton and C. R. Martin, *Nano Letters*. **2003**, *3*, 815.
8. C. R. Sides and C. R. Martin, *Advanced Materials*. **2005**, *17*, 125.
9. M. Wirtz, S. A. Miller and C. R. Martin, *International Journal of Nanoscience*. **2002**, *1*, 255.
10. C. J. Brumlik and C. R. Martin, *Journal of the American Chemical Society*. **1991**, *113*, 3174.
11. C. J. Brumlik, C. R. Martin and K. Tokuda, *Analytical Chemistry*. **1992**, *64*, 1201.
12. G. Che, B. B. Lakshmi, E. R. Fisher and C. R. Martin, *Nature*. **1998**, *393*, 346.
13. C. A. Foss, Jr., G. L. Hornyak, J. A. Stockert and C. R. Martin, *Journal of Physical Chemistry*. **1992**, *96*, 7497.
14. P. Kohli, M. Wirtz and C. R. Martin, *Electroanalysis*. **2004**, *16*, 9.
15. B. B. Lakshmi, C. J. Patrissi and C. R. Martin, *Chemistry of Materials*. **1997**, *9*, 2544.
16. S. B. Lee, D. T. Mitchell, L. Trofin, T. K. Nevanen, H. Soederlund and C. R. Martin, *Science*. **2002**, *296*, 2198.
17. S. A. Miller, V. Y. Young and C. R. Martin, *Journal of the American Chemical Society*. **2001**, *123*, 12335.
18. B. T. Holland, C. F. Blanford and A. Stein, *Science*. **1998**, *281*, 538.
19. J. P. Hoogenboom, C. Retif, E. de Bres, M. van de Boer, A. K. van Langen-Suurling, J. Romijn and A. van Blaaderen, *Nano Letters*. **2004**, *4*, 205.
20. P. Jiang, J. F. Bertone and V. L. Colvin, *Science*. **2001**, *291*, 453.
21. J. C. Lytle, H. Yan, N. S. Ergang, W. H. Smyrl and A. Stein, *Journal of Materials Chemistry*. **2004**, *14*, 1616.
22. A. Stein and R. C. Schroden, *Current Opinion in Solid State and Materials Science*. **2001**, *5*, 553.
23. O. D. Velev, A. M. Lenhoff and E. W. Kaler, *Science*. **2000**, *287*, 2240.
24. C. J. Brumlik, V. P. Menon and C. R. Martin, *Journal of Materials Research*. **1994**, *9*, 1174.
25. G. L. Hornyak, C. J. Patrissi and C. R. Martin, *Journal of Physical Chemistry B*. **1997**, *101*, 1548.
26. V. P. Menon and C. R. Martin, *Analytical Chemistry*. **1995**, *67*, 1920.
27. M. Nishizawa, V. P. Menon and C. R. Martin, *Science*. **1995**, *268*, 700.
28. C. R. Martin and L. S. Van Dyke, in *Molecular Design on Electrode Surfaces*, (Ed. R. W. Murray), John Wiley & Sons, Inc., New York **1992**, 403–424.
29. C. R. Martin, *Accounts of Chemical Research*. **1995**, *28*, 61.
30. R. M. Penner, L. S. Van Dyke and C. R. Martin, *Journal of Physical Chemistry*. **1988**, *92*, 5274.
31. R. M. Penner, L. S. Van Dyke and C. R. Martin, *Solid State Ionics*. **1989**, *32–33*, 553.
32. B. B. Lakshmi, P. K. Dorhout and C. R. Martin, *Chemistry of Materials*. **1997**, *9*, 857.

33. V. M. Cepak, J. C. Hulteen, G. Che, K. B. Jirage, B. B. Lakshmi, E. R. Fisher, C. R. Martin and H. Yoneyama, *Chemistry of Materials*. **1997**, *9*, 1065.
34. G. Che, B. B. Lakshmi, C. R. Martin and E. R. Fisher, *Langmuir*. **1999**, *15*, 750.
35. G. Che, K. B. Jirage, E. R. Fisher and C. R. Martin, *Journal of the Electrochemical Society*. **1997**, *144*, 4296.
36. N. Li and C. R. Martin, *Journal of the Electrochemical Society*. **2001**, *148*, A164.
37. N. Li, C. J. Patrissi, G. Che and C. R. Martin, *Journal of the Electrochemical Society*. **2000**, *147*, 2044.
38. C. R. Sides, F. Croce, V. Young, C. R. Martin and B. Scrosati, *Electrochemical and Solid-State Letters*. **2005**, *8*, A483.
39. C. G. Wu and T. Bein, *Science*. **1994**, *264*, 1757.
40. G. A. Ozin, *Advanced Materials*. **1992**, *4*, 612.
41. C. R. Martin, in *Handbook of Conducting Polymers, 2nd Ed.*, (Ed. J. R. Reynolds, T. Skotheim and R. Elsenbaumer), Marcel Dekker, Inc, **1997**, Chap. 16, p. 409.
42. GE Osmotics, Inc., *Product Guide*, http://www.osmolabstore.com/documents/1220457-OEMMembranes.pdf.
43. R. L. Fleisher, P. B. Price and R. M. Walker, *Nuclear Tracks in Solids*, University of California Press, Berkeley, CA, **1975**.
44. C. R. Martin, M. Nishizawa, K. Jirage and M. Kang, *Journal of Physical Chemistry B*. **2001**, *105*, 1925.
45. C. Schoenenberger, B. M. I. van der Zande, L. G. J. Fokkink, M. Henny, C. Schmid, M. Krueger, A. Bachtold, R. Huber, H. Birk and U. Staufer, *Journal of Physical Chemistry B*. **1997**, *101*, 5497.
46. P. Apel, *Radiation Measurements*. **2001**, *34*, 559–566.
47. C. A. Foss, Jr., G. L. Hornyak, J. A. Stockert and C. R. Martin, *Advanced Materials*. **1993**, *5*, 135.
48. C. A. Foss, Jr., G. L. Hornyak, J. A. Stockert and C. R. Martin, *Journal of Physical Chemistry*. **1994**, *98*, 2963.
49. A. Despic and V. P. Parkhutik, *Modern Aspects of Electrochemistry*. **1989**, *20*, 401.
50. N. Li, D. T. Mitchell, K.-P. Lee and C. R. Martin, *Journal of the Electrochemical Society*. **2003**, *150*, A979.
51. H. Masuda, H. Yamada, M. Satoh, H. Asoh, M. Nakao and T. Tamamura, *Applied Physics Letters*. **1997**, *71*, 2770.
52. H. Asoh, K. Nishio, M. Nakao, A. Yokoo, T. Tamamura and H. Masuda, *Journal of Vacuum Science and Technology B*. **2001**, *19*, 569.
53. T. Yanagishita, K. Nishio, M. Nakao, A. Fujishima and H. Masuda, *Chemistry Letters*. **2002**, *10*, 976.
54. J. D. Joannopoulos, R. D. Meade and J. N. Winn, *Photonic Crystals: Molding the Flow of Light*, University Press, Princeton, NJ, **1995**.
55. M. E. Turner, T. J. Trentler and V. L. Colvin, *Advanced Materials*. **2001**, *13*, 180.
56. A. Stein, *Microporous and Mesoporous Materials*. **2001**, *44–45*, 227.
57. O. D. Velev, P. M. Tessier, A. M. Lenhoff and E. W. Kaler, *Nature*. **1999**, *401*, 548.
58. O. D. Velev, A. M. Lenhoff and E. W. Kaler, *Science*. **2000**, *287*, 2240.
59. P. Jiang, J. F. Bertone and V. L. Colvin, *Science*. **2001**, *291*, 453.
60. J. C. Hulteen and R. P. Van Duyne, *Journal of Vacuum Science and Technology A*. **1995**, *13*, 1553.
61. L. Xu, L. D. Tung, L. Spinu, A. A. Zakhidov, R. H. Baughman and J. B. Wiley, *Advanced Materials*. **2003**, *15*, 1562.
62. G. E. Possin, *Review of Scientific Instruments*. **1970**, *41*, 772.
63. R. D. Patel, M. G. Takwale, V. K. Nagar and V. G. Bhide, *Thin Solid Films*. **1984**, *115*, 169.
64. S. Kawai, in *Symposium on Electrochemical Technology in Electronics*, (Eds. L.T. Romankiw and T. Osaka), Electrochemical Society, Pennington, NJ, **1987**, 389.

65. S. K. Chakarvarti and J. Vetter, *Nuclear Instruments & Methods in Physics Research, Section B: Beam Interactions with Materials and Atoms*. **1991**, *B62*, 109.
66. T. M. Whitney, J. S. Jiang, P. C. Searson and C. L. Chien, *Science*. **1993**, *261*, 1316.
67. Y. Wang, C. Ye, X. Fang and L. Zhangy, *Chemistry Letters*. **2004**, *33*, 166.
68. L. S. Van Dyke and C. R. Martin, *Langmuir*. **1990**, *6*, 1118.
69. R. M. Penner and C. R. Martin, *Journal of the Electrochemical Society*. **1986**, *133*, 2206.
70. C. R. Martin, *Advanced Materials*. **1991**, *3*, 457.
71. Z. Cai, J. Lei, W. Liang, V. Menon and C. R. Martin, *Chemistry of Materials*. **1991**, *3*, 960.
72. *Electroless Plating: Fundamentals and Applications* (Eds. G. O. Mallory and J. B. Hajdu). William Andrew Publishing, **1990**.
73. M. A. Aegerter, R. C. Mehrota, I. Oehme, R. Reisfeld, S. Sakka, O. Wolfbeis and C. K. Jorgensen, *Optical and Electronic Phenomena in Sol-Gel Glasses and Modern Applications*, Springer-Verlag, Berlin **1996**.
74. S. A. Miller and C. R. Martin, *Journal of the American Chemical Society*. **2004**, *126*, 6226.
75. T. Kyotani, T. Nagai, S. Inoue and A. Tomita, *Chemistry of Materials*. **1997**, *9*, 609.
76. A. A. Zakhidov, R. H. Baughman, Z. Iqbal, C. Cui, I. Khayrullin, S. O. Dantas, J. Marti and V. G. Ralchenko, *Science*. **1998**, *282*, 897.
77. R. A. Saraceno, C. E. Engstrom, M. Rose and A. G. Ewing, *Analytical Chemistry*. **1989**, *61*, 560.
78. G. Che, B. B. Lakshmi, C. R. Martin, E. R. Fisher and R. S. Ruoff, *Chemistry of Materials*. **1998**, *10*, 260.
79. S. M. George, A. W. Ott and J. W. Klaus, *Journal of Physical Chemistry*. **1996**, *100*, 13121.
80. M. A. Cameron, I. P. Gartland, J. A. Smith, S. F. Diaz and S. M. George, *Langmuir*. **2000**, *16*, 7435.
81. H. Shin, D.-K. Jeong, J. Lee, M. M. Sung and J. Kim, *Advanced Materials*. **2004**, *16*, 1197.
82. J. W. Elam, D. Routkevitch, P. P. Mardilovich and S. M. George, *Chemistry of Materials*. **2003**, *15*, 3507.
83. Y.-S. Min, E. J. Bae, K. S. Jeong, Y. J. Cho, J.-H. Lee, W. B. Choi and G.-S. Park, *Advanced Materials*. **2003**, *15*, 1019.
84. E. K. Seo, J. W. Lee, H. M. Sung-Suh and M. M. Sung, *Chemistry of Materials*. **2004**, *16*, 1878.
85. L. J. LeGore, O. D. Greenwood, J. W. Paulus, D. J. Frankel and R. J. Lad, *Journal of Vacuum Science and Technology A*. **1997**, *15*, 1223.
86. M. Ritala and M. Leskela, *Nanotechnology*. **1999**, *10*, 19.
87. J. C. Badot, S. Ribes, E. B. Yousfi, V. Vivier, J. P. Pereira-Ramos, N. Baffier and D. Lincot, *Electrochemical and Solid-State Letters*. **2000**, *3*, 485.
88. A. Russell, K. Repka, T. Dibble, J. Ghoroghchian, J. J. Smith, M. Fleischmann and S. Pons, *Analytical Chemistry*. **1986**, *58*, 2961.
89. Z. J. Karpinski and R. A. Osteryoung, *Journal of Electroanalytical Chemistry*. **1993**, *349*, 285.
90. S. M. Drew, R. M. Wightman and C. A. Amatore, *Journal of Electroanalytical Chemistry and Interfacial Electrochemistry*. **1991**, *317*, 117.
91. I. F. Cheng, L. D. Whiteley and C. R. Martin, *Analytical Chemistry*. **1989**, *61*, 762.
92. J. C. Giddings, *Unified Separation Science*, John Wiley & Sons, Inc., New York **1991**, Chap. 4, p. 73.
93. S. H. Joo, S. J. Choi, I. Oh, J. Kwak, Z. Liu, O. Terasaki and R. Ryoo, *Nature*. **2001**, *412*, 169.
94. B. D. Bath, R. D. Lee and H. S. White, *Analytical Chemistry*. **1998**, *70*, 1047.
95. V. Srinivasan and W. I. Higuchi, *International Journal of Pharmaceutics*. **1990**, *60*, 133.
96. A. B. Garcia, A. Cuesta, M. A. Montes-Moran, A. Martinez-Alonso and J. M. D. Tascon, *Journal of Colloid and Interface Science*. **1997**, *192*, 363.
97. A. H. Tullo, *Chemical and Engineering News*. **2002**, *80*, 25.
98. B. A. Johnson and R. E. White, *Journal of Power Sources*. **1998**, *70*, 48.

99. C. R. Sides, N. Li, C. J. Patrissi, B. Scrosati and C. R. Martin, *MRS Bulletin*. **2002**, *27*, 604.
100. C. J. Patrissi and C. R. Martin, *Journal of the Electrochemical Society*. **2001**, *148*, A1247.
101. N. Li, C. R. Martin and B. Scrosati, *Electrochemical and Solid-State Letters*. **2000**, *3*, 316.
102. L. Fransson, T. Eriksson, K. Edstrom, G. T. and J. G. Thomas, *J. Power Sources*. **2001**, *101*, 1.
103. C. Delmas, H. Cognac-Auradou, J. M. Cocciantelli, M. Menetrier and J. P. Doumerc, *Solid State Ionics*. **1994**, *69*, 257.
104. *The Cell – A Molecular Approach* (Ed. G. M. Cooper). Sinauer Associates, Inc, Sunderland, MA **2000**.
105. Z. Siwy, E. Heins, C. C. Harrell, P. Kohli and C. R. Martin, *Journal of the American Chemical Society*. **2004**, *126*, 10850.
106. Z. Siwy, P. Apel, D. Baur, D. D. Dobrev, Y. E. Korchev, R. Neumann, R. Spohr, C. Trautmann and K.-O. Voss, *Surface Science*. **2003**, 532.
107. Y. Jiang, A. Lee, J. Chen, V. Ruta, M. Cadene, B. T. Chait and R. MacKinnon, *Nature*. **2003**, *423*, 33.
108. C. C. Harrell, P. Kohli, Z. Siwy and C. R. Martin, *Journal of the American Chemical Society*. **2004**, *126*, 15646.
109. Y. Jiang, A. Lee, J. Chen, M. Cadene, B. T. Chait and R. MacKinnon, *Nature*. **2002**, *417*, 515.
110. S. Lee, Y. Zhang, H. S. White, C. C. Harrell and C. R. Martin, *Analytical Chemistry*. **2004**, *76*, 6108.
111. G. Bonnet, S. Tyagi, A. Libchaber and F. R. Kramer, *Proceedings of the National Academy of Sciences of the United States of America*. **1999**, *96*, 6171.
112. M. Kang, S. Yu, N. Li and C. R. Martin, *Small*. **2005**, *1*, 69.
113. C. R. Martin, *Chemistry of Materials*. **1996**, *8*, 1739.
114. C. R. Martin and D. T. Mitchell, *Electroanalytical Chemistry*. **1999**, *21*, 1.
115. J. C. Hulteen and C. R. Martin, *Journal of Materials Chemistry*. **1997**, *7*, 1075.

Electroless Fabrication of Nanostructures

T. Osaka

1 Fundamental (with Contributions from A. Sugiyama)

This chapter reviews recent electroless deposition technologies mainly on the basis of the researches in the fields of interconnect materials for ultralarge-scale integration (ULSI) device and magnetic nanodot arrays for magnetic recording device, metal and metal oxide nanoparticles.

Various techniques are available for depositing metal films on substrates, such as chemical vapor deposition (CVD), sputtering, evaporating, electrodeposition, and electroless deposition. Among these processes, electro- and electroless deposition, that is, electrochemical technique, have attracted much attention because of its simplicity in operation and sufficient mass productivity. Especially, electroless deposition is advantageous for forming thin film on large-scale substrate, and for depositing films uniformly on the complicated shapes or nonconductive substance [1, 2].

Electroless deposition technique was first reported by Brenner and Riddell in 1946 [3, 4]. In the electroless deposition process, metal ions in the solution were reduced by means of a chemical-reducing agent, resulting in the metal formation [5]. Nowadays, electroless deposition techniques are widely used in electronics application [1, 2]. In semiconductor device industry, attempts of formation of Ni and Ni alloys have been made to utilize the technique for delineating semiconductor junctions, making ohmic contacts, and micropattering integrated circuits [6–9]. In the late 1980s, filling via holes and trenches with Cu have been attempted for producing ULSI interconnects [10–12]. Recently, metal dots of Ni and Co alloys with smaller than 100-nm diameter can be fabricated on Si wafers by electroless deposition, which has a great potential for realization of ultrahigh-density read-only memory (ROM) or random-access memory (RAM) devices. Ion implantation apparatus was found to modify the silicon surface and initiate electroless gold

T. Osaka (✉)
Department of Applied Chemistry, School of Science and Engineering, Waseda University, 3-4-1 Okubo, Shinjuku-ku, Tokyo, 169-8555 Japan
e-mail: osakatets@waseda.jp

P. Schmuki, S. Virtanen (eds.), *Electrochemistry at the Nanoscale,* Nanostructure Science and Technology, DOI 10.1007/978-0-387-73582-5_8,

deposition. In magnetic recording field, the application of electroless cobalt film was first proposed by Fisher et al. in 1962 [13], and the electroless Co film had already been successfully applied in the fabrication of practical longitudinal disk media [14]. Furthermore, soft magnetic films of electroless CoNiFe alloy are developed for soft magnetic underlayer of a perpendicular magnetic recording media [14–17].

Some plating equipment such as paddle-plating system was first invented for electrodeposition technique by IBM group [18–20]. Recently, we developed it not only for electroplating but also for electroless plating in such a system, which is now absolutely essential for electroless plating. In conventional plating bath with screw stirring device or jet-stream impingement, uniformity of mass-transfer rate for plating metal was insufficient over the entire substrate surface [21, 22]. In addition, paddle-plating system with filtering unit was developed for both electrodeposition and electroless deposition. Also, the control of the plating bath composition and bath temperature was significant. Figure 1 shows the scanning electron microscopy (SEM) images of electroless CoFeB on photoresist/Cu substrate with line patterns [23]. In the case of electroless deposition using dimethylamine-borane (DMAB) as a reducing agent, the CoFeB alloy was deposited not only on the catalytic surface but also on the photoresist; extraneous deposition was observed at the edge of the pattern. The abnormal phenomenon is originally caused by the use of DMAB as a reducing agent. For the improvement of deposition selectivity, we developed the novel plating bath containing some organic additives and plating apparatus with the paddle unit for agitation of the bath solution, resulting in the suppression of extraneous deposition, as shown in Fig. 1b.

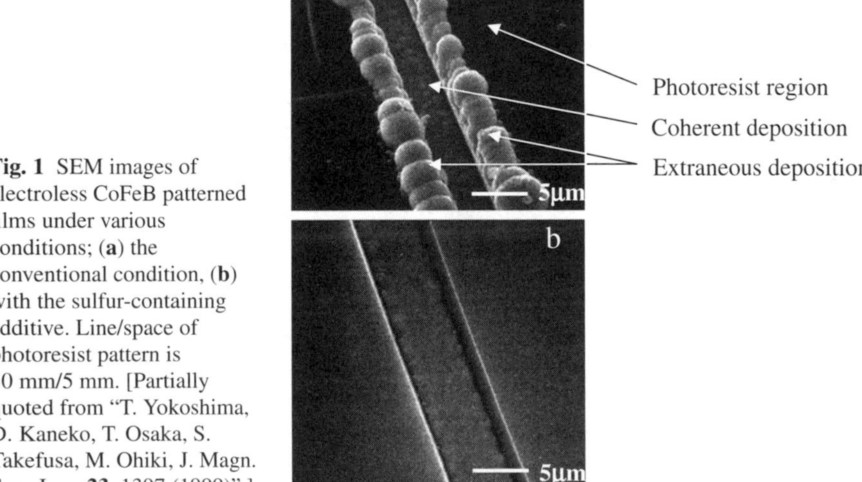

Fig. 1 SEM images of electroless CoFeB patterned films under various conditions; (**a**) the conventional condition, (**b**) with the sulfur-containing additive. Line/space of photoresist pattern is 50 mm/5 mm. [Partially quoted from "T. Yokoshima, D. Kaneko, T. Osaka, S. Takefusa, M. Ohiki, J. Magn. Soc. Jpn., **23**, 1397 (1999)".]

2 Superfilling of Cu into Patterned Substrates by Electroless Plating (with Contributions from J. Sasano)

2.1 Introduction

Since the finding of the effective diffusion barrier layers such as TaN, Cu electrical wiring has become the main stream for the interconnection in ULSI, instead of Al wiring deposited by the physical vapor deposition (PVD). Cu wiring is performed by well-known "damascene process [24]" which is the technique for filling trenches or vias patterned in insulating layers with Cu metal by electrodeposition. Because Cu is deposited by electrodeposition, this finding opened the research field to a large number of electrochemists. Inclusion of some additives in the plating bath enables Cu filling from the bottom of the patterns without voids or seams which may deteriorate the quality of the electrical and mechanical performance of the ULSI chip. This bottom-up filling phenomenon, the so-called superfilling, is one of the most important key factors to realize the damascene process. Conductive seed layers needed for electrodeposition are normally formed by dry processes such as sputtering. This is one drawback in productivity, because equipment of dry processes operated in a vacuum chamber is generally expensive. On the other hand, wet processes have the advantage of simplicity of the facility for metal deposition and the mass production. Therefore, wiring technique without including dry processes is of great worth. Electroless plating is the most promising among wet processes, because it does not need external power sources and formation of conductive seed layers. Actually, Cu wiring by electroless plating has been practiced in the manufacturing of printed circuit boards (PCBs). However, the copper-plating solution formulated for this purpose is not applicable to the wiring for ULSI as it is, because the critical dimension of the Cu wiring for PCBs is 10 μm or more and that for ULSI is submicrometer or less. To fill patterned substrates with Cu directly, specially formulated plating solution must be explored. Normally, conformal deposition is the special feature of electroless plating. Hence, the research on electroless plating for ULSI had been devoted mainly to the formation of conformal layers over the patterned substrates [25–26]. The conformally grown layers are aiming at the applications to conductive seed layers for successive damascene electrodeposition or barrier layers against Cu diffusion into interlevel dielectrics. Recently, however, a few papers about bottom-up filling of Cu into patterned substrates by electroless plating have been published [27–29]. In this chapter, the results are introduced and compared with the general damascene process by electroplating.

2.2 Superfilling by Electrodeposition

As mentioned in the introduction, Cu wiring of ULSI is performed by damascene process. Bottom-up filling of Cu into trench-patterned substrates is achieved by putting several additives in $CuSO_4$ solution. The most common additives are chloride ions (Cl^-), polyethylene glycol (PEG), bis(3-sulfopropyl)disulfide (SPS), and Janus green B (JGB) [30–32]. Despite the successful use of the process in the semiconductor fabrication technology, the behavior of these additives during the copper electrodeposition process is still incompletely understood because of the complexity resulting from interactions between the effects of the multiple numbers of additives [30–33]. Therefore, investigations for understanding additive effects on the copper electrodeposition inside submicrometer trenches have been carried out by many researchers [31–39]. It has been reported [34, 37, 38] that bottom-up growth occurred in the bath containing Cl^-, PEG, and SPS or its monomeric derivative. Many reports have shown that the addition of PEG with Cl^- inhibits copper deposition [31–34, 40–46], whereas SPS accelerates the deposition when it is added together with PEG and Cl^- [47, 48]. It has been suggested that the competitive adsorption between the accelerator and the inhibitor results in the bottom-up effect [34, 27, 49–51]. The bottom-up effect achieves the superfilling of copper in trenches, while it also brings about the "overfill" phenomenon, in which copper bumps are formed above the copper-filled trenches [31, 34]. Moffat et al. [52, 53] and West et al. [37] have independently proposed models in which the coverage of accelerator at the bottom of trenches is assumed to increase during the filling of trenches. Their models predict not only the superfilling but also the overfill phenomenon, explaining successfully the fundamental aspects of the bottom-up growth. However, from a practical point of view, the overfill phenomenon itself is disadvantageous for the planarization step (chemical mechanical polishing, CMP), which follows the copper deposition in the damascene process. JGB is regarded to serve as a leveling agent, flattening these bumps of copper deposits on the surface [31, 32], or to influence the filling properties [30, 35]. However, a high concentration of JGB inhibits the bottom-up growth resulting in void formation. Hasegawa et al. [54] investigated the agitation effects on the overfill phenomenon in the presence of JGB. Bath agitation inhibited the overfill phenomenon significantly without producing voids in copper-filled trenches. The effect of agitation was considered to result from the enhanced transfer of JGB at the exterior of trenches. A cross-sectional SEM image of trenches filled with Cu deposited from the electroplating bath containing the four additives written above is shown in Fig. 2.

2.3 Superfilling by Electroless Deposition

There are few reports on superfilling of Cu into patterned substrates by electroless plating. As long as we know, two research works are found in this field. One is reported by Nawafune et al. [27] and the other by Shingubara et al. [28, 29].

Nawafune et al. filled trench patterns with the width of 250–1000 nm. The point of their research is that $Co(NO_3)_2$ was used as a reducing agent and pre-adsorption of additives on substrates was performed before the electrodeposition of Cu. The electroless plating solution for Cu deposition with $Co(NO_3)_2$ as a reducing agent was originally found by Vasikelis et al. [55]. The solution has advantages of avoiding

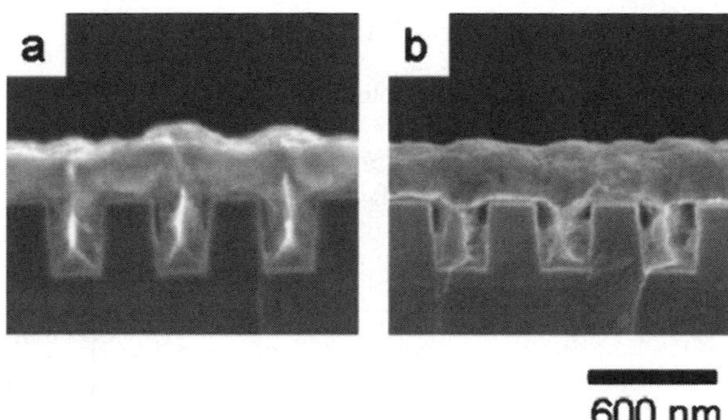

Fig. 2 Cross-sectional SEM images of copper deposited in trenches from the Cl-PEG-SPS-JGB bath with agitation of the bath. Both SPS and JGB concentrations were 2 ppm. Agitation speeds were (**a**) 0 and (**b**) 100 rpm. [Reprinted from "M. Hasegawa, Y. Negishi, T. Nakanishi, and T. Osaka, J. Electrochem. Soc., **152**, C221 (2005)". Copyright 2005, The Electrochemical Society.]

hydrogen evolution and the almost neutral bath pH. However, the solution produces conformal deposition as it is. Therefore, Nawafune et al. utilized the pre-adsorption of polyethyleneglycol (PEG) as a suppressor for Cu deposition. The pre-adsorption process is to dip a substrate into a solution containing 100 mg/L PEG and 20 mg/L Cl^- ions before dipping into an electroless plating bath. The bath composition of the electroless copper plating bath is described in Table 1.

They investigated the local anodic and cathodic polarization behavior of Cu electrode with or without the pre-adsorption process in the solutions got rid of $Cu(NO_3)_2$ and $Co(NO_3)_2$, respectively.(Fig. 3) Then, they found that the onset potential of cathodic current with pre-adsorption process was shifted toward more negative potential than that without pre-adsorption process. Pre-adsorption process did not change the onset potential of anodic current, leading to the negative shift of the

Table 1 Basic composition of electroless copper plating bath and pre-adsorption solution of additive. [Reprinted from H. Nawafune, S. Higuchi, K. Akamatsu, and E. Uchida, J. Surface Finish. Soc. Jpn., **54**, 683 (2003). Copyright 2003, The Surface Finishing Society of Japan.]

Basic bath composition	
$Cu(NO_3)_2$	0.050 mol/L
$Co(NO_3)_2$	0.15 mol/L
Ethylenediamine	0.60 mol/L
Ascorbic acid	0.010 mol/L
HCl	0.0050 mol/L
2,2′-Bipyridyl	20 mol/L
pH (adjusted with HNO_3)	6.6
Bath temperature	50 °C
Pre-adsorption solution of additive	
Polyethylene glycol 4000 (PEG)	100 mg/L
Cl^-	20 mg/L

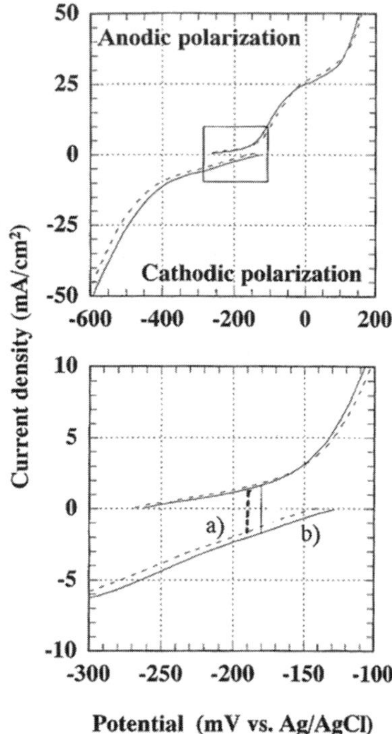

Fig. 3 Effect of pre-adsorbed PEG and chloride ion on polarization curves of copper electrode: (**a**) pre-adsorbed PEG and chloride ion, (**b**) basic bath. [Reprinted from "H. Nawafune, S. Higuchi, K. Akamatsu, and E. Uchida, J. Surface Finish. Soc. Jpn., **54**, 683 (2003)". Copyright 2003, The Surface Finishing Society of Japan.]

mixed potential with pre-adsorption. This indicates that PEG adsorbed on a Cu electrode under the presence of Cl^- ions hardly concerns the oxidation of the reducing agent, Co(II) ions, and merely inhibits the cathodic reaction, Cu deposition.

They tried to deposit Cu into contact holes formed in a SiO_2 layer. They found complete filling of Cu into contact holes with diameter of 1 μm without pre-adsorption process. However, seams and voids were observed for Cu deposited into contact holes with diameters smaller than 450 nm. Meanwhile, superfilling of Cu was achieved even in the case of contact holes with the diameter of 250 nm by electroless plating after pre-adsorption process. They suggested that PEG suppresses Cu deposition mainly at the opening of the contact holes because it is difficult for PEG to diffuse into the bottom of the holes, leading to the higher deposition rate at the bottom of the holes.

According to the results from X-ray diffraction, Cu deposited by this method showed a preferential orientation to (111) faces. Since Cu wiring which has preferential orientation to (111) faces is known to have superior resistance against electromigration to (200) orientation [56], the Cu wiring is advantageous for ULSI applications. Moreover, the specific resistance of this Cu was 1.85 μΩ cm at 20 °C, which is very close to the value of pure Cu, that is, 1.67 μΩ cm.

On the other hand, Shingubara et al. [28, 29] was successful in superfilling of Cu by electroless plating utilizing SPS as a key additive similar to the ordinary damascene process by electroplating. However, SPS acts as a suppressor for Cu deposition, and not as an accelerator in this case. They investigated the dependence of Cu thickness on SPS concentration. The samples were prepared by electroless deposition for 15 min. Cu thickness decreased with an increase in SPS concentration, and electroless deposition of Cu did not proceed when the SPS concentration in the plating bath was greater than 1.5 mg/L.

The dependence of plating time on hole-filling characteristics was investigated in the plating bath containing 0.5 mg/L SPS. They used SiO_2 substrates with holes whose diameter and depth were 0.5 μm and 2.3 μm, respectively (Fig. 4). According to the resulting cross-sectional SEM images, the extent of inhibition at the top surface was greater than that at the bottom of the holes, although Cu deposition rates both at the top surface and the bottom of the holes were inhibited by addition of SPS. At a plating time of 4 min, Cu thickness at the top surface (T_s) and at the bottom of the holes (T_b) was approximately 32 nm and 129 nm, respectively, while 87 nm of T_s and 709 nm of T_b were obtained for 10 min of plating. This indicates that the Cu deposition rate at the bottom of the holes was higher than the rate on the surface. The holes were almost filled with Cu at a plating time of 20 min. This suggests that bottom-up filling of Cu could be obtained by addition of SPS into the

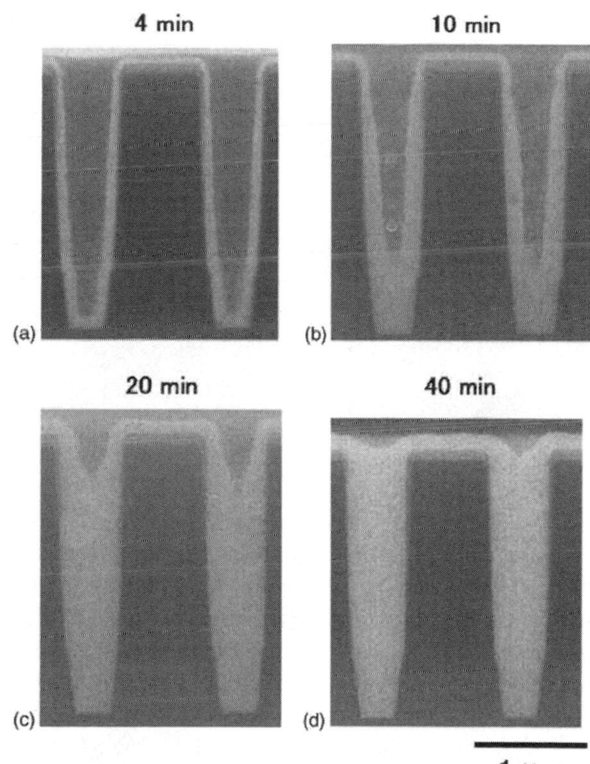

Fig. 4 Cross-sectional SEM images of holes with plating time. Hole diameter, 0.5 mm; hole depth, 2.3 mm; SPS concentration, 0.5 mg/L. [Reprinted from "Z. Wang, O. Yaegashi, H. Sakaue, T. Takahagi, S. Shingubara, J. Electrochem. Soc., **151**, C781 (2004)". Copyright 2004, The Electrochemical Society.]

Fig. 5 Dependence of bottom-up ratio: Tb /Ts, of electroless-plated Cu on hole diameter. SPS concentration, 0.5 mg/L; plating time, 10 min. [Reprinted from "Z. Wang, O. Yaegashi, H. Sakaue, T. Takahagi, S. Shingubara, J. Electrochem. Soc., **151**, C781 (2004)." Copyright 2004, The Electrochemical Society.]

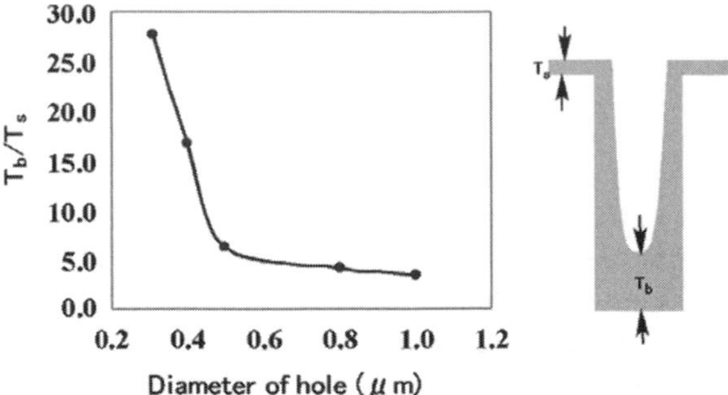

bath for electroless plating. They also investigated the effects of the hole diameter on the bottom-up ratio (T_b/T_s) and indicated that the Cu deposition rate increased with a reduction in the hole diameter (Fig. 5). The bottom-up ratio increased slowly with a decrease in the hole diameter from 1.0 to 0.5 μm. However, the bottom-up ratio increased rapidly with a decrease in the hole diameter from 0.5 to 0.31 μm.

They proposed that bottom-up fill of Cu during electroless plating was achieved through the following mechanism. SPS adsorbing on the Cu film surface decreased the active area for self-catalytic deposition of Cu. Although the deposition rates both at the top surface of the substrate and the bottom of the holes are the same initially, the rate at the bottom is gradually decreasing because of the consumption of SPS. SPS is incorporated in the Cu film during electroless deposition. Because it is difficult for SPS to diffuse to the bottom of the holes, difference in concentration of SPS at the bottom of the holes and the surface of the substrate becomes large during electroless plating. Then, bottom-up filling of Cu is induced.

We duplicated the results of Shingubara's group. However, the deposition rate was too slow to be in practical use, because SPS works as a suppressor. Therefore, we propose a new system to achieve superfilling using an additive which acts as an accelerator in Cu electroless deposition [57].

2.4 Summary

In this section, Cu filling into patterned substrates, which is important for ULSI technology, by electrochemical methods was introduced. Damascene process is achieved by electrodeposition of Cu. Success of bottom-up filling of Cu is due to the selection of appropriate additives. Recently, this bottom-up filling of Cu could be achieved by electroless plating. Two examples of that are introduced in this chapter. Both of them achieved bottom-up filling of Cu by electroless plating by selecting suitable additives which induce the difference in the deposition rates at the bottom of the holes and at the top surface of the substrate. Although the deposition rate of Cu in electroless plating is much slower than that in electroplating, electroless plating is the more promising method. The improvement of the deposition rate can be achieved by selecting appropriate additives and bath conditions.

3 A Novel Process for Fabrication of ULSI Interconnects (with Contributions from M. Yoshino)

3.1 Introduction

In the field of ULSI technology, Cu has been used instead of Al for the interconnection material because of its lower electrical resistivity and higher electromigration resistance. The Cu interconnect technology has been developed through the implementation of "damascene process" by IBM [24]. The damascene process shows the usefulness of wet processes, including electrochemical deposition (plating), in ULSI technology. Although only Cu wiring structures are filled currently by a wet process in practical ULSI manufacturing, it is expected that several steps in conventional dry processes will also be replaced by wet processes.

For a successful achievement of Cu interconnections, it is essential to fabricate an effective barrier layer as well as a capping layer because copper diffuses readily into the interlevel dielectric such as SiO_2, and also because it is oxidized easily, both of which cause the degradation of electrical performance. Even though several materials such as TaN and TiN are employed in practice, these materials are deposited by dry process. With an increase in the degree of integration, difficulties have been experienced in depositing thin films uniformly over fine-patterned structures with high aspect ratios because of the intrinsic limitation of the sputtering technique itself. Recently, a process of forming a barrier layer by electroless deposition has been proposed [58–61]. The electroless deposition has several advantages compared with the sputtering method, for example, the simplicity of operation and the low cost. In particular, its potential to produce uniform deposits independently of the size and geometry of the structure is attractive to overcome the above-mentioned problem of coverage, normally accompanying the sputtering method. However, the electroless deposition reaction requires a seed layer to be deposited by sputtering to initiate its reaction [58–62], resulting in partial negation of the advantage of the electroless deposition method.

We have proposed a new concept for the Cu wiring technology. In this proposal, the capping and the barrier layers are both fabricated by electroless deposition, and particularly, we proposed a novel process for fabricating a barrier layer without sputtered seed layer by using a self-assembled monolayer (SAM). The schematic illustration of our proposal is shown in Fig. 6. The SAM followed by treatment in a Pd solution were utilized as the adhesion promoter/catalyzed layer for the electroless deposition on SiO_2 substrate [62–64], The SAM formation of 3-aminopropyltrimethoxysilane (APS) on SiO_2 is schematically illustrated in Fig. 7. APS molecules are bonded to the SiO_2 surface by silane-coupling reaction, leading to the formation of SAM. Amino groups exposed at the top surface of the SAM trap

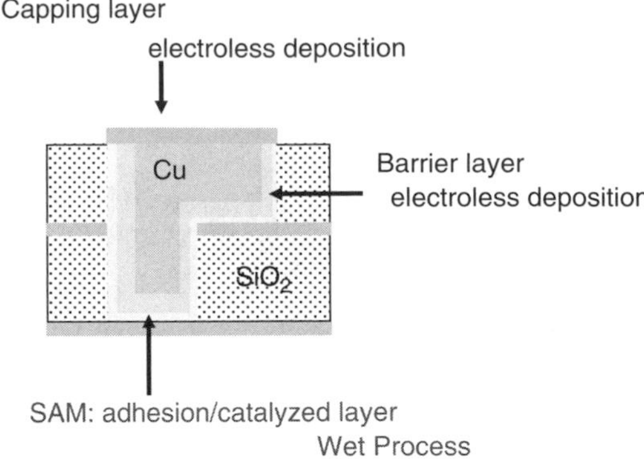

Fig. 6 Schematic illustration of all-wet fabrication process for ULSI interconnects technologies

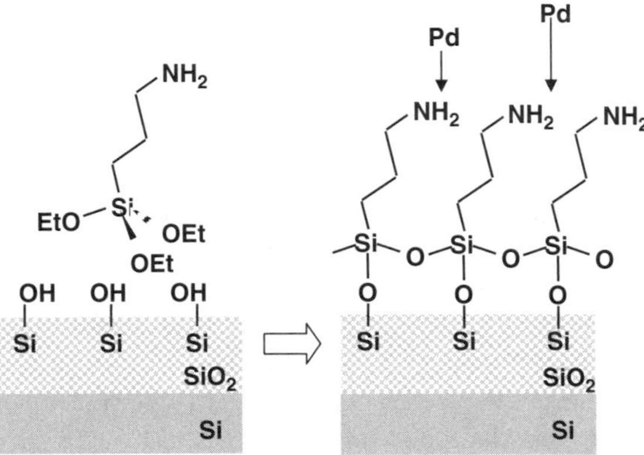

Fig. 7 Schematic illustration of SAM formation and catalyzation processes. [Quoted from "T. Osaka, Chem. Rec., **4**, 346 (2004)".]

Pd catalysts needed for activating electroless deposition. By using this process, an all-wet barrier process can be realized with electroless plated films.

In the following sections, the results obtained with various electroless Ni-alloy films deposited on Cu will be described first from the viewpoint of application as a barrier layer and a metal cap in the interconnect technology [62, 63, 65]. Then, the use of organosilane SAM instead of sputtered seeds as an adhesion and catalyst layer for the subsequent electroless deposition of a Ni alloy barrier layer will be described. The procedure used for the formation of SAM was similar to that described by Calvert, et al. [66–71].

3.2 Thermal Stability of Electroless Ni-Alloy Diffusion Barrier Layer

For the evaluation of thermal stability of the electroless Ni-alloy films for preventing Cu diffusion, the effect of annealing on sheet resistance was measured (Fig. 8(a)).

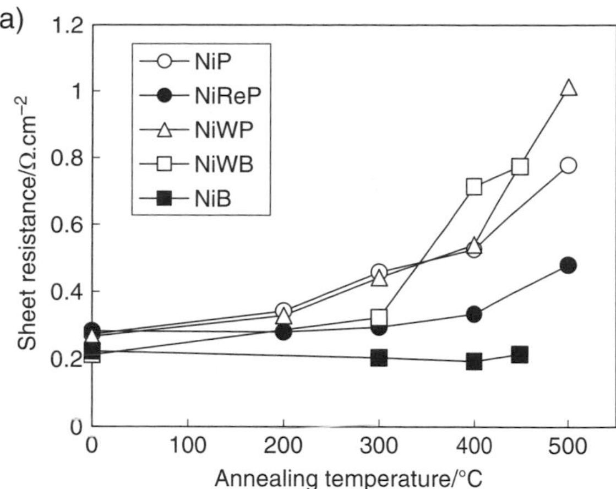

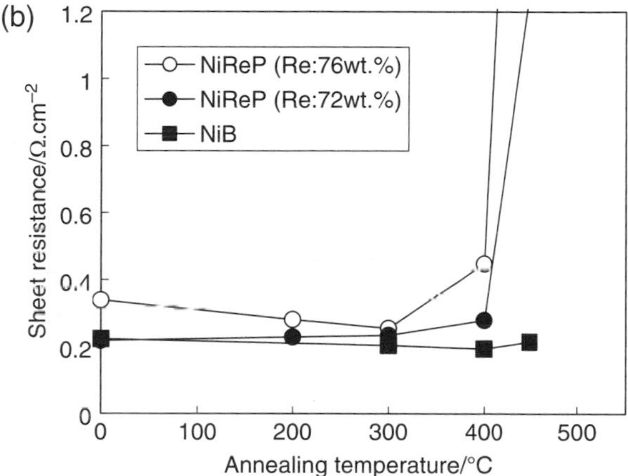

Fig. 8 (**a**) Variation of sheet resistance with temperature of Cu deposited on NiP, NiReP NiWP, NiWB, and NiB films formed on Cu substrate. (**b**) Variation of sheet resistance with temperature of Cu (100 nm) on NiReP and NiB films, formed on SAM/SiO_2/Si substrate. [Quoted from "T. Osaka, Chem. Rec., 4, 346 (2004)".]

Details of the dependence of film composition on the bath composition and plating conditions are described in the literature [62, 63, 65]. The sheet resistance of each film tended to increase with annealing temperature. The large increase in sheet resistance observed with the NiP film at 300 °C suggests that interdiffusion occurred at this temperature. On the other hand, the sheet resistances of all NiReP films were stable up to 400 °C irrespective of the Re content of the film. The XRD pattern of NiReP contained small peaks attributable to Ni_3P phase, and their intensity did not change upon heating up to 450 °C. From these results, it is apparent that the electroless NiReP films have a sufficiently high thermal stability against interdiffusion. Although we also investigated NiWP films [63], further improvement of thermal stability was difficult to achieve perhaps because only a small amount of W (∼5 wt% at most) could be incorporated regardless of plating conditions. On the other hand, the sheet resistance of the NiWB film was found to begin to increase at 300 °C, while that of the NiB film was stable up to 450 °C, suggesting that the electroless NiB film has an excellent thermal stability in the interfacial region of Cu/barrier layer. The XRD patterns of NiB and NiWB showed that annealing took place at different temperatures for the two different films [65]. Before annealing, only the peaks attributable to the substrate were seen for both films. At elevated temperatures, the peaks corresponding to Ni_3B phase appeared and became stronger in intensity for the NiB film, whereas the diffraction pattern of NiWB film hardly changed, showing an amorphous-like structure. The NiB film with an excellent thermal stability showed an evidence of crystal growth caused by annealing, while the NiWB film showing thermal degradation of the barrier property retained the amorphous-like structure even after annealing.

3.3 Fabrication of Barrier Layer on SiO_2 Substrate Without Sputtered Seed Layer

As mentioned in the introductory section, exclusion of sputtered seed layer is considered to be highly advantageous for the future ULSI technology. In general, however, a seed layer formed by sputtering is essential to initiate the electroless deposition reaction of a barrier layer on the interlevel dielectric, that is, SiO_2. Therefore, the fabrication of electroless NiP films on SiO_2 without sputtered seed layer was first investigated to obtain fundamental information on the formation of barrier layers, and then this method was applied to other Ni-alloys.

SAMs of organosilane with a functional group such as amino and pyridyl groups were utilized as the adhesion promoter/catalyzed layer, because these SAMs are known to be highly effective for electroless metallization of SiO_2 surface [68]. After optimization to shorten the treatment time, we decided to use a toluene solution of APS [63]. The SAM treatment was found to yield a uniform and glossy NiP film on the surface, whereas on SiO_2 without SAM, only a small area was coated with NiP deposit. This deposit seemed to have formed on a few catalytic Pd particles which remained barely attached on the surface.

For the formation of NiReP films on the SAM–SiO_2 substrate, a strike plating in the acid NiP bath must be performed before immersion in the alkaline NiReP bath [62, 63]. The predeposited NiP nuclei protect the SAM from possible damage triggered by the alkaline solution [71]. By adopting this two-step process, nearly perfect and uniform NiReP films with a bright appearance could be formed. As shown in Fig. 8(b), the NiReP films formed on SAM–SiO_2 was confirmed to be stable up to 400°C, which was also proved by AES analysis. Although the sudden rise of sheet resistance at 500 °C is suggestive of a lower stability of the NiReP barrier layer on SAM–SiO_2 than that on Cu/Ta/SiO_2/Si, the thermal stability up to 400 °C is sufficient for practical use.

As shown in the preceding section, the electroless NiB film is another superior candidate for the barrier material. Electroless NiB film was deposited on the

SAM–SiO_2 substrate. Continuous films having smooth surfaces with R_a of 1.842 nm were obtained. The result of peel test with a tape showed an excellent adhesion of the NiB film with a thickness of 20–60 nm. This film has also a good thermal stability as shown in Fig. 8(b). Compared to the other Ni-alloys, the rate of increase in sheet resistance of this film was very low. This suggests the viability of the NiB layer as a diffusion barrier for the Cu wiring.

3.4 All-Wet Fabrication of Cu Wiring

The NiB barrier layer was formed on a SiO_2–Si substrate with submicron trench patterns. The width and the aspect ratio of the trench were 130 nm and 2.5 nm, respectively. The surface of the substrate was modified by Pd-activated SAM before the electroless deposition of the NiB layer. Figure 9(a) and (b) show cross-sectional SEM images of the NiB layer deposited on a patterned substrate. It is seen that the shape of the deposited NiB layer is highly conformal. The thickness of the NiB layer increased with deposition time. The adhesion of the film was as good as that of NiB on a nonpatterned substrate.

Fig. 9 Cross-sectional SEM images of patterned substrate coated with NiB film (**a, b**). The substrate surface was modified by Pd-activated SAM. Trench pattern after Cu electrodeposition on NiB (40 nm)/SAM/SiO_2/Si substrate(c). [Quoted from "T. Osaka, Chem. Rec., **4**, 346 (2004)".]

Cu filling in the trench pattern by electroplating was performed without sputtered seed layer. A conventional electroplating bath for the damascene process was used in this experiment. Figure 9(c) shows a cross-sectional SEM image of the specimen after Cu electroplating. Cu is filled in the trenches with no voids or seams, demonstrating the success of the "all-wet" Cu wiring process. Our ongoing investigation of fundamental aspects of Cu superfilling [54] indicates that this technique can be made even more sophisticated. We believe that our proposal is a promising technology in advanced ULSI applications. Such a technique is now successfully applied to the low *k* (*k*: dielectric constant) materials [72] and organic materials such as polyimides [73].

4 Magnetic Nanodot Arrays for Patterned Media (with Contributions from J. Kawaji)

4.1 Introduction

A patterned magnetic recording medium, in which minute magnetic dot arrays are aligned periodically, is expected to realize a higher-density magnetic recording than a conventional thin film media (or called "continuous media") [74]. Figure 10 shows the structure of patterned media together with that of thin film media [75]. In patterned media, information was recorded as the direction of magnetization of each dot array, that is, one dot functions as one bit. On the other hand, one bit in thin film media consists of a large number of magnetic grains; it was described that about 100 grains should be present in one bit in order to obtain a sufficient signal-to-noise ratio [75]. Preparation of magnetic nanodot arrays with perpendicular anisotropy has been extensively studied with various processes. Physical processes, such as sputter deposition of magnetic film and the patterning of the deposited film by reactive ion etching or focused ion beam lithography [76, 77], or sputter deposition of magnetic film on prepatterned substrate [78], have been conventionally used. On the other hand, chemical process attracts much attention as an alternative candidate for preparation method of magnetic dot arrays featuring its mass productivity and the ability to form metallic fine structures with high aspect ratio, excellent area selectivity, and uniformity. For example, magnetic dot arrays such as fct-CoPt and FePt

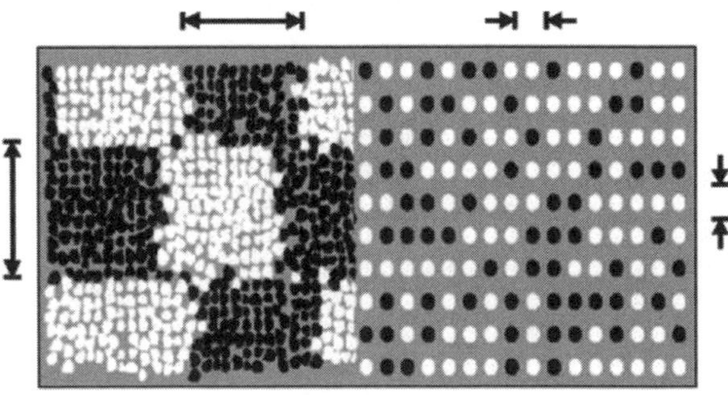

Fig. 10 Comparison of thin film medium (continuous medium, 100 crystal per bit) and patterned medium (one dot = one bit). [Quoted from "J. C. Lodder, J. Magn. Magn. Mater., **272–276**, 1692 (2004)".]

are formed by filling the magnetic metals into patterned substrate with nanoscopic pores by using electrodeposition [79, 80].

In this section, electroless deposition process was focused on as a fabrication process of dot arrays. Nonmagnetic NiP dot arrays were reported to be successfully formed on the patterned Si substrate with SiO_2 resist by utilizing the difference in a reactivity of the electroless deposition between the surfaces of Si and SiO_2 [81–84]. On the other hand, CoNi-alloy magnetic thin films with perpendicular anisotropy, such as CoNiMnP [85], CoNiReMnP [86], CoNiReP [87], and CoNiP [88–90], were prepared by electroless deposition. In this section, the study on fabrication of CoNiP magnetic dot arrays on patterned Si wafers is overviewed.

4.2 Fabrication of Magnetic Nanodot Arrays on Si Wafer

Figure 11 shows illustration of the typical process steps for the fabrication of patterned substrate and the formation of CoNiP dot arrays. A Si (100) wafer covered with a SiO_2 was used as a substrate. The wafer was coated with an electron beam (EB) resist, and then patterning was carried out with EB irradiation. The resist pattern was transferred to the SiO_2 layer by reactive ion etching. The height and diameter of the patterned pore was expressed as h_{pore} and d_{pore}. After being cleaned properly [91], the specimen was immersed into CoNiP electroless bath [88]. By using this bath, CoNiP continuous films whose *c*-axis is preferably orientated in the direction perpendicular to the film surface are formed on Si wafers. By only immersing the specimen into the bath without activation process, CoNiP alloy was deposited into the patterned pores. Also, no deposits were observed onto the SiO_2 layer, demonstrating a high area selectivity of the CoNiP deposition. (Fig. 12) On the contrary, it was observed that, when the d_{pore} decreased, the deposition did not occur, which indicated that the decrease in the diameter and the increase in aspect ratio of patterned pores suppressed the activity of the Si surface for the deposition reaction. Figure 13(a) shows cross-sectional SEM image of patterned pores with 100 nm in diameter, where the CoNiP deposition was not seen. Inversed-pyramid-shaped Si structure is observed, which is thought to be <111> oriented facet formed by an anisotropic etching of Si during the immersion in an alkaline CoNiP bath (pH ~9.5). The etching reaction of Si seems to proceed more rapidly than the

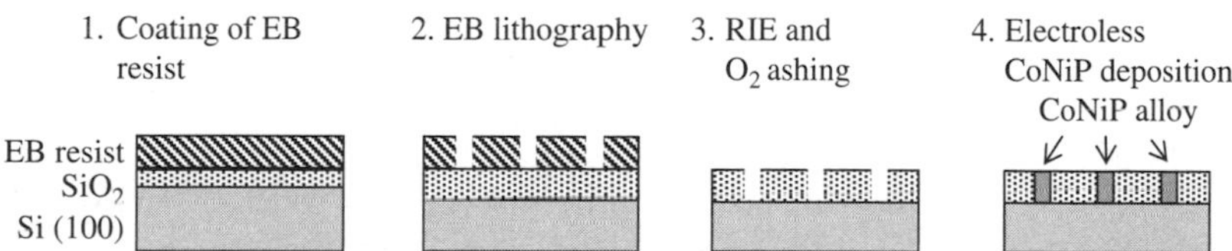

Fig. 11 Schematic illustration of process steps for the fabrication of electroless CoNiP dot arrays. [Quoted from "J. Kawaji, F. Kitaizumi, H. Oikawa, D. Niwa, T. Homma, T. Osaka, J. Magn. Magn. Mater., **287**, 245 (2005)."]

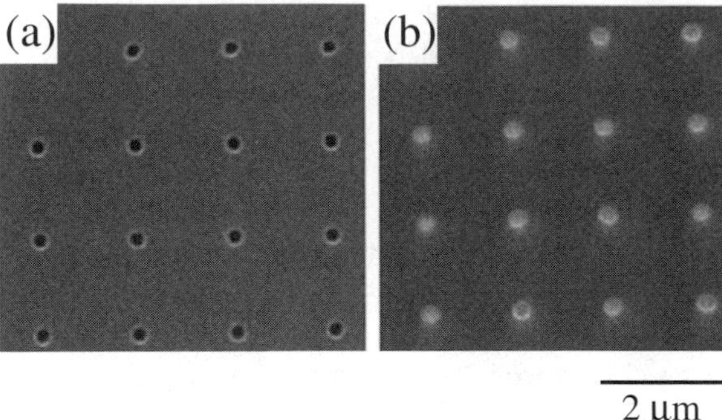

Fig. 12 SEM images of the patterned surface before (**a**) and after (**b**) the CoNiP deposition. The height and diameter of dot patterns were 500 nm and 200 nm, respectively. [Quoted from "J. Kawaji, F. Kitaizumi, H. Oikawa, D. Niwa, T. Homma, T. Osaka, J. Magn. Magn. Mater., **287**, 245 (2005)."]

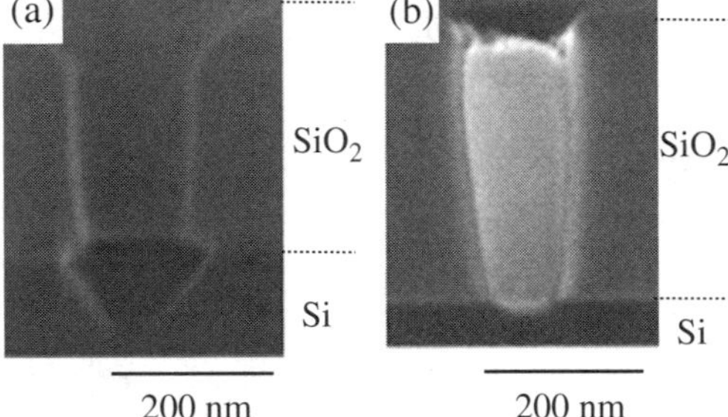

Fig. 13 Cross-sectional SEM images of the patterned surface after the immersion into CoNiP bath without any activation processes(a) and with commercial Pd activation solution (OPC-AC, Okuno Chemical Industries Co. Ltd.). [Quoted from "J. Kawaji, F. Kitaizumi, H. Oikawa, D. Niwa, T. Homma, T. Osaka, J. Magn. Magn. Mater., 287, 245 (2005)."]

deposition reaction when the smaller patterns were used. Therefore, the surface should be activated more effectively for the deposition reaction. Pd is well-known catalyst to initiate the electroless deposition reaction, and by using commercial Pd catalyst solution, CoNiP alloy was deposited into the patterned pores with d_{pore} of 100 nm without significant chemical etching of Si substrate (Fig. 13(b)). The Pd complex in the solution was expected to preferably adsorb to Si surface, rather than to the SiO_2 surface, and the electrochemical substitution of Pd was also believed to take place on Si surface only. In addition, any voids caused by the influence of the hydrogen gas generated by an oxidation reaction of hypophosphite ion as a reducing agent were not observed.

4.3 Magnetic Properties of CoNiP Nanodot Arrays

Figure 14(a) and (b) show the MFM images of the CoNiP continuous thin film and dot arrays (h_{pore}, d_{pore} = 450 nm and 270 nm) at dc-magnetized state. In Fig. 14(b), the CoNiP dot arrays show a clear magnetization state in contrast to that of the CoNiP film where numerous reversed domains were seen (Fig. 14(a)). Such a difference originated from the effect of shape anisotropy of dot arrays with high aspect

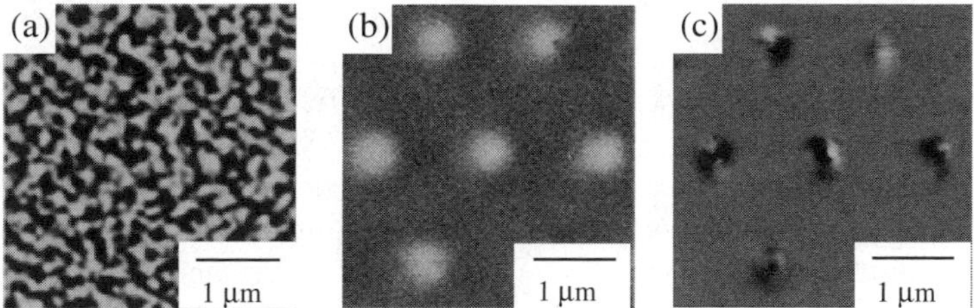

Fig. 14 MFM images of CoNiP continuous film at dc-magnetization state (**a**) and CoNiP dot arrays with 270 nm in diameter at dc-magnetization state (**b**) and ac-demagnetization state (**c**). [Quoted from "J. Kawaji, F. Kitaizumi, H. Oikawa, D. Niwa, T. Homma, T. Osaka, J. Magn. Magn. Mater., **287**, 245 (2005)."]

Table 2 Magnetic properties of CoNiP continuous film and dot arrays

Shape of CoNiP deposits	Perpendicular coercivity [Oe]	Perpendicular squareness ratio [–]
CoNiP continuous film	800	0.15
Dot arrays (200 nm diameter, 500 nm height)	440	0.44

ratio (in this case, *ca.* 1.7). Table 2 shows the magnetic properties of the CoNiP dot arrays on substrate A with an aspect ratio of 2.5 (h_{pore}, d_{pore} = 500 nm, 200 nm) measured with a vibrating sample magnetometer. The squareness ratio was high compared with the CoNiP continuous film, which could be also due to the shape anisotropy of the dot arrays. Figure 14(c) shows MFM image for the CoNiP arrays at ac-demagnetized state. This image indicates that the dot arrays have multidomain structure, where each magnetic domain was about 100 nm in diameter. For realizing magnetic dot array with single-domain structure, the decrease in dot-pattern diameter and enhancement of magnetic properties of the deposits both are essential, which is desirable for achieving extremely higher read–write performance of the patterned media.

In summary, electroless CoNiP deposition demonstrated sufficient area selectivity on the SiO_2/Si patterned substrate, and the control of the activation condition of the Si surface at the bottom of patterned pores led to successful deposition of CoNiP alloy into the pores with smaller than 100 nm diameter and aspect ratio higher than 5. Moreover, the perpendicular squareness ratio was improved by forming the CoNiP deposits with a high aspect ratio. By developing the deposition condition to satisfy the two requirements, the control of the surface activity for metal deposition and the enhancement of magnetic anisotropy of the deposits, the electroless deposition could be a promising candidate for the fabrication method of patterned magnetic recording media.

5 Nanoparticles

5.1 Oxide Nanoparticles (*with Contributions from T. Nakanishi*)

Oxide nanoparticles have received much attention, as well as metal nanoparticles, because of their applications in various fields depending on their optical or magnetic properties, and thus, a lot of methods for synthesis of oxide nanoparticles in solution phase have been developed and reported. In particular, a synthetic method using a water-in-oil microemulsion including a reverse micelle system has been used widely [92]. As for the nanoparticles of iron oxides, such as maghemite (γ-Fe_2O_3) and magnetite (Fe_3O_4), which are the representative magnetic materials, the precipitation technique, or alkalization of metal salts solution and hydrolysis in microemulsions [93–96], are commonly used. The synthesis of iron oxide nanoparticles has been also achieved by sonochemical decomposition of iron pentacarbonyl [97], thermal decomposition of iron complexes [98, 99], and thermal decomposition of iron pentacarbonyl followed by oxidation [100]. Especially, the synthetic method of high crystalline maghemite nanoparticles by mild chemical oxidation of iron nanoparticles, which were generated from the thermal decomposition of iron–surfactant complex [100], is a remarkable method for the synthesis of oxide nanoparticles. The iron–surfactant complex was prepared by the thermal decomposition of iron pentacarbonyl in the presence of surfactant. The intermediate formation of metallic state is considered to be one of the keys for the high crystallinity of nanocrystalline oxide [100, 101a], and this procedure has been applied to the syntheses of cobalt ferrite ($CoFe_2O_4$) [101b] and manganese ferrite ($MnFe_2O_4$) [101c] nanocrystals.

The formation of intermediate metallic state, or assembled state of iron atoms, should be realized not only by the thermalysis of organometallic compounds such as iron–surfactant complex but also by a chemical reduction of metal ions, and the following oxidation is expected to result in the formation of iron oxide nanoparticles (Fig. 15). The deposition of metal on initial nuclei from a metal ion with a reducing agent is considered to be exactly an electroless deposition process. The knowledge of electroless plating or electroless metal deposition appears very useful

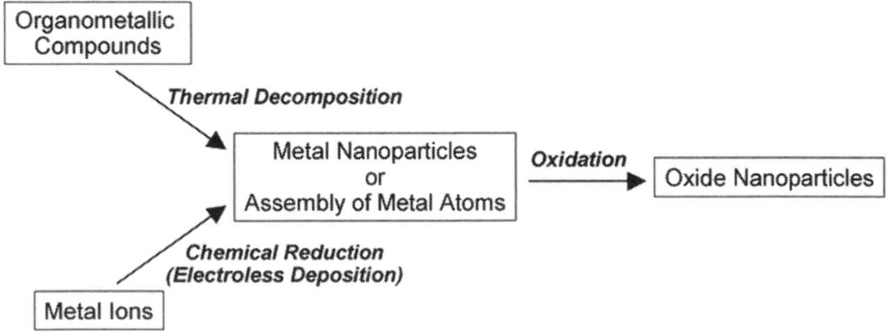

Fig. 15 Synthetic procedure for oxide nanoparticles via intermediate metallic state

to control the formation of metal nanoparticle as a precursor of oxide nanoparticle, even though spontaneous formation of particles in elecroless plating solutions brings about the decomposition of plating solutions. To control the growth of particles in solutions, the use of reverse micelle system is effective.

The synthesis of iron oxide nanoparticles via successive reduction–oxidation in the reverse micelles, in which metal oxides are formed by the reduction of metal ions with a reducing agent followed by air oxidation (Fig. 16), has been achieved [102]. Two reverse micellar solutions consisting of heptane, sodium bis(2-ethylhexyl)sulfosuccinate (Aerosol OT; AOT), and an aqueous solution of $FeSO_4$ or $NaBH_4$ were prepared. After the addition of $NaBH_4$-containing solutions to the $FeSO_4$-containing solution, yellowish solution turned light brown, via dark-blue intermediate state. This is suggestive of the formation of iron oxide within the nanometer-sized reverse micelles accompanied by the reduction of Fe^{2+} or Fe^{3+} ions, followed by the immediate oxidation due to the frequent contacts with oxygen in the air. X-ray diffraction pattern of the dark-brown powder collected from the light-brown micellar solution with the addition of 1,6-hexanediamine (Fig. 17) indicated formation of γ-Fe_2O_3 with relatively high crystallinity, and the mean diameter of obtained γ-Fe_2O_3 nanoparticles was estimated as $\sim$4 nm by using Scherrer formula for the (311) reflection peak observed at 2θ of 36°. From the temperature dependence of the magnetization measured with a SQUID magnetometer under zero-field cooling (ZFC) and field cooling (FC) conditions, superparamagnetic behavior, which is characteristic of magnetic nanoparticles, was observed (Fig. 18). The sharp maximum of the ZFC curve at T_{max} of $\sim$25 K and the splitting between ZFC and FC curves just above T_{max} indicate a narrow particle size distribution.

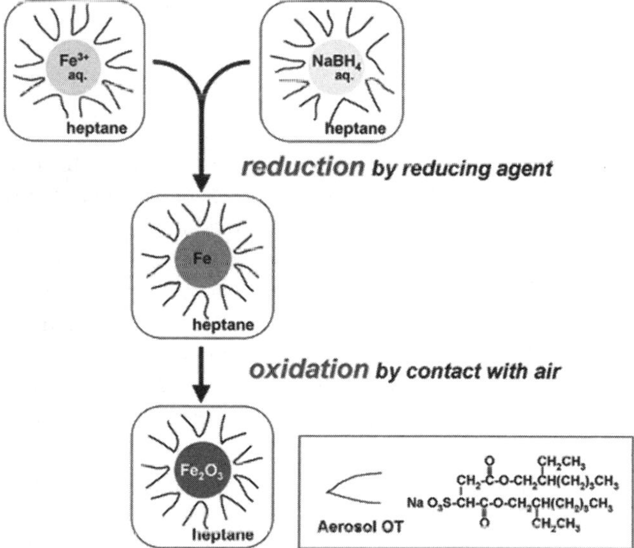

Fig. 16 Schematic illustration of the formation of iron oxide nanoparticles via successive reduction–oxidation in reverse micelles

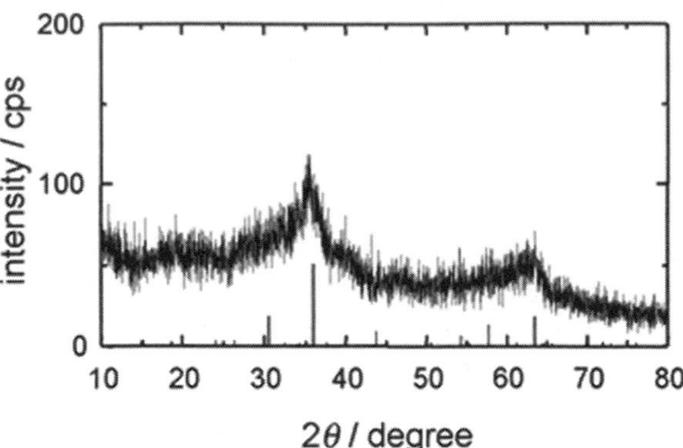

Fig. 17 XRD pattern of the sample collected with hexanediamine. Standard pattern for γ-Fe_2O_3 (JCPDS#39-1346) is also shown. [Quoted from "T. Nakanishi, H. Iida, T. Osaka, Chem. Lett., **32**, 1167 (2003).]

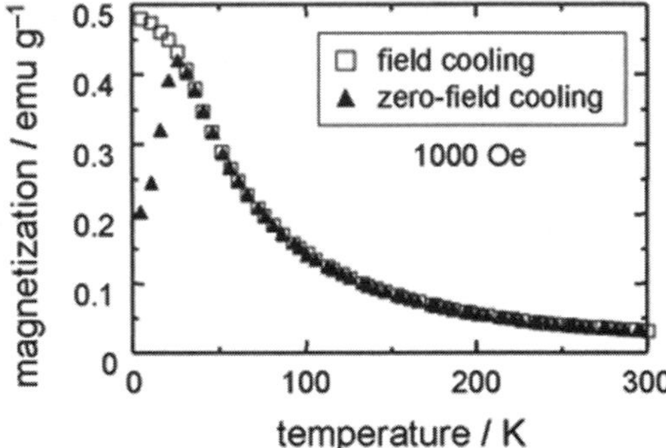

Fig. 18 Temperature dependence of the magnetization curves under ZFC (*solid triangles*) and FC (*open squares*) for the sample collected with hexanediamine. [Quoted from "T. Nakanishi, H. Iida, T. Osaka, Chem. Lett., **32**, 1167 (2003).]

In the successive reduction–oxidation in the reverse micelles, the reducing agents such as N_2H_4 and NaH_2PO_2 were revealed to affect on the final oxidized products differently from $NaBH_4$, suggesting that an intermediate metallic state controls the formation of oxide nanoparticles [103]. Two electrochemical factors, that is, the oxidation potential of the reducing agent being more negative than the redox potential of the metal to be deposited and the catalytic activity of the depositing metal for the anodic oxidation of the reducing agent are essential for electroless deposition [104]. As for the case of chemical reduction in reverse micelles, they are also the key to realize successful metallization, and the formation of highly crystalline γ-Fe_2O_3 nanoparticles was considered to result from air oxidation of well-reduced and properly ordered iron, or assembled state of iron atoms, obtained by reduction with $NaBH_4$ [102, 103]. It is suggested that the knowledge of electroless plating is useful to control the formation of metal nanoparticle or assembled state of atoms as a precursor of oxide nanoparticles with an optimum combinations of metal salt and reducing agent.

5.2 *Metallic Mesoporous Particles* (*with Contributions from T. Momma*)

Starting the first discovery of mesoporous silica from a polysilicate kanemite [105], mesoporous materials, which have ordered structure of aligned pores with the diameter in meso-scale (a few to 50 nm), attract many researchers for their potential of applications with its promising feature of high surface area, ordered porous structure, and so on [106]. In 1995, Attard, et al. proposed the novel synthetic method of mesoporous silica by "*Direct Physical Casting*" using lyotropic liquid crystal (LLC) formed by highly concentrated nonionic surfactants solution [107]. There are many advantages of the "*Direct Physical Casting*". This method has the ability to predict the meso-structures of the final products by those of the initial mesophases of LLC. Moreover, the range of materials which can be produced is beyond silica to other inorganic oxides, metals, semiconductors, and polymers.

Mesoporous metals are also fascinating objects from the viewpoint of applications, such as electronic devices, magnetic recording media, metal catalysts, photonic devices, etc., due to their expected electric conductivity, catalytic activity, etc. In 1997, G. S. Attard, et al. proposed one-pot process of mesoporous metal powder by the reduction of metallic ions in the presence of LLC made of nonionic surfactants at high concentrations [108], with expanding their technique on preparation of mesoporous silica using LLC. Since then, many types of mesoporous metals with various compositions have been reported up to date, while most of them were prepared by electroplating in the LLC media [109]. Although electrodeposition by an external power source is widely known as a general deposition method of metals, by its nature of electrodeposition, the products are the films adhered on conducting substrates. As for the variety of reducers, combination of noble metal ions and base metals was proposed. With forming the local cell, mesoporous Pt or Pt-Ru powders were reported from the LLC containing metallic ions with addition of Zn powder [110]. With this method, metallic spherical powders with widely distributed diameter of 90–500 nm were reported. Also, preparation of mesoporous Pt layer without outer electric source was proposed by using "contact plating", which utilize the formed local cell with hexachloro platinum and Al metal [111]. With changing the area of Al contacting with LLC, the ordering structure of resulting mesoporous Pt was influenced with the change in the deposition rate.

In order to extend the possible applications of mesoporous metals synthesized by wet processing, chemical reduction method and electroless deposition of mesoporous particles are also proposed. By using chemical reduction and/or electroless plating, particles of mesoporous metal and mesoporous metal layers formed on insulating surfaces, can be prepared, in addition to some unique features of electrolessly plated metals. Preparation of porous Pt powder was reported to be synthesized from the LLC media with the hexachloro platinum and coexisting reactive reducer, such as sodium borohydride (SBH), in the water phase of LLC [112]. With the self-decomposition of SBH in water phase, spherical Pt powder was grown with incorporating the rod-shaped micelles forming LLC. After washing out the micelles, mesoporous Pt powder was obtained. The alloy of Pt and Ru was also realized by the introduction of ruthenium trichloride in the water phase of synthetic LLC [113].

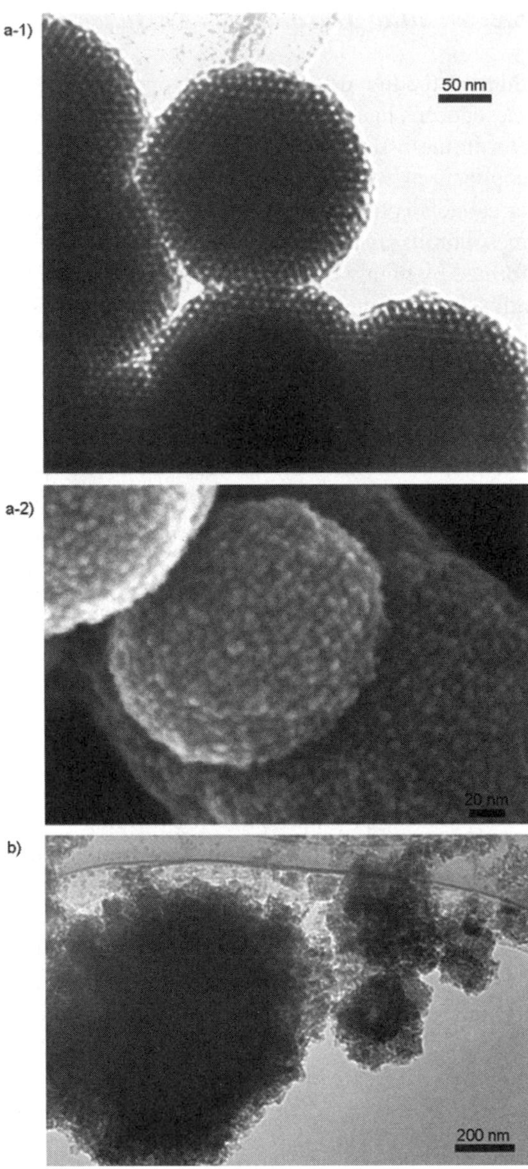

Fig. 19 TEM and HR-SEM images of (**a**) the Ni particles and (**b**) the Co particles prepared by using both SBH and DMAB. a-2): HR-SEM image. [Quoted from "Y. Yamauchi, T. Momma, T. Yokoshima, K. Kuroda, and T. Osaka, J. Mater. Chem., **15**, 1987–1994 (2005)."]

The quality of ordering structure or mesopores in the mesoporous metallic powder was demonstrated to be strongly influenced by the system, which was applied to synthesize the mesoporous materials. Using the electroless plating, the combination of metallic ions, which are the source of product; reducer, which is the electron source and the operation conditions are dominating the reaction mechanism, and rate of metallic phase formation. Mesoporous structure is achieved by the growth of metallic phase with incorporation of the rod-shaped micelles formed by concentrated surfactant and existing in the LLC medium. Generation of fine metallic nuclei in LLC medium results in lower ordering structure, while the growth of metal thrusting the micelles aside would result in nonporous structure. Sophisticated combination of reducers was reported to obtain highly ordered mesoporous Ni powder [114] as shown in Fig. 19. Dimethylamineborane (DMAB) is well known as a reducer for Ni electroless plating. DMAB is oxidized at the surface having catalytic activity for oxidation of DMAB, while dissolved DMAB in water is chemically stable. The selective oxidation of DMAB at the certain surface results in the controlled deposition of Ni in the electroless plating systems [115].

Mesoporous alloy particle prepared by electroless deposition was also proposed. Alloys with various compositions should possess new solid-state properties and contribute to the development of novel materials with desired functions, such as magnetic, electronic, and optical properties, etc. The fabrication of metal-based meso-structured materials with high ordering and controlled compositions is essential for the development of mesoporous metals. Nickel and cobalt, both of which are suitable for preparation of alloys by electroless deposition, were examined to prepare mesoporous alloy by electroless deposition. Ni–Co binary alloy is known as a magnetic material [116], and magnetic mesoporous Ni–Co alloy having high surface area and periodic pore arrangement can be very useful for the applications including a novel reactor. Meso-structured Ni–Co binary alloys with various compositions were prepared in the electroless deposition bath containing both metal ions (Ni^{2+} and Co^{2+}) with varied compositions and nonionic surfactants with high concentrations. Simply controlling the bath conditions varied the composition of the metal framework.

References

1. "New Trends and Approaches in Electrochemical Technology", ed. N. Masuko, T. Osaka, Y. Fukunaka, Kodansha, Tokyo (1993).
2. "Electrochemical Technology Innovation and New Developments", ed. N. Masuko, T. Osaka, Y. Ito, Kodansha, Tokyo (1996).
3. A. Brenner, G. E. Riddell, J. Res. Natl. Bur. Std., **37**, 1 (1946).
4. A. Brenner, G. E. Riddell, Proc. Am. Electroplaters Soc., **34**, 156 (1947).
5. F. A. Lowenheim, "Modern Electroplating", John Wiley & Sons Inc., New York, (1974).
6. H. Iwasa, M. Yokozawa, I. Teramoto, J. Electrochem. Soc., **115**, 485 (1986).
7. H. Cachet, M. Froment, E. Souteyrand, J. Electrochem. Soc., **139**, 2920 (1992).
8. M. V. Sullivan, J. H. Eiger, J. Electrochem. Soc., **104**, 226 (1957).
9. J. M. Calvert, G. S. Calabrese, J. F. Bohland, M. Chen, W. J. Dressick, C. S. Dulcey, J. H. George, Jr., E. K. Pavelcheck, K. W. Rhee, L. M. Shirey, J. Vac. Sci. Technol. B, **12**, 3884 (1994).
10. C. H. Ting, M. Paunovic, P. L. Pai, G. Chin, J. Electrochem. Soc., **136**, 462 (1989).
11. M. J. Desilva, Y. Shacham-Diamand, J. Electrochem. Soc., **143**, 3512 (1996).
12. V. M. Dubin, Y. Shacham-Diamand, B. Zhao, P. K. Vasudev, C. H. Ting, J. Electrochem. Soc., **144**, 898 (1996).
13. R. D. Fisher, W. H. Chilton, J. Electrochem. Soc., **109**, 485 (1962).
14. T. Osaka, H. Nagasaka, F. Goto, J. Electrochem. Soc., **128**, 1686 (1981).

15. H. Uwazumi, N. Nakajima, M. Masuda, T. Kawata, S. Takenoiri, S. Watanable, Y. Sakai, K. Enomoto, IEEE Trans. Magn., **40**, 2392 (2004).
16. T. Asahi, T. Yokoshima, J. Kawaji, T. Osaka, H. Ohta, M. Ohmori, H. Sakai, IEEE Trans. Magn., **40**, 2356 (2004).
17. T. Osaka, T. Asahi, T. Yokoshima, J. Kawaji, M. Ohmori, H. Sakai, J. Magn. Magn. Mater., **287**, 292 (2005).
18. L. T. Romankiw, U.S. Patent 3, **853**, 715 (1974).
19. L. T. Romankiw, Electrochem. Soc. Proc., **PV90-8**, 59 (1990).
20. P. C. Andricacos, K. G. Berridge, J. O. Dukovic, M. Flotta, J. Ordonez, H. R. Poweleit, J. S. Richter, L. T. Romankiw, O. P. Schick, K. H. Wong, U.S. Patent 5, **516**, 412 (1996).
21. C. Karakus, D. T. Chin, "Metal Distribution in Jet Plating", J. Electrochem. Soc. **141**, 691 (1994).
22. S. A. Amadi, D. R. Gabe, M. R. Goodenough, "Air Agitation for Electrodeposition Process II. Experimental", Trans. Inst. Metal Finish., **72**, 66 (1994).
23. T. Yokoshima, D. Kaneko, T. Osaka, S. Takefusa, M. Oshiki, J. Magn. Soc. Jpn., **23**, 1397 (1999).
24. P. C. Andricacos, C. Uzoh, J. O. Dukovic, J. Horkans, H. Deligianni, IBM J. Res. Dev., **42**, 567 (1998).
25. Y. Shacham-Diamand, A. Inberg, Y. Sverdlov, V. Bogush, N. Croitoru, H. Moscovich, A. Freeman, Electrochim. Acta, **48**, 2987 (2003).
26. V. Bogush, A. Inberg, N. Croitoru, V. Dubin, Y. Shacham-Diamand, Microelectronic Eng., **70**, 489 (2003).
27. H. Nawafune, S. Higuchi, K. Akamatsu, M. Uchida, J. Surface Finish. Soc. Jpn., **54**, 683 (2003).
28. S. Shingubara, Z. Wang, O. Yaegashi, R. Obata, H. Sakaue, T. Takahagi, Electrochem. Solid-State Lett., **7**, C78 (2004).
29. Z. Wang, O. Yaegashi, H. Sakaue, T. Takahagi, S. Shingubara, J. Electrochem. Soc., **151**, C781 (2004).
30. J. J. Kelly, A. C. West, Electrochem. Solid-State Lett., **2**, 561 (1999).
31. P. Taephaisitphongse, Y. Cao, A. C. West, J. Electrochem. Soc., **148**, C492 (2001).
32. S. Miura, K. Oyamada, Y. Takada, H. Honma, Electrochemistry (Tokyo, Jpn.), **69**, 773 (2001).
33. M. Kang, A. A. Gewirth, J. Electrochem. Soc., **150**, C426 (2003).
34. T. P. Moffat, J. E. Bonevich, W. H. Huber, A. Stanishevsky, D. R. Kelly, G. R. Stafford, D. Josell, J. Electrochem. Soc., **147**, 4524 (2000).
35. T. P. Moffat, D. Wheeler, C. Witt, D. Josell, Electrochem. Solid-State Lett., **5**, C110 (2002).
36. T. Haba, T. Itabashi, H. Akahoshi, A. Sano, K. Kobayashi, H. Miyazaki, Mater. Trans., JIM, **43**, 1593 (2002).
37. A. C. West, S. Mayer, J. Reid, Electrochem. Solid-State Lett., **4**, C50 (2001).
38. D. Josell, B. Baker, C. Witt, D. Wheeler, T. P. Moffat, J. Electrochem. Soc., **149**, C637 (2002).
39. M. Kang, M. E. Gross, A. A. Gewirth, J. Electrochem. Soc., **150**, C292 (2003).
40. M. Yokoi, S. Konishi, T. Hayashi, Denki Kagaku oyobi Kogyo Butsuri Kagaku, **52**, 218 (1984).
41. D. Stoychev, C. Tsvetanov, J. Appl. Electrochem., **26**, 741 (1996).
42. J. J. Kelly, A. C. West, J. Electrochem. Soc., **145**, 3472 (1998).
43. J. J. Kelly, A. C. West, J. Electrochem. Soc., **145**, 3477 (1998).
44. J. P. Healy, D. Pletcher, M. Goodenough, J. Electroanal. Chem. Interfacial Electrochem., **338**, 155 (1992).
45. V. D. Jovic, B. M. Jovic, J. Serb. Chem. Soc., **66**, 935 (2001).
46. Z. V. Feng, X. Li, A. A. Gewirth, J. Phys. Chem. B, **107**, 9415 (2003).
47. M. Tan, J. N. Harb, J. Electrochem. Soc., **150**, C420 (2003).
48. J. J. Kim, S.-K. Kim, Y. S. Kim, J. Electroanal. Chem., **542**, 61 (2003).

49. Y. Cao, P. Taephaisitphongse, R. Chalupa, A. C. West, J. Electrochem. Soc., **148**, C466 (2001).
50. K. Kondo, N. Yamakawa, Z. Tanaka, K. Hayashi, J. Electroanal. Chem., **559**, 137 (2003).
51. K. Kondo, T. Matsumoto, K. Watanabe, J. Electrochem. Soc., **151**, C250 (2004).
52. T. P. Moffat, D. Wheeler, W. H. Huber, D. Josell, Electrochem. Solid-StateLett., **4**, C26 (2001).
53. D. Josell, D. Wheeler, W. H. Huber, T. P. Moffat, Phys. Rev. Lett., **87**, 016102 (2001).
54. M. Hasegawa, Y. Negishi, T. Nakanishi, T. Osaka, J. Electrochem. Soc., **152**, C221 (2005).
55. A. Vaskelis, H. Norkus, G. Rozovskis, H. Vinkevicius, Trans. Inst. Metal Finishing, **75**, 1 (1997).
56. K. Ueno, J Surface Finish. Soc. Jpn., **49**, 1177 (1998).
57. M. Hasegawa, Y. Shacham-Diamand, T. Osaka, Electrochem. Solid-State Lett., **9**, C138 (2006).
58. Y. Shacham-Diamand, S. Lopatin, Microelectron. Eng., **37/38**, 77 (1997).
59. Y. Shacham-Diamand, S. Lopatin, Electrochim. Acta, **44**, 3639 (1999).
60. Y. Shacham-Diamand, Y. Sverdlow, Microelectron. Eng., **50**, 525 (2000).
61. E. J. O'Sullivan, A. G. Schrott, M. Paunovic, C. J. Sambucetti, J. R. Marino, P. J. Bailey, S. Kaja, K. W. Semkow, IBM J. Res. Develop., **42**, 607 (1998).
62. T. Osaka, N. Takano, T. Kurokawa, K. Ueno, Electrochem. Solid State Lett., **5**, C7 (2002).
63. T. Osaka, N. Takano, T. Kurokawa, T. Kaneko, K. Ueno, J. Electrochem. Soc., **149**, C573 (2002).
64. M. Yoshino, T. Yokoshima, T. Osaka, Proceedings of 204th Meeting of the Electrochemical Society, PV2003-10, The Electrochem Soc Inc: New Jersey, p. 137 (2004).
65. T. Osaka, N. Takano, T. Kurokawa, T. Kaneko, K. Ueno, Surf. Coat. Technol., **169/170**, 124 (2003).
66. C. S. Dulcey, J. H., Jr. Georger, V. Krauthamer, D. A. Stenger, T. L. Fare, J. M. Calvert, Science, **252**, 551 (1991).
67. W. J. Dressick, J. M. Calvert, Jpn. J. Appl. Phys., **32**, 5829 (1993).
68. J. M. Calvert, W. J. Dressick, C. S. Dulcey, M.-S. Chen, J. H. Georger, D. A. Stenger, T. S. Koloski, G. S. Calabrese, Polymers for Microelectronics: Resists and Dielectrics, Thompson, L. F., Willson, C. G., Tagawa, S., Eds. American Chemical Society: Washington, DC, p. 210 (1994).
69. S. L. Brandow, M.-S. Chen, T. Wang, C. S. Dulcey, J. M. Calvert, J. F. Bohland, G. S. Calabrese, W. J. Dressick, J. Electrochem. Soc., **144**, 3425 (1997).
70. S. L. Brandow, W. J. Dressick, C. R. K. Marrian, G.-M. Chow, J. M. Calvert, J. Electrochem. Soc., **142**, 2233 (1995).
71. M.-S. Chen, S. L. Brandow, C. S. Dulcey, W. J. Dressick, G. N. Taylor, J. F. Bohland, J. H., Jr. Georger, E. K. Pavelchek, J. M. Calvert, J. Electrochem. Soc., **146**, 1421 (1999).
72. M. Yoshino, T. Masuda, J. Sasano, T. Yokoshima, I. Matsuda, T. Osaka, A. Hahimoto, Y. Hagiwara, I. Sato, Abstract for 206 th Meeting of The Electrochemical Society, Hilton Hawaiian Village, Hawaii, USA, October 2004, Abst. No. 95 (2004).
73. T. Masuda, M. Yoshino, J. Sasano, I. Matsuda, T. Osaka, Abstracts for 110th Meeting of The Surface Finishing Society of Japan, Hotel Matsushima, Miyagi, Japan, September 2004, p. 59 (2004).
74. C. A. Ross, H. I. Smith, T. Savas, M. Schattenburg, M. Farhoud, M. Hwang, M. Walsh, M. C. Abraham, R. J. Ram, J. Vac. Sci. Technol. B, **17**, 3168 (1999).
75. J. C. Lodder, J. Magn. Magn. Mater., **272–276**, 1692 (2004).
76. T. Aoyama, S. Okawa, K. Hattori, H. Hatte, Y. Wada, K. Uchiyama, T. Kagotani, H. Nishio, I. Sato, J. Magn. Magn. Mater., **235**, 174 (2001).
77. C. T. Rettner, M. E. Best, B. D. Terris, IEEE Trans. Magn., **37**, 1649 (2001).
78. J. Moritz, S. Landis, J. C. Toussaint, P. Bayle-Guillemaud, B. Rodmacq, G. Casali, A. Lebib, Y. Chen, J. P. Nozières, B. Diény, IEEE Trans. Magn., **38**, 1731 (2002).
79. N. Yasui, A. Imada, T. Den, Appl. Phys. Lett., **83**, 3347 (2003).

80. Y. H. Huang, H. Okumura, G. C. Hadjipanayis, D. Weller, J. Appl. Phys., **91**, 6869 (2002).
81. T. Osaka, N. Takano, S. Komaba, Chem. Lett, **657** (1998).
82. N. Takano, N. Hosoda, T. Yamada, T. Osaka, J. Electrochem. Soc., **146**, 1407 (1999).
83. D. Niwa, N. Takano, T. Yamada, T. Osaka, Electrochim. Acta, **48**, 1295 (2003).
84. D. Niwa, T. Homma, T. Osaka, J. Phys. Chem. B, **108**, 9900 (2004).
85. T. Osaka, N. Kasai, I. Koiwa, F. Goto, Y. Suganuma, J. Electrochem. Soc., **130**, 790 (1983).
86. T. Osaka, I. Koiwa, Y. Okabe, H. Matsubara, A. Wada, F. Goto, N. Shiota, J. Nakashima, IECEJ Tech. Group Meeting on Magnetic Recording, **MR84-15** (1984).
87. I. Koiwa, H. Matsubara, T. Osaka, Y. Yamazaki, T. Namikawa, J. Electrochem. Soc., **133**, 685 (1986).
88. T. Homma, K. Inoue, H. Asai, K. Ohrui, T. Osaka, IEEE Trans. Magn., **27**, 4909. (1991).
89. T. Homma, T. Nakamura, J. Shiokawa, T. Osaka, J. Magn. Soc. Jpn., **18-S1**, 73 (1994).
90. K. Itakura, T. Homma, T. Osaka, Electrochim. Acta, **44**, 3707 (1999).
91. H. Morinaga, M. Suyama, M. Nose, S. Verhaverbeke, T. Ohmi, IEICE Trans. Electron., **E79-C**, 343 (1996).
92. D. Ganguli, M. Ganguli, Eds., *Inorganic Particle Synthesis via Macro- and Microemulsions*; Kluwer Academic/Plenum Publishers: New York, 2003.
93. M. Gobe, K. Kon-no, K. Kandori, K. J. Kitahara, Colloid Interface Sci., **93**, 293–295 (1983).
94. K. M. Lee, C. M. Sorensen, K. J. Klabunde, G. C. Hadjipanayis, IEEE Trans. Magn., **28**, 3180–3182 (1992).
95. (a) M. A. López-Quintela, J. J. Rivas, Colloid Interface Sci., **158**, 446–451 (1993). (b) J. A. López Pérez, M. A. López Quintela, J. Mira, J. Rivas, S. W. Cherles, J. Phys. Chem. B, **101**, 8045–8047 (1997).
96. T. Hirai, J. Mizumoto, S. Shiojiri, Komasawa, I. J. Chem. Eng. Jpn., **30**, 938–943 (1997).
97. (a) X. Cao, R. Prozorov, Y. Koltypin, G. Kataby, I. Felner, A. Gedanken, J. Mater. Res., **12**, 402–406 (1997). (b) K. V. P. M. Shafi, A. Ulman, X. Yan, N.-L. Yang, C. Estournes, H. White, M. Rafailovich, Langmuir, **17**, 5093–5097 (2001).
98. J. Rockenberger, E. C. Scher, A. P. Alivisatos, J. Am. Chem. Soc., **121**, 11595–11596 (1999).
99. S. Sun, H. Zeng, J. Am. Chem. Soc., **124**, 8204–8205 (2002).
100. T. Hyeon, S. S. Lee, J. Park, Y. Chung, H. B. Na, J. Am. Chem. Soc., **123**, 12798–12801 (2001).
101. T. Hyeon, Chem. Commun., 927–934 (2003). (b) T. Hyeon, Y. Chung, J. Park, S. S. Lee, Y.-W. Kim, B. H. Park, J. Phys. Chem. B, **106**, 6831–6833 (2002). (c) E. Kang, J. Park, Y. Hwang, M. Kang, J.-G. Park, Hyeon, T. J. Phys. Chem. B, **108**, 13932–13935 (2004).
102. T. Nakanishi, H. Iida, T. Osaka, Chem. Lett., **32**, 1166–1167 (2003).
103. H. Iida, T. Nakanishi, H. Takada, T. Osaka, Electrochim. Acta, **52**, 292–296 (2006).
104. (a) Y. Okinaka, T. Osaka, In *Advances in Electrochemical Science and Engineering*; Gerischer, H., Tobias, C. W., Eds., VCH: Weinheim, Vol. 3, pp. 55–116 (1994). (b) I. Ohno, Mater. Sci. Eng., A, **146**, 33–49 (1991).
105. Y. Yanagisawa, T. Shimizu, K. Kuroda, C. Kato, Bull. Chem. Soc. Jpn. 63, 988–992 (1990).
106. (a) C. T. Kresge, M. E. Leonowicz, W. J. Roth, J. C. Vartuli, J. S. Beck, Nature, 359, 710–712 (1992); (b) S. Inagaki, Y. Fukushima, K. Kuroda, J. Chem. Soc., Chem. Commun., 680–682 (1993); (c) A. Monnier, F. Schüth, Q. Huo, D. Kumar, D. Margolese, R. S. Maxwell, G. D. Stucky, M. Krishnamurty, P. Petroff, A. Firouzi, M. Janicke, B. F. Chmelka, Science, 261, 1299–1303 (1993); (d) Q. S. Huo, D. I. Margolese, U. Ciesla, P. Y. Feng, T. E. Gier, P. Sieger, R. Leon, P. M. Petroff, F. Schüth, G. D. Stucky, Nature, 368, 317–321 (1994); (e) S. A. Bagshaw, E. Prouzet, T. J. Pinnavaia, Science, 269, 1242–1244 (1995); (f) Y. Sakamoto, M. Kaneda, O. Terasaki, D. Y. Zhao, J. M. Kim, G. D. Stucky, H. J. Shim, R. Ryoo, Nature, 408, 449–453 (2000); (g) S. Che, Z. Liu, T. Ohsuna, K. Sakamoto, O. Terasaki, T. Tatsumi, Nature, 429, 281–284 (2004).
107. G. S. Attard, J. C. Glyde, C. G. Göltner, Nature, 378, 366–368 (1995).
108. G. S. Attard, J. M. Corker, C. G. Göltner, S. Henke, R. H. Templer, Angew. Chem. Int. Ed., **36**, 1315(1995).

109. A. M. Whitehead, J. M. Elliott, J. R. Owen, G. S. Attard, Chem. Commun., **331** (1999); I. S. Nandhakumar, J. M. Elliott, G. S. Attard, Chem. Mater., **13**, 3840 (2001); G. S. Attard, S. A. A. Leclerc, S. Maniguet, A. E. Russell, I. Nandhakumar, P. N. Bartlett, Chem. Mater., **13**, 1444 (2001); P. A. Nelson, J. M. Elliott, G. S. Attard, J. R. Owen, Chem. Mater., **14**, 524 (2002); T. Gabriel, I. S. Nandhakumar, G. S. Attard, Electrochem. Commun., **4**, 610 (2002); P. N. Bartlett, J. Marwan, Chem. Mater., **15**, 2962 (2003); I. S. Nandhakumar, T. Gabriel, X. Li, G. S. Attard, M. Markham, D. C. Smith, J. J. Baumberg, Chem. Commun. 1374 (2004); G. S. Attard, P. N. Bartlett, N. P. B. Coleman, J. M. Elliott, J. R. Owen, J. H. Wang, Science, **278**, 838 (1997).
110. (a) J. Jiang, A. Kucernak, J. Electroanal. Chem., **520**, 64–70 (2002); (b) J. Jiang, A. Kucernak, J. Electroanal. Chem., **533**, 153–165 (2002); (c) J. Jiang, A. Kucernak, J. Electroanal. Chem., **543**, 187–199 (2003); (d) A. Kucernak, J. Jiang, Chem. Eng. J., **93**, 81–91 (2003).
111. Y. Yamauchi, T. Yokoshima, T. Momma, T. Osaka, K. Kuroda, Chem. Lett., **33**, 1576 (2004).
112. G. S. Attard, C. G. Göltner, J. M. Corker, S. Henke, R. H. Templer, Angew. Chem. Int. Ed., **36**, 1315–1317 (1997).
113. G. S. Attard, SAA. Leclerc, S. Maniguet, AE. Russell, I. Nandhakumar, PN. Bartlett, Chem. Mater., **13**, 1444 (2001).
114. (a) Y. Yamauchi, T. Yokoshima, H. Mukaibo, M. Tezuka, T. Shigeno, T. Momma, T. Osaka, K. Kuroda, Chem. Lett., **33**, 542–543 (2004); (b) Y. Yamauchi, T. Momma, T. Yokoshima, K. Kuroda, T. Osaka, J. Mater. Chem., **15**, 1987–1994 (2005).
115. Y. Yamauchi, T. Yokoshima, T. Momma, T. Osaka, K. Kuroda, Electrochem. Solid-State. Lett., **8**, C141 (2005).
116. Y. Yamauchi, T. Yokoshima, T. Momma, T. Osaka, K. Kuroda, J. Mater. Chem. **14**, 2935–2940 (2004); Y. Yamauchi, S. S. Nair, T. Yokoshima, T. Momma, T. Osaka, K. Kuroda, Stud. Surf. Sci. Catal., **156**, 457 (2005).

Electrochemical Fabrication of Nanostructured, Compositionally Modulated Metal Multilayers (CMMMs)

S. Roy

1 Introduction

Compositionally modulated (CM) materials started attracting attention when it was found that they had unusual mechanical [1, 2], magnetic [3–5], electronic [5], and corrosion properties [6–9]. Nanostructured, compositionally modulated materials usually consist of stacks of two or three different metals, metal oxides, ceramics, as shown in Fig. 1, which have significantly different properties. The composition and thickness of each individual layer can be manipulated to optimize the desired property of the deposit or coating. In addition, due to the increased effect of surface or interface arising from the exceptional thinness of the layers, there can be large deviations from bulk behavior, which raises the possibility of totally new properties for compositionally modulated materials (CMMs).

The overall property of the deposit is, then, a function of the individual components, the modulation thickness, as well as the structure of the interface. Since the need and requirement of a compositionally modulated structure depends on its application, researchers have produced different processes and technologies to fabricate these materials. Although sputtering, evaporation, and chemical vapor deposition have been used to fabricate these materials, in this chapter the discussion is on electrodeposited CMM. In addition, since the focus will be on materials which exhibit unusual properties due to the constriction of the modulation to two dimensions, that is, surface area versus thickness, the concentration is on modulation thicknesses <100 nm, which is smaller than a single grain. The fabrication schemes covered here stress the techniques and issues related to the electrodeposition of compositionally modulated metal multilayers (CMMMs). Although authors have fabricated nanostructured layers [10] as well as nanowires [11] by electrochemical means, the electrochemical principles are similar for both cases, and, therefore, this chapter covers techniques to fabricate either item.

S. Roy (✉)
School of Chemical Engineering and Advanced Materials, Newcastle University, New Castle, Upon Tyne NE1 7RU, UK
e-mail: s.roy@newcastle.ac.uk

P. Schmuki, S. Virtanen (eds.), *Electrochemistry at the Nanoscale,* Nanostructure Science and Technology, DOI 10.1007/978-0-387-73582-5_9,

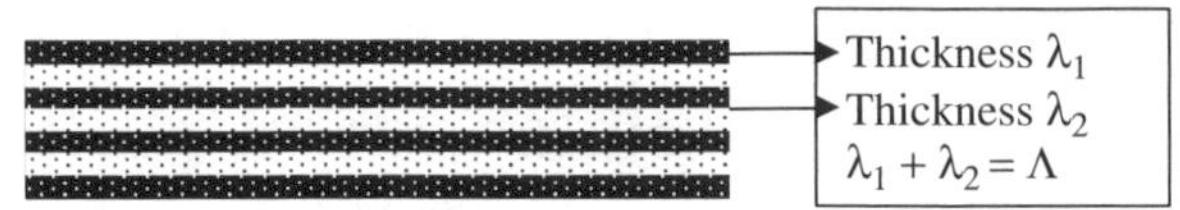

Fig. 1 A Schematic of a compositionally modulated material consisting of two different components. The individual thickness of each layer is λ_1 and λ_2. The overall modulation thickness is the sum of the individual thicknesses, Λ

The first motivation to write this chapter is that, although there have been several excellent reviews on the properties of nanostructured CMMs [12, 13], there have been relatively few articles reviewing the state-of-then-art fabrication techniques. There is some information on combining electrochemical and nonelectrochemical techniques for coatings [14], with their respective merits and demerits, or short critiques on a single electrochemical technique [15], but there is no comprehensive review on the historical development and understanding of the issues related to electrochemical fabrication of CMMMs. In this regard, this work is a critical review of the important developments in this area, and highlights the peculiarities encountered during the electrochemical fabrication of CMMMs. Although the total number of papers published in this field are numerous, for the sake of brevity, as well as keeping in mind the availability, most of the reference material has been gleaned from standard journals (with very few articles from symposium proceeding's volumes).

The second motivation is that there is some confusion regarding the "best" method or technique for fabrication, perhaps due to the claims of different groups pursuing different avenues for electrodeposition. Since the fabrication of each nanostructured CMM system will critically depend on the constituents, modulation thickness, and desired property, there is no "best" alternative. This was demonstrated and discussed by various groups. For example, collaboration between Denmark and Loughborough Universities on the deposition of CMMMs found that there was no single fabrication scheme which provided a "best practice" scheme. While Zn–Fe CMAs plated for stress relief in Zn–Fe deposits [7] required optimization of modulation thickness, other Zn–Fe alloys that were developed for corrosion protection [8, 9], required optimization of electrochemical current pulse forms.

In the following sections, we present the different methods for the electrochemical fabrication of nano structured compositionally modulated multilayers, mainly metals and alloys, with some allusion to ceramics and semiconductors. The Chapter aims to provide the readers with the information which allows them to design and choose electrochemical process and technology to fabricate CMMs. As there is yet no standard industrial process plate to fabricate CMMMs, it is hoped that the readers will enable this vision in the near future.

2 Experimental Apparatus and Schemes

Electrodeposition of nanostructured metal multilayers can be broadly categorized as: (1) electrodeposition from separate electrolytes, (2) electrodeposition from a single electrolyte using pulse currents, and (3) electrodeposition from a single electrolyte using pulsing flows. Some researchers have used a combination of these methods to achieve the desired deposit composition or thickness. The following sections provide a description of the different techniques.

2.1 Dual-Bath Electrodeposition

The obvious technique to plate a variety of metal multilayers is to use different electrolytes to plate the individual metals. This can be achieved by plating a substrate in a particular bath, and then transferring the substrate to another electrolyte where a second metal is plated, and so on. If two metals are plated alternately, then the substrate can be transferred back and forth between the two electrolytes to build up a multilayer structure. The earliest report of such a technique was provided by Blum [16] to fabricate coarse multilayer structures of Cu/Ni, Cu/Ag, and Cu/Cu, the latter consisting of copper layers of different grain sizes. Micron-scale Cu/Ni structures were prepared by Ogden [17] by transferring the substrate from one electrolyte to another. At the nanoscale level, two groups, one in the US [18–21] and another in Belgium [22–24], developed different mechanical apparatus to carry out the fabrication of nanostructured multilayers.

The simplest apparatus needed to carry out dual-bath electrodeposition is shown in Fig. 2. As shown in the figure, the cathode is transferred between the two baths. Each bath is equipped with a separate counter (and reference) electrode. The composition of each electrolyte can consist of only the metal which is to be plated in that bath, and often standard electrolytes are used. The power supply is a simple direct current plating unit (galvanostat) or a potentiostat, or even a pulse rectifier, depending on the need. In essence, the principle of plating is no different from a standard dc deposition technique. The crucial step, however, is of rinsing between the electrolytes, which is not shown in this figure. There may be a need for a single-stage or

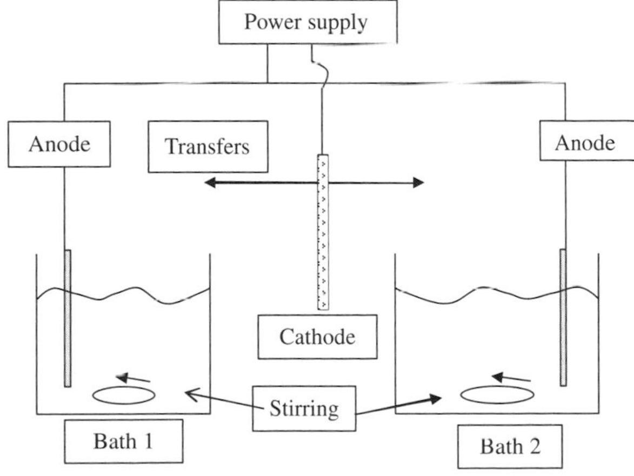

Fig. 2 Dual-bath electrodeposition apparatus and scheme. The electrode is transferred between the two electrolytes with rinses in between. The two electrolytes can have different compositions and agitation schemes

double-stage rinsing [24] and even a prerinse [24, 25] before the substrate is introduced into the main electrolyte. This needs a high degree of automation [25] and tends to increase the capital cost needed for deposition.

Although the underlying principle is the same, the equipment developed by the different groups differ considerably. Ross and co-workers developed a dedicated apparatus to deposit thin metal and amorphous structures. Initially the group devised an apparatus, which consisted of a plating tank, which was subdivided into two compartments [18]. The two compartments had cut-out windows through which the substrate could be exposed to the plating solution. The substrate was placed in a circular cut-out section and was fixed to a rotating brass disk. As the substrate rotated, it was exposed to one electrolyte and subsequently to another. As the substrate was exposed to one electrolyte, it was plated by that metal. Then the second metal was plated over it when it was exposed to the second solution. Therefore, this arrangement provided a suitable method to plate a multilayers structure of alternating metals. This arrangement has similarities with the equipment used in sputter deposition techniques.

The researchers used this apparatus to plate Ni/Ni–P between 2.1 and 4.0 nm thick. They used pure nickel-plating electrolyte and nickel–phosphorus electrolyte solutions. Although they obtained layered structures, the Ni film was found to contain phosphorus as an impurity, which was attributed to cross-contamination. Consequently, a new apparatus was designed [19, 20], where a substrate was suspended above nozzles (or jets) of electrolytes on three ball bearings, and rotated by a motor. The nozzles were wedge shaped and the substrate–nozzle distance was maintained at 0.5 mm in order to ensure that plating took place only in the area where the substrate contacted the individual electrolyte. Fine capillaries were placed just inside the perimeter of the nozzle which sucked the electrolyte back, so that electrolyte leakage was minimized. Adequate arrangement was made to wash and dry the substrate in nitrogen in between the plating periods. This reduced the problems related to cross-contamination of the individual electrolytes.

Using this apparatus, the researchers plated Ni/Ni–P deposits using nickel chloride and nickel–phosphorus electrolytes. The workers were successful in electrodepositing Ni/Ni–P multilayers of thicknesses ranging between 2 and 8 nm [20], which was verified by low-angle X-ray diffraction (XRD) and transmission electron microscopy (TEM). However, the authors had significantly more difficulty in plating other metallic systems, that is, Co/Ni–P. They found that individual layers less than 20 nm repeat length were indistinct and no periodicity was detectable via low-angle XRD. It was found that although Ni–P grew uniformly on Co, cobalt nucleation was uneven and faceted, which prevented the formation of a layer of uniform thickness. Electrodeposition of Cu/Ni and Cu/Co were only partially successful because the less noble material dissolved when it contacted the electrolyte containing the more noble metal, thereby indicating that galvanic coupling was a major problem in this system [20, 21].

Celis and co-workers carried out electrodeposition of Sn/Ni–P [22, 23] and Cu/Ni [24] multilayers and attempted to solve some of the problems encountered by Ross et al. The Sn/Ni–P multilayer was chosen especially, because it represents a system where a hard and a soft material are interlaced, leading to some interesting mechanical properties. From the electrodeposition point of view, the challenge was that the two materials are difficult to nucleate on each other. The researchers examined electrolyte chemistry which could promote layer-by-layer growth of Sn on Ni–P. They clearly showed the necessity of using leveling and brightening agents and demonstrated the possibility of growing individual alternating layers of Sn/Ni–P of 500 nm thickness. They showed that a combination of potential modulation along with a dual-bath principle was necessary to grow Sn/Ni–P CMMMs.

A second system, that is, the Cu/Ni system was chosen, mainly to demonstrate the feasibility of using a dual-bath technique to produce metal multilayers, which were <5 nm in thickness [24]. The scheme for electrodeposition was slightly different from that shown in Fig. 2. The electrode was rinsed in water after plating in each electrolyte, and before introducing it into the next plating bath, it was rinsed in the same electrolyte. The authors demonstrated that modulation thickness down to 20 nm was achievable using such a system, but formation of faceted structures could not be avoided. Several layers of Cu/Ni formed a part of a single columnar grain – in all cases, the layers were found to be continuous.

An automated dual-bath plating system was developed by the Centre of Advanced Electroplating [25, 26], who carried out copper–nickel deposition in it. Their apparatus included an activation step as well as a water rinse before each layer was plated. They found that although the dual bath could be used to fabricate pure metal modulations, the technique was incapable of producing layers which were much thinner than 25 nm. This is because the washing and activation routine needed the same amount of time, even when the time required for electroplating was reduced. If the total thickness was kept constant, the number of modulations has to increase as the modulation thickness was reduced. This resulted in an exponential increase in process time (as vs. plating time) which made the process unviable. They found that practical advantages of a dual-bath system was enhanced when the layers were close to the micron scale [25], but operating costs and degree of automation needed higher sophistication as the thickness of the individual layers is decreased.

2.2 Single-Bath Electrodeposition

The most common technique for electrodeposition of CMMMs involves the use of a single electrolyte containing two or more metal ions. Pulsed currents or potentials are applied to preferentially reduce one of the metals, so as to form alternating layers of (nearly) pure metallic layers. The more noble metal ion is usually present in a solution in very small quantities, typically in mM concentration, which limits the plating rate of the noble component. The less noble component, on the other hand, is kept close to its solubility limit (or at least at a high concentration). The molar ratio of the two materials in the electrolyte is typically 1:40–1:100. By maintaining such a difference in metal-ion concentration, the noble material is plated under mass-transfer-controlled conditions, and the less noble component is deposited under kinetic control. During the period when the less noble component is discharged, the noble component is also co-deposited. However, since the noble metal is always reduced at the mass-transfer-limiting current, a nearly pure layer of the less noble component is obtained. In the literature, CMMMs are usually described by the layer compositions, such as Cu/Ni(Cu) or Cu/Co(Cu), with (Cu) implying the copper impurity in the nickel or cobalt layer. In this chapter, the nomenclature is Cu/Ni or Cu/Co, with the implicit understanding that some amount of the noble material is present in the nickel or cobalt layer.

The apparatus used for this technique is simple, such as shown in Fig. 3. The cell can be a two- or three-electrode system, with a pulse rectifier as a power supply. A reference electrode is needed if plating is carried out by potential pulses. In order to deposit multiple layers, the current or potential is switched repeatedly between a low value, where only the noble metal is reduced, and a high value, where a near-pure layer of the less noble material is deposited. The advantage of this technique is that there are no moving parts; consequently, there is no requirement for rinsing, are no drag-out problems, and minimum probability of oxidation in air.

The important electrochemical control parameter for composition modulation is the choice of current or potential pulse trains. A typical pulse train is described by Fig. 4. If the time for applying a low current i_L is t_L, then the charge of the noble metal discharged is given by $Q_L = i_L{\cdot}t_L$, provided that the current efficiency is 100%. If the high current is i_H and is applied for a time t_H, then the charge corresponding to the layer rich in the less noble material is $Q_H = i_H{\cdot}t_H$, again, provided that the current efficiency is 100%. The thickness of each layer can be calculated from the charge balances and molar density using Faraday's law. The content of the noble material in the layer containing the less noble component is i_L/i_H.

Although the electrochemical cell is simple, the requirement of a complex current or potential pattern necessitates the use of more complicated power supplies or potentiostats. Depending on the metal or alloys to be plated, one may need to switch between high and low currents, high and low potentials, or a combination, that is, switch between currents and potentials to achieve the desired layer thickness and

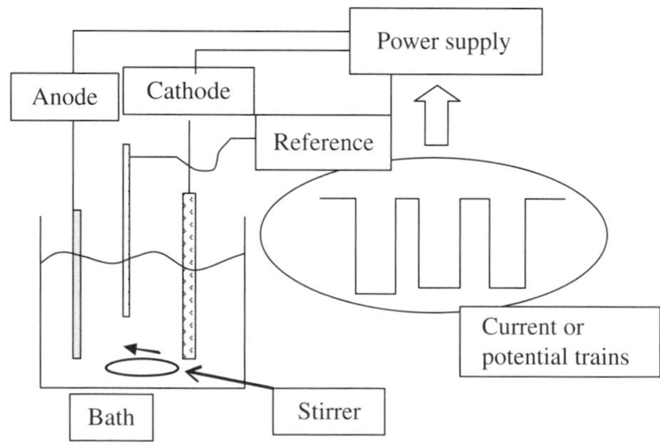

Fig. 3 Electrochemical apparatus used during CMMM plating from a single electrolyte. Solution composition as well as current or potential pulses ensure that one of the layers is deposited preferentially, thereby producing nearly pure layers of each metal

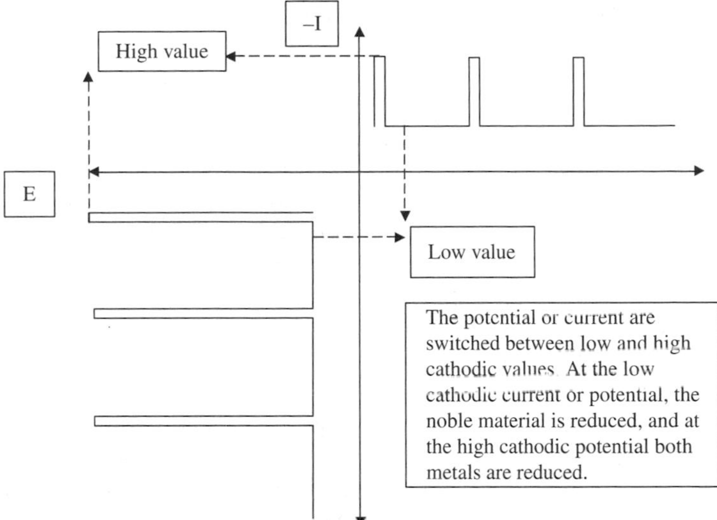

Fig. 4 Typical pulse current or potential trains used for CMMM deposition from a single electrolyte. Since a metal-deposition rate is fixed either by applying an appropriate current or potential, switching between currents, potentials, or current and potential can be used to deposit CMMMs

composition. The instrumentation which allows one to plate with a high degree of accuracy and control can be expensive. Process monitoring also requires high-speed data-acquisition cards.

Second, the choice of electrolyte is problematic because one needs to solubilize two different metal ions into a single solution. It is well known that some baths can precipitate due to slow homogeneous reactions. In the case of ternary metal systems, such as those needed to deposit Cu/Ni–Fe, or Cu/Ni–Co multilayers, the choice of electrolyte becomes even more convoluted, since three different metal ions have to be dissolved in the same electrolyte. For instance, Cu can be readily plated from acid–copper sulphate electrolytes, but the necessity of plating nickel eliminates the use of such acid electrolytes, because H_2 evolution needs to be minimized. In this case, the issue is resolved by using either citrate-based electrolytes, or by adding $CuSO_4$ in a nickel sulphamate electrolyte.

In addition to these practicalities, the choice of anode is not simple. Since the less noble material is present in very small quantities, it is consumed as the layers are plated and needs replenishment. This can be critical, because as the metal-ion concentration is depleted, the limiting current drops. If the noble metal is plated galvanostatically, then a decrease in the limiting current can incorporate nickel into the copper layer during the deposition of Cu layer. Alternatively, if copper were plated under potentiostatic control, then a feedback loop to calculate the total charge plated is required to ensure desired layer thicknesses. Therefore, despite the apparent simplicity of the procedure, fairly sophisticated know-how is needed to fabricate modulated structures of correct thickness and composition.

The earliest use of a single-bath deposition method to fabricate modulated alloys was recorded by Brenner [27]. The same method for fabrication of micron and submicron-scale modulated metals by Cohen et al. was reported in 1983 [28]. In their work, Cohen et al. demonstrated that charge-balance calculations could be used to determine the thickness of each layer. In the mid- to late 1980s, Yahalom and Zadok [29] prepared metal multilayers using this technique and filed a US patent [30]. In the paper and the patent [29, 30] it was suggested that the noble material should be deposited under mass-transport-controlled conditions, and the less noble metal should be deposited under kinetic control. In a subsequent paper, the technique was used to deposit Cu/Ni metal multilayers of 10-nm modulation thickness [31], which were characterized by TEM. It was found that the layers deposited coherently, were lamellar, and the thickness of the layers correlated well with the ones predicted from charge-balance calculations.

Lashmore and co-workers, in parallel, deposited Cu/Ni [32] and Cu/Co multilayers [33] of modulation thicknesses >5 nm. In their experiments, they stirred the solution during copper deposition, but stopped all agitation during the electrodeposition of the less noble metal. They justified this on the basis that since the copper-deposition current was lower in unstirred solutions, they could minimize the amount of copper which was incorporated in the Ni or Co layer [32, 33]. They also found that for galvanostatic pulses, when the deposition current was changed from Ni to Cu deposition, the potential change was sluggish. They argued that this, in principle, could produce a diffuse interface and incorporated a short dissolution current to provide a sharper interface. The layer thickness was calculated from the total charge passed and verified using XRD [32]. Cu/Ni layers were found to possess a (111) texture [32], and Cu/Co layers exhibited an fcc structure [33], which showed evidence of superlattice formation.

Tench and White fabricated Cu/Ni metal multilayers and studied their mechanical properties [34]. A thorough electrochemical analysis and critical consideration for multilayer deposition was presented in a subsequent paper [35]. They suggested that copper should be deposited under potential control because the copper-limiting current decreased with increasing plating time. Since many researchers deposited copper at the same (limiting) current throughout the experiment, they risked incorporating nickel into the copper layer. In addition, they tried to determine if nickel could passivate during copper deposition. They carried out cyclic voltammetry to show that passivation did not occur. Since the nickel-covered substrate experienced an anodic potential during the period when copper was being deposited, they estimated that about 2 nm of nickel was dissolved from the deposit due to displacement of Ni by Cu. Based on this finding, they proposed that such a displacement reaction could be exploited to plate Cu–Ag alloys [36], where Cu, the less noble component, was displaced by silver. They showed that the displacement reaction was controlled by mass transfer for a certain period of time, and speculated that it was likely that Ag grew epitaxially on copper during the process.

Despic and Jovic had independently suggested that a single current-pulse electrodeposition technique could be used to fabricate layered structures of controlled composition by using simple concentration polarization methods [37]. Their technique involved applying a pulse current greater than the limiting current of the noble component in solution. In this case, the noble metal would reduce first, and a layer containing only the noble component would be deposited until the metal at the vicinity of the electrode was entirely consumed. Thereafter, the less noble component would begin to discharge. This would produce an alloyed layer, rich in the less noble component. After forming such a layer, the current was switched off in order to regain equilibrium conditions. This amounts to using single-pulse currents, where, instead of a high and low current pulsing, only a high current, followed by a null (or zero) current is used.

Interestingly, they chose to use the same concentration of the noble and base metal ion in their electrolyte, which was different from their peers. The feasibility of their technique was demonstrated using two model metal-plating systems – (1) copper–lead and (2) copper–nickel [37]. These two systems were chosen because copper and lead are immiscible, whereas copper and nickel are completely miscible. Therefore, they represent two extreme systems for forming layered structures. They developed the theoretical basis to predict the thickness and composition of the modulated layers, which showed fair agreement with their experimental results.

Their method suffered from three different practical problems: (1) the layer corresponding to the less noble metal included significant amounts of copper, (2) the deposited copper was rough due to high-concentration polarization, and (3) there was a significant problem related to lead dissolution from the copper–lead system due to Pb displacement by Cu. In order to overcome some of these problems, in a subsequent work [38], they refined their earlier idea and used a dual-current pulse technique (which was, in essence, similar to the traditional single-bath technique used by other researchers). They demonstrated its use by fabricating Cu/Ni multilayer structures with a modulation thickness of the order 100–200 nm and developed the theoretical equations to calculate the thickness and composition of the layers.

Based on the experiences of the researchers mentioned earlier, at the beginning of the 1990s, it was reasonably established that a single-bath electrolyte was useful for electrodepositing nanostructured CMMMs. However, outstanding issues related to multilayer fabrication remained. The foremost of this was the ability to predict layer composition and thickness. Although the original work of Cohen et al. [28] showed that a simple charge-balance analysis could be used for the calculation of layer thickness, it has to be remembered that they used it only for predicting micron-scale layer thickness. Tench and White [36] clearly showed that a displacement reaction was occurring, which influenced the thickness of the individual layer thickness, but did not provide a method to correct the problem. Although Despic and Jovic's [37, 38] work provided some useful mathematical tools, it involved many assumptions and could not be used in a production-worthy fashion.

The second issue was related to the loss of the less noble component by displacement, and the effect it had on the deposit properties and microstructure. This was quite critical, because Co, Ni, Fe (base metals) showed unusual properties in layered structures, and their loss compromised the functionality of the material. This also raised the spectre of producing diffuse interfaces, which could compromise the quality of the materials, which were dependent on the interface properties (such as giant magnetoresistance properties).

Finally, since the noble metal was being deposited under mass-transfer-controlled conditions, rough and dendritic structures were produced. Strategies for the optimization of process conditions were required to reduce these problems. In fact, a much greater effort was necessary to iron out these issues before the single-bath technique could be adopted in a production situation, which became the main focus of research in the early to mid-1990s.

2.3 Electrodeposition by Flow Modulation

A less-explored technique for inducing composition modulation in electrodeposited materials is via changes in mass-transfer conditions near the cathode surface. This method induces composition modulation by increasing or decreasing the limiting current of the noble material. If the total current is constant, then the composition of the deposit follows the changing limiting current of the noble metal. For most practical purposes, this method can only produce composition modulated alloys rather than pure metallic layers.

Electrodeposition of compositionally modulated alloys by flow modulation was proposed by Schwartz and co-workers [39, 40], using a theoretical analysis at a rotating disk electrode. Although analysis for oscillating flows for a variety of electrochemical systems had been explored by other groups, its application for depositing compositionally modulated multilayers was proposed and pursued mainly by Schwartz and co-workers [39–42]. The analysis showed that if two ions, such as Cu(II) and Ni(II) were co-deposited in a fluctuating flow field; and if Cu(II) was deposited under mass-transfer limitations, then, inducing a flow modulation would change the partial current for copper reduction, creating a composition modulation. Their analysis showed that the thickness of modulation would be inversely dependent on the oscillation frequency of the flow field [39, 40], which meant that slow oscillations would produce larger modulation thicknesses, and fast oscillations would lead to smaller modulation thicknesses.

The theoretical proposition was tested experimentally by depositing Ni–Fe alloys at a rotating ring disk electrode (RRDE) [41]. The rotation speed of the RRDE was varied using a function generator to create oscillating flows, thereby changing the momentum and concentration boundary layers. In these experiments, the quantity of Fe(II) was maintained at very low concentrations, thereby controlling Fe(II) reduction by mass transfer. Ni(II) concentration was 40 times higher in solution, and therefore controlled by electrode kinetics [41]. The composition of the deposit plated at the disk was then analyzed by anodic stripping voltammetry. Appreciable composition modulation was detected for low hydrodynamic oscillations; where modulation thickness was in the range of 25–100 nm. The modulation thickness for fast oscillations was less difficult to detect; possibly due to the fact that a significant time was needed before the hydrodynamic changes induced a change in the mass-transfer-limiting current. The authors changed their electrodeposition system to an uniform-flow-injection system and induced velocity changes by controlling the pump speed, and again found detectable composition modulation only for slower oscillations [42].

The problem regarding CMA deposition by oscillating flow fields are twofold: (1) that the flow variation can at most produce alloys of different composition and (2) that for appreciable composition modulation and thinner modulation thickness, the concentration and momentum boundary layers have to be comparable. Since most aqueous electroplating systems are characterized by a high Schmidt number of ~1000, the concentration boundary layer is much thinner than the momentum boundary layer. Any oscillation induced in the flow field requires a certain time before the concentration boundary layer responds to it – this eliminates the possibility of using high-frequency oscillations, which defines the lower limit of modulation thickness.

3 Issues Related to Electrodeposition of Nanostructured Metal Multilayers

As mentioned earlier, from the early 1990s researchers concentrated their attention mainly to obtain a better understanding of the underlying phenomena during electrodeposition from a single electrolyte. These issues are discussed in this section.

3.1 Estimation of Layer Thickness and Composition

One of the most important aspects related to the fabrication of modulated materials reliably and reproducibly is to be able to predict individual layer thicknesses. Earlier works on dual-bath electrodeposition [25, 28] as well as single-bath electrodeposition [5, 29, 30] used charge balances to estimate the thickness of the individual layers. In this approach, if a partial current of a species is given by i_k, then the total deposition current, i_p, is the summation of the partial currents for all species, m, which may be the individual metals, as well as oxygen and proton reduction (i.e., other parasitic reactions),

$$i_p = \sum_{1}^{m} i_k \tag{1}$$

This approach was reasonably successful for large modulation thicknesses, that is, greater than 50 nm. Researchers could determine partial current densities from direct current polarization data. By simply computing the charge of a metal plated, and using the Faraday equation, and the molar density of a metal, the individual layer thickness and the overall modulation thickness could be calculated.

However, Eq. (1) does not predict small modulation thickness, that is, $\Lambda < 20$ nm, adequately. This is mainly because nonsteady state effects, which influence metal deposition and dissolution during the beginning and end of the application of a current or potential step have a strong effect when the layer thicknesses are small. Since this requires solution of nonsteady state currents, the prediction of deposit composition as well as thickness becomes more difficult. In order to determine the partial current for each species, a numerical approach is used. The kinetics of each species are given by Butler–Volmer equation

$$i_k = i_0, k \left\{ \exp\left(\frac{\alpha n F}{RT}\eta_k\right) - \exp\left(-\frac{(1-\alpha)nF}{RT}\eta_k\right)\right\} \tag{2}$$

where i_0 is the exchange current density, α is the charge transfer coefficient, n is the number of electrons transferred, and η is the electrode overpotential, and k is the species in question. For the noble component, the reaction rate is controlled by diffusion, and hence, mass transfer to the electrode surface has to be included. This is usually described by Fickian diffusion in one dimension through a Nernst boundary layer. Finally, the equations are solved with appropriate boundary conditions, and the condition described in Eq. (1).

However, if certain conditions are fulfilled, simple analytical equations can also be used to compute modulation thickness as well as composition. For example, if a single current pulse, such as that suggested by Despic and co-workers [37], is used to deposit metal multilayers using short current pulses, then analytical equations can be used to describe the partial current of metallic species. For this analysis to hold, there are two crucial assumptions: (1) that the current pulses are so short that the "pulsed" diffusion boundary layer is much smaller than the steady-state Nernst diffusion layer [43] and (2) that the time period for displacement reactions is short, and the amount of metal displaced is so small that it can be neglected [43].

When a single-pulse current of value i_p, which is greater than the limiting current of the noble species in solution, is applied, then the Sand equation can be used to describe the concentration changes of the noble metal ion in the diffusion layer [37, 43]. The Sand equation allows one to calculate the time period, t_1, during which the applied current, i_p, is completely consumed by the reduction of the noble metal, identified in the following equation as species 1, thereby creating a noble metal rich layer:

$$t_i \frac{\pi D_1 n^2 F^2 C_{1,b}^2}{4 i_p^2} \tag{3}$$

Where D is the diffusion coefficient, F is Faraday's constant, and C is concentration. The subscript 1 refers to the noble metal and b denotes bulk. The corresponding charge, Q_1, for the noble material deposited during this time is given by:

$$Q_1 = \frac{\pi D_1 n^2 F^2 C_{1,b}^2}{4i_p} \tag{4}$$

Thereafter, the concentration profile of the noble metal has to relax and the less noble metal (and possibly proton reduction) begins. During this period, the Cottrell equation (with certain assumptions [43]) can be applied. The Cottrell equation shows that the noble metal current decays with time:

$$i(t) = nFC_{1,b}\sqrt{\frac{D_1}{\pi t}} \tag{5}$$

This analysis clearly indicates that if a single-pulse current is used to plate-modulated layers, then the interface between the noble and less noble material will be diffused, although the thickness of this diffuse layer could be minimized by maintaining a low concentration of noble metal ions.

The above equations were tested at a rotating disk [37] and rotating cylinder systems [43], and show that simple fundamental charge balance analysis can be used to compute modulation thicknesses even in unsteady–state systems. However, Despic and co-workers [37] recognized that if there were a displacement reaction between the noble and less noble components, then the calculations would be severely compromised. However, earlier [20, 25, 28, 32, 36, 37] investigations had already pointed out the significant effect of displacement reactions which merits a more detailed discussion.

3.2 Effect of Displacement Reaction

When modulated materials are plated from a single electrolyte a displacement reaction, such as Cu(II) + Ni → Ni(II) + Cu, can occur. This can be understood if one considers Fig. 5. Consider a modulated deposit is being produced by switching a current or potential between the points *H* and *L*, where the base metal rich alloy is plated at the point *H*, and noble metal layer is deposited at *L*. Then, as soon as the

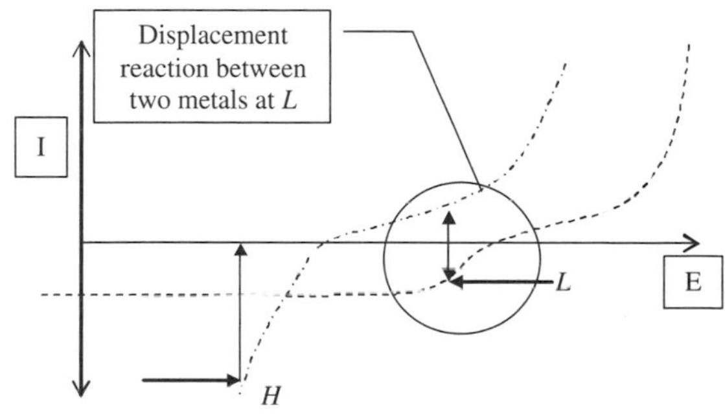

Fig. 5 Schematic of displacement or corrosion reaction during electrodeposition of CMMMs from a single electrolyte by switching between *H* and *L* values Polarization data for - - - noble component; -.-.-.- less noble component

current is switched to L, the base metal potential swings to an anodic potential. This should cause it to dissolve, as shown by the anodic current corresponding to the less noble metal indicated in the figure.

The effect on the deposit composition and individual layer thicknesses are somewhat different depending on whether a constant current or constant potential is applied. In the former the system may adopt a mixed potential, usually encountered in corrosion systems, whereas in the latter, the potential of the electrode determines the partial current of the dissolving species. In any case, corrosion or displacement reaction is inevitable in the absence of passivation, inhibition, or other schemes which slow down the kinetics of the dissolution reaction. This process modifies the behavior of the overall system to such an extent that the analytical expressions to compute layer thicknesses using charge balances for single- or dual-pulse current or potential can be invalid (unless proven otherwise). This process is not confined to a single-bath electrolyte, and has been observed in dual-bath systems [20, 25, 28], as well as a modulated flow systems [41]. Therefore, a through consideration of galvanic coupling during metal multilayer fabrication is vital.

Theoretical models to show the effect of galvanic corrosion during pulse plating of alloys was proposed by Roy et al. [44]. The effect was demonstrated by developing two extreme models for pulse plating of alloys. The first model was called the no-corrosion model, where two metals, one considerably noble than the other (such as copper and nickel) were plated at a rotating cylinder electrode. In the no-corrosion model, copper was reduced according Butler–Volmer kinetics, whereas nickel kinetics were described by cathode Tafel kinetics. This meant that nickel was reduced at the cathode, but not allowed to dissolve. The second model was called the corrosion model, where the kinetics of both metals was described by Butler–Volmer equations, which allowed nickel to dissolve from the deposit. These models also included the effect of mass transfer to the electrode through a Nernst diffusion layer. The models were solved using appropriate boundary conditions and Eq. (1) to fulfill the condition that the total imposed current was a summation of the partial currents of the two metals.

The models were solved numerically and tested for Cu–Ni alloy deposition from a citrate electrolyte [44]. It was found that (1) when displacement reactions were ignored then the partial current for the two metals pulsated with time, as did the concentration at the vicinity of the electrode and (2) when there was an unimpeded displacement reaction, then the partial current of the noble component remained at the diffusion limiting current and its concentration near the electrode surface was virtually zero. The less noble component simply filled the additional current so as to fulfill Eq. (1). This phenomenon took place because the noble metal was reduced at the electrode surface throughout the pulse time; in effect, a displacement reaction caused the system to behave like a direct current plating system [44]. Since the plating of nanostructured CMMMs is similar to the time periods encountered for alloy deposition, the effect of displacement reaction in nanostructured CMMMs can be expected to be similar to that observed in pulse plating of alloys.

This is borne out in a subsequent work, which included the enrichment of the surface of nickel by copper due to displacement [45], which showed that for a critical time the displacement reaction was unimpeded and controlled by the liquid-phase limiting current of copper. Thereafter, the reaction rate was hampered by diffusion of nickel to the surface, that is, nickel diffusing through the copper layer formed at the surface, which finally stopped the displacement altogether. Beyond this point, the no-corrosion model was applicable. Clearly, the faster the surface-enriched layer was formed the sooner the displacement reaction stopped. A refinement of the model was suggested by Bradley and Landolt [46], which included the effect of gradual surface coverage by the noble metal, and allowed one to assess the effects of two-dimensional versus three-dimensional nucleation effects.

Chaissang and co-workers studied the effect of displacement reaction on Cu–Co deposits using a quartz crystal microbalance (QCM) which allowed one to monitor weight changes as well as internal stresses accompanying current or potential modulation [47, 48]. This was achicvcd by using a double-quartz crystal microbalance with an AT and BT crystals. The AT and BT crystals have identical responses to weight gain but opposite responses to stress. Since a considerable amount of stress is generated due to lattice mismatch between the different components of a multilayer, this method allowed one to observe if the dissolution of the less noble component was a response to relieve stresses within the deposit. The researchers deposited Cu/Fe-Ni multilayers by potentiostatic cycling and did not detect significant mass changes during the deposition of the more noble component, although they detected significant stress changes [47]. A similar behavior was observed when Cu/Co multilayers were deposited – although stress changes were observed, Co dissolution was not observed [48]. However, since a displacement reaction would result in very small weight changes (because the atomic weights of the metals deposited are similar), a QCM measurement may not be an appropriate method to detect corrosion reactions.

Bradley and Landolt [49], on the other hand, found that the displacement reaction during the deposition of Cu–Co multi- layers and alloys took place for a longer period of time than that for Cu/Ni. Indeed, they estimated that several monolayers of copper had to be deposited before the displacement rate decreased from the diffusion limiting rate, and found that it continued indefinitely, albeit slowly. This explained earlier findings of Ross et al. [20], who found that they could not deposit modulatcd layers of Cu/Co in a dual-bath system, where the microstructure of the deposit was significantly rougher than Cu/Ni deposit. Bradley and Landolt [49] also indicated that their multilayer deposits consisted of three phases; Cu, Co, and Cu–Co. This indicated that Cu/Co multilayers could be more porous than Cu/Ni, and this could be due to the difference in the miscibility and phase structures of the two systems.

3.3 Interfacial and Phase Instability

The above discussion shows that CMMs may have electrochemical behavior which is influenced by the deposit microstructure itself. The interest in crystalline and microstructural stability of CMMMs arose from the effort to form high-quality strained superlattices [50]. Moffat [50] showed that a Cu/Ni single crystal could be grown on Cu(100), but due to stress, growth defects, and twinning, this was not feasible for Cu(111) and Cu(110). Cziráki et al., through TEM, showed that the multilayer structures mutate significantly along the growth direction in Cu/Co multilayers [51]. Near the electrode surface the grain size remains small due to unrelieved stresses resulting from the mismatch of copper and cobalt lattices; but as the deposit grows, some of these stresses are relieved by "tilting" of grains, thereby increasing the roughness of the deposits. Since higher roughness causes the overall current density to decrease (due to the increase in surface area), modulation thicknesses were found to decrease with increasing deposit thickness [51].

Bonhôte and Landolt [52], on the other hand, examined the changes in deposit microstructure during the growth of Cu/Ni multilayers. They examined the development of morphological features as a function of the copper deposition current, since it is plated close to its mass-transfer-limiting, current (nickel is plated under kinetically controlled conditions and does not contribute to roughness). They found that at low currents the deposit was characterized by columnar growth with somewhat curved multilayers, as is shown in Fig. 6 (A). There were discontinuities at the grain boundaries, showing that the layers grew within each grain. At high deposition currents, the grain size approached the modulation thickness and the multilayers stretched across grains. They also found that Cu/Ni structures showed preferred texture when the copper deposition rate was low and, that the texture was lost when copper was plated very close to its limiting current. This corroborated earlier observations of Lashmore and co-workers [32, 33], as well as Moffat [50], and clearly showed that electrochemically fabricated multilayer structures were more complex than the simple pervasive view of horizontally stacked layers being formed during electrodeposition.

Cu/Co metal mutilayers, however, showed a different structural evolution. Shima et al. [53] found that the dissolution of cobalt from a Cu/Co multilayer was dependent on the modulation thickness as well as the thickness of the individual layers. When the Co thickness was small (equivalent to 12 monolayers), then the number of copper layers which completely stopped Co dissolution was equivalent to six

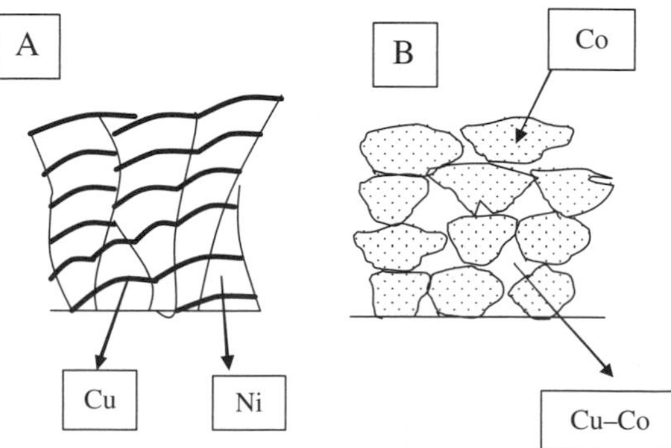

Fig. 6 (**A**) Deposit microstructure for Cu/Ni multilayers, when the current for Cu plating is much lower than the mass-transfer-limiting current. (**B**) The growth of Cu–Co alloys where separate phases are observed

monolayers. However, when the Co layer was thicker, that is, 51 monolayers, then more than 50 monolayers of copper were required to stop the cobalt from dissolving. This indicated that the structure had defects and instabilities which allowed the cobalt to dissolve, similar to the findings reported by Bradley and Landolt [49].

A later investigation by Dulal et al. [54], using a RRDE showed that cobalt was dissolved from beneath a copper layer, even while the electrode potential was held at a potential where copper was plated. Even when the copper layer was relatively thick, (about 50–80 atomic layers), cobalt dissolution continued. An SEM inspection revealed pores through which the underlayer was contacting the electrolyte. The authors surmised that due to the lack of miscibility of Co and Cu, and the fact that the nucleation potential for Co on Cu was higher than Co on Co (and vice versa), Co nuclei grew one over the other, thereby providing "channels" or "columns" for the less noble material to remain in contact with the electrolyte. Bradley and Landolt [49] had observed that Co and Co–Cu phase separation took place when Cu–Co alloys were deposited. In effect, therefore, a Cu–Co structure could be visualized as a "granular" material such a shown in Fig. 6 (B).

Kelley et al. carried out a more detailed examination of Cu/Co multilayers and alloys [55, 56]. They found that Cu–Co alloys developed a phase structure which contained Cu, Cu–Co, and Co. For low modulation thicknesses the Co grains were small, but as the modulation thickness increased the grains took on a fan-like structure. These authors stated that the Cu phase was most likely to have deposited during the period where galvanic corrosion occurred. The Cu–Co intermediate phase was unstable, but could be stabilized by annealing. In effect, the lack of miscibility of copper and cobalt resulted in a microstructure which was considerably more complex than a simple multilayer system.

Based on the discussions presented in this section, it is clear that a simple "layer-by-layer growth" view of electrodeposited CMMs is not applicable. The changes in grain structure, passivation (or lack of it), as well as modulation thickness and changes in composition and phases can lead to deposits which are not only different from those prepared by vacuum techniques, but also vary from those fabricated by different electrochemical methods. These findings show that for fabrication of metal multilayers, considerable effort should be made to obtain reproducible and reliable deposit characteristics.

4 Optimization of Electrodeposition Processes

A more detailed understanding of metal multilayer systems and fabrication led to a search for optimized processes and technology for their fabrication. Process optimization covers different aspects which spread across chemistry, electrochemistry, and engineering and is therefore the realm of interdisciplinary research. In order to progress from science to a production-worthy technology, there are several aspects of CMMM electrodeposition that need to be considered. They are (1) electrolyte stability and electrochemical cell design, (2) control of electrochemical parameters, and (3) methods to control grain growth and interface structure.

4.1 Electrolyte Stability and Electrochemical Cells

One of the main problems for an exploitable electrochemical technology is the development of stable and reliable electrolytes from which two or more metals can be deposited. Although the obvious choice is to use electrolytes which are generally used to plate the alloys of the two metals, there is a subtle but important difference for CMMM deposition. Alloy plating baths are mainly *designed to deposit both metals together, and not individually*. This means that most alloy deposition solutions bring the deposition potentials of the individual metals closer, rather than move them apart. A minimum degree of separation of potentials of the two metals is essential to obtain any degree of composition modulation, and therefore some commercially available electrolytes are unsuitable for electrodepositing CMMMs. A second consideration is that the more noble material, such as copper or silver, which is plated under diffusion controlled conditions, can give dendritic growth. Often, while plating an alloy of such a metal, such as Cu–Ni or Ag–Pd, dendrite growth can be suppressed due to the leveling action of the second component. This luxury is not afforded during the plating of individual layers and, therefore, it is unclear if standard alloy plating electrolytes can be used for multilayer manufacture.

An empirical engineering approach has been presented by Validzadeh et al. [57, 58], who searched for an electrochemical system to fabricate Au/Co or Ag/Co nanostructured multilayers. They used a variety of electrolytes and combined them (without applying a combinatorial approach) to determine if they were suitable for modulated alloy deposition [57]. Thereafter, Taguchi methodology was employed to design experiments in order to obtain CMAs and CMMs [58]. The virtue of such an approach is useful in an industrial context, but the disadvantage is that it does not help to provide a fundamental understanding of the electrochemical system [58].

Roy and co-workers [59, 60] suggested an alternative approach to design stable electrolytes for CMMM deposition. They undertook a thermodynamic analysis of citrate-based electrolytes, which has been used by numerous researchers to electrodeposit Cu/Ni, or Cu/Co multilayers, to identify the complexes in solution. Their analysis was used to determine the species responsible for precipitation, and then choosing the solution parameters to eliminate that species [59]. An in-depth analysis was carried out by determining the limiting current of each species such that the electroactive complexes in solution were also identified [60].

Péter et al., on the other hand, undertook a step-by-step approach to fabricate Cu/Co multilayers [61] from an electrolyte containing metal sulphates and sodium chloride to demonstrate that deposit properties are critically dependent on the chosen electrolyte. The same group deposited a variety of compositionally modulated systems [62]. They carried out detailed experiments with a variety of electrolytes, with and without additives, and measured the property of the deposit obtained in each case, and chose the one with the best performance [61, 62]. Although this approach allows one to examine thoroughly the effect of each electrolyte, it is slow.

The electrochemical cells used in laboratory-scale experiments are, typically, simple coupons placed in an unstirred beaker. However, this is not a desirable situation because passing a current induces natural convection. The problem with such an experimental apparatus is that there is no control over the limiting current of the noble component. Although some authors suggest that using a quiescent solution (i.e., using natural convection) to plate the noble component is better [32, 33], this actually exacerbates the problem, because the degree of natural convection *varies along the electrode length*. The limiting current of the noble material is different at different parts of the electrode, and if the metal were to be plated potentiostatically, different thicknesses (arising from different limiting currents) would be deposited. If the deposition were carried out galvanostatically, the composition of the CMMM would vary with the location on the electrode. Therefore, it is better to control the limiting current by controlling the hydrodynamic conditions.

Since the RDE provides uniform mass transfer, many researchers used an RDE to deposit metal multilayers [37, 38]. However, the problem is that an RDE has nonuniform current distribution which would form alloys of different composition at different parts of the electrode. This was recognized by some researchers [41], who then proposed the uniform injection cell [42]. Other researchers eliminated these effects by choosing to plate at a rotating cylinder electrode [43–45], where current density is uniform and turbulent mass transfer controls the hydrodynamic conditions. Bradley and Landolt [49] and Kelly et al. [55, 56] used an RDE with the disk set in a recess which was facing upward. This provided uniform current density and eliminated problems of gas entrapment. Péter et al. used a horizontal electrochemical cell with an upward-facing cathode [62] to obtain a uniform current distribution.

The problem with all these cells is that they are applicable only for laboratory-scale research. For any industrial exploitation, it is imperative that the electrochemical cell has to be scalable and can be used to plate strip metal, sheet metal, wafers, and other electrical components, which have industrial relevance. Barring very complex shapes, industrial processes require plating on flat surfaces of large diameters (or length). In this regard, the most appropriate plating systems are those described by the Centre for Advanced Electroplating [25, 26], the uniform injection cell [41] and a cell proposed by Roy et al. [63], which has been used for metal deposition on wafers in an industrial production facility [64].

A scaled-down version of the Roy cell was used to deposit Ni/Cu [63, 65], Co/Cu [66], and Co–Ni/Cu [66, 67] multilayers and is presented in Fig. 7. The figure shows the essential concept that the electrode is flush against the wall of the channel, and that the liquid flows past the electrode under forced convection laminar conditions – which is effected by maintaining a narrow channel gap. The proximity of the anode and cathode ensures that the current density is nearly uniform at the cathode. The use of force convection flow ensures that the natural convection effects are negligible. This allows one to exploit the full range of current densities and convective flow regime to fabricate compositionally modulated layers under a variety of conditions. Laboratory- [66] and industrial (-) scale testing [64] have shown that electrodeposited alloys, CMMMs, and metals can be deposited reproducibly.

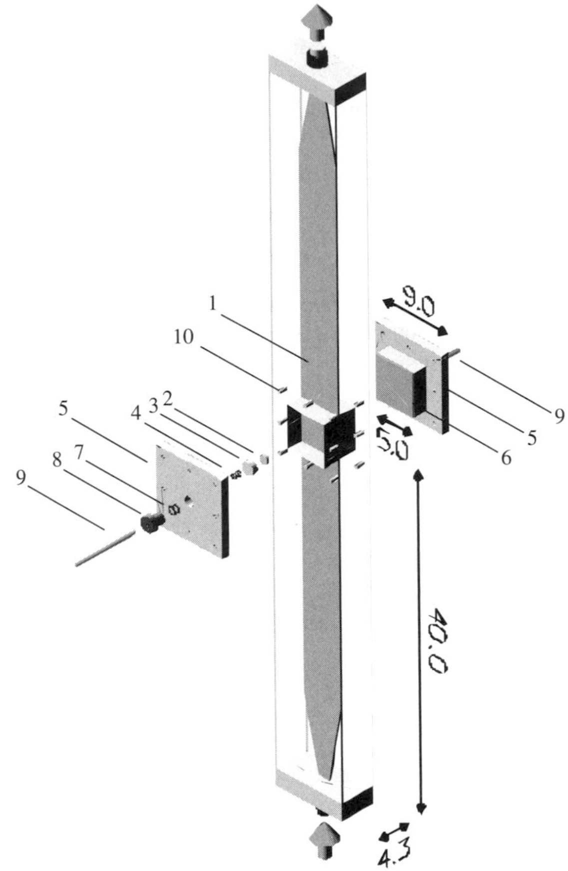

Fig. 7 A schematic of the rectangular flow system used to electrodeposit CMMMs. The system is scalable. The components are (1) flow channel, (2) glass disk, (3) PTEE cup, (4) and spring contact, (5) perspex holder, (6) copper plate, (7) O-rings, (8) PTFE screw, (9) copper electrode contact, and (10) screw holes. Arrows indicate the direction of flow

4.2 Control of Electrochemical Parameters

An issue of even greater importance than electrochemical system is the choice of electrochemical parameters. In earlier work, when CMA and (nearly) pure metals were produced, there was some debate as to whether these structures should be produced under constant current (galvanostatic) or constant potential (potentiostatic) conditions. This arose due to differing deposit properties obtained by the two different schemes [4]. However, such a debate is not valid for any steady-state system, since according to the Butler–Volmer equation, only one single current can exist at a particular imposed potential. The current and potential are similar to state variables, and there should not be a difference whether one imposes a current or a potential. However, as mentioned earlier, it is the dynamic current or potential profile which contributes greatly to the deposit property (since it is the interface that influences deposit properties), and it is the dynamics of the current or potential application that is responsible for these changes.

This question was answered by a systematic analysis by Meuleman et al. [63, 65]. In order to determine the effect of electrochemical parameters on CMMM composition and modulation thickness, they deposited Cu/Ni multilayers using two different schemes: (1) copper and nickel were plated by applying a constant potential and constant current, respectively, and (2) Cu and Ni were plated by applying a constant current, but a relaxation period was inserted in-between the current applications. These researchers developed a charge-balance model to compute the amount of nickel dissolved when the potential or current was switched to plate a copper layer (immediately after nickel plating) and monitored the electrode potential. The model was compared to Cu/Ni multilayers plated in a flow channel. This allowed the researchers to investigate the effect of formation diffuse interface layers and the role of displacement reactions under well-characterized conditions.

For the constant current deposition, a displacement reaction (cf. Fig. 5) proceeded when the current was changed from the high to the low value. The displacement reaction continued until the surface was covered with copper and the nickel dissolution was stifled. The electrode potential, which was monitored during these experiments, showed a gradual change, indicating the formation of a diffuse layer during this process. These results were similar to earlier observations during CMMM deposition [32] or alloy deposition [44]. For the case of constant potential deposition, however, nickel dissolution was stopped within 10 m. During this period less than a monolayer of copper covered the electrode. This indicated that nickel dissolution was stopped due to passivation of the surface, due to the high anodic potential, eliminating the formation of a diffuse layer. Based on deposit analysis, oxygen content was found to be small [32, 45]. This may be because passivation takes place only on particular parts of the Ni surface layer, rather than throughout the deposit.

Péter and co-workers carried out a chronoamperometric analysis for Co/Cu multilayers to calculate the charge of the less noble metal lost by displacement [68] in their optimized deposition cell. They determined the charge of Co dissolved from the deposit when different potentials for copper deposition were applied. Based on this, accurate prediction of the thickness of the Co layer was possible, which allowed them to improve the deposit properties.

4.3 Control of Interface Structure and Properties

Since optimization needs to take into account the roughness induced at the interface, studies have also concentrated on determining the roughness at the interface as a function of deposit growth [66]. Figure 8 shows the roughness of the deposit at the end of copper deposition and cobalt–nickel deposition cycles in the mentioned reference. The figure shows that at the end of the copper deposition cycle the surface is rougher than at the end of the Ni–Co deposition cycle. This is because copper plates close to its limiting current, which promotes roughness, and Ni–Co, which is controlled by reaction kinetics, smoothes the surface. In this fashion the Ni–Co layer counteracts the roughness developed during Cu deposition.

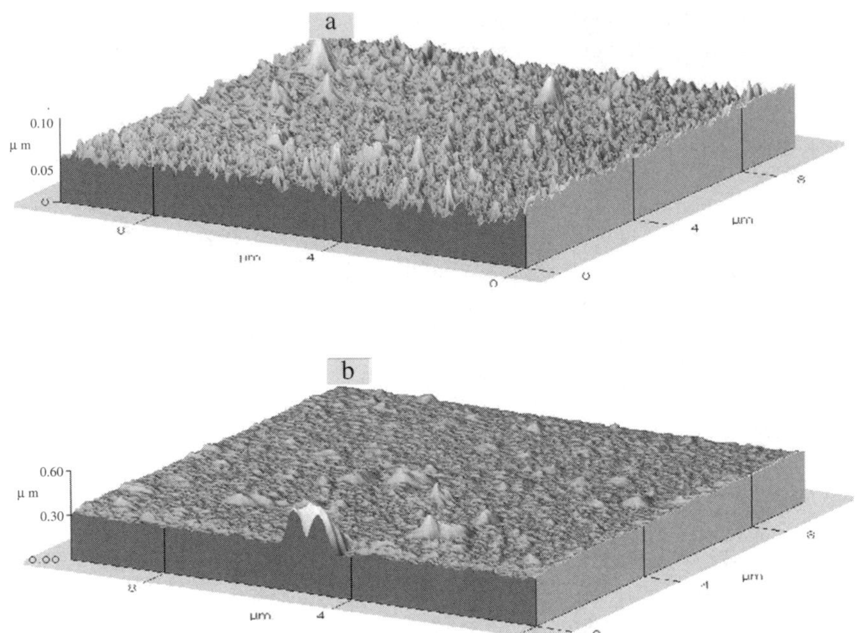

Fig. 8 10 μm × 10 μm AFM images of 50 × Co–Ni/Cu multilayers which have been obtained after the termination of (**a**) the Cu deposition cycle and (**b**) Ni–Co deposition cycle

AFM measurements by Cziráki et al. showed that for Cu–Co alloys, the roughness on the substrate side was small (due to the smaller grain size) and increased in the growth direction [12]. In an another study, Péter et al. showed that the thickness of copper required to entirely cover the cobalt surface was greater than that required to cover cobalt during Co/Cu deposition [69]. This indicated that the roughness of the deposit increased during the copper plating cycle, as is shown in Fig. 8.

The use of additives for control of interfacial structure is less definitive. While citrate has been extensively used to plate Cu–Ni CMAs due to its leveling and grain-refining action, it has been found that the magnetic properties of nickel plated from these electrolytes is poor, probably due to the smaller grain structure [65]. Kelley [55] has shown that using different additives can change the phase and grain structure of Co–Cu alloys and multilayers, which provides a good method for controlling the deposit structure. He demonstrated that saccharin, SDS, and their combination produce different effects on the deposit structure. Less is known about the role of adsorbed intermediates, H^+, OH^-, adsorbed hydrogen (H) and other chemical species, which accompanies metal deposition, although pH has been shown to affect deposit properties [70].

5 Other Compositionally Modulated Systems: Ceramics and Seminconductors

It is important to mention that electrodeposition techniques have also been used to deposit semiconductor and metal-oxide-modulated structures. Although this discussion has mostly been restricted to the deposition for metallic and alloyed CMMs, it is important to mention that the same fabrication techniques and issues dominate the case for nonmetallic systems. Switzer and co-workers [71] reported that they had successfully deposited $Tl_aPb_bO_c/Tl_dPb_eO_f$ structures, that is, two different compositions of Tl-Pb-O ceramic using electrochemical technique. They deposited these ceramic-modulated materials using galvanostatic pulses and used simple charge balance calculations to compute the thickness of modulation. The modulation thickness ranged between 5.9 nm and 8.4 nm and was verified using XRD.

In a later study they used galvanostatic and potentiostatic waveforms to deposit Tl-Pb-O compositionally modulated ceramic superlattices [72]. They carried out scanning tunneling microscopy (STM) to determine the modulation thickness. Interestingly, this technique could be applied to discern between the two different layers because the resistance of the two layers were different. Their analysis showed that potentiostatic waveforms produced square modulations, whereas galvanostatic methods resulted in a gradual change on composition modulation. However, it is unclear if the difference is due to displacement reactions (as would be the case for metallic systems), inherent instrumental problems or other systematic factors. In essence, however, the authors claimed that their ceramic superlattices were comparable in quality to that attained by vacuum techniques.

Interestingly, a variation of the dual-bath electrodeposition technique has been used to deposit semiconductors using underpotential deposition (UPD). UPD is defined as the phenomenon whereby an element deposits on another material below (or under) the potential where it would deposit on itself. This process is due to the favorable energetics leading to formation of a compound. Stickney and co-workers [73] proposed this method to fabricate CdTe using electrochemical atomic layer epitaxy (EC-ALE).

The researchers first deposited Cd from an electrolyte solution containing $CdSO_4$. This was followed by the oxidative deposition of Te from Te^{2}. For this method of deposition, careful consideration for deposition potential has to be made. In addition, judgment related to oxidative versus reductive deposition is essential. Since the formation of a compound (such as CdTe) demands that metals remain at a fixed ratio, a compositional analysis can be made to test if the individual elements are undergoing UPD (where the ratio will be maintained) or depositing at "over" potentials (OPD). Although this method does not produce compositionally modulated structures, it modulates the structures at the atomic scale. The advantage of this method is that it is self-terminating, that is, once the compound is formed, the reaction stops. Unless a second cycle is deposited using the next UPD cycle, the structure does not grow. This allows one to grow a layer-by-layer structure.

Stickney and co-workers also developed a specialized apparatus for EC-ALE deposition of semiconductors [74, 75]. They developed a modified electrochemical cell consisting of a six-way delivery valves, which would control the flow of reactants, rinsing agents (which were blanks of the metal), and waste solution. These reagents were introduced in a sequential fashion into the electrochemical cell and the compound semiconductors were plated. The researchers even adapted the system so that the deposit could be analyzed in a vacuum chamber.

6 Summary

In summary, it has to be noted that electrochemical methods for the fabrication of nanostructured modulated deposits pose good opportunities for exploitation. Electrochemical techniques offer advantages related to low capital cost, good thickness control, and ability to control the texture and interfaces. There are different techniques and apparatus to deposit CMMMs and a choice of an appropriate technique for fabrication should be made. Overall, a dual-bath technique is useful for relatively thick CMMMs (>50-nm layer thickness), whereas a single electrolyte is more useful for depositing thinner modulation thicknesses.

Researchers have shown that a picture of CMMMs being "stacked layers of individual metals" is an oversimplification of a multilayer system. In fact, electrolyte chemistry, displacement reactions, miscibility, and phase structure in the solid state influence the deposit composition, structure, and properties. The ability to obtain high-quality deposits requires attention to various chemical and physical aspects such as control of mass-transfer-limiting current, displacement reactions, additive effects, and proper use of well-designed reactors. For electrochemical processes to truly succeed in depositing CMAs in a large industrial scale, all these aspects need to be considered and designed into the development of any process. This may be one of the reasons that there is still no standard "industrial-scale" process to deposit nanostructured CMMMs. In future, one would expect to see a development in that direction, and the results from all these studies should feed into it.

References

1. T. Tsalakos and A. F Jankowski, "Mechanical Properties of Composition-Modulated Metallic Foils" *Ann. Rev. Mater. Sci.*, **16**, (1986) 293–313.
2. T. Foecke and D. S. Lashmore, "Mechanical Behaviour of Compositionally Modulated Alloys" *Scripta Metallurgica et Materialia*, **27**, (1992) 651–656.
3. R. D. Schull and L. H. Benett, "Nanocomposite Magnetic Materials" *Nanostructured Materials*, **1** (1992) 83–88.
4. M. Alper, K. Attenborough, R. Hart, S. J. Lane, D. S. Lashmore, C. Younes and W. Schwarzacher, "Giant Magnetoresistance in Electrodeposited Superlattices", *Appl. Phys. Lett.*, **63–15** (1993), 2144–2146.
5. J. Tóth, L. F. Kiss, E. Tóth-Kádár, A. Dinia, V. Pierron-Bohnes and I. Bakonyi, "Giant Magnetoresistance and Magnetic Properties of Electrodeposited Ni81Cu19/Cu Multilayers" *J. Magnetism Magnetic Mater.*, **198–199**, (1999) 243–245.

6. M. E. Bahrololoom, D. R. Gabe and G. D. Wilcox, "Development of a Bath for Electrodeposition of Zinc-Cobalt Compositionally Modulated Alloy Multilayered Coatings", *J. Electrochem. Soc.*, **150–3** (2003) C144–C151.
7. J. D. Jensen, "Engineering Metal Microstructures: Process-Microstructure-Property Relationships for Electrodeposits", *Dissertation no. 784*, Linkoping Studies in Science and Technology, Linkoping Universitet, SE-581 83 Linkoping, Sweden. ISBN: 91-7373-458-6.
8. J. D. Jensen, D. R. Gabe and G. D. Wilcox, "The Practical Realisation of Zinc-Iron CMA Coatings", *Surf. Coatings Tech.*, **105** (1998) 240–250.
9. J. D. Jensen, G. W. Critchlow and D. R. Gabe, "A Study on Zinc-Iron Alloy Electrodeposition from a Chloride Electrolyte" *Trans. Inst. Met. Finish.*, **76–5** (1998) 187–191.
10. S. K. J. Lenczowski, C. Schönenberger, M. A. M. Gijs and W. J. M. Jonge, "Giant Magnetoresistance of Electrodeposited Co/Cu Coatings" *J. Magnetism Magnetic Mater.*, **148** (1995) 455–465.
11. A. Blondel, J. P. Meier, B. Doudin and J.-Ph. Ansermet, "Giant Magnetoresistance of Nanowires of Multilayers" *Appl. Phys. Lett.* **65** (1994) 3019–3021.
12. Á. Cziraki, L. Péter, V. Weihnacht, J. Tóth, E. Simon, J. Pádár, L. Pogány, C. M. Schneider, T. Gemming, K. Wetzig, G. Tichy and I. Bakoyi, "Structure and Giant Magnetoresistance Behaviour of Co-Cu/Cu Multilayers Electrodeposited Under Various Conditions", *J. Nanosci. Nanotechnol.*, **6–7**, (2006) 2000–2012.
13. W. Schwarzacher and D. S. Lashmore, "Giant Magnetoresistance in Electrodeposited Films" *IEEE Trans. On Magnetics*, **32** (1996) 3133–3153.
14. J. P. Celis, D. Drees, M. Z. Huq, P. Q. Wu and M. De Bonte, "Hybrid Processes – a Versatile Technique to Match Process Requirements and Coating Needs" *Surf. Coating. Tech.*, **113** (1999) 165–181.
15. Y. D Gamburg, "Electrodeposition of Alloys with Composition Modulated over Their Thickness: A Review", *Transl. from Electrokhimiya*, **37–6**, (2001) 686–692.
16. W. Blum, *Trans. A. Electrochem. Soc.*, **40** (1921) 307–320.
17. C. Ogden, "High Strength Composite Copper-Nickel Electrodeposits" *Plat. Surf. Finishing*, **73**, (1986) 130–134.
18. L. M. Goldman, B. Blancpain, F. Spaepen, "Short Wavelength Compositionally Modulated Ni/Ni-P Films Prepared by Electrodeposition", *J. Appl. Phys.*, **60**, (1986) 1374–1376.
19. L. M. Goldman, C. A. Ross, W. Ohashi, D. Wu and F. Spaepen, "New Dual Bath Technique for Electrodeposition of Short Repeat Length Multilayers", *Appl. Phys. Lett.*, **55–21**, (1989) 2182–2184.
20. C. A. Ross, L. M. Goldman and F. Spaepen, "An Electrodeposition Technique for Producing Multilayers of Nickel-Phosphorus and Other Alloys", *J. Electrochem. Soc.*, **140**, (1993), 91–98.
21. C. Ross, "Electrodeposited Multilayer Thin Films", *Annu. Rev. Mater. Sci.*, **24**, (1994) 159–188.
22. G. Wouters, J.-P. Celis and J. R. Roos, "The Electrocrystallisation of Compositionally Modulated Multilayers of Tin and Amorphous Nickel-Phosphorus", *J. Electrochem. Soc.*, **140**, (1993) 3639–3643.
23. G. Wouters, M. Bratoeva, J.-P. Celis and J. R. Roos, "Electrochemical Diagnostic of Bright Tin Deposition in View of the Electrolytic Synthesis of Ni-P/Sn Multilayers", *J. Electrochem. Soc.*, **141**, (1994) 397–401.
24. A. S. M. A. Haseeb, J.-P. Celis and J. R. Roos, "Dual Bath Electrodeposition of Cu/Ni Compositionally Modulated Multilayers", *J. Electrochem. Soc.*, **141**, (1994) 230–237.
25. P. T. Tang, P. Leisner, P. Møller, C. Neilsen and D. M. Nabirani, "Dual Bath Plating Composition Modulated Alloys (CMA) Based on a Newly Developed Computer Controlled Plating System", *SUR/FIN'94*, June 20–23 (1994), Indianapolis.
26. D. M. A, Nabirani, P. T. Tang and P. Leisner, "The Electrolytic Plating of Compositionally Modulated Alloys and Laminated Metal Nano-structures Based on an Automated Computer-Controlled Dual-Bath System", *Nanotechnology*, **7**, (1996) 134–143.

27. A. Brenner, "Electrodeposition of Alloys" Vol. II, p. 589, Academic Press, New York (1963).
28. U. Cohen, K. R. Walton and R. Sard, "Electroplating of Cyclic Multilayered Alloy Plating for Electrical Contact Applications", *J. Electrochem. Soc.*, 130, (1983). 1987–1995.
29. J. Yahalom and O. Zadok, "Formation of Composition-Modulated Alloys by Electrodeposition" *J. Mater. Sci.*, **22**, (1987) 499–503.
30. J. Yahalom and O. Zadok, "Method for the Production of Alloys Possessing High Elastic Modulus and Improved Magnetic Properties by Electrodeposition" US Patent No. 4652348 (1987).
31. J. Yahalom, D. F. Tessier, R. S. Timsit, A. M. Rosenfeld, D. F. Mitchell and P. T. Robinson, "Structure of Composition-Modulated Cu/Ni Thin Films Prepared by Electrodeposition" *J. Mater. Res.*, **4**, (1989) 755–758.
32. D. S. Lashmore and M. P. Dariel, "Electrodeposited Cu-Ni Textured Superlattices" *J. Electrochem. Soc.*, **135**, (1988) 1218–1221.
33. M. Dariel, L. H. Bennett, D. S. Lashmore, P. Lubitz, M. Rubinstein, W. L. Lechter and M. Z. Hartford, "Properties of Electrodeposited Co-Cu Multilayer Structures", *J. Appl. Phys.*, **61–18**, (1987), 4067–4069.
34. D. Tench and J. White, "Tensile Properties of Nanostructured Ni-Cu Multilayered Materials Prepared by Electrodeposition" *J. Electrochem. Soc.*, **138** (1991) 3757–3758.
35. D. M. Tench and J. T. White, "Considerations in Electrodeposition of Compositionally Modulated Alloys" *J. Electrochem. Soc.*, **137**, (1990) 3061–3066.
36. D. M. Tench and J. T. White, "A New Periodic Displacement Method Applied to Electrodeposition of Cu-Ag Alloys" *J. Electrochem. Soc.*, **139**, (1992) 443–446.
37. A. R. Despic and V. D. Jovic, "Electrochemical Formation of Laminar Deposits of Controlled Structure and Composition: 1. Single Current Pulse Galvanostatic Technique" *J. Electrochem. Soc.*, **134**, (1987) 3004–3011.
38. A. R. Despic, V. D. Jovic and S. Spaic, "Electrochemical Formation of Laminar Deposits of Controlled Structure and Composition: 1. Dual Current Pulse Galvanostatic Technique" *J. Electrochem. Soc.*, **136**, (1989) 1651–1657.
39. D. T. Schwartz, P. Stroeve and B. G. Higgins, "Electrodeposition of Composition-Modulated Alloys in Fluctuating Flow Field" *AIChE J.*, **35**, (1989) 1315–1327.
40. D. T. Schwartz, "Multilayered Alloys Induced by Fluctuating Flow" *J. Electrochem. Soc.*, **138**, (1989) 53C–56C.
41. S. D. Leith and D. T. Schwartz, "Flow-Induced Composition Modulated Ni-Fe Thin Films with Nanometer-Scale Wavelengths" *J. Electrochem. Soc.*, **143** (1996) 873–878.
42. J. A. Medina and D. T. Schwartz, "Electrodeposition of Flow Induced Composition Modulated NiFe Alloys in the Uniform Injection Cell" *Electrochim. Acta*, **42**, (1997) 2679–2684.
43. S. Roy, "An Analytical Equation to Compute the Composition of Pulse Plated Binary Alloys" *Plat. Surf. Finish.*, **76** (1999) 202–205.
44. S. Roy, M. Matlosz and D. Landolt, "Effect of Corrosion on the Composition of Pulse-Plated Cu-Ni Alloys", *J. Electrochem. Soc.*, **141**, (1994) 1509–1517.
45. S. Roy and D. Landolt, "Effect of Off-Time on the Composition of Pulse Plated Cu-Ni Alloys", *J. Electrochem. Soc.*, **142**, (1995) 3021–3027.
46. P. E. Bradley and D. Landolt, "A Surface Coverage Model for Pulse Plating of Binary Alloys Exhibiting a Displacement Reaction" *J. Electrochem. Acta*, **42**, (1997) 993–1003.
47. E. Chaissang, "In-Situ Mass Changes and Stress Measurements in Cu/$Fe_{20}Ni_{80}$ Electrodeposited Multilayers" *J. Electrochem. Soc.*, **144**, (1997) L328–L330.
48. E. Chaissang, A. Morrone and J. E. Schmidt, "Nanometric Cu-Co Multilayers Electrodeposited on Indium-Tin Oxide Glass" *J. Electrochem. Soc.*, **146**, (1999) 1794–1797.
49. P. E. Bradley and D. Landolt, "Pulse Plating of Copper Cobalt Alloys" *Electrochim. Acta*, **45**, (1999) 1077–1087.
50. T. P. Moffat, "Electrochemical Production of Single-Crystal Cu-Ni Strained-Layer Superlattices on Cu(100)", *J. Electrochem. Soc.*, **142** (1995) 3767–3770.

51. A. Cziráki, L. Péter, B. Arnold, J. Thomas, H. D. Bauer, K. Wetzig and I. Bakonyi, "Structural Evolution During Growth of Electrodeposited Co-Cu/Cu Multilayers with Giant Magnetoresistance" *Thin Solid Films*, **424**, (2003) 229–238.
52. Ch. Bonhôte and D. Landolt, "Microstructure of Ni-Cu Multilayers Electrodeposited from a Citrate Electrolyte", *Electrochim. Acta*, **42**, (1997) 2407–2417.
53. M. Shima, L. Salamanca-Riba and T. P. Moffat, "Dissolution of Artifically Structured Materials", *Electrochem. Solid-State Lett.*, **2** (1999) 271–274.
54. S. M. S. I. Dulal, E. A. Charles and S. Roy, "Dissolution from Electrodeposited Copper-Cobalt-Copper Sandwiches" *J. Appl. Electrochem.*, **34**, (2004) 151–158.
55. J. J. Kelley, P.E. Bradley and D. Landolt, "Additive Effects during Pulsed Deposition of Cu-Co Nanostructures", *J. Electrochem. Soc.*, **147**, (2000) 2975–2980.
56. J. J. Kelley, M. Cantoni and D. Landolt, "Three Dimensional Strcuturing of Electrodeposited Cu-Co Multilayer Alloys" *J. Electrochem. Soc.*, **148**, (2001) C620–C626.
57. S. Validzadeh, G. Holmbom and P. Leisner, "Electrodeposition of Cobalt-Silver Multilayers" *Surf. Coatings Technol.*, **105** (1998) 213–217.
58. S. Valizadeh, E. B. Svedberg and P. Leisner, "Electrodeposition of Compositionally Modulated Au/Co Alloy Layers" *J. Appl. Electrochem.*, **32** (2002) 97–104.
59. Todd Green, A. E. Russell and S. Roy, "The Development of a Stable Citrate Electrolyte for the Electrodeposition of Copper-Nickel Alloys" *J. Electrochem. Soc.*, **145**, (1998) 875–881.
60. W. R. A. Meuleman and S. Roy, "Electrochemical Characterisation of Copper Deposition from Citrate Solutions" **PV 99-33**, Ed M. Matlosz and D. Landolt, The Electrochemical Society, Inc., (2000) Pennington, N.J., USA, pp 61–70.
61. L. Péter, Z. Kupay, A. Cziráki, J. Pádár, J. Tóth and I. Bakonyi, "Additive Effects in MultilayerElectrodepositon: Properties of Co-Cu/Cu Multilayers Deposited with NaCl Additive", *J. Phys. Chem., B*, **105**, (2001) 10867–10873.
62. V. Weinacht, L. Péter, J. Tóth, J. Pádár, Zs Kerner, C. M. Schneider and I Bakonyi, "Giant Magnetoresistance in Co-Cu/Cu Multilayers Prepared by Various Electrodeposition Modes", *J. Electrochem. Soc.*, **150**, (2003) C507–C515.
63. W. R. A. Meuleman, S. Roy, L. Péter and I. Varga, "Effect of Current and Potential Waveforms on Sublayer Thickness of Electrodeposited Copper Nickel Multilayers" *J. Electrochem. Soc.*, **151**, (2002) C479–C486.
64. S. Roy, Y. Gupte and T. A. Green, "Flow Cell Design for Metal Deposition at Recessed Circular Electrodes and Wafers, *Chem. Eng. Sci.*, **56**, (2001) 5025–5035.
65. W. R. A. Meuleman, S. Roy, L. Péter and I. Bakonyi, "Effect of Current and Potential Waveforms on GMR Characteristics of Electrodeposited Ni(Cu)/Cu Mutlilayers", *J. Electrochem. Soc.*, **149**, (2002) C479–C486.
66. S. M. S. I Dulal, E. A. Charles and S. Roy, "Characterisation of Co-Ni(Cu)/Cu Multilayers Deposited from a Citrate Electrolyte in a Flow Channel Cell" *Electrochim. Acta*, **49** (2004) 2041–2049.
67. G. Nabiyouni, O. I. Kasyutich, S. Roy and W. Schwarzacher, "Co-Ni-Cu/ Cu Multilayers Electrodeposited Using a Channel Flow Cell" *J. Electrochem. Soc.* **149**, (2002) C218–C222.
68. L. Péter, Q. Liu, Z. Kerner and I. Bakonvi, "Relevance of Potentiodynamic Method in Parameter Selection for Pulse Plating of Co-Cu/Cu Multilayers", *Electrochim. Acta*, **49** (2004) 1513–1526.
69. L. Péter, A. Cziráki, L. Pogány, Z. Kupay, I. Bakonyi, M. Uhlemann, M. Herrich, B. Arnold, T. Bauer and K. Wetzig, "Microstructure and Giant Magnetoresistance of Electrodeposited Co-Cu/Cu Multilayers" *J. Electrochem. Soc.*, **148** (2001) C168–C176.
70. M. Alper and W. Schwarzacher, "The Effect of pH Changes on the Giant Magnetoresistance of Electrodeposited Superlattices" *J. Electrochem. Soc.*, **144** (1997), 2346–2352.
71. J. A. Switzer, M. J. Shane, R. J. Phillips, "Electrodeposited Ceramic Superlattices" *Science*, **247**, (1990) 444–446.

72. J. A. Switzer, R. P. Raffaelle, R. J. Phillips, C.-J. Hung and T. D. Golden, "Scanning Tunnelling Microscopy of Electrodeposited Ceramic Superlattices" *Science*, **258**, (1992) 1918–1921.
73. B. W. Gregory, D. W. Suggs and J. L. Stickney, "Conditions for the Deposition of Cd-Te by Electrochemical Atomic Layer Epitaxy", *J. Electrochem. Soc.*, **138**, (1991) 1279–1284.
74. L. P. Colletti and J. L. Stickney, "Optimization of the Growth of CdTe Thin Films Formed by Electrochemical Atomic Layer Epitaxy in an Automated Deposition System" *J. Electrochem. Soc.*, **145**, (1998) 3594–3602 (1998).
75. L. P. Colletti, B. H. Flowers Jr. and J. L. Stickney, "Formation of Thin Films of Cd-Te, Cd-Se, and Cd-S by Electrochemical Atomic Layer Epitaxy" *J. Electrochem. Soc.*, **145**, (1998) 1442–1449.

Corrosion at the Nanoscale

Vincent Maurice and Philippe Marcus

1 Introduction

In a variety of new technologies, materials with nanoscale architecture or thin functional layers on their surfaces are being used and will be developed in the near future. Like all materials, these new materials can be used only if they are stable against environmental degradation (corrosion). However, the loss of material due to corrosion which is accepted for conventional materials (i.e., mm/year) is no longer acceptable and must be controlled and limited to values less than 10–100 nm. The concept of corrosion at the nanoscale is not restricted to metallic materials in the classic sense, but applies to all products incorporating small-size metallic parts (e.g., contacts, interconnects, displays, implants, and sensors), and ultrathin layers conferring the relevant property (or function) to the material surface. To obtain good stability, protective films of nanometer thickness can be used, but the protective effect must not be obtained at the expense of other surface functionalities. To increase the durability of the protection, self-repair of the ultrathin protective films is a key factor.

To control the stability of these materials, it is therefore essential to develop our knowledge of corrosion and its prevention at the nanoscale. The usual chemical or structural heterogeneities that cause corrosion in conventional materials (e.g., inclusions and microstructure) must not be overlooked, but are not necessarily the major issue. Defects related to the nanostructure of the materials and their protective ultrathin layers become the major causes of corrosion at the nanoscale. Furthermore, if appropriately controlled, the process of corrosion at the nanoscale can be turned to an advantage, leading to novel ways of manufacturing nanostructured materials.

V. Maurice (✉)
Laboratoire de Physico-Chimie des Surfaces, CNRS-ENSCP (UMR 7045), Ecole Nationale Supérieure de Chimie de Paris, Université Pierre et Marie Curie, 11, rue Pierre et Marie Curie, 75231 Paris Cedex 05, France
e-mail: vincent maurice@enscp.fr

P. Schmuki, S. Virtanen (eds.), *Electrochemistry at the Nanoscale,* Nanostructure Science and Technology, DOI 10.1007/978-0-387-73582-5_10,

The methods that are necessary for investigating the corrosion mechanisms must operate at the appropriate scale (the nanoscale) and in the relevant environment (aqueous solution): these are in situ nanoprobes (scanning tunneling microscopy (STM) and atomic force microscopy (AFM)). They can be combined with ex situ surface-sensitive analytical tools (X-ray photoelectron spectroscopy (XPS), Auger electron spectroscopy (AES), secondary ion mass spectrometry (SIMS)), some of them reaching nanometer-scale resolution (AES, SIMS) and other nonlocal in situ techniques (e.g., infrared reflection absorption spectroscopy (IRRAS)).

The purpose of this chapter is to highlight recent advances in the understanding of the stability and degradation of metallic materials in electrochemical environments based on studies of corrosion and corrosion protection performed at the nanometer and subnanometer (i.e., molecular and atomic) scale. Results obtained on model single-crystal metallic surfaces with in situ scanning probe microscopies (STM, AFM) are emphasized. The (nano)structural aspects of anodic dissolution, protection by corrosion inhibitors, passivation by ultrathin anodic oxide films, passivity breakdown, and initiation of localized corrosion are presented and discussed.

2 Corrosion and Protection of Materials for the Nanoelectronics: The Case of Copper

To meet the demand for decreasing dimensions, aluminum has been replaced by copper as an interconnect metal in the newly developed integrated circuits for the microelectronic industry. The production technology of interconnects involves a planarization process where metal, after deposition, is removed by combined electrochemical dissolution and mechanical polishing, leaving the polished copper interconnects in the trenches and vias only. Conditions have to be such that the metal remains passive (i.e., protected from dissolution by an ultrathin anodic oxide film) when not in contact with the polishing pad, and repassivates quickly after activation (i.e., dissolution). Due to the nanometer dimensions of the final interconnects which are expected to be reached within the next few years, this technology requires to control the corrosion and protection of the metal at the (sub)nanometer scale. Therefore, understanding fundamental corrosion processes for Cu such as active dissolution and protection by inhibitors or passive films grown on the surface at the subnanometer scale becomes not only interesting from the academic point of view, but also essential to the development of new technologies.

2.1 Active Dissolution

The active dissolution of copper has recently been studied at the subnanometer (i.e., molecular and atomic) scales with scanning probe microscopes implemented with electrochemical control (mostly ECSTM).[1, 2, 3, 4, 5, 6, 7]. These studies provide striking examples of the recent advances made in the understanding of the fundamental mechanisms of corrosion processes. Investigations have focused on the structural modifications of atomically flat surfaces resulting from the adsorption of anions in the double-layer potential region preceding the onset of dissolution and on the dissolution of the metal atoms at specific sites. The dynamics of these processes have been studied in some cases. Results obtained on Cu(001) single-crystal surfaces are chosen here to illustrate this topic.

A major result evidenced by nanometer scale studies is that dissolution proceeds layer by layer via a step-flow mechanism at moderate potentials (i.e., slow etching rate). This process is illustrated in Fig. 1 for Cu(001) in H_2SO_4 and HCl [5]. The preferential etching of the surface takes place at the defects corresponding to the preexisting step edges. The selective etching of atoms at these sites results from their lower coordination to nearest neighbor atoms. The superstructure formed by strongly adsorbed ions in the double-layer region influences the anisotropy of the etching as illustrated in Fig. 1. On Cu(001), a disordered and possibly mobile sulfate adlayer is formed in sulfuric acid solution and the steps are etched isotropically without any preferential direction [4, 5]. In contrast, a highly ordered c(2×2) adlayer is formed in hydrochloric or hydrobromic acid solutions, that stabilizes the step edges along the close-packed <100> directions of the superstructure [2, 3, 4, 5, 7].

In the absence of strongly adsorbed anions, the etching process, at moderate overpotentials, stabilizes the step edges oriented along the close-packed crystallographic directions of the crystal, where the nearest-neighbor coordination of the

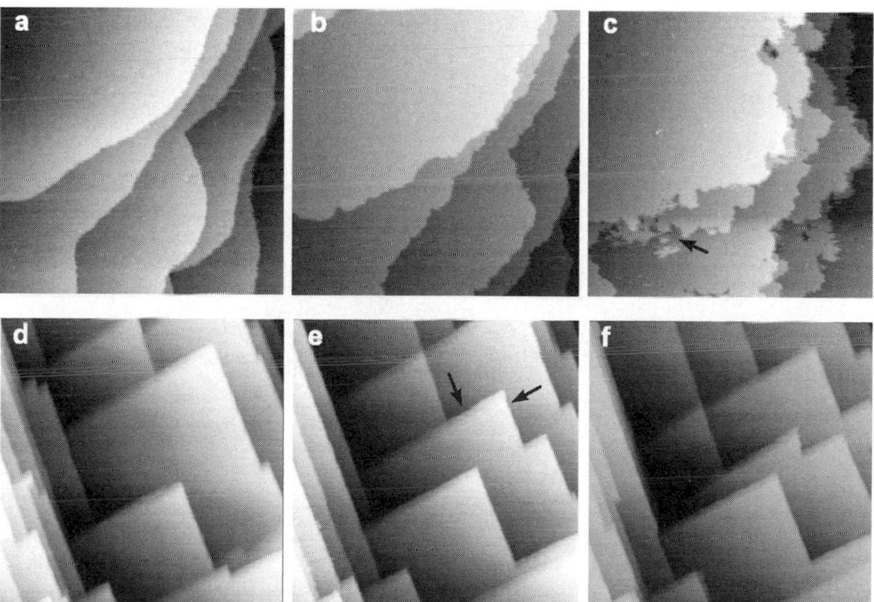

Fig. 1 ECSTM images of Cu(001) recorded (**a**), (**b**), (**c**) in 10 mM H_2SO_4 (180 nm × 180 nm) and (**d**), (**e**), (**f**) in 1 mM HCl (160 nm × 160 nm), showing the isotropic and anisotropic step flow caused by anodic dissolution in H_2SO_4 and HCl, respectively. From ref. [5]

atoms is maximum, for example, along the atomically smooth <1-10> directions as observed on Cu(111) [1]. This is also observed on Ni(111) [8] and Ag(111), [9] and, accordingly, along <10-10> on Co(0001) [10].

Figure 2a shows the observed square lattice corresponding to the adsorbed Cl layer on Cu(001). Atomic-scale imaging shows that the dissolution proceeds at structurally well-defined kinks of the [100]- and [010]-oriented steps *via* removal of the primitive unit cells of the Cl adlayer containing two Cu atoms and one adsorbed

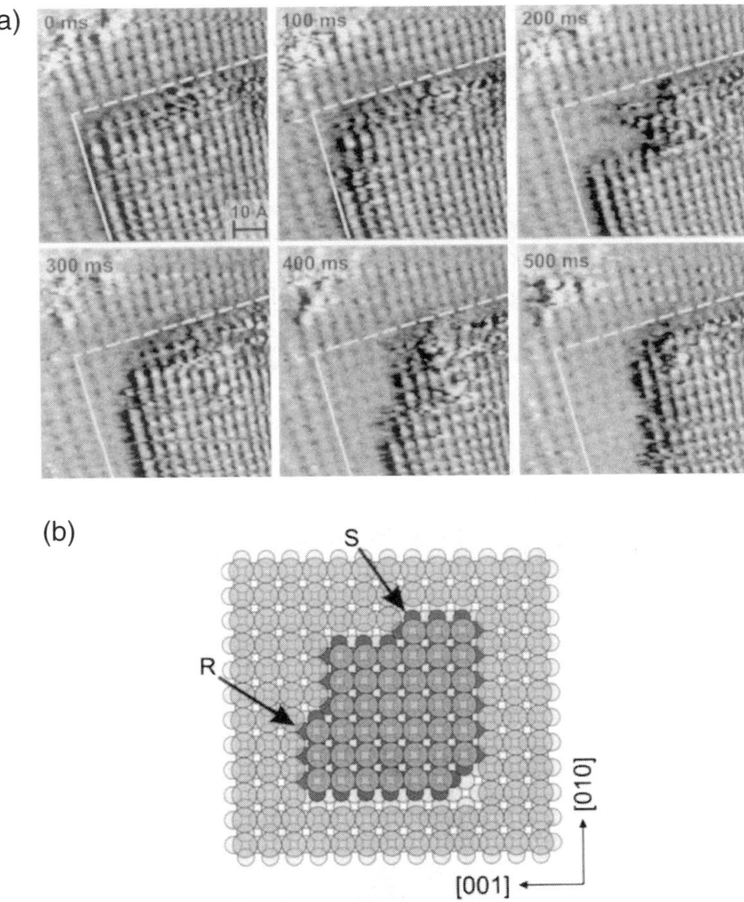

Fig. 2 (**a**) Series of fast-scan ECSTM images (100 ms/image) recorded on Cu(001) in 0.01 M HCl at −0.17 V/SCE showing the progressing dissolution of a Cu terrace starting at the outer terrace corner. The observed lattice is the c(2×2) ad-lattice formed by adsorbed chlorine. The markedly different dissolution behavior at the active and stable steps whose initial position is marked by the solid and dashed lines, respectively, is clearly visible. (**b**) Model of the (001) surface of a face-centered cubic metal (e.g. Cu) covered by an ordered c(2×2) ad-layer (e.g., Cl) showing the different structure of steps and kinks along the [010] and [001] directions. R and S mark the reactive and stable Cu atoms at the outer kinks, respectively. From ref. [7]

Cl atom. In the double-layer potential range, equilibrium fluctuations of these step edges are observed that result from local removal/redeposition processes. These fluctuations are also illustrated by the sequence of fast-san images shown in Fig. 2 [4, 7]. At more anodic potentials, the dissolution process prevails, and a net removal is observed, but the redeposition still occurs.

Another effect of the c(2×2) adlayer is an induced anisotropy of the dissolution process along the symmetrical <100> directions of the step edges [4, 5, 7, 14]. This is also visible in Fig. 2a where one orientation of the step edges dissolves more rapidly than the other. This is related to the structural anisotropy of the step edges induced by the c(2×2) adlayer, as illustrated by the model shown in Fig. 2b. This has been tentatively explained by the coordination of the outmost Cu atom forming the dissolving kink to the adjacent Cl adsorbates [7]. At the reactive kink site, the Cu atom is closely coordinated to two adsorbed Cl atoms, whereas at the stable kink site it is closely coordinated to one adsorbed Cl atom. Hence the formation of the $CuCl_2^-$ dissolving complex would be already initiated at the reactive kink site, generating a preferential dissolution along one of the symmetrical <100> directions. A similar anisotropic effect of the c(2×2) ordered adlayer on the dissolution of the (001)-oriented metal substrate has been reported for Br on Cu [5], S on Ni [11], I on Ag [12] and Pd [13, 14].

2.2 Protection by Corrosion Inhibitors

Subnanometer scale data on the mechanisms of protection by corrosion inhibitors have also been obtained from studies performed below and at or near the onset of dissolution on well-prepared single-crystal surfaces exhibiting a terrace and step topography. The effect of benzotriazole (BTAH, $C_6H_4N_3H$) has been studied in sulfuric, hydrochloric, and perchloric acid solutions on the Cu(001) surface [5, 15, 16, 17]. The effect of the surface structure has been addressed by a comparative study on Cu(111) in sulfuric acid solution [18] and on Cu(111) and Cu(110) in perchloric acid solution [17]. The effect of strongly bonded self-assembled monolayers (alkanethiol), modeling corrosion inhibitors, has also been reported for Cu(001) in HCl solution [19].

Figure 3 illustrates at the nanometer scale the modified mechanism of dissolution in the presence of a corrosion inhibitor. Instead of proceeding uniformly by a step-flow process as described above, the Cu dissolution in the presence of the inhibitor is observed to proceed locally, leading to the nucleation and growth of monoatomic pits resulting in the roughening of the surface. The effect has been observed for submonolayer coverage of the inhibitor. It can be understood by the stabilization or "passivation" of the steps against dissolution which occurs even for submonolayer coverage, assuming that the inhibiting species adsorb preferentially at these defect sites. The pits presumably form at nonprotected sites corresponding to the defects of the corrosion inhibiting adlayer. An anodic shift of the onset of dissolution ranging

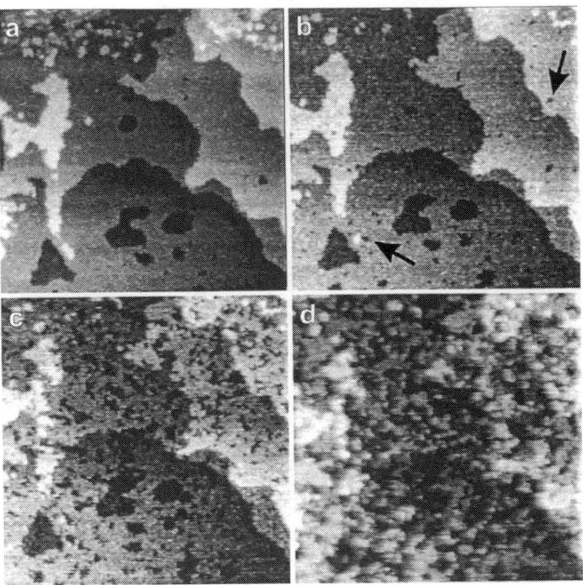

Fig. 3 Series of ECSTM images (150 nm × 150 nm) showing the anodic dissolution of Cu(111) in 0.01 M H_2SO_4 + 10^{-4} M BTAH recorded at −0.33 V/SCE (**a, b**) and −0.06 V (**c, d**). The arrows in (**b**) point the formation of etch pits. From ref. [18]

from 0.05 to 0.3 V has been observed with ECSTM, depending on the concentration of the BTAH in the electrolyte.

The reduced inhibiting efficiency of the BTAH on the dissolution of Cu in the presence of chloride has been addressed by comparative observations made in the double-layer potential region in sulfuric and hydrochloric acid solutions on Cu(001). In both solutions, commensurate ordered adlayers of the BTAH have been observed for potentials $<-0.6\ V_{SCE}$. They form a (5×5)R26.6° structure assigned to a parallel stacking of BTAH molecules oriented perpendicular (or slightly tilted) to the Cu(001) surface. At potentials $>-0.6\ V_{SCE}$ in HCl solutions, the structure as well as the dynamic behavior of the surface are identical to that found in the absence of inhibitor and described above. Figure 4 illustrates the displacement of the ordered structure of adsorbed BTAH species by adsorbed chloride ions forming the same c(2×2) structure at the same potential as in BTAH-free solutions.

In contrast, in sulfuric acid solutions, the BTAH adlayer is observed in the entire underpotential region with formation of a chain-like structure with increasing potential assigned to the polymerization of the BTAH adlayer. These differences in the initial stages of growth of the inhibiting Cu(I)BTA film that forms at higher anodic potential in the Cu(I) oxidation region have been invoked to explain the lower inhibition efficiency in chloride-containing solution. In HCl solution, the direct conversion of the chemisorbed BTAH layer into the polymerized Cu(I)BTA phase is hindered by the formation of the c(2×2) adlayer of adsorbed chloride, which is not the case in H_2SO_4 solution. Consequently, in HCl solution, BTAH or Cu(I)BTA is not present on the surface in the dissolution potential range and the dissolution proceeds by the same step-flow mechanism. The dissolving species ($CuCl_2^-$) reacts in the solution

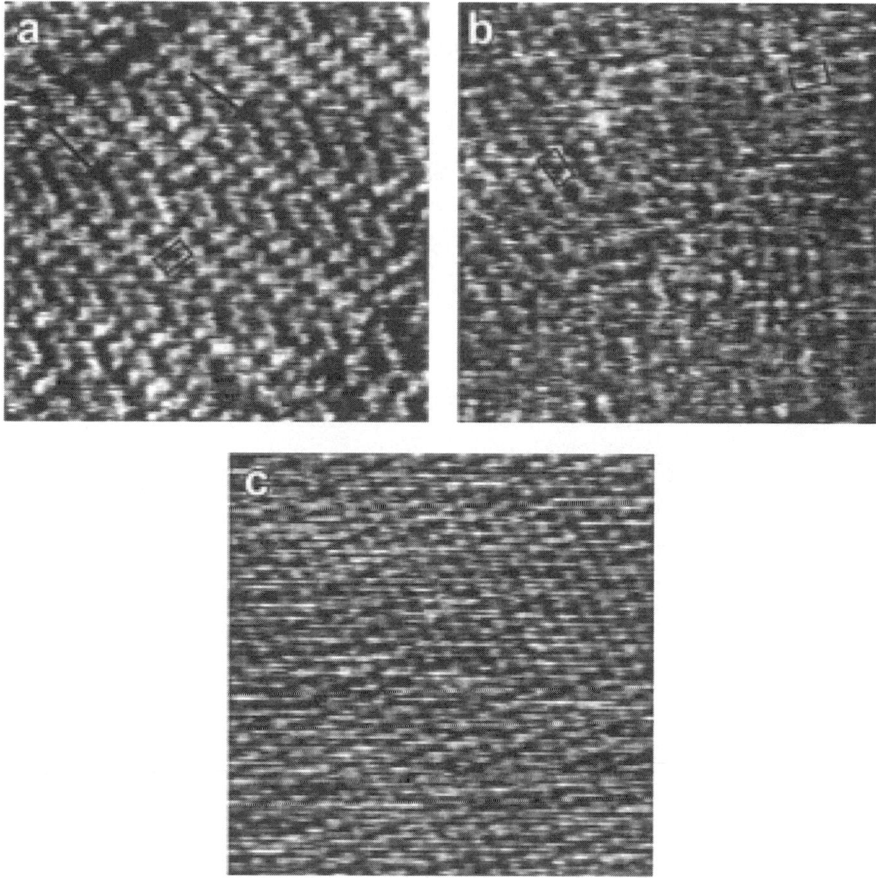

Fig. 4 ECSTM images of Cu(001) in 0.1 M HCl + 0.075 M BTAH showing large (**a**) and smaller (**b**) ordered domains of the chemisorbed BTAH adlayer at – 0.7 V/SCE (10 nm × 10 nm) and (**c**) the c(2×2) ordered adlayer at – 0.45 V/SCE (7.5 nm × 7.5 nm). From ref. [16]

with BTAH to form Cu(I)BTA, which subsequently precipitates on the Cu surface. However, this dissolution/redeposition mechanism produces an inhibiting layer less protective against corrosion than the inhibiting layer directly produced in the Cu(I) oxidation range by a surface reaction between adsorbed polymerized BTAH and Cu(I) ions in sulfuric acid solutions.

On copper, nanoscale ECAFM or ECSTM observations have also been combined with other techniques like quartz-crystal microbalance (QCM), electrochemical impedance spectroscopy (EIS), XPS, AES, Fourier-transform infrared (FTIR), and/or infrared reflection absorption spectroscopy (IRRAS) with the objective of optimizing the corrosion resistance. For example, ECSTM was used to study the corrosion inhibition of copper by 2-mercaptobenzoxazole (MBO) [20]. The topographs showed the rapid formation of a compact three-dimensional film on the surface, optimizing inhibition, and containing cuprous ions and S and N with a modified chemical environment determined by XPS. It was concluded that the inhibitor reacts with Cu(I) species, resulting from dissolution, to produce a water-insoluble copper complex constituting the inhibiting film. ECAFM was used to optimize the corrosion protection by sodium heptanoate, a nontoxic "green" inhibitor [21]. It was observed that in the optimal conditions of inhibition (pH 8, $[C_7H_{13}O_2Na]$=0.08 M), a continuous metallic soap thin film was formed, mainly constituted of copper heptanoate acting as a blocking barrier, whereas, at pH 5.7 or 11, a discontinuous and therefore poorly protective film was formed, constituted of crystals of copper heptanoate or copper oxides.

2.3 Passivation

The structural aspects of the passivation of copper have also been recently studied at the nanometer and subnanometer scales on well-defined single-crystal Cu(111) and Cu(001) surfaces [22, 23, 24, 25, 26, 27], bringing new insight into the mechanisms of self-protection of the metal against corrosion in aqueous solutions. Figure 5 shows typical polarization curves characterizing the growth of the passive films on copper in sodium hydroxide solution. The surface structures obtained in the Cu(I) and Cu(II) oxidation potential ranges as well as in the potential range preceding the oxide formation have been clarified. Epitaxial crystalline Cu_2O films with a nanostructured surface resulting from faceting are formed in the Cu(I) oxidation range. They grow a few degrees off the (111) and (001) directions on Cu(111) and Cu(001), respectively. Their growth is preceded by a surface reconstruction of the copper surface induced by the adsorption of hydroxide species, forming a structural precursor for the growth of the Cu(I) oxides. Crystalline duplex passive films are formed in the Cu(II) oxidation range. On both substrates, the structure of the upper layer corresponds to CuO(001).

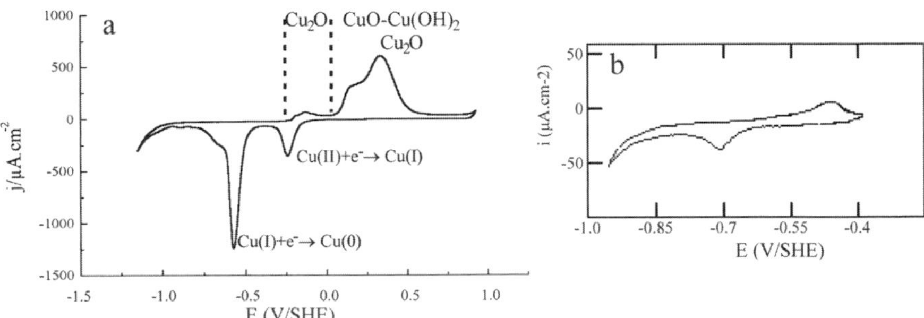

Fig. 5 Voltammograms recorded for Cu(111) in 0.1 M NaOH between the hydrogen and oxygen evolution limits (**a**) and in the potential range below oxidation (**b**). The passive film consists of a single Cu(I) oxide layer of a duplex Cu(I)/Cu(II) oxide layer depending on the potential region

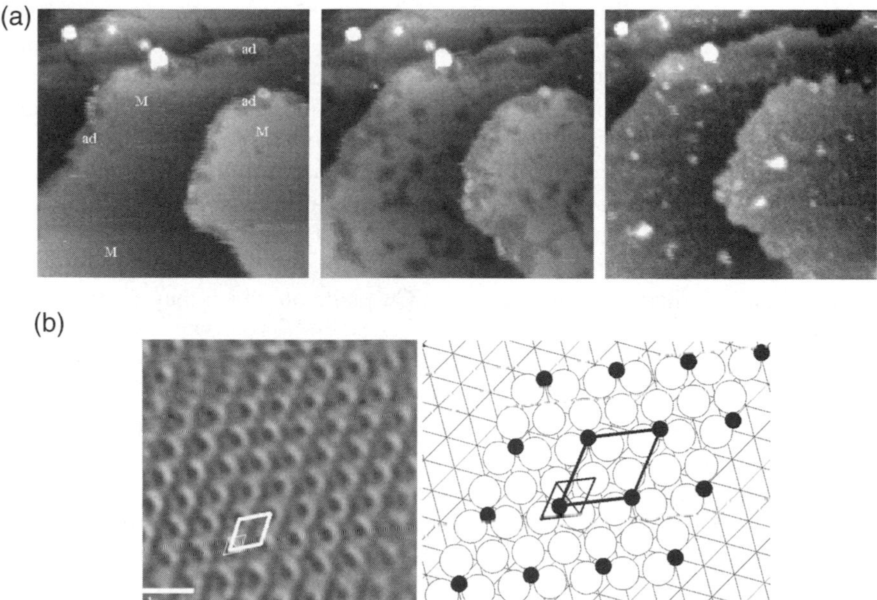

Fig. 6 (**a**) Sequence of ECSTM images (40 nm × 40 nm, 38 s/image) showing the growth of an adlayer of OH groups at $-0.6\ V_{SHE}$ at the onset of the anodic peak in Fig. 5b. (**b**) ECSTM image and model of the ordered structure of adsorbed OH groups formed on Cu(111) in 0.1 M NaOH in the potential region below the potential of formation of copper oxide. The large and small cells mark the lattice of adsorbed OH and of reconstructed copper, respectively. From refs. [23, 25]

Figure 6 illustrates the growth of the adsorbed hydroxide layer on Cu(111) at potentials below the potential at which copper oxide is formed. The sequence of images shown in Fig. 6a was obtained after stepping anodically the potential to the onset of the anodic peak observed at $-0.46\ V_{SHE}$ (see Fig. 5b) and assigned to the adsorption of OH groups. The atomically smooth terraces of the surface (marked M) become progressively covered by darker-appearing islands (marked ad) that grow laterally and coalesce to cover completely the terraces. The adsorbed layer preferentially grows at the step edges, confirming the preferential reactivity of these defect sites of the surface. The terraces grow laterally due to displacement of the step edges, and monoatomic ad-islands are formed at the end of the growth process. These two features are indicative of the reconstruction of the topmost Cu plane induced by the adsorption of the OH groups. The reconstruction causes the ejection of Cu atoms above the surface. The ejected atoms diffuse on the surface and aggregate at step edges, which cause the observed lateral displacement of the step edges. In the final stages of the adsorption process where most of the surface is already covered by the adlayer, the ejected atoms have a reduced mobility on the OH-covered terraces and aggregate to form the observed monoatomic ad-islands.

Figure 6b shows an atomically resolved image and the model of the ordered $OH_{ads}/Cu(111)$ surface that confirm the reconstruction of the Cu(111) topmost atomic plane. A hexagonal lattice with a parameter of 0.6 ± 0.02 nm is measured. Each unit cell contains one minimum and four maxima of intensity. The interatomic distance between the maxima (assigned to Cu atoms) is ~0.3 nm, which is larger than the interatomic spacing of 0.256 nm in Cu(111) from bulk parameter values and confirms the reconstruction of the topmost Cu plane into a plane of lower density. A coverage of ~0.2 OH per Cu(111) atom is deduced from the density of the intensity minima, in excellent agreement with the coverage of 0.19 obtained from the electrochemical charge transfer measurements. In addition, the STM data show that the position of the adsorbed OH groups corresponds to threefold hollow sites of the reconstructed Cu plane on which they form a (2×2) structure.

On Cu(001), a partially ordered OH adsorption layer is formed in the potential range below oxidation (see Fig. 7). It is assigned to OH-stabilized dimers of Cu atoms ejected from the topmost plane and superimposed above it with the OH presumably adsorbed in bridge sites. The orientation of the dimers along the <100> substrate directions causes the reorientation of the step edges along these directions. By alternating the dimer orientation along the [100] and [010] directions on the terraces, zig-zag arrangements are generated with a main orientation along the <110>directions. Long-range ordering of the zig-zag chains is observed in areas limited in width with $c(2\times6)$ and $c(6\times2)$ superstructure domains.

Figure 8 illustrates the influence of the oversaturation potential on the nucleation, growth, crystallization, and structure of the Cu_2O oxide film formed in the potential range of Cu(I) oxidation [22, 24, 26]. At low oversaturation (see Fig. 8a obtained on Cu(111)), poorly crystallized and one monolayer thick islands covering partially the substrate are formed after preferential nucleation at step edges. They are separated by islands of the ordered hydroxide adlayer. At higher oversaturation (see Fig. 8b obtained on Cu(111)), well-crystallized and several monolayer thick films are formed, and the step edges are not preferential sites of nucleation. The equivalent thickness of the oxide layer can be deduced from subsequent measurements of the charge transfer during cathodic reduction scans. It was ~0.5 and 7 equivalent monolayers (ML; one ML corresponds to one (111)-oriented O^{2-}–Cu^{+}– O^{2-} slab)

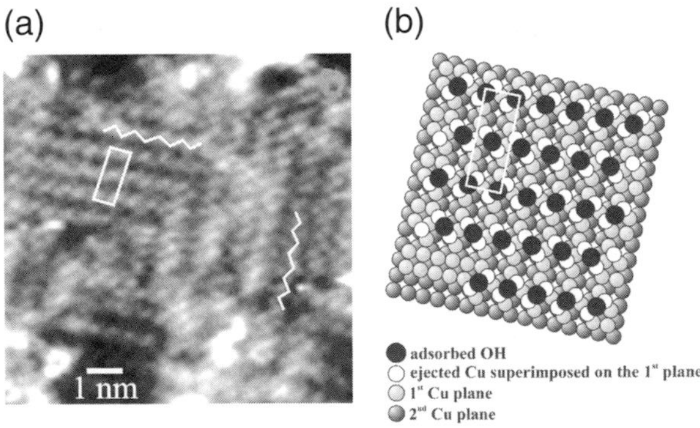

Fig. 7 ECSTM image and model of the locally ordered structure of the adsorbed OH on Cu(001) in 0.1 M NaOH in the potential region below the potential of formation of copper oxide. From ref. [26]

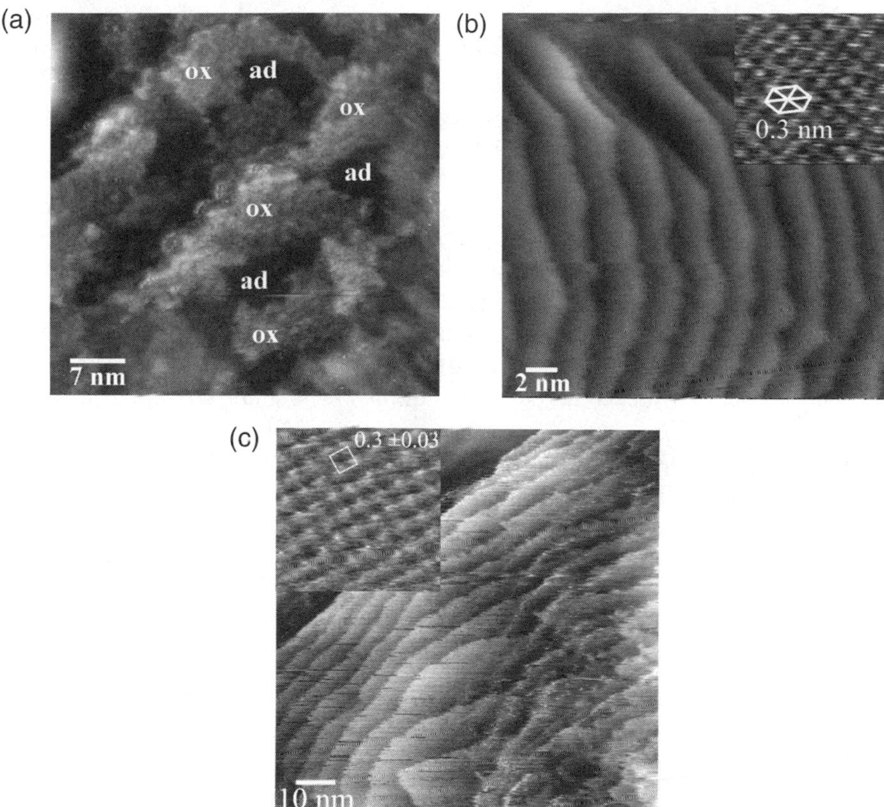

Fig. 8 ECSTM images of the Cu(I) oxide formed on Cu(111) at -0.25 V_{SHE} (**a**) and -0.20 V_{SHE} (**b**), and on Cu(001) at -0.11 V_{SHE} (**c**) in 0.1 M NaOH. At low oversaturation (**a**), noncrystalline 2D oxide islands (ox.) separated by the adsorbed OH layer (ad) cover partially the substrate. At higher oversaturation (**b, c**), a 3D crystalline oxide layer fully covers the substrate. Its atomic lattice, shown in the inset, corresponds to that of $Cu_2O(111)$ (**b**) and $Cu_2O(001)$ (**c**) on Cu(111) and Cu(001), respectively. From ref. [24, 26]

after growth at -0.25 and -0.2 V_{SHE}, respectively. The observed lattice of the oxide layer is hexagonal with a parameter of ~0.3 nm, consistent with the Cu sublattice in the (111)-oriented cuprite. The oxide grows in parallel (or antiparallel) epitaxy

$$(Cu_2O(111)[1\bar{1}0] \parallel Cu(111)[1\bar{1}0] \text{ or } [\bar{1}10]).$$

On Cu(001) (see Fig. 8c), the oxide layer has a square symmetry and the same periodicity of 0.3 nm, consistent with the Cu sublattice of (001)-oriented Cu_2O. The rotation of 45° of the close-packed $[1\bar{1}0]$ direction of the oxide lattice with respect to the close-packed direction of the Cu(001) lattice gives an epitaxial relationship noted as $Cu_2O(001)[1\bar{1}0] \parallel Cu(001)[100]$. The 45° rotation is initiated by the formation of the adsorbed layer of OH groups described above, where the dimers of Cu atoms stabilized by bridging OH groups are aligned along the [100] and [010] directions.

On both substrates, the crystalline Cu(I) oxide layers have a nanostructured and facetted surface as shown in Fig. 8b and c. The surface faceting results from a tilt of a few degrees of the orientation of the oxide lattice with respect to the Cu lattice (see Fig. 9). The tilt is assigned to the relaxation of the epitaxial stress in the metal/oxide interface resulting from the large mismatch between the two lattices. The height of the surface steps of the oxide layers corresponds to 1 ML of cuprite, indicating an identical chemical termination of the $Cu_2O(111)$ and $Cu_2O(001)$ oxide terraces. It is thought that the surface of the oxide layer is hydroxylated in the aqueous solution and that the measured lattice corresponds to OH and/or OH^- groups forming a (1×1) layer on the Cu^+ planes of the (111)- and (001)-oriented cuprite layers.

Crystalline Cu(I)/Cu(II) duplex passive films are formed in the potential range of Cu(II) oxidation in 0.1 M NaOH [27]. On both substrates, a terrace and step topography of the passivated surfaces are observed with terraces extending up to 20 nm in width and with a step height of ~0.25 nm corresponding to the thickness of one equivalent monolayer of CuO(001). Accordingly, the atomic lattices observed are consistent with a bulk-terminated CuO(001) surface which is characterized by a distorted hexagonal symmetry with in-plane nearest-neighbor distances of ~0.28 nm along the closed-packed directions (see Fig. 10). The epitaxial relationships for the duplex layers are:

$$CuO(001)[\bar{1}10] \parallel Cu_2O(111)[1\bar{1}0] \parallel Cu(111)[1\bar{1}0] \text{ or } [\bar{1}10]$$

and

$$CuO(001)[\bar{1}10] \parallel Cu_2O(001)[1\bar{1}0] \parallel Cu(001)[100],$$

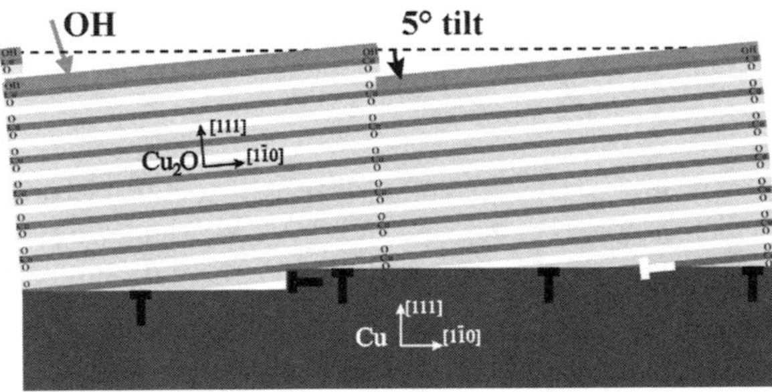

Fig. 9 Model (section view) of the 7-ML thick Cu(I) oxide grown in tilted epitaxy on Cu(111). The stacking sequence of the O^{2-} and Cu^+ planes in the oxide is illustrated by the gray lines. The oxide surface is terminated by a monolayer of hydroxyl/hydroxide groups. The faceted surface corresponds to a tilt of ~5° between the two lattices. Interfacial misfit dislocations and misorientation dislocations are illustrated by the T and |— symbols, respectively. White and black symbols correspond respectively to dislocations of the oxide and metal lattices

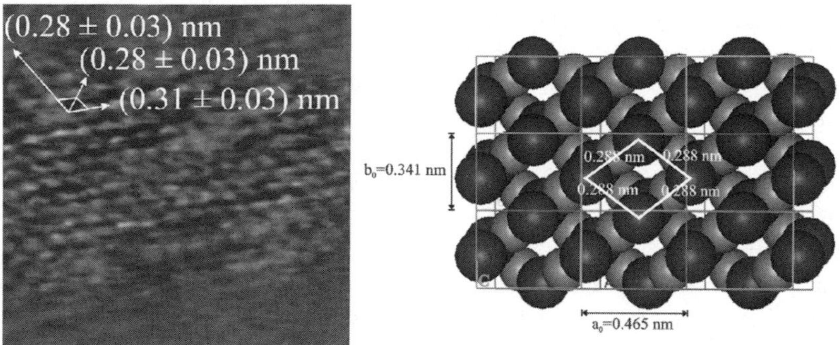

Fig. 10 ECSTM image (5 nm × 5 nm) of the atomic lattice recorded on the terraces of the 3D Cu(II) anodic oxide formed on Cu(111) in 0.1 M NaOH at 0.83 V_{SHE}, and top view of the (001) face of CuO. The dark disks represent the O^{2-}, the bright disks Cu^{2+}. From ref. [27]

corresponding in both cases to the parallel alignment of the closed-packed directions of the CuO and Cu_2O lattices.

The common (001) orientation of the CuO outer layers of the duplex films on the two substrates is assigned to their surface hydroxylation at the passive film/electrolyte interface, necessary to stabilize the bulk-like termination of CuO(001) which is polar and unstable when anhydrous. This assignment is supported by the step height measurements that indicate an identical chemical termination of all terraces. It is proposed that the surface is terminated by an OH^- (or OH) layer in (1×1) registry on the topmost plane of the Cu^{2+} sublattice, then O^{2-} and Cu^{2+} planes alternate toward the bulk of the oxide layer.

3 Nanostructure of Passive Films

Other metals like Ni, Fe, Cr, and iron-based and nickel-based alloys are self-protected against corrosion in aqueous environments by the growth of passivating anodic oxide layers on their surface. The anodic oxide films consist of hydroxylated oxides whose thickness does not exceed a few nanometers, making these passive films nanomaterials by nature. Their anticorrosion properties have been recently reviewed [28, 29, 30, 31]. Nanometer scale studies performed with scanning probe microscopies (STM and AFM) have focused on the growth and structure of these passive films. Besides the Cu described above, Ni [8, 32, 33, 34, 35, 36, 37, 38, 39, 40], Fe [41, 42, 43, 44, 45], Cr [46, 47], Al [48], Co [10, 49], and ferritic and austenitic stainless steels [50, 51, 52, 53, 54] have been studied. Prior to these studies, it was generally considered that the ultrathin passive films were amorphous, and that the absence of crystallinity was favorable to the corrosion resistance. This view has now changed, because the (sub)nanometer scale studies have revealed that in many cases the passive films are crystalline. Data obtained on nickel and stainless steels are selected to illustrate this topic.

Nickel has been the most-studied substrate. It was on Ni that the first high-resolution images of nanometer thick passive films were obtained [33, 34], confirming by ex situ STM measurements the crystallinity of the passivated surface previously observed by razing high energy electron diffraction (RHEED) [55]. The ex situ results were later confirmed by in situ ECSTM [8, 35, 36, 38, 39] and ECAFM [37] measurements, and by grazing X-ray diffraction (GXRD) measurements [38, 56]. The initial stages of growth of the passive film have been investigated in acid [32, 38, 39] and alkaline [35, 40] solutions. In acid solutions, extensive anodic dissolution occurs prior to blocking by the passive film, as shown by an intense active peak in the polarization curve. Since studying the nucleation and growth of the passive films requires to maintain the potential in the passivation peak, it produces a large amount of dissolved cations that can precipitate on the sample surface to form a relatively thick layer of corrosion products, not characteristic of the passive film and precluding the observation of the nucleation and growth of the barrier layer of the passive film [32]. Some data have nevertheless been obtained by ECSTM showing severe roughening of the surface at the nanoscale with the formation of 3D nanopits caused by the intense ongoing anodic dissolution occurring in the not-yet passivated sites [38]. This severe roughening prevents any detailed (sub)nanometer scale study of the nucleation of the passive film in acid solutions, but can be minimized in alkaline solutions.

Figure 11 illustrates the nucleation of the passive film observed on Ni(111) in a sodium hydroxide solution (pH $\sim$11) at $E = -0.3\ V_{SHE}$, that is, at the onset of the anodic peak corresponding to the growth of the Ni(II) passivating oxide [40]. A slow dissolution of the nickel surface is observed. As in the case of Cu described above, the dissolution takes place at the step edges, leading to the consumption of the metal terraces by a step-flow mechanism. The nucleation of the passivating oxide occurs preferentially at the step edges at this potential. It is characterized by the formation of 2D grains with a lateral size of $\sim$2 nm. The ongoing dissolution at the nearby step edges not yet passivated leads to the formation of isolated 2D islands of the passivating oxide.

At $E = -0.28\ V_{SHE}$, dissolution of the metal terraces by step flow is still observed, but the 2D growth of the passivating oxide is faster. It forms extended islands of 2D nanograins blocking the dissolution at the step edges. These 2D islands extend laterally with time to completely cover the surface. This leads to the formation of a 2D passivating oxide with a nanogranular morphology (see Fig. 12a). The crystalline areas have an hexagonal lattice with a parameter of 0.306 $\pm$ 0.004 nm assigned to 2D nanocrystals of $Ni(OH)_2$(0001) in strained epitaxy on Ni(111) [40]. Similar strained epitaxial 2D films have been observed on Ni(001) in 0.1 M NaOH(aq) [35]. On Ni(111) in sulfuric acid (pH $\sim$3) [39], crystalline 2D islands with a lattice parameter consistent with $Ni(OH)_2$(0001) have also been reported in the initial stage of growth prior to complete coverage of the surface, suggesting that in both acid and alkaline solutions, nickel hydroxide predominates in the submonolayer regime of growth of the passive film.

At $E \geq -0.13\ V_{SHE}$, that is, at the top of the anodic peak corresponding to the growth of the passivating oxide, the 3D growth of the passive film is observed (see

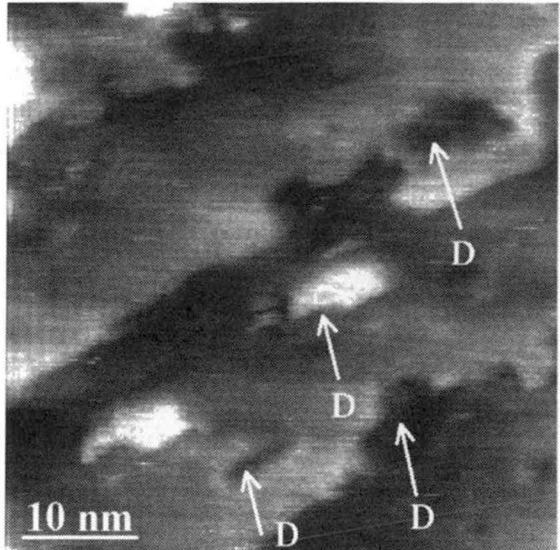

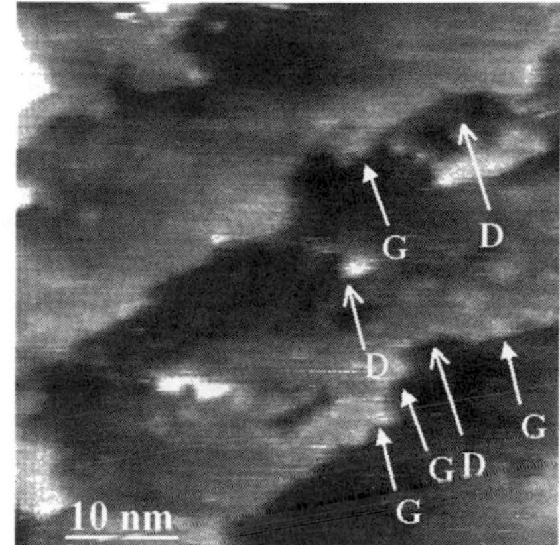

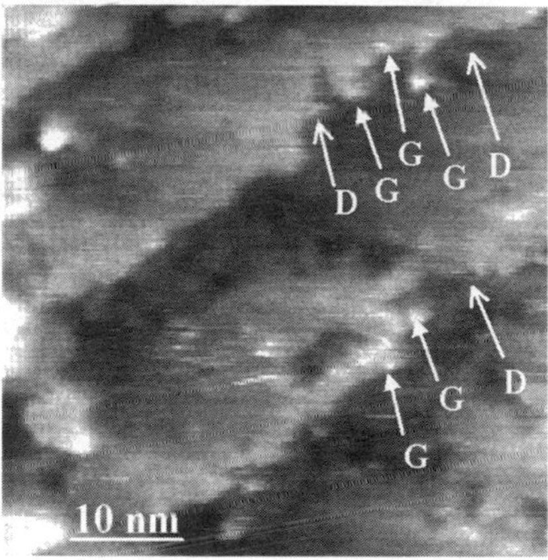

Fig. 11 Sequence of ECSTM images of the Ni(111) surface recorded in 1 mM NaOH(aq) 30 s, 73 s, and 164 s after polarization at E = –0.3 V_{SHE}, showing the step-flow process generated by dissolution (sites marked D) and the nucleation of 2D grains of the passive film (marked G) at the step edges. From ref. [40]

Fig. 12b). It is characterized by the formation of facetted topography indicative of the tilt between the oxide lattice of the passive film and the lattice of the substrate. The hexagonal lattice formed at −0.13 V and 0.22 V has a parameter of 0.318 ± 0.015 nm, slightly larger than that measured at −0.28 V when a 2D passivating oxide is formed. The parameter is in excellent agreement with the value of 0.317 nm expected for an unstrained 3D layer assigned to β-$Ni(OH)_2$(0001). This is consistent with a crystalline 3D outer hydroxide layer of the passive film formed in alkaline electrolytes, as opposed to the 2D outer hydroxide layer formed in the whole passive range in acid electrolytes.

Figure 12c illustrates the typical topography of a Ni(111) single-crystal surface polarized in the middle of the passive domain in a sulfuric acid solution (pH ~2.9)

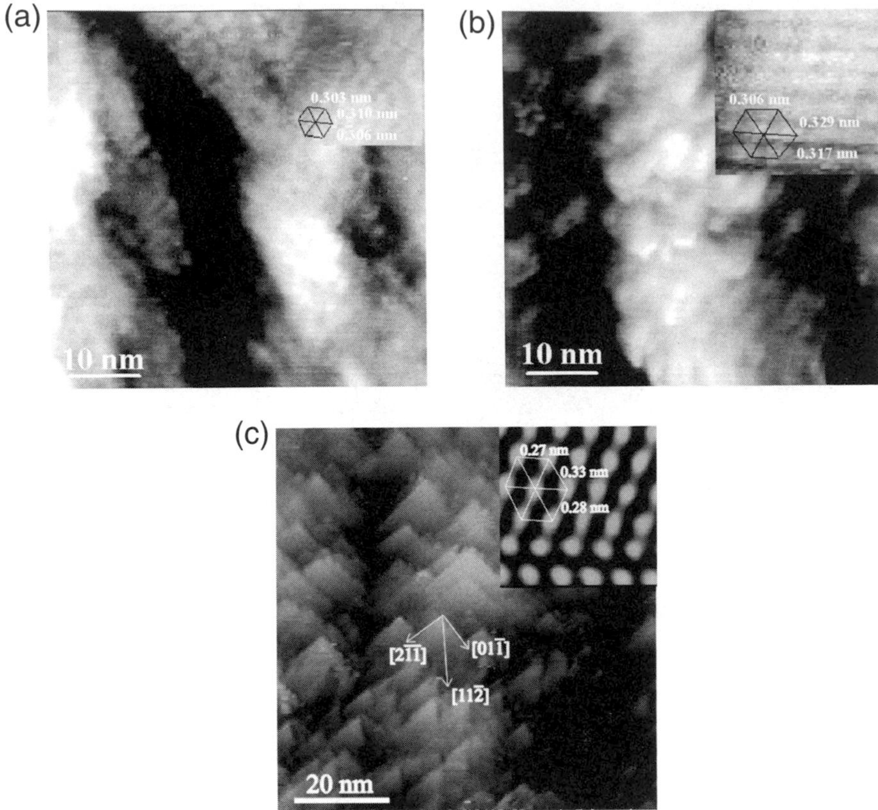

Fig. 12 ECSTM images of Ni(111) surfaces passivated (**a**) in 1 mM NaOH(aq) at E = –0.28 V_{SHE}, (**b**) in 1 mM NaOH(aq) at E = –0.13 V_{SHE}, (**c**) in 0.05 M H_2SO_4 + 0.095 M NaOH(aq) (pH~2.9) at E = +0.95 V_{SHE}. The inserts show the atomic lattices. From refs. [40, 36]

[36]. The passivated surface is also facetted exhibiting terraces and steps. The presence of terraces at the surface of the crystallized passive film is indicative of a slightly tilted epitaxy between the NiO(111) lattice forming the inner part of the passive film and the Ni(111) substrate terraces. It has been proposed that this tilt partly relaxes the interfacial stress associated with the mismatch of 16% between the two lattices. This tilt has been confirmed by GXRD measurements on Ni(111) [56]. A similar surface faceting is observed for the Cu(I) oxide layer grown on copper (as described above) for which the lattice misfit between Cu_2O and Cu lattices is similar [22, 24]. The lattice measured on the terraces is hexagonal with a parameter of 0.3 ± 0.02 nm assigned to NiO(111), NiO being the constituent of the inner part of the passive film. It must be pointed out that the (111) surface of NiO which has the NaCl structure is normally polar and unstable. It is however the surface which is obtained by passivation. The reason for this is that the surface is stabilized by adsorption of a monolayer of hydroxyl groups or by the presence of a monolayer of

β-Ni(OH)$_2$ in parallel epitaxy with the NiO surface. These data show that the direction of growth of the oxide film is governed, at least in part, by the minimization of the oxide surface energy by the hydroxyl/hydroxide groups. The presence of water is thus a major factor for the structural aspects of the growth mechanism.

The structure of passive films has also been investigated on chromium [46, 47] and on ferritic and austenitic stainless steels [50–51, 52, 53, 54]. A major finding for these systems is that potential and ageing under polarization are critical factors for the development of crystalline passive films and for the dehydroxylation of the passive film as evidenced by combined XPS measurements. On chromium, the crystals of Cr_2O_3 can be very small at low potential where the oxide inner layer is not fully developed and where the passive film is highly hydrated and consists mainly of hydroxide (as shown by combined XPS measurements). This supports the view

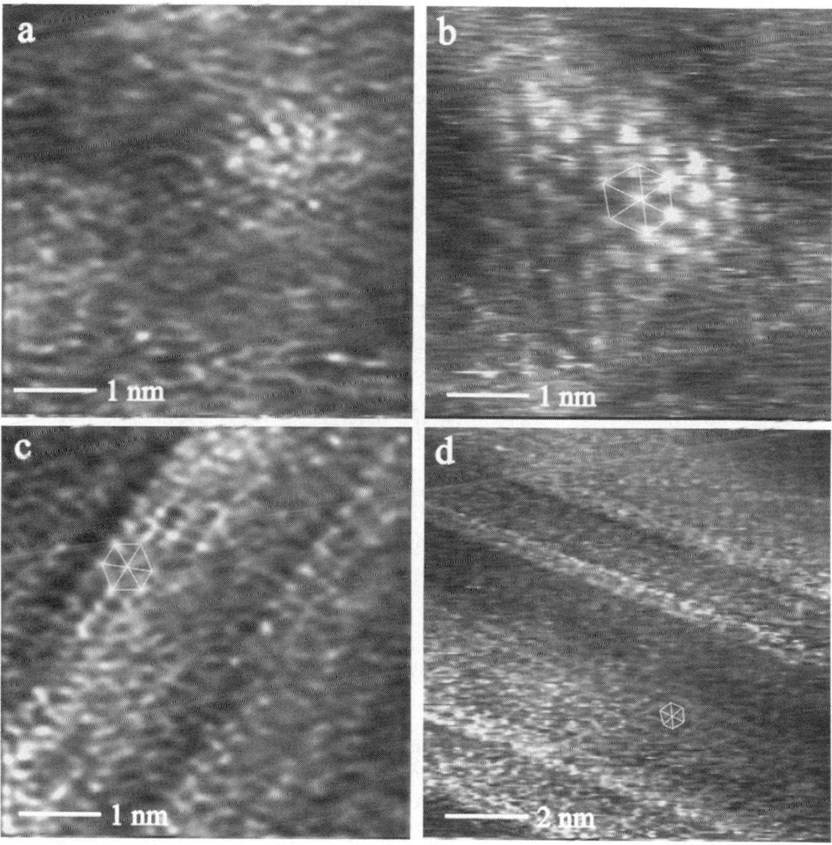

Fig. 13 STM images of the Fe–22Cr(110) (*left*) and Fe–18Cr–13Ni(100) (*right*) surfaces recorded after passivation in 0.5 M H_2SO_4 at +0.5 V_{SHE} for 2 h (**a**), (**b**) and for 22 h (**c**), (**d**). The nearly hexagonal lattice is marked. The effect of ageing under polarization is evidenced by the extension of the observed crystalline areas. From refs. [52, 54]

that the passive film on chromium can have a nanocrystalline structure. At high potential where the inner part of the passive film is dehydrated and consists mostly of chromium oxide, larger crystals are formed. The nanocrystals and the larger crystals have a lattice consistent with α-$Cr_2O_3(0001)$. The basal plane of the oxide is parallel to Cr(110). A special feature of the passive film on chromium is that the oxide nanocrystals are cemented together by the chromium hydroxide layer. It has been suggested that the role of cement between grains played by chromium hydroxide, and, of course, the high stability of chromium oxide and chromium hydroxide make this passive film extremely protective against corrosion.

Structural changes also occur during ageing under anodic polarization of the chromium-rich passive layers formed on stainless steels in aqueous solution. The major modification is an increase of the crystallinity of the film and the coalescence of Cr_2O_3 islands in the inner oxide as observed on Fe–22Cr and Fe–18Cr–13Ni alloys studied over time periods of up to 65 h. This is illustrated by the images shown in Fig. 13. For short polarization times ($\leq$2 hours), the crystallinity of the passive films decreases with increasing Cr content of the alloy [50, 51]. The comparison of the rates of crystallization of Fe–22Cr(110) and Fe–18Cr–13Ni(100) revealed that the rate of crystallization is more rapid on the austenitic stainless steel than on the ferritic one [52, 54]. This is tentatively explained by a regulating effect of Ni on the supply of Cr on the alloy surface, a lower rate of Cr enrichment being in favor of a higher degree of crystallinity [57].

4 Nanostructural Aspects of Passivity Breakdown and Localized Corrosion

4.1 Dissolution in the Passive State and Passivity Breakdown

Localized corrosion by pitting in aggressive environments (e.g., with Cl^-) is a major cause of degradation of passivated metal surfaces. Its understanding and control require to study at the (sub)nanometer scale the mechanisms of initiation, including dissolution in the passive state and passivity breakdown. Recent STM and AFM results have addressed these issues on nickel [58, 59, 60, 61]. In these model studies, Ni(111) surfaces were first passivated in a chloride-free sulfuric acid solution (pH 2.9) to produce the characteristic surface described above. Chlorides (0.05 M NaCl) were subsequently introduced in the electrolyte without changing the pH.

Figure 14 shows that the presence of crystalline defects at the surface of passive films plays a key role in the dissolution in the passive state. The sequence of three images shows that the passive film dissolves at the edges of the facets resulting from the tilted epitaxy between the NiO(111) oxide and the Ni(111) substrate lattice. The observed dissolution is a 2D process leading to a progressively decreasing size of the dissolving facets by a step-flow process. The process is similar to that of active dissolution of metal surfaces at moderate potential with no pit forming, as shown above. The 2D step-flow process is dependent on the step orientation:

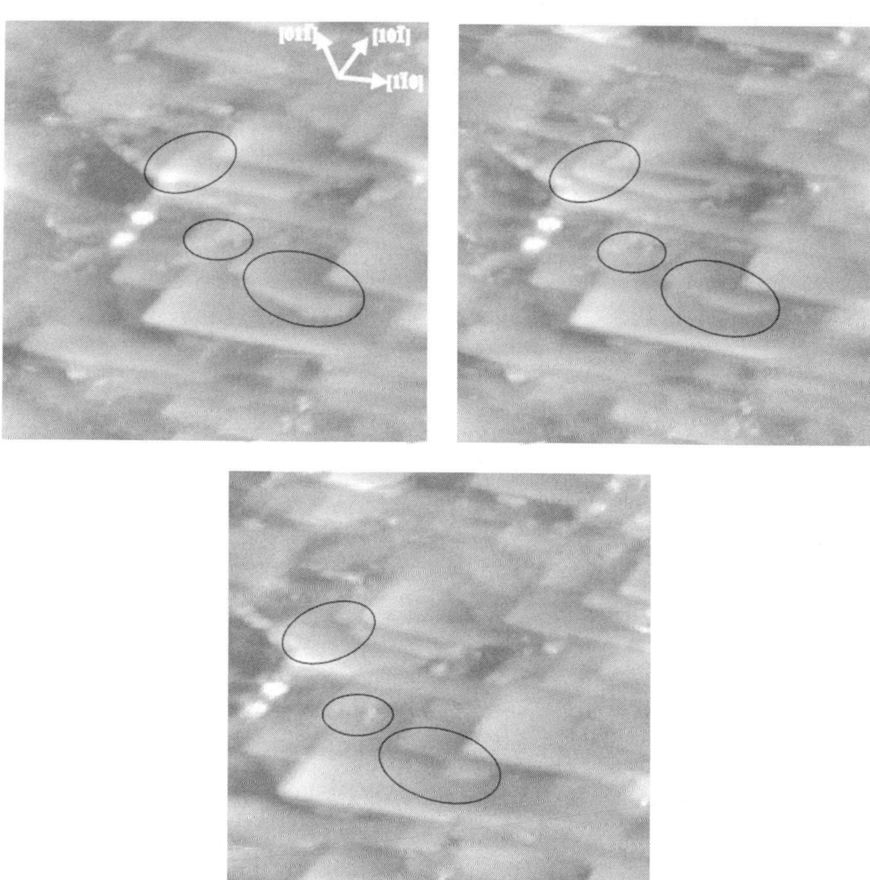

Fig. 14 Sequence of ECSTM images (75 nm × 75 nm, 63 s/image) showing the localized dissolution of the passivated Ni(111) surface at +0.85 V_{SHE} in 0.05 M H_2SO_4 + 0.095 M NaOH + 0.05 M NaCl (pH 2.9). The oxide crystallographic directions are indicated. The circles show the areas of localized dissolution. From ref. [60]

the step edges oriented along the closed-packed directions of the oxide lattice dissolve much less rapidly due to the higher coordination of their surface atoms. This process leads to the stabilization of the facets, with edges oriented along the close-packed directions of the oxide lattice, and produces steps that are oriented along the $\{100\}$ planes, the most stable orientations of the NiO structure. The average dissolution rate of the facets, measured to be $0.44 \pm 0.25\ nm^2s^{-1}$ from such measurements, does not vary significantly with applied potential below or above the pitting potential value (~0.9 V_{SHE}) [60]. This indicates that the potential drop at the passive film/electrolyte interface, associated with the surface reaction of dissolution of the oxide film, remains constant. Also, the average dissolution rate does not significantly increase in the presence of chlorides in the electrolyte [61], showing that the surface reaction of dissolution at the regular step edges at the surface of the nanograins of the well-passivated surface is not accelerated by the presence of chloride. The chlorides may however prevent the stabilization of the dissolving oxide film in the more disoriented steps forming the grain boundaries of the passive film where metastable pits are formed (see below). Consistently with the absence of effect of chloride on the dissolution rate, the atomic lattice of the passive film formed in the absence of chloride is unmodified after the addition of chloride [59].

Figure 15 shows the modifications of the nanostructure of the passivated Ni(111) surface resulting from the increase of the potential in the passive range and the effect of the presence of chlorides. After passivation at 0.55 V_{SHE} (Fig. 15a), the surface is completely covered with platelets assigned to the grains of the passive film. This is the facetted surface of these grains that is revealed by the typical higher magnification STM images shown above. The lateral size of the platelets,

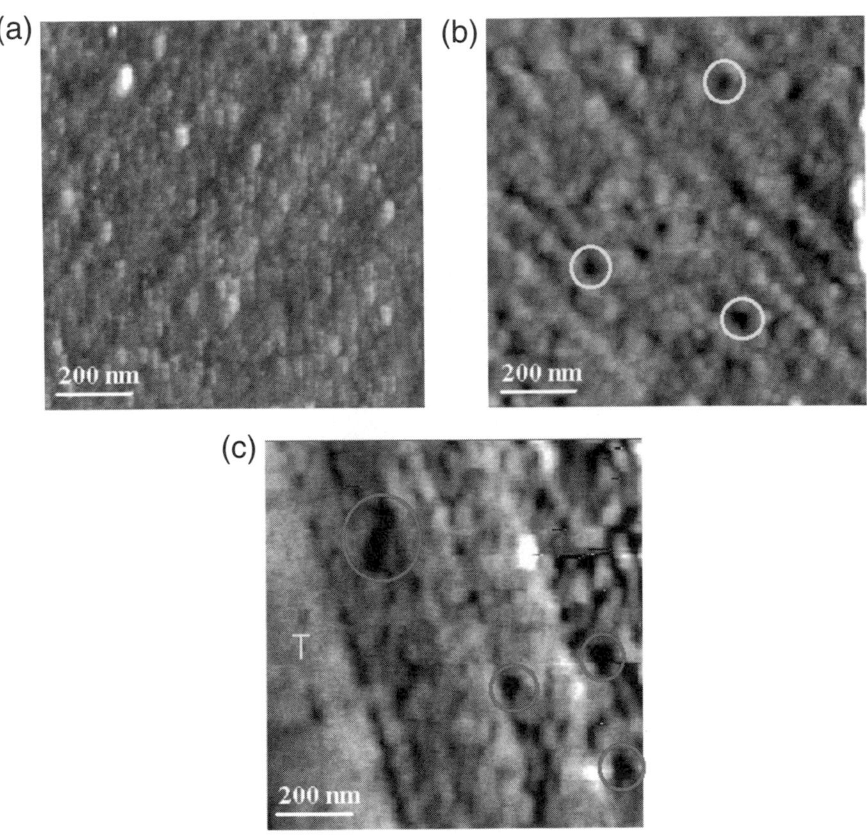

Fig. 15 AFM images (1000×1000 nm^2, Z range = 3 nm) of the Ni(111) surface after prepassivation for 30 min. in 0.05 M H_2SO_4 + 0.095 M NaOH (pH 2.9) at +0.55 V_{SHE} (**a**), and subsequent increase of the potential stepwise (steps of 0.1 V) every 30 min up to +1.05 V_{SHE} in the absence (**b**) or presence (**c**) of chloride (0.05 M NaCl). From ref. [61]

varying from 50 to 230 nm, suggests a varying degree of advancement of the coalescence of the oxide grains. Between the platelets, depressions with a depth varying from 0.4 to 1.4 nm are measured. Their formation is assigned to the dissolution occurring on the non- yet passivated (or less-protected) sites that are formed in the transient process of growth of the oxide film, prior to complete passivation of the surface.

Increasing the potential in the passive range by successive potential steps up to 1.05 V_{SHE} causes the roughening of the prepassivated surface. In the absence of chlorides, the formation of local depressions of nanometer dimensions (20–30 nm at the surface) is observed with a density of $(3 \pm 2) \times 10^{10}$ cm^{-2} (marked in Fig. 15b). The depth of these depressions ranges from 2.2 to 3.8 nm. This is larger than after prepassivation at 0.55 V_{SHE} and than the thickness of the passive film formed in these conditions (<2 nm), indicating the locally enhanced corrosion of the substrate in these sites. This implies that the surface, prepassivated at +0.55 V_{SHE}, has been locally modified (depassivated) with a local enhancement of the corrosion of the substrate, and subsequently repaired (repassivated). It cannot be concluded, however, if these events correspond to the exposition of the bare metal surface to the electrolyte (complete passivity breakdown) or only results from a temporary enhancement of dissolution without complete removal of the passive film (transient passive film thinning). However, these local modifications of the passive film occur in the absence of chloride, showing that the key role of the chloride in pitting may not be related to the initiation of these modifications.

In the presence of chlorides (Fig. 15c), the depressions observed between the grains have, for the most part, the same dimensions as those described above. However, significantly larger depressions are also observed (marked by circles in Fig. 15c). Their lateral dimension ranges between 40 and 50 nm at the surface and their depth between 5 and 6 nm. Their density is $(2 \pm 1) \times 10^9$ cm^{-2}. Figure 16 shows localized attacks of similar dimensions of the passivated surface that were observed by STM after exposing to chlorides the Ni(111) surface prepassivated at +0.9 V_{SHE} [58, 59]. These localized attacks were assigned to metastable pits on the basis of the absence of variations of the current characteristic of stable pitting during the corrosion test. The observed atomic structure inside these metastable pits (see Fig. 16) is crystalline with lattice parameters similar to those measured on the passivated surface prior to exposure to chlorides, showing a similar structure of the surface repassivated in the presence of chlorides and confirming the metastable character of the pit.

Figure 15c also shows an area, marked T, corresponding to an extended terrace of the substrate, that exhibits much less depressions than the rest of the imaged surface where narrower terraces are observed due to a higher density of substrate step edges. This role of the defects of the substrate (monoatomic step edges) in passivity breakdown is confirmed in Fig. 16, which shows a preferential alignment of the metastable pits along the substrate step edges. Figures 15 and 16 also show that the pits are located at the boundaries between the oxide crystals of the passive film, showing that the grain boundaries of the passive film are preferential sites of passivity breakdown.

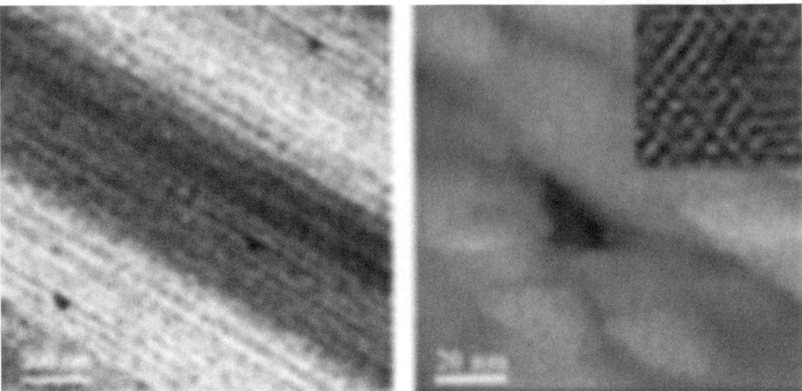

Fig. 16 STM images of Ni(111) passivated in 0.05 M H_2SO_4 + 0.095 M NaOH (pH~2.9) at 0.9 V_{SHE} for 30 min and subsequently exposed to 0.05 M NaCl for 90 min, showing metastable pits at two magnifications. The inset in the right images shows the crystalline lattice observed inside the metastable pit. From refs. [58, 59]

These nanostructural data on metastable pitting indicate that modifications of passivity initiate by dissolution of the oxide in the more defective grain-boundary sites of the passive film and enhance the corrosion of the substrate. The increased size and lower density of the metastable pits observed in the presence of chloride suggest that the role of the chloride is to modify and/or retard the stabilization of the oxide dissolving in these specific sites, and thus to sustain the transient localized corrosion of the substrate initiated by the dissolution of the oxide. A retarding effect on repair following complete passivity breakdown is also suggested.

4.2 Initiation of Pitting Corrosion

The initiation of pitting corrosion by, at, or near inclusions has been studied at the (sub)micrometer scale with AFM on Al alloys [62, 63] and stainless steels [64, 65, 66, 67]. The images in Fig. 17 illustrate the development of a corrosion trench observed around an iron-rich inclusion on the Al–6061–T6 alloy. It was found that corrosion was initiated in regions where the ratio of inclusions to host matrix surface area was high. After formation of the trench, the dissolution of the matrix was uniform and proceeded radially from the inclusion to form a circular pit [63]. While the extent of dissolution around the inclusion was at first independent of the inclusion size, it became distinctly dependent in later stages with larger cavities being formed around the larger inclusions. The formation of the trench was promoted by a −500 mV cathodic overpotential as shown in Fig. 17. Based on comparative measurements in aerated and de-aerated solutions, it has been proposed that the intermetallic inclusions act as cathodic site for the reduction of oxygen [63] causing the formation of an alkaline environment in the immediate vicinity of each inclusion,

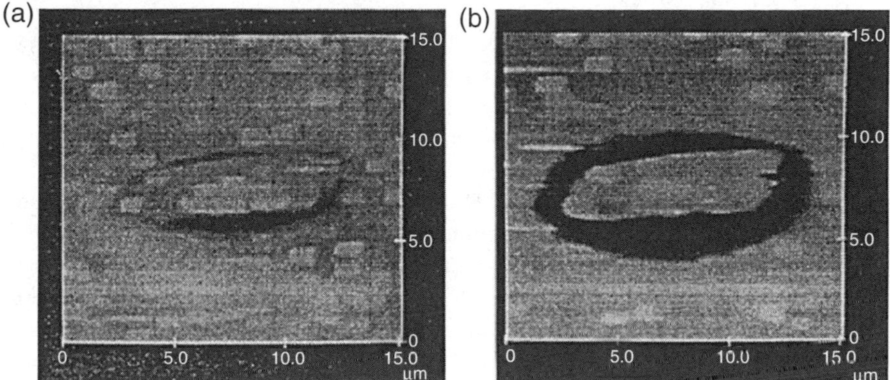

Fig. 17 ECAFM images (15 µm × 15 µm) of corrosion around an Al_3Fe inclusion in Al–6061 alloy in 0.6 M NaCl at (**a**) open-circuit potential, 3 h, and (**b**) −0.5 V cathodic overpotential, 3 min. From ref. [63]

confirmed by local pH measurements. This local increase of pH is thought to promote the local dissolution of the Al matrix.

On stainless steels, recent studies have focused on the initiation of pitting near or at MnS inclusions. On 304 SS, no pit initiation was observed around or at the MnS inclusions. Morphological changes were assigned to the buildup of corrosion products resulting from dissolution of the inclusion itself [62]. As pits were initiated elsewhere on the surface, it was suggested that scanning with the AFM probe tip may affect the pit initiation. A more recent study [65] combining ex situ AFM with scanning electrochemical microscopy (SECM) and EDX analysis has shown that certain inclusions (sulfides) concentrate chloride, by electromigration, under a sulfur crust as a result of their dissolution evidenced by AFM. It has been proposed that the high local current density (evidenced by SECM), the electromigration of chloride and the sulfur crust (evidenced by EDX) generate an occluded extreme environment in which the stainless steel depassivates.

On 316 SS, the activation of MnS inclusions was initiated by injecting an aggressive solution locally with microcapillaries and subsequently studied in situ using the scanning vibrating electrode technique (SVET) and ex situ using Auger electron spectroscopy (AES) and AFM [67]. It was observed that a single pit could be initiated when hydrochloric acid was injected, whereas sulfuric acid only partially dissolved the inclusion. No changes of surface morphology were observed when a sodium chloride solution was injected. A significant enrichment in sulfur was detected around the inclusion by AES, and micropits were observed in the metal at the edge of the inclusion after HCl activation. Anodic zones were detected by SVET around the inclusion, whereas a cathodic current flowed from the inclusion. The anodic current was ascribed to the breakdown of passivity induced by the adsorption of sulfur released from the MnS inclusion by dissolution.

Nanoscale chemical analysis performed by secondary ion mass spectroscopy (SIMS) and TEM/EDX have been recently applied to measure the chemical composition of the boundary between matrix and MnS particles in 316 SS [68, 69]. It was found by Ryan et al. that the stainless steel matrix around the particles was characterized by a significant reduction of the Cr:Fe atomic ratio. Figure 18a shows the variation of the Cr:Fe atomic ratio normalized to the matrix value as a function of the distance from a sulfide inclusion. The analysis was reported for four inclusions, and it can be seen that the Cr concentration in the 200-nm-wide region surrounding the edge of the inclusion was measured to be 9–10 at% instead of 18 at% in the nonmodified areas forming the matrix of the alloy. On this basis, the authors concluded that these zones were Cr depleted below the well-known critical value of ~13 at% required to make stainless steels corrosion resistant. As a result, it was

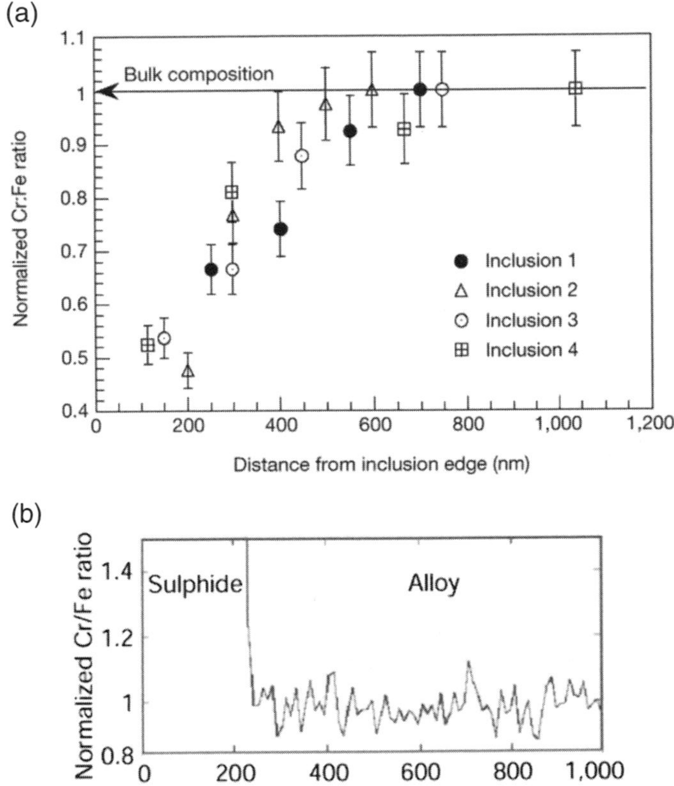

Fig. 18 Variation of the Cr:Fe normalized ratio obtained from nanometer scale SIMS analysis of the MnS/matrix interface in 316 stainless steel. Contrasting results reported by Ryan et al. (**a**) and Meng et al. (**b**) regarding Cr depletion of the matrix near the inclusions are shown. From refs. [68, 69]

proposed that high-rate metal dissolution events in these Cr-depleted areas would trigger localized corrosion by pitting and by increasing the acidity of the local environment. This would initiate the dissolution of the inclusion itself, with the sulfur corrosion products forming a crust around the dissolving inclusion and providing the occluded environment of strongly aggressive, modified chemistry in which the steel matrix would also be unstable, thus leading to stable pit growth. However, in subsequent measurements performed by Meng et al. on several stainless steels samples, including the same 316F grade analyzed by Ryan et al., Cr depletion around the MnS particles was not confirmed. Figure 18b shows the variation of the Cr:Fe atomic ratio normalized to the matrix value as a function of the distance from a sulfide inclusion reported by Meng et al. In contrast with Ryan et al., the authors found that the matrix surrounding the particles does not show any change of local concentration and that the interface between sulfide inclusion and matrix shows a sharp transition to the higher Cr:Fe in the sulfide inclusion. This led Meng et al. to conclude that further (nanoscale) analysis was needed to clarify the mechanism triggering localized corrosion near second-phase particles.

Apart from inclusions, localized corrosion of stainless steels can be initiated at defects such as dislocations or grain boundaries. Dislocations are similar to surface steps which have been observed to be preferential sites of metastable pitting on pure nickel as described in Section 4.1. The dissolution of carbide precipitates at grain boundary in the early stage of intergranular corrosion was recently reported for 304 SS in NaCl solution [66]. Selective dissolution is another form of localized corrosion for duplex stainless steels which have a two-phase microstructure. It was observed with low-resolution ECSTM for 2205-type duplex SS having large separate volumes of ferrite and austenite in approximately equal fractions with a grain size of ~5 μm [70]. In 0.05 M H_2SO_4 + 1 M NaCl, no significant active dissolution was observed, but some selective dissolution of the ferrite phase revealed the phase-boundary region by forming a step for potentials higher than E_{corr} + 1 V. In 4 M H_2SO_4 + 1 M HCl, the selective dissolution of ferrite grains was observed at E_{corr} + 0.05 V. At E_{corr} + 0.15 V, the dissolution of the austenite phase was also observed. These results show a stepwise process starting from the phase-boundary region, which had been previously revealed by selective dissolution of the ferrite grains.

4.3 Tip-Induced Controlled Localized Corrosion

Using an STM or AFM tip to form nanostructures by locally induced corrosion represents a potential means of nanostructuring metallic surfaces. Recently, it has been shown that ECSTM can be used to induce metal dissolution spatially confined to a region underneath the tip at electrode potentials lower than the M/M^{z+} equilibrium potential, provided that the tip potential is positive with respect to the equilibrium value [71, 72, 73]. The result obtained on Cu(111) in sulfuric acid is shown on Fig. 19. The tip-induced dissolution of the electrode was attributed to the role of counter-electrode played by the tip in the close vicinity of the electrode. The

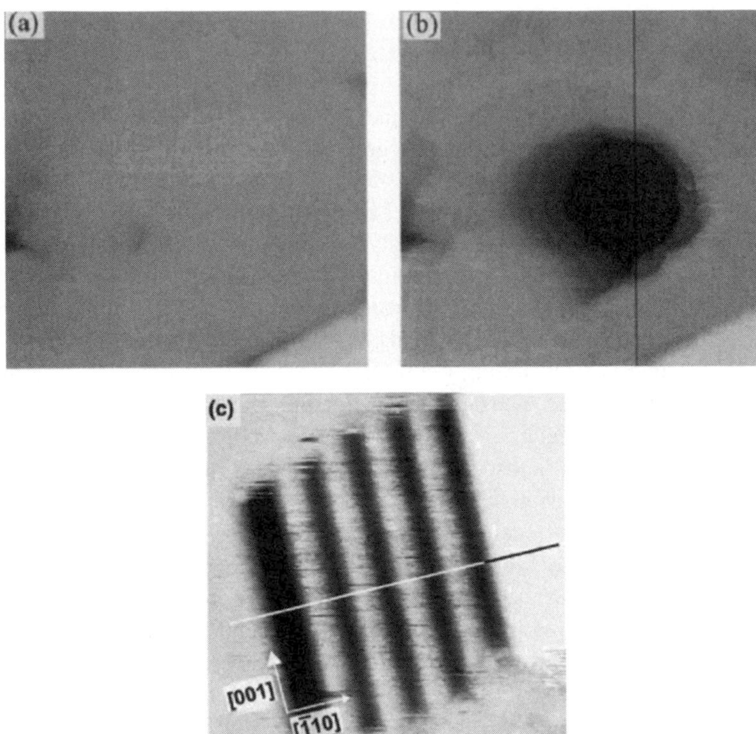

Fig. 19 ECSTM images of a Cu(111) surface in a Cu^{2+}-containing sulfuric acid solution at −0.05 V vs. Cu/Cu^{2+}, E_{tip}= 0 V vs. Cu/Cu^{2+}, I = 2 nA, prior to (**a**) and after (**b**) the tip has been scanned over a 25-nm^2 area at −0.05 V vs. Cu/Cu^{2+}, E_{tip}= +0.02 V vs. Cu/Cu^{2+}, I= 2 nA. From ref. [71] (**c**) ECSTM image of a Cu(110) surface in 10 mM HBr showing the reconstruction of a tip-induced corrosion hole

proposed mechanism involves electron transfer from the surface Cu atoms, not possible to the empty states of the electrode, to the empty states of the tip in regard. This requires the Fermi level of the tip to be lower than that of the electrode and below the M/M^{z+} equilibrium potential, and the Fermi level of the electrode to be above the M/M^{z+} equilibrium potential. The phenomenon is inhibited for electrode potential far negative (<0.1 V) to the equilibrium potential.

For Cu(110) in bromic acid solution, the tip-induced localized dissolution of the Cu surface has also been observed. However, in this case, the strongly adsorbed Br adlayer has been observed to promote highly ordered and stable nanostructures owing to a self-organizing reconstruction process. As a result, trenches are formed along preferential directions in the nanoscale area initially dissolved (Fig. 19c). The driving force for the reconstruction has been proposed to be the formation of Cu(100) facets on which the adsorption of Br would be energetically favored compared to the (110) orientation.

Tip-induced localized corrosion was also observed for Ag(111) in sulfuric and perchlorate acid solutions [72]. It was observed for polarization potentials of the electrode and tip negative and positive with respect to the Ag/Ag^+ equilibrium potential, respectively. In this case, the authors proposed that the local tip-induced dissolution of the electrode was related to a cathodic shift of the Ag/Ag^+ equilibrium potential resulting from a local decrease of the Ag^+ concentration under the tip that would be caused by an electrostatic repulsion of Ag^+ by the positively charged tip and by mass-transport limitations arising from shielding effects. This explanation differs from that proposed for the similar tip-induced localized dissolution observed on Cu and suggests that further studies are required to clarify the mechanism leading to tip-induced localized corrosion.

AFM tips can also be used to induce localized corrosion at the nanometer scale. It was observed that the AFM tip–surface interaction can be used to locally increase the dissolution rate of Al [74, 75]. This requires a moderate corrosivity of the solution (0.1 M NaCl), since the effect was not observed in a noncorroding solution (demineralized water), and the thin film was corroding too fast to observe the effect of the scanning tip in 1 M NaCl. The enhanced dissolution of Al was attributed to the energy dissipation from frictional forces arising from scanning the tip over the surface (contact mode AFM). Local heating of the substrate may provide enough energy to overcome the activation-energy barrier for the chemical reaction of the chloride ion with aluminum, thereby accelerating its dissolution in the aqueous solution.

AFM tips have also been used to perform high-resolution scratching experiments [76, 77]. It was found that the rastering with the AFM tip in contact mode in chloride solution resulted in accelerated dissolution of pure Al and alloy AA2024-T3. Rastering of pure Al resulted in enhanced uniform dissolution, while on alloy AA2024-T3 it was dependent on the exact location in the microstructure that was scratched. The abrasion associated with AFM in contact mode resulted in the immediate dissolution of the Al–Cu–Mg particles, otherwise stable for hours. This behavior was attributed to the influence of unstable surface films that provided some protection when present but could be easily destabilized by the tip [76]. In the presence of dichromate concentrations of 0.005 M or more, localized attack was not observed on pure Al, perhaps owing to the formation of a harder film more resistant to AFM scratching. On AA2024-T3, the addition of 0.0005 M dichromate was sufficient to protect the Al–Cu–Mg particles from destabilization by the rastering tip. At higher applied tip forces, the Al–Cu–Mg particles could be destabilized by the rastering tip, but not the Al matrix [77].

5 Conclusion

In this chapter, the structural aspects of corrosion, corrosion protection by inhibitors, and passive film and passivity breakdown have been examined at the nanoscale. Recent results obtained on metal surfaces mostly with in situ nanoprobes (STM and

AFM) implemented to electrochemistry have been reviewed. Nanochemical aspects of the initiation of localized corrosion have also been addressed based on SIMS and TEM/EDX data. The selected examples demonstrate the following:

- the anisotropic effect of the surface atomic structure of strongly adsorbed layers of anions on the mechanism of active dissolution of metals, and of the surface atomic structure of the passive film on the mechanism of dissolution in the passive state,
- the role of the competitive adsorption between inhibitors and aggressive anions on the structure of the inhibitor layer and its properties of corrosion protection,
- the existence of a precursor of adsorbed hydroxide in the growth of passive films,
- the crystallinity of passive films on numerous substrates and the effects of substrate structure and surface hydroxylation on orientation of the crystalline structure,
- the role of surface defects related to the nanostructure of passivated surfaces in passivity breakdown and the effect of chlorides on passivity repair, and
- the correlation between the near-matrix region around surface chemical heterogeneities and pit initiation.

The obtained data greatly contribute to the advances in the understanding and control of corrosion of metals and alloys at the nanoscale. In the future, the impact of combined structural and chemical characterization at the nanoscale of the nanostructural defects and heterogeneities of surfaces protected by ultrathin layers should allow a major breakthrough in the design and control of new materials, corrosion resistant at the nanoscale. In situ nanoprobes should also emerge as powerful tools, using corrosion in a controlled manner, to produce nanostructured surfaces.

References

1. D.W. Suggs, A.J. Bard, J. Am. Chem. Soc. 116: 10725, 1994.
2. D.W. Suggs, A.J. Bard, J. Phys. Chem. 99: 8349, 1995.
3. T.P. Moffat, Mat. Res. Soc. Symp. Proc. 451: 75, 1997.
4. M.R. Vogt, A. Lachenwitzer, O.M. Magnussen, R.J. Behm, Surf. Sci. 399: 49, 1998.
5. O.M. Magnussen, M.R. Vogt, J. Scherer, R.J. Behm, Appl. Phys. A 66: S447, 1998.
6. P. Broekmann, M. Anastasescu, A. Spaenig, W. Lisowski, K. Wandelt, J. Electroanal. Chem. 500: 241, 2001.
7. O.M. Magnussen, L. Zitzler, B. Gleich, M.R. Vogt, R.J. Behm, Electrochim. Acta 46: 3725, 2001.
8. T. Suzuki, T. Yamada, K. Itaya, J. Phys. Chem. 100: 8954, 1996.
9. M. Dietterle, T. Will, D.M. Kolb, Surf. Sci. 327: L495, 1995.
10. S. Ando, T. Suzuki, K. Itaya, J. Electroanal. Chem. 431: 277, 1997.
11. S. Ando, T. Suzuki, K. Itaya, J. Electroanal. Chem. 412: 139, 1996.
12. T. Teshima, K. Ogaki, K. Itaya, J. Phys. Chem. B 101: 2046, 1997.
13. K. Sashikata, Y. Matsui, K. Itaya, M.P. Soriaga, J. Phys. Chem. 100: 20027, 1996.
14. K. Itaya, in Interfacial Electrochemistry – Theory, Experiments and Applications, A. Wieckowski, Editor, p. 187, Marcel Dekker, New York, 1999.
15. M.R. Vogt, W. Polewska, O.M. Magnussen, R.J. Behm, J. Electrochem. Soc. 144: L113, 1997.

16. M.R. Vogt, R.J. Nichols, O.M. Magnussen, R.J. Behm, J. Phys. Chem. B 102: 5859, 1998.
17. M. Sugimasa, L.-J. Wan, J. Inukai, K. Itaya, J. Electrochem. Soc. 149: E367, 2002.
18. W. Polewska, M.R. Vogt, O.M. Magnussen, R.J. Behm, J. Phys. Chem. B 103: 10440, 1999.
19. J. Scherer, M.R. Vogt, O.M. Magnussen, R.J. Behm, Langmuir 13: 7045, 1997.
20. C.W. Yan, H.C. Lin, C.N. Cao, Electrochim. Acta 45: 2815, 2000.
21. E. Rocca, G. Bertrand, C. Rapin, J.C. Labrune, J. Electroanal. Chem. 503: 103, 2001.
22. N. Ikemiya, T. Kubo, S. Hara, Surf. Sci. 323: 81, 1995.
23. V. Maurice, H.-H. Strehblow, P. Marcus, Surf. Sci. 458: 185, 2000.
24. J. Kunze, V. Maurice, L.H. Klein, H.-H. Strehblow, P. Marcus, J. Phys. Chem. B, 105: 4263, 2001.
25. J. Kunze, V. Maurice, L.H. Klein, H.-H. Strehblow, P. Marcus, Electrochim. Acta 48: 1157, 2003.
26. J. Kunze, V. Maurice, L.H. Klein, H.-H. Strehblow, P. Marcus, J. Electroanal. Chem. 554–555, 113–125, 2003.
27. J. Kunze, V. Maurice, L.H. Klein, H.-H. Strehblow, P. Marcus, Corrosion Sci. 46: 245–264, 2004.
28. G.S. Frankel, J. Electrochem. Soc. 145: 2186, 1998.
29. P. Marcus, V. Maurice, Passivity of metals and alloys in *Corrosion and Environmental Degradation,* M. Schütze, (Ed.), (Wiley-VCH, Weinheim, 2000), pp. 131–169.
30. J.W. Schultze, M.M. Lohrengel, Electrochim. Acta 45: 2499, 2000.
31. P. Marcus (Ed.), Corrosion Mechanisms in Theory and Practice, 2nd edition (Marcel Dekker Inc., New York, 2002).
32. O. Lev, F.-R. Fan, A. J. Bard, J. Electrochem. Soc. 135: 783, 1988.
33. V. Maurice, H. Talah, P. Marcus, Surf. Sci. 284: L431, 1993.
34. V. Maurice, H. Talah, P. Marcus, Surf. Sci. 304: 98, 1994.
35. S.-L. Yau, F.-R. Fan, T. P. Moffat. A. J. Bard, J. Phys. Chem. 98: 5493, 1994.
36. D. Zuili, V. Maurice, P. Marcus, J. Electrochem. Soc. 147: 1393, 2000.
37. N. Hirai, H. Okada, S. Hara, Transaction JIM 44: 727, 2003.
38. J. Scherer, B.M. Ocko, O.M. Magnussen, Electrochim. Acta 48: 1169, 2003.
39. M. Nakamura, N. Ikemiya, A. Iwasaki, Y. Suzuki, M. Ito, J. Electroanal. Chem. 566: 385–391, 2004.
40. A. Seyeux, V. Maurice, L.H. Klein, P. Marcus, J. Solid State Electrochem. 9: 337–346, 2005.
41. R.C. Bhardwaj, A. Gonzalez-Martin, J.O'M. Bockris, J. Electroanal. Chem. 307: 195, 1991.
42. M.P. Ryan, R.C. Newman, G.E. Thompson, J. Electrochem. Soc. 142: L177, 1995.
43. J. Li, D.J. Meier, J. Electroanal. Chem. 454: 53, 1998.
44. I. Diez-Pérez, P. Gorostiza, F. Sanz, C. Müller, J. Electrochem. Soc. 148: B307, 2001.
45. E.E. Rees, M.P. Ryan, D.S. MacPhail, Electrochem. Solid-State Lett. 5: B21, 2002.
46. V. Maurice, W. Yang, P. Marcus, J. Electrochem. Soc. 141: 3016, 1994.
47. D. Zuili, V. Maurice, P. Marcus, J. Phys. Chem. B 103: 7896, 1999.
48. R.C. Bhardwaj, A. Gonzalez-Martin, J.O'M. Bockris, J. Electrochem. Soc. 138: 1901, 1991.
49. A. Foelske, J. Kunze, H.-H. Strehblow, Surf. Sci. 554: 10–24, 2004.
50. M.P. Ryan, R.C. Newman, G.E. Thompson, Philosophical Mag. B 70: 241, 1994.
51. M.P. Ryan, R.C. Newman, G.E. Thompson, J. Electrochem. Soc. 141: L164, 1994.
52. V. Maurice, W. Yang, P. Marcus, J. Electrochem. Soc. 143: 1182, 1996.
53. H. Nanjo, R.C. Newman, N. Sanada, Appl. Surf. Sci. 121: 253, 1997.
54. V. Maurice, W. Yang, P. Marcus, J. Electrochem. Soc. 145: 909, 1998.
55. J. Oudar, P. Marcus, Appl. Surf. Sci. 3: 48, 1979.
56. O.M. Magnussen, J. Scherer, B. M. Ocko, R. J. Behm, J. Phys. Chem. B 104: 1222, 2000.
57. P. Marcus, V. Maurice, *Passivity and Its Breakdown, Joint ECS/ISE meeting*, Paris, France, 1997, P. Natishan, H.S. Isaacs, M. Janik-Czachor, V.A. Macagno, P. Marcus, M. Seo, (Eds.), The Electrochemical Society Proceedings Series, PV 97-26, Pennington, NJ, 1998, pp. 254–265.

58. V. Maurice, V. Inard, P. Marcus, *Critical Factors in Localized Corrosion III,* P.M. Natishan, R.G. Kelly, G.S. Frankel, R.C. Newman (Eds.), The Electrochemical Society Proceedings Series, PV 98-17, Pennington, NJ, 1999, pp. 552–562.
59. V. Maurice, L. Klein, P. Marcus, Electrochem. Solid-State Lett. 4: B1, 2001.
60. V. Maurice, L. Klein, P. Marcus, Surf. Interf. Anal. 34: 139, 2002.
61. V. Maurice, T. Nakamura, L. Klein, P. Marcus, Proceedings of Eurocorr'2004, EFC Series, in press.
62. R.M. Rynders, C.-H. Paik, R. Ke, R.C. Alkire, J. Electrochem. Soc. 141: 1439, 1994.
63. J.O. Park, C.-H. Paik, Y.H. Huang, R.C. Alkire, J. Electrochem. Soc. 146, 517, 1999.
64. G. Gugler, J.D. Neuvecelle, P. Mettraux, E. Rosset, D. Landolt, *Modifications of Passive Films*, P. Marcus, B. Baroux, M. Keddam, (Eds.), EFC publications n°12, The Institute of Materials, p. 274, 1994.
65. D.E. Williams, T.F. Mohiuddin, Y.Y. Zhu, J. Electrochem. Soc. 145: 2664, 1998.
66. R.E. Williford, C.F. Windisch Jr., R.H. Jones, Mat. Sci. Engin. A288: 54–60, 2000.
67. B. Vuillemin, X. Philippe; R. Oltra, V. Vignal, L. Coudreuse, L.C. Dufour, E. Finot, Corros. Sci. 45: 1143–1159, 2003.
68. M.P. Ryan, D.E. Williams, R.J. Chater, B.M. Hutton, D.S. McPhail, Nature 415: 770–774, 2002.
69. Q. Meng, G.S. Frankel, H.O. Colijn, S.H. Goss, Nature 424: 389–390, 2004.
70. M. Femenia, J. Pan, C. Leygraf, P. Luukkonen, Corrosion Sci. 43: 1939, 2001.
71. Z.-X. Xie, D.M. Kolb, J. Electroanal. Chem. 481: 177, 2000.
72. S.G. Garcia, D.R. Salinas, C.E. Mayer, W.J. Lorenz, G. Staikov, Electrochim. Acta, 48: 1279–1285, 2003.
73. B. Obliers, M. Anastasescu, P. Broeckmann, K. Wandelt, Surf. Sci. 573: 47–56, 2004.
74. L. Chen, D. Guay, J. Electrochem. Soc. 141: L43, 1994.
75. L. Roue, L. Chen, D. Guay. Langmuir 12: 5818, 1996.
76. P. Schmutz, G.S. Frankel, J. Electrochem. Soc. 145: 2295, 1998.
77. P. Schmutz, G.S. Frankel, J. Electrochem. Soc. 146: 4461, 1999.

Nanobioelectrochemistry

A.M. Oliveira Brett

1 Overview

Bioelectrochemistry is an interdisciplinary subject that deals with the aspects of electrochemistry and electroanalysis characterizing biological processes at the molecular level and relevant to the mechanisms of biological regulation of cells [1]. Nanotechnology has been first used to refer to the ability to engineer materials precisely at the scale on nanometers [2]. Nanotechnology is thus defined as the design and fabrication of materials, devices, and systems with control at nanometer dimensions. So its essence is therefore size and control at the nanometer scale.

The nanometer, one-millionth of a millimeter, is the characteristic scale of molecules, groups of atoms bound covalently together. Atoms are of the order of a tenth of a nanometer. Consequently a nano-object contains tens or hundreds of atoms in order to achieve the required size.

Nanobiotechnology deals with materials and processes at the nanometer scale, which are based on biological, biomimetic or biologically relevant molecules, and nanotechnological devices, which are used to monitor or control biological processes in medicine, examples being nanoswitches incorporating biomolecules and biochips.

Chemistry, physics, and biology have always dealt with atoms and molecules, their behavior, reactions, and manipulation, and quantum mechanics is the science of the absolutely small. Quantum mechanics defines as absolutely small a system that is perturbed by that act of observing it, and most nanosystems are not small enough. However, quantum effects are needed to explain nano-objects, tiny spheres of a solid, the small clusters of atoms called quantum dots, nanodots, or nanoparticles. Chemists are the ultimate nanotechnologists, always attaching individual atoms to other atoms. If it is necessary to cover a very small area of a surface with molecules,

A.M.O. Brett (✉)
Departamento de Química, Universidade de Coimbra,
3004-535 Coimbra, Portugal
e-mail: brett@ci.uc.pt

P. Schmuki, S. Virtanen (eds.), *Electrochemistry at the Nanoscale,* Nanostructure Science and Technology, DOI 10.1007/978-0-387-73582-5_11,

a very powerful way to do it is using the idea of self-assembly of molecules. The development of synthesis technology to make nanometer-scale, three-dimensional structures on surfaces with prechosen properties is materials by design.

Biological structures at macromolecular and supramolecular scales are assembled naturally using principles of self-assembly, and most of these protein-based structures combine lightness with tremendous strength in miniature complex mechanisms, as in the proton-gradient mechanism across the cellular membrane. Many biomolecules are of nanometer dimensions and can exhibit similar properties. As one simple example, the dimeric glucose oxidase enzyme has the dimensions of $7 \times 5.5 \times 8$ nm.

In the 1980s, instruments were invented which allow the examination of solids and surfaces down to the atomic scale in both air and liquids. G. Binnig and H. Rohrer won the Nobel Prize for physics in 1986 for their design of the scanning tunneling microscope, which led to the development of many other forms of scanning local probe microscopes [3]. Techniques such as scanning electron microscopy and transmission electron microscopy require the use of a high vacuum. The new family of microscopies – collectively known as scanning probe microscopy (SPM) – encompasses techniques such as scanning tunneling microscopy (STM), atomic force microscopy (AFM), scanning Kelvin microscopy (SKM) and scanning electrochemical microscopy (SECM) [3, 4]. All of these require accurate positioning of a nanometric-sized probe, often within less than 0.1 nm. SPM manipulation can be used to accurately and reliably position molecular-sized biocomponents.

Nanotechnology brings a different, new approach imposing conditions, engineering materials, and trying to control the systems, although chemistry may impose fundamental limitations on the freedom of nanotechnologies to manipulate matter at the atomic and molecular scales.

Nanotechnology can be divided into fabrication of materials and devices in complex three-dimensional structures, and metrology, the means to quantitatively assess the quality of the manufactured nano-objects. Two possible approaches for obtaining structures within the size range of 1–100 nm are generally discussed. The first one is the "top-down approach" based on lithography, which is currently used to fabricate integrated circuits. The second is the "bottom-up (or chemist's) approach" in which complex structures are assembled from single atoms and molecules into supramolecular structures.

Nanotechnology is both a technology for fabricating ultra-small materials and devices, and a concept in which everything is considered from the viewpoint of atomic or molecular building blocks. An immediate consequence of miniaturization of materials is the great increase in specific surface area, and a possible advantage is that the intrinsic properties may be changed for the better when they are finely divided. Techniques that enable large assemblage of nanocomponents are a manifestation of nanotechnology.

Progressive miniaturization of existing technologies leads to new areas of application. The conceptual aspect, at atomic or molecular level, leads us to the living systems where proteins are broken down into their amino-acid constituents that will be used as templates for synthesis of new proteins. Biological molecular motors are being intensively studied as a source of design inspiration for truly nanoscale motors.

Applications in medicine are concerned with "smart medicine" or "magic bullet" in which the nanostructured drug-delivery particles target solely the organ of interest, avoiding all the deleterious effects on the whole body due to the higher doses otherwise necessary to be effective. In nanosurgery, the ultimate development of surgical tools are the still-to-be-realized quasi-autonomous nanosized robots, nanobots, which would be released into the blood, through which they would travel to the site needing intervention, where, controlled by external commands given by a surgeon, they would carry out the required repair action. The implantable nanosensor diagnostic device, for continuously monitoring physiological parameters, is an ultimate goal that requires advances in miniaturization and surpasses the biocompatibility drawback.

Nanorobotics encompasses the design, fabrication, and programming of robots with overall dimensions below a few micrometers, and the programmable assembly of nanoscale objects, expected to have revolutionary applications in environmental monitoring for micro-organisms and healthcare. Nanorobots have overall dimensions comparable to those of biological cells and organelles. Nanorobots are only one example of nanoelectromechanical systems (NEMSs), which represent a new frontier in miniaturization [5].

In addition, continuous monitoring of food, in a faster, safer, and cheaper fashion, is required, and nanosensor devices will be developed to assess packaged and fresh food condition. The "intelligent" nanosensor will be responding to a tiny fraction, or one, of the total complex mixture of molecules present, and the target chemicals have to be specifically chosen for each food.

Electrochemical research in nanotechnology [6, 7] is of great relevance to explain many biological mechanisms, and the electrode interface is a very good model for simulating interaction with cell membranes and to clarify the mechanisms of action. Electrochemical biosensors have the advantage of being rapid, sensitive, and cost-effective and enable *in situ* generation of reactive intermediates and their detection and are easy to miniaturize.

Comprehensive descriptions of research on electrochemical biosensors [8] show the great possibilities of using electrochemical transduction in diagnostics. The electrochemical transduction is dynamic, in that the electrode is itself a tuneable charged reagent as well as a detector of all surface phenomena, which greatly enlarges the electrochemical biosensing capabilities. However, it is necessary that the analyte is electroactive, that is, capable of undergoing electron-transfer reactions, in order to use an electrochemical transducer. To design heterogeneous electrochemical biosensors, it is essential to understand the surface structures of the modified surfaces.

Nanobioelectrochemistry is a word that refers to different aspects of electrochemical research of biological compounds at nanoscale, such as bioelectroanalysis at and below nanomolar concentration, and characterization and applications of nanofilms for biologically modified electrode-surface processes.

2 Electrochemistry of Biomolecules at the Nanoscale

Nowadays, detection of biomolecules at the nanoscale is very important and atto- and femtomole concentrations are not possible to achieve with many analytical methods. Voltammetric studies of proteins that are tightly bound to an electrode surface reveal interesting and intricate chemistry that is not detected easily or at all by other methods. Electroanalytical methods are inexpensive, faster, and can be automated, and appropriate voltammetric methods can allow direct measurements on biological samples with very little or no pretreatment of the samples. Electrochemistry, using ultramicroelectrodes and of strongly adsorbed biofilms on electrode surfaces, is enabling the study of biologically relevant events and compounds at nanoscale concentrations.

Living cells exchange information through emission of chemical messengers. The importance of such messengers has been widely recognized by biologists. A significant number of exocytosis measurements of neurotransmitters have been correlated with messengers released via a transitory fusion pore before full exocytosis. However, what is less understood is how these chemical messengers are released by cells in their outer-cytoplasmic fluids. This difficulty is easily understood when one becomes aware that most of these releases occur in the atto–femtomole ranges which prevent the use of classical detection methods. Ultramicroelectrodes have proved extremely useful for monitoring such events, as shown in Fig. 1. Electrochemical data allow the whole process of exocytosis to be described with a precision that has never been achieved before by other techniques [9].

The physicochemical properties of vesicles are key factors in the control of the dynamics of release through the fusion pore, and the high and variable frequency of this release makes it highly significant. The release of chemical messengers by dense, core vesicles, such as chromaffin cells, located above the kidneys, have also been quantitatively explained [10].

Adriamycin is an antibiotic of the family of anthracyclines with a wide spectrum of chemotherapeutic applications and antineoplastic action, but that causes cardiotoxicity ranging from a delayed and insidious cardiomyopathy to irreversible heart failure. Detecting and quantifying adriamycin on organic fluids or cells are very important. The strong and irreversible adsorption of adriamycin onto carbon electrodes allows detection limits 10,000 times lower than the usual limits for voltammetric methods. In fact, the detection limit of the order of 10^{-11} M (10 pM) for adriamycin is the lowest reported in the literature for this substance using voltammetric methods [11]. The adsorption of adriamycin onto glassy carbon and highly oriented pyrolytic graphite electrodes was studied by voltammetric techniques and MAC Mode AFM, Fig. 2. *In situ* AFM images show quick and spontaneous adsorption of the adriamycin on HOPG surfaces.

The oxidation of the adsorbed adriamycin is pH dependent and corresponds to a two-electron/two-proton mechanism. The total surface concentration of adriamycin adsorbed onto glassy carbon, from a 50 nM adriamycin solution during 3 min, was calculated to be 2.57×10^{-12} mol/cm^2. The strong adriamycin adsorption enables the development of a voltammetric method for direct detection and quantification of

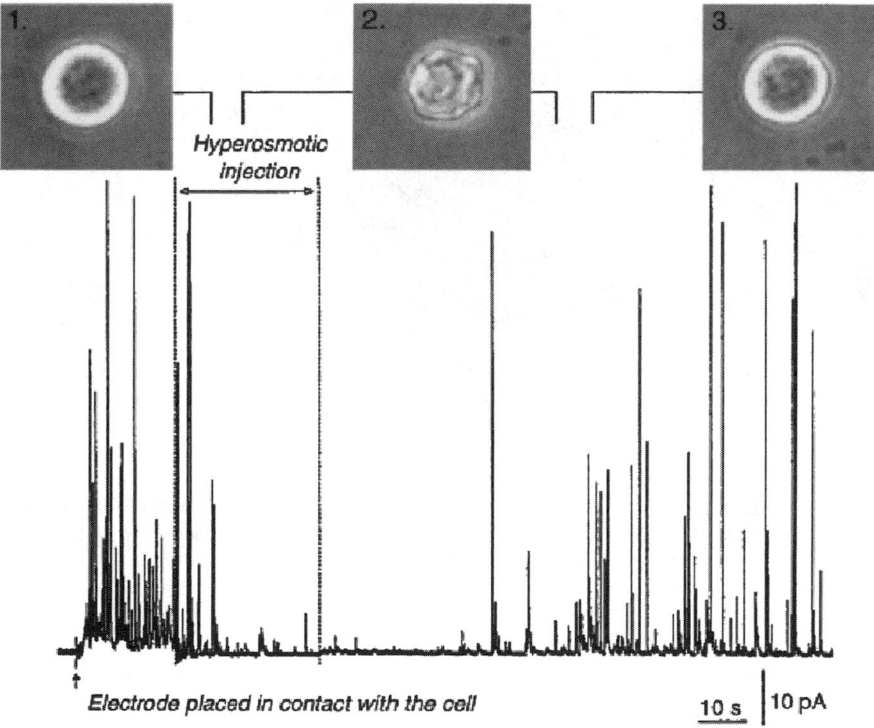

Fig. 1 Effect of transient exposures (30–60 s) to hyperosmotic media (630, 750, or 970 mOsm, obtained by adjusting the NaCl concentration of Locke buffer supplemented with 0.7 mM $MgCl_2$ without carbonates) on adrenaline release by chromaffin cells. [Reprinted with permission by Wiley, from Chem. Phys. Chem., Vol. 4, pp. 101–108, (2003), C. Amatore, S. Arbault, I. Bonifas, Y. Bouret, M. Erard, M. Guille, "Dynamics of Full Fusion During Vesicular Exocytotic Events: Release of Adrenaline by Chromaffin Cells.]

adriamycin in biological fluids and makes adriamycin a very interesting substance for studying adsorption and interfacial phenomena.

Protein film voltammetry (PFV) provides detailed and integrated kinetic and thermodynamic information on the electron-transfer properties of redox proteins [12, 13]. PFV examines a film of protein molecules, adsorbed directly onto the electrode surface. The protein sample is adsorbed on an electrode, up to monolayer coverage, in such a way that electron exchange with the active sites is fast. Unmodified graphite electrodes were used, providing a less well-defined surface than SAMs, onto which many proteins spontaneously adsorb in a redox-active state. Direct adsorption leads to a protein mono/submonolayer which is tightly adsorbed on the electrode surface. The electrode is effectively a redox partner, delivering or removing electrons, but with the added advantage of being able to vary and control the potential (driving force) and measure the corresponding rates. The kinetic parameters of the redox reaction are readily extracted. In achieving this aim, there

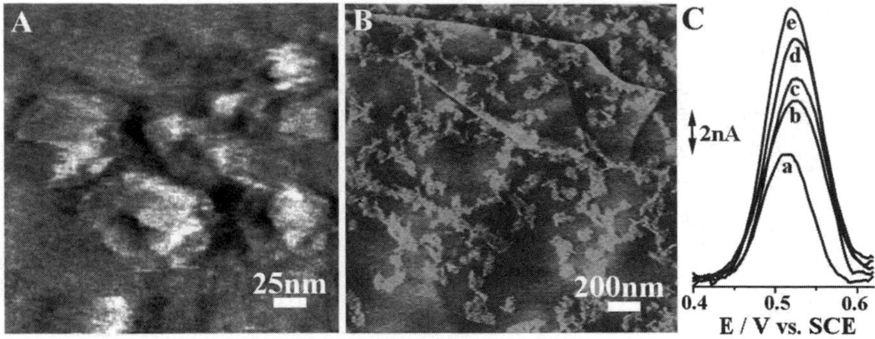

Fig. 2 MAC Mode AFM images obtained in pH 4.5 0.1 M acetate buffer electrolyte of adriamycin adsorbed onto HOPG, after free adsorption from a pH 4.5 0.1 M acetate buffer electrolyte 1 nM adriamycin solution: (**A**) 30 s (**B**) 3 min. (**C**) Differential pulse voltammograms in pH 4.5 0.1 M acetate buffer solution of adriamycin adsorbed on GCE, free adsorption during 3 min from pH 4.5 0.1 M acetate buffer adriamycin solutions: a – 1×10^{-10} (100 pM), b – 3×10^{-10} (300 pM), c – 5×10^{-10} (500 pM), d – 9×10^{-10} (900 pM), e – 1×10^{-9} M (1 nM). [Adapted with permission from J. Electroanalytical Chem., Vol. 267–276, A.M. Olivera Brett, J.A.P. Piedade, A.M. Chiorcea, "Voltammetry and AFM imaging of picomoles of adsorbed adriamycin onto glassy carbon and HOPG electrode surfaces", pp. 538–539, Copyright Elsevier (2002).]

is considerable interest in electrodes (particularly Au) that are modified with a self-assembled monolayer (SAM) of thiol-containing adsorbates. The terminal functionalities on these adsorbates are usually designed to provide a surface that will give strong noncovalent interactions with the protein molecules and orient them for fast electron transfer.

Metallophthalocyanine (MPc) and its derivatives are well known as organic nanosized electrocatalytic materials that catalyze oxidation or reduction of some species, such as cysteine and nitric oxide. Bulk nanosized particles of metallomacrocyclic compounds (phthalocyanines and phorphyrins) have been used in biosensing. Carbon-paste electrodes (CPE) modified with immobilized nanosized cobalt phthalocyanine (nano-CoPc) particles (nano-CoPc-CPE) were fabricated [14]. The electrocatalytic oxidation of theophylline (1,3-dimethylxanthine, THP), a member of xanthine-based alkaloids, with a stimulating effect on respiration widely used in the treatment of asthma and chronic bronchitis in adults, on the nano-CoPc-CPE was investigated and the current obtained was enhanced compared to bulk CoPc-modified CPE (B-CoPC-CPE).

3 Electrochemistry of Self-Assembled Nanostructured Biomolecules

Bionanotechnology adapts not only the results, functional proteins, and nucleic acids but also the processes, supramolecular aggregates, and metabolosome constructions of molecular evolution. Molecular addressing systems, based on deoxyribonucleic acid (DNA) complementarity and DNA–protein interaction selectivity, are emerging [15]. Molecular motors, DNA-based switches, DNA-based oscillators, enzymes, ribozymes, gold nanoparticles, antibodies, aptamers, and nucleic-acid-binding proteins can be ordered along the nucleic acid backbone. Strands of DNA interact in the most programmable way. Their enormous variability provides ample scope for designing molecules; DNA is more than just the secret of life. It is also a versatile component for making nanoscopic structures and devices [16]. Designs for extended 3D crystals based on DNA cages could be used for crystallization of macromolecules and enable their structure determination by X-ray diffraction.

Self-assembly tries to develop nano- and microstructures following the bottom-up procedure, from simple molecules to more complicated systems. Self-assembled structures using liposomes, polymerized lipid vesicles, or pseudo-cellular membranes are the most widely studied.

The interaction of biologically derived peptides with phospholipid monolayers and bilayers has been of interest for some time. The results have relevance to biological mechanisms, since peptide and protein interactions with biological membranes are of great significance in many aspects of physiology, such as cell signaling and toxicology. Gramicidin is a well-known membrane-active peptide consisting of 15 amino acid residues. It forms channels in phospholipid bilayers of two β-helices joined together through the hydrogen bonding of the terminal formyl groups within the hydrocarbon environment of the bilayer. Due to these properties, gramicidin is the most used channel model. In phospholipid monolayers, gramicidin forms monomolecular-conducting channels and the activity of gramicidin is unique, in that the inside of the gramicidin β-helix acts as the channel lumen. The properties of gramicidin derivatives in bilayers experiments using electrochemical impedance spectroscopy have shown that gramicidin interacts with dioleoyl phosphatidylcholine (DOPC) monolayers, increasing the surface "roughness" and introducing an extra capacitative element [17].

The tethering of molecular analogs of biological membranes to solid surfaces has been used in a variety of biomimetic systems. Because of the simplicity of sulfur–gold tethering chemistry, thiol- and disulphide-labeled compounds have been the basis of most of these studies. A generic immunosensing device, in which the polypeptide ion channel, gramicidin A, was assembled into a tethered lipid membrane and was coupled to an antibody targeting a compound of diagnostic interest was described [18]. This gramicidin-based biosensor operates as an ion-channel switch amperometric nanobiosensor in which the binding of the target molecule shifts the conformation of the gramicidin channels from predominantly conducting dimers to nonconducting monomers. Applying a small alternating potential between the gold electrode and a reference electrode in the test solution generates a charge at the gold surface and causes electrons to flow in an external circuit. The device provides a rapid, quantitative measure of the concentration of the target compound in the test solution. The test solution can be any electrolyte containing biological fluid, including blood, serum, urine, or saliva. The specificity of the response is dependent on the specificity of the receptor and is equivalent to a conventional sandwich immunoassay. The receptors most extensively studied have been antibody Fab fragments.

Synthetic micropore and gold nanotubule membranes that mimic the function of a ligand-gated ion channel have been described [8]. These membranes can be switched from an "off" state, no or low ion-current though the membrane to an "on" state, higher ion-current, in response to a chemical stimulus, such as a drug. These ion-channels mimetic membranes were based on both gold-nanotube and microporous alumina membranes. The gold nanotube membranes were prepared via electroless deposition of Au onto the pore walls of a polycarbonate membrane, the pores acting as templates for the nanotubes. The current is forced through the nanotubes, and analyte molecules present in a contacting solution phase modulate the value of the transmembrane current.

There is an increasing interest in the concept of using nanopores as the sensing elements in biosensors. The nanopore most often used is the α-hemolysin protein channel, and the sensor consists of a single channel embedded within a lipid bilayer membrane. An ionic current is passed through the channel, and analyte species are detected as transient blocks in this current associated with translocation of the analyte through the channel-stochastic sensing. While this is an extremely promising sensing paradigm, it would be advantageous to eliminate the very fragile lipid bilayer membrane and perhaps to replace the biological nanopore with an abiotic equivalent. A new family of protein biosensors that are based on conical-shaped gold nanotubes embedded within a mechanically and chemically robust polymeric membrane has been described [19]. Three different molecular-recognition agents, and correspondingly three different protein analytes, were investigated: (i) biotin/streptavidin, (ii) protein-G/immunoglobulin, and (iii) an antibody to the protein ricin, with ricin as the analyte.

Through-mask electrodeposition has been used extensively to pattern metals, semiconductors, and polymers on conductive substrates. The mask, a patterned arrangement of solvent-accessible openings, limits the regions of the substrate where material growth occurs, thereby allowing synthesis of a high-fidelity negative replica with nanometer-thickness control. Emerging techniques for electrochemical nanofabrication frequently exploit the masks that self-organize, such as anodic alumina and molecular crystals such as amphiphilic surfactants and block-copolymers, but these self-assembling masks form a limited number of unit cell geometries, precluding the creation of complex patterns. Proteins can organize into homo- or heterostructures representing all possible two-dimensional (2D) space groups built from chiral molecules. Electrodeposition through crystalline protein masks has not yet been demonstrated as a robust nanofabrication strategy. In electrodeposition, material growth proceeds from the substrate outward and need not follow a line-of-sight path through the mask. Thus, electrodeposition offers the unique prospect of being able to grow dense materials through a tortuous multilayer crystalline protein mask, or other complex protein structure, thereby easing the difficult task of optimizing the protein–surface interface to attain perfect monolayer coverage. Of particular interest for bio-inspired nanofabrication are surface layer (S-layer) proteins, a class of 2D crystalline proteins that encapsulate certain bacterial cells, protecting them from extracellular enzymes and regulating molecular trafficking. S-layers are highly resistant to conditions that normally denature proteins (e.g., low pH, chaotropic agents, and heat) and can assemble into all 2D rotational symmetries. They have 1–4 nm solvent-accessible openings organized with typical lattice parameters ranging from 10 to 20 nm.

A simple and robust method to fabricate nanoarrays of metals and metal oxides over macroscopic substrates using the crystalline surface layer (S-layer) protein of *Deinococcus radiodurans* as an electrodeposition mask was developed [20]. The hexagonally packed intermediate (HPI) layer from *Deinococcus radiodurans* was selected because of the potential of S-layer proteins to through-mask electrodeposition based on its chemical resistance and ease of purification in the crystalline state. S-layers are highly resistant to conditions that normally denature proteins. Substrates are coated by adsorption of the S-layer from a detergent-stabilized aqueous protein extract, producing insulating masks with 2–3 nm diameter solvent-accessible openings to the deposition substrate. Because this characteristic is shared by many S-layer proteins from other organisms, and possibly other self-assembling proteins, will certainly prove useful for the electrodeposition of ordered nanostructures with a variety of superlattice symmetries. The coating process can be controlled in order to achieve complete or fractional surface coverage. The general applicability of the technique by forming arrays of cuprous oxide (Cu_2O), Ni, Pt, Pd, and Co exhibiting long-range order with the 18-nm hexagonal periodicity of the protein openings was demonstrated [20]. This protein-based approach to electrochemical bio-inspired nanofabrication should permit the creation of a wide variety of 2D inorganic structures.

Two-dimensional crystalline bacterial cell S-layers on solid substrates are of fundamental and technological interest in biotechnology and bottom-up nanostructuring technologies. The mechanism of the recrystallization of nanoscale bacterial surface protein layers (S-layer proteins) on solid substrates is of fundamental interest in the understanding and engineering of biomembranes and biosensors [21].

The electrochemical behavior of the S-layers on gold electrodes has been investigated by *in situ* electrochemical quartz crystal microbalance (EQCM) measurements, AFM, and small-spot X-ray photoelectron spectroscopy (SS-XPS) of potentiostatiscally emersed substrates [21]. It was shown that the negatively charged bonding sites of the S-layer units (e.g., carboxylates) can bond with positively charged Au surface atoms in the positively charged electrochemical double-layer region positive of the point of zero charge (~ -0.8 V vs. saturated mercury–mercurous sulphate electrode). Surface conditions in other potential regions decelerated the recrystallization and fixation of S-layers. Time-resolved *in situ* and *ex situ* measurements demonstrated that 2D S-layer crystal formation on gold electrodes with potential control can occur within a few minutes, in contrast to the hours, common in open-circuit SAM generation. These results proved that the recrystallization and fixation of 2D-crystalline S-layers on an electronic conductor can be influenced and controlled by direct electrochemical manipulation.

4 Electrochemistry and AFM of Nanoscale DNA Surface Layers on Conducting Surfaces

DNA is a very important biomolecule that plays a crucial role in all living organisms, being responsible for storage, duplication, and genetic information [22]. Due to its important chemical and biophysical characteristics, which establish a high specificity of recognition and binding to other molecules, nanotechnology and biosensor technology have been using DNA molecules for construction of DNA-based biosensor devices [23–27]. In particular, DNA electrochemical biosensors have received special attention due to numerous successful applications in medicine and in environmental and food control [25–31].

A DNA electrochemical biosensor is an integrated receptor–transducer device: an electrochemical transducer (the electrode) coupled to an immobilized DNA nanofilm on its surface (the probe) as biological matrix-recognition element in order to detect both DNA damage and DNA damaging agents. Interaction of DNA with the damaging agent is converted, via changes in the electrochemical properties of the DNA recognition film, into measurable electrical signals. The most important step in the development and manufacture of a sensitive DNA-biosensor for the detection of DNA–drug interactions is the immobilization and stabilization procedure of the nucleic acid probe on the electrode surface.

Many factors can influence the efficiency of DNA immobilization on the electrode and on the nature of DNA–surface interactions: the electrode characteristics and electrode pretreatment conditions, the DNA adsorption procedure, the pH and ionic strength of the DNA solution, the sequence, length, and concentration of the DNA molecules. Many of the biophysical properties of DNA and its flexibility are influenced by these factors. Therefore, it is important to have a good understanding of all key factors that influence the immobilization of the nucleic acid probe onto the transducer surface.

The different structures and conformations that DNA molecules can adapt at the electrode surface lead to different types of interaction and to the modification of the accessibility of the chemical compounds to the DNA grooves. Consequently, the understanding of DNA biosensor surface morphological characteristics is essential for its practical application and for a better understanding of the voltammetric results obtained.

In recent years, magnetic A/C mode AFM (MAC Mode AFM) has proven to be, at the same time, a powerful technique to investigate the interfacial and conformational proprieties of biological samples [31, 32]. It is a gentle technique that permits the direct visualization of biomolecules that are softly bound to the electrode surface.

MAC Mode AFM has been used to investigate the overall surface topography of DNA-based biosensors obtained by adsorption of dsDNA molecules on the electrode surface, because it can bring important information concerning the internal morphology of DNA–electrochemical biosensors. It also enables visualization of individual DNA molecules and DNA–protein complexes immobilized onto different surface materials, with extraordinary resolution and accuracy.

The nature of the DNA interactions with carbon electrode surfaces and the topographical conformations that short-length DNA molecules can adopt are still not clearly understood. Short-chain 10-mer oligodeoxynucleotides (ODNs) molecules, with various specific sequences, represent particularly attractive models to study nucleic acid adsorption, because they allow a clear interpretation of the experimental data.

The free adsorption of 10-mer synthetic oligodeoxynucleotides (ODNs) onto highly oriented pyrolytic graphite (HOPG) surfaces was studied using AFM [33]. The mechanism of interaction of nucleic acids with carbon electrode surfaces was elucidated, using 10-mer synthetic homo- and hetero-ODNs sequences of known base sequences, because they allow clear interpretation of the experimental data. AFM images in air revealed different adsorption patterns and degree of HOPG surface coverage for the ODNs and the correlation with the individual structure and base sequence of each ODN molecule. The hydrophobic interactions with the HOPG hydrophobic surface explain the main adsorption mechanism, although other effects such as electrostatic and van der Waals interactions may contribute to the free-adsorption process. The ODNs interacted differently with the HOPG surface, according to the ODN sequence hydrophobic characteristics, being directly dependent on the molecular mass, the hydrophobic character of the individual bases, and on the secondary structure of the molecule.

The adsorption of homo-ODNs containing only adenines, guanines, thymines, and cytosines showed that the degree of HOPG surface coverage by A10, T10, and C10 sequences followed the decrease of the ODN molecular mass and the hydrophobicity of the constituent bases A > T > C Fig. 3. The homo-ODNs A10 and T10 containing single-stranded species interact and adsorb strongly onto the HOPG surface, when compared with double-helical C10 and quadruple G10 ODNs. The A10 and G10 sequences adsorb strongly onto HOPG due to the superior hydrophobicity of the purinic bases at the chain extremities, while the adsorption of pyrimidinic C10 sequences was almost completely inhibited [33]. The hydrophobic interactions of the ODN molecules with the surface represent the main adsorption mechanism, although other effects such as electrostatic and van der Waals interactions may also contribute to the adsorption process.

The 10-mer ODN molecules adsorbed spontaneously on the HOPG surface, through hydrophobic interactions with the hydrophobic surface, and the adsorption is strongly influenced by the base type in the ODN sequence. The importance of the type of base at the sequence extremities on the adsorption process was evaluated. The differences in HOPG surface coverage obtained with ODN molecules with mixed-base composition and the same number of each base, in different orders, demonstrated that the type of base at the molecule extremities plays the most important role in the adsorption process.

The molecular mass and the presence of secondary structures in the ODNs influence the hydrophobic character of the molecules and consequently the adsorption onto HOPG. Single-stranded ODN molecules interacted and adsorbed strongly onto HOPG, since the bases are more exposed and free to undergo hydrophobic interactions with the HOPG surface. The formation of double-helical or more complex

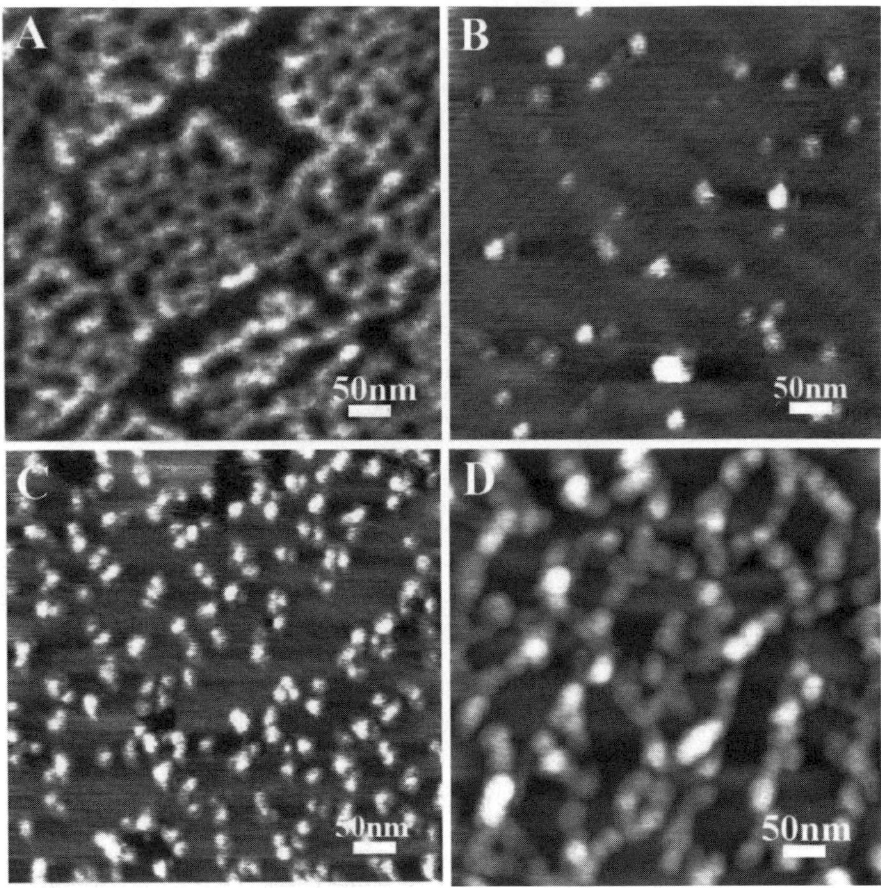

Fig. 3 MAC Mode AFM topographical images in air of ODN molecules, immobilized onto HOPG by free adsorption during 3 min, from 0.3 μM (**A**) A10, (**B**) C10, (**C**) G10, and (**D**) T10 in pH 4.5 0.1 M acetate buffer. [Adapted with permission from Biophys. Chem., Vol. 121, A.-M. Chiorcea Paquim, T.S. Oretskaya, A.M. Olivera Brett, "Adsorption of synthetic homo- and hetero-oligodeoxynucleotides onto highly oriented pyrolytic graphite. Atomic force microscopy characterization, pp. 131–141, Copyright Elsevier (2006).]

structures reduced the interaction of the ODN molecules with the surface, because the molecules had the bases protected in the interior of the secondary structure. The importance of the type of base existent at the ODN chain extremities on the adsorption process was investigated [33] and different adsorption patterns were obtained (Fig. 4) with ODN sequences composed by the same group of bases aligned in a different order. The adsorption of ODNs corresponding to pairs of complementary sequences: homo-ODNs sequences A10 with T10, G10 with C10 (Fig. 3), and hetero-ODNs sequences A6G4 with T6C4, A8G2 with T8C2, and MIX 1 with MIX 2 (Fig. 4) showed in the AFM images that each of these complementary ODN struc-

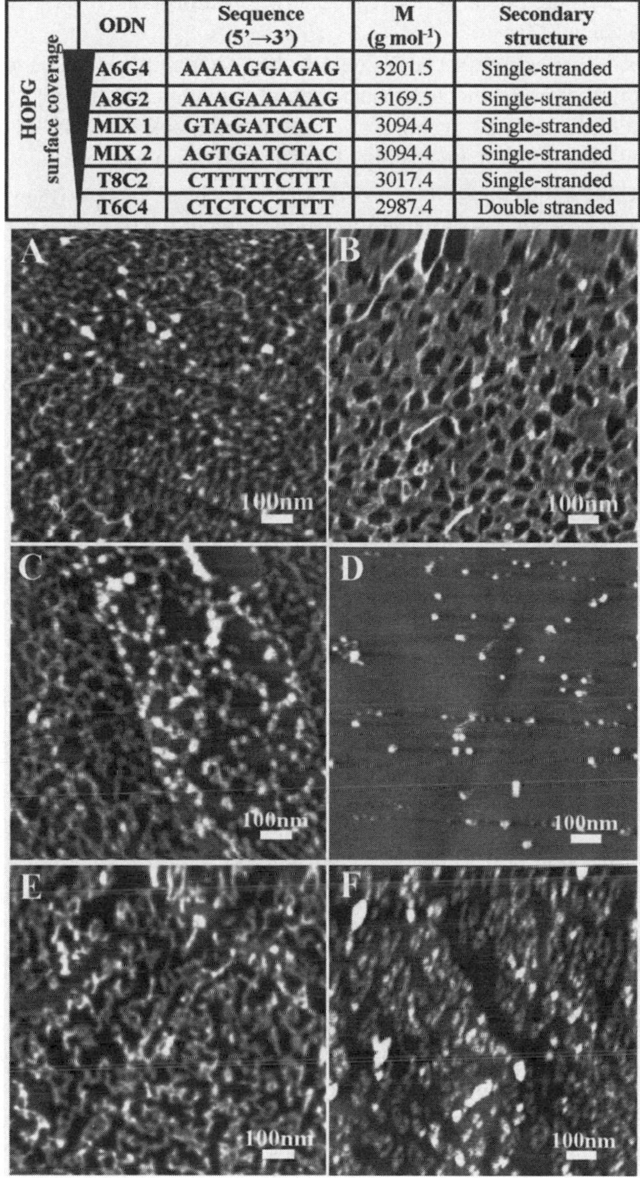

HOPG surface coverage	ODN	Sequence (5'→3')	M (g mol^{-1})	Secondary structure
	A6G4	AAAAGGAGAG	3201.5	Single-stranded
	A8G2	AAAGAAAAAG	3169.5	Single-stranded
	MIX 1	GTAGATCACT	3094.4	Single-stranded
	MIX 2	AGTGATCTAC	3094.4	Single-stranded
	T8C2	CTTTTTCTTT	3017.4	Single-stranded
	T6C4	CTCTCCTTTT	2987.4	Double stranded

Fig. 4 MAC Mode AFM topographical images in air of ODN molecules, immobilized onto HOPG by free adsorption during 3 min, from 0.3 μM (**A**) A6G4, (**B**) A8G2, (**C**) T8C2, (**D**) T6C4, (E) MIX 1, and (**F**) MIX 2 in pH 4.5 0.1 M acetate buffer. [Adapted with permission from Biophys. Chem., Vol. 121, A.-M. Chiorcea Paquim, T.S. Oretskaya, A.M. Olivera Brett, "Adsorption of synthetic homo- and hetero-oligodeoxynucleotides onto highly oriented pyrolytic graphite. Atomic force microscopy characterization, pp. 131–141, Copyright Elsevier (2006).]

tures always presented a very different adsorption morphology and coverage onto the HOPG surface [33].

This aspect must be carefully considered when choosing one sequence as a probe to be immobilized on the electrode surface and the other as a target for hybridization. The electrode surface coverage, orientation, and packing of the adsorbed ODN probe layer are very important. Ideally, the probe ODNs must be connected with the electrode surface at one point only, oriented with the sequences perpendicular to the electrode surface in such a way that the target sequence will have a total access to the immobilized probe, the molecules of which are sufficiently separated from one another to be able to enable hybridization. Besides this, they must present a uniform coverage of the surface in order to minimize nonspecific adsorption of target molecules on the uncovered areas of the electrode. Additionally, the target ODNs should be chosen in order to have a reduced adsorption on the carbon electrode surface, in order to minimize nonspecific adsorption. Consequently, the sequences' dynamic secondary structure always has to be considered.

The AFM results clearly show the importance of characterizing and understanding the adsorption process of small ODNs on carbon electrode surface and evaluation of all the factors influencing the correct manufacture of DNA electrochemical biosensors for hybridization detection.

To increase the understanding of how a DNA biosensor should be designed in the context of the choice of sensing layer, the interaction of the four bases with a thin polycrystalline gold film was investigated. The structure and desorption dynamics of mono- and multilayer samples of adenine, cytosine, guanine, and thymine on polycrystalline gold thin films was studied using temperature-programmed desorption-infrared reflection absorption spectroscopy (TPD-IRAS) and temperature-programmed desorption-mass spectroscopy (TPD-MS) [34].

It was shown that the purines, adenine and guanine, adsorb onto gold in a complex manner and that both adhesive (adenine) and cohesive (guanine) interactions contribute to the apparent binding energies to the substrate surface. Adenine displays at least two adsorption sites, including a high-energy site (210°C, ~136 kJ/mol), wherein the molecule coordinates to the gold substrate via the NH_2 group in a sp^3-like, strongly perturbed, nonplanar configuration. The situation is very different for guanine. Two different phases coexist simultaneously on gold for a situation corresponding to approximately one monolayer of guanine: a dispersed phase of oriented guanine molecules that is structurally very different from solid guanine, and a phase consisting of aggregated, solid-like guanine. The dispersed phase of oriented molecules desorbs first, at about 180°C (127 kJ/mol), that is, at approximately the same temperature as the adenine molecules, whereas aggregated guanine desorbs at a higher temperature, ~220°C (139 kJ/mol). This observation shows that cohesive forces are more important than adhesive forces for obtaining a stable overlayer of guanine on gold.

The pyrimidines, cytosine and thymine, display a less complicated adsorption/desorption behavior [34]. The desorption energy of cytosine (160°C, ~122 kJ/mol) is similar to those obtained for adenine and guanine, but desorption occurs from a single site of dispersed, nonaggregated cytosine. Thymine also desorbs from a single site but at a significantly lower energy (100°C, ~104 kJ/mol).

Infrared data reveal that the monolayer architectures are structurally very different from those observed for the bases in the bulk crystalline state. It is also seen that both pyrimidines and purines adsorb on gold with the plane of the molecule in a nonparallel orientation with respect to the substrate surface. These results should aid in improving the understanding of strategies for capturing oligonucleotides or DNA strands for bioanalytical applications, and in particular, for gold nanoparticle-based assays.

From the perspective of biosensor applications, this detailed information about how DNA bases interact with gold surfaces will enable researchers to improve the stability and activity of oligonucleotide capture layers. The need for fast, cheap, and precise detection of DNA in samples of hair, blood, saliva, and other body fluids is increasing, both in medical and pharmaceutical applications as well as in forensic science.

The DNA molecule has proven to be a powerful technological building block. Several ways of combining DNA molecules that allow them to carry out computations in test tubes and create 2D patterns and 3D structures at the nanoscale are being investigated.

The programmed self-assembly of rationally designed molecular structures into patterned superstructures is a major challenge in nanotechnology. Bottom-up techniques for the molecular assembly of electronic circuits is using self-assembling DNA nanostructures as scaffolds for constructing and positioning molecular-scale electronic devices and wires [35]. Such DNA nanostructures provide programmable methodology for nanoscale construction of patterned structures, utilizing macromolecular building blocks ("DNA tiles") based on branched DNA, that self-assemble into periodic and aperiodic lattices [36]. DNA nanostructures do not appear in themselves to have good properties as electrical conduits or devices [37]. The programmability is due to the highly specific hybridization of complementary DNA strands used to specify interactions both within and between DNA tiles. The process of self-assembly by hybridization entails simply heating a solution of properly designed oligonucleotides above their dissociation temperature and then cooling them down very slowly to allow annealing with low error rates.

In recent years, predictable self-assembly of DNA "smart tiles" to construct periodically patterned lattices has been demonstrated [38, 39]. The artificial DNA used in DNA computing and nanotechnology has a fundamental characteristic: sticky ends. Sticky ends are short sections of single-stranded DNA that extend beyond an end of a double-stranded DNA molecule. Matching sticky ends serve to join two pieces of double-stranded DNA. Structure formation using DNA smart tiles begins with the chemical synthesis of single-stranded DNA oligonucleotides, which, when properly annealed, self-assemble into DNA branched-junction building blocks through Watson–Crick base pairing. DNA tiles can carry sticky ends that preferentially match the sticky ends of other particular DNA tiles, thereby facilitating further assembly into tiling lattices, Fig. 5.

One major application of DNA nanotechnology is to use self-assembled DNA lattices as scaffolds for the assembly of other molecular components. The 4×4 lattice in Fig. 5, which displays a square aspect ratio, can be useful for forming

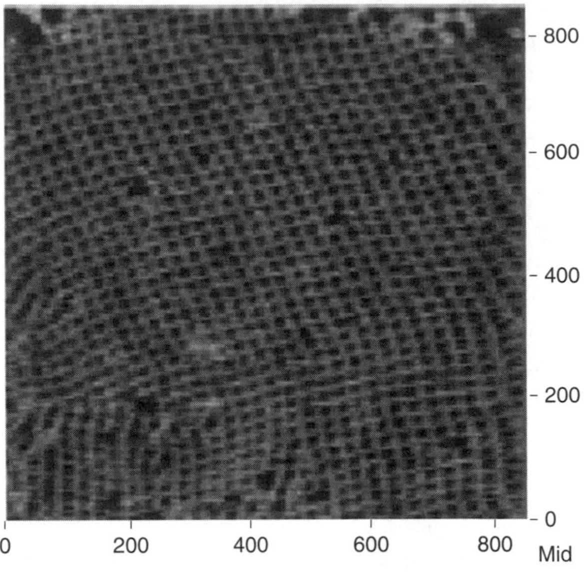

Fig. 5 An AFM image of the self-assembled lattice based on novel 4 × 4 DNA tiles. [Reprinted with permission by IOP from Nanotechnology, Vol. 15 (2004), S525–S527, S.H. Park, H. Yan, J.H. Reif, T.H. Labean, G. Finkelstein, "Electronic nanostructures templated on self-assembled DNA scaffolds"]

regular arrays of other nanoscale "building blocks". It was modified by incorporating a relatively small biotin molecular group into one of the T_4 loops at the tile center. By adding streptavidin to the solution of the self-assembled 4 × 4 DNA nanogrids, the interaction of streptavidin with biotin led to the formation of periodic streptavidin arrays.

The precise control of periodic spacing between individual protein molecules by programming the self-assembly of DNA tile templates was demonstrated (Fig. 5). In particular, the application of two self-assembled periodic DNA structures, 2D nanogrids, and one-dimensional (1D) nanotrack as template for programmable self-assembly of streptavidin protein arrays was reported [40].

These programmable protein assemblies utilize a two-tile system (A tile and B tile) as selectable templates for protein binding, where A tile and B tile associate with each other alternatively through rationally designed sticky ends and self-assemble into either 2D nanogrids or 1D nanotracks. A tile and B tile can be selectively modified such that either one or both tile types carry biotin groups (Fig. 6). To make a template of the assembly of streptavidin molecules, the loops at the center of the A and B tiles were selectively modified to incorporate a biotin group. Biotinylated A and B tiles are denoted as A* and B*. Streptavidin has a diameter of ~4 nm. The AFM images of the bare 2D A*B DNA nanogrids (Fig. 6a) and A*B* DNA nanogrids (Fig. 6b) show the two distinct forms of protein nanoarrays resulting from streptavidin binding to lattices A*B (Fig. 6c) and A*B* (Fig. 6d) and clearly demonstrate the regular periodicity of the streptavidin molecules templated on the 2D DNA nanogrids.

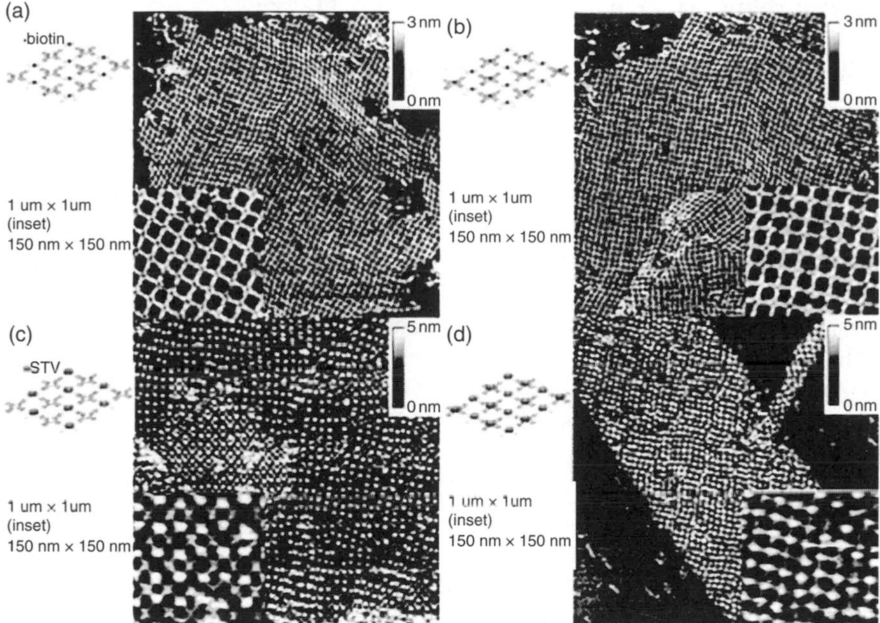

Fig. 6 Atomic-force microscopy images of the programmed self-assembly of streptavidin on 2D DNA nanogrids. (**a** and **b**) AFM images of bare A*B and A*B* nanogrids before streptavidin attachment. (**c** and **d**) AFM images obtained after binding of streptavidin to the bare DNA nanogrids A*B and A*B*, respectively. Scan sizes of all AFM images are 1 μm × 1 μm with 150 nm × 150 nm zoom-in insets. [Reprinted with permission by American Chemical Society ACS from Nano Lett., Vol. 5 (2005), pp. 729–733, S.H. Park, Y. Yin, Y. Liu, J.H. Reif, T.H. Labean, H. Yan, "Programmable DNA self-assemblies for nanoscale organisation of ligands and proteins"]

The streptavidin binding to the self-assembled A*B and A*B* arrays generates topographical features on the mica surface that are higher than bare DNA lattices and are visualized as brighter bumps at the appropriate tile centers. The binding of streptavidin to the biotin sites in A*B lattice or A*B* lattice results in two distinct forms of streptavidin nanoscale arrays with different periodic spacing between adjacent protein molecules. Consequently, the combination of selectively modified A and B tiles in the self-assembly and subsequent binding of streptavidin to biotin leads to varied periodic spacing of the protein molecules on the DNA lattices (Fig. 6) forming two distinct protein arrays.

The successes in constructing self-assembled 2D DNA-tiling lattices composed of tens of thousands of tiles may lead to potential applications, including nanoelectronics, biosensors, and programmable/autonomous molecular machines. The diversity of materials with known DNA attachment chemistries considerably enhances the attractiveness of DNA tiling assembly, which can be used to form superstructures upon which other materials may be assembled.

5 Nanoscale Electrochemical Biosensor Devices

The use of biological components: enzymes, antibodies, DNA, etc., to detect specific compounds has led to the development of biosensors. Electrochemical biosensors are devices that couple a biological recognition element to an electrode transducer. Electrochemical biosensors have great advantages over spectrophotometric biosensors in the relatively little time consumed, ease of operation, sensitivity, high selectivity, and, in principle, low cost. Nanotechnology is enabling the development of highly sensitive and innovative electrochemical biosensing devices [41]. Most enzymatic electrochemical nanoscale biosensors have been developed using gold nanoparticles. Colloidal gold is a metallic colloid, whose particles are formed by little octahedral units, called primary particles. The size of gold particles essentially depends on the way colloidal gold is formed. Gold nanoparticles carry a negative net charge. When a protein or another biological molecule is added to colloidal gold, this negative charge is neutralized by physical adsorption of these molecules on the surface of the gold nanoparticles, but most macromolecules adsorb on to colloidal gold with retention of their bioactivity.

Metal nanoparticles provide three important functions for electroanalysis. These are the roughening of the conductive sensing interface, the catalytic properties of the nanoparticles permitting the amplified electrochemical detection of the metal deposits, and the conductivity properties of nanoparticles at nanoscale dimensions that allow the electrical contact of redox centers in proteins with electrode surfaces [42, 43]. Dissolution of the nanoparticle labels and the electrochemical collection of the dissolved ions on the electrode followed by the stripping-off of the deposited metals represents a general electroanalytical procedure. These unique functions of nanoparticles were employed for developing electrochemical gas sensors, electrochemical sensors based on molecular- or polymer-functionalized nanoparticle sensing interfaces and for the construction of different biosensors, including enzyme-based electrodes, immunosensors, and DNA sensors.

5.1 Template-Synthesized Biomolecule Nanotubes

Since 1991, when multiwalled nanotubes (MWNTs) were accidentally synthesized using the arc process [44], much progress has been made in learning more about them and how they are formed. Carbon nanotubes (CNTs) are nanostructures with dimensions of $\sim$1 nm diameter and can be metallic or semiconductor depending on the diameter and orientation of the hexagonal lattice (Fig. 7). The electronic properties of single-walled nanotubes (SWNTs) were found to depend on their atomic structures, which are described by their chiral angle and diameter.

Nanotubes are a special type of nanoporous material that may produce unique behavior in fluids confined to the nanotube interiors relative to fluids confined to the interiors of zeolites and other molecular sieves. Similar to other carbon materials used in electrochemistry, such as glassy carbon, graphite, and diamond, CNTs are

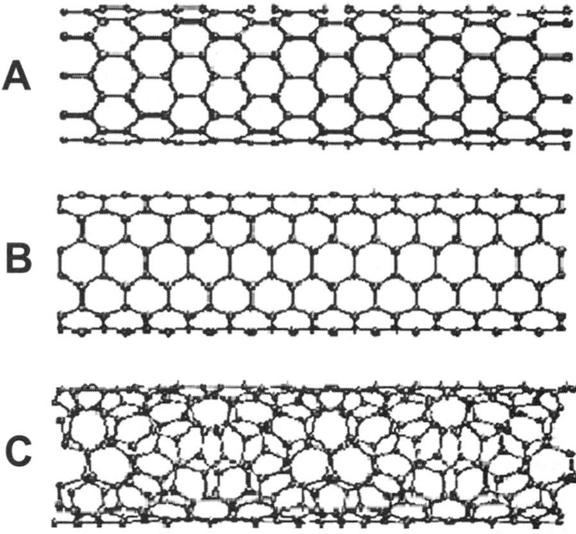

Fig. 7 Side view of nanotubes. (**A**) Zigzag, (**B**) Armchair, and (**C**) Chiral

electrochemically inert and have a good potential window [45]. The side walls and tube ends have been chemically modified by biomolecular components [46].

The combination of the catalytic properties of biomaterials with single-walled carbon nanotubes (SWCNTs), 1D molecular wires (~1–2 nm), which are capable of accommodating biomolecules, enables the development of nanobiosensor devices [46, 47]. The unique chemical and physical properties of CNT have paved the way for new and improved sensing devices, in general, and electrochemical biosensors, in particular. CNT-based electrochemical transducers offer substantial improvements in the performance of amperometric enzyme electrodes, immunosensors, and nucleic-acid-sensing devices. The greatly enhanced electrochemical reactivity of hydrogen peroxide and NADH at CNT-modified electrodes makes these nanomaterials extremely attractive for numerous oxidase- and dehydrogenase-based amperometric biosensors. Aligned CNT "forests" can act as molecular wires to allow efficient electron transfer between the underlying electrode and the redox centers of enzymes. Bioaffinity devices utilizing enzyme tags can greatly benefit from the enhanced response of the biocatalytic-reaction product at the CNT transducer and from CNT amplification platforms carrying multiple tags. The successful realization of CNT-based biosensors requires proper control of their chemical and physical properties, as well as their functionalization and surface immobilization. However, research is proceeding toward better chemical functionalization of the CNTs for achieving biomolecule–CNT hybrid system applications.

Vertically aligned arrays of SWCNTs forests on pyrolytic graphite surfaces were developed for amperometric enzyme-linked immunoassays [48]. Improved fabrication of these SWNT forests utilizing aged nanotube dispersions provided higher nanotube density and conductivity. Biosensor performance enhancement was monitored using nanotube-bound peroxidase enzymes showing a 3.5-fold better sensitivity for H_2O_2 than when using fresh nanotubes to assemble the forests and improved detection limits. Absence of improvements by electron mediation for detection of H_2O_2 suggested very efficient electron exchange between nanotubes and enzymes attached to their ends. Protein immunosensors were made by attaching antibodies to the carboxylated ends of nanotube forests. Utilizing casein/detergent blocking to minimize nonspecific binding, a detection limit of 75 pmol mL^{-1} (75 nM) was achieved for human serum albumin (HSA) in unmediated sandwich immunosensors using horseradish peroxidase (HRP) labels. Mediation of the immunosensors dramatically lowered the detection limit to 1 pmol mL^{-1} (1 nM), providing significantly better performance than alternative methods. In the immunosensor case, the average distance between HRP labels and nanotube ends is presumably too large for efficient direct electron exchange, but this situation can be overcome by electron mediation.

A layer-by-layer deposition strategy for preparing protein nanotubes within the pores of a nanopore alumina template membrane was described [49]. This method entails alternately exposing the template membrane to a solution of the desired protein and then to a solution of glutaraldehyde, which acts as cross-linking agent to hold the protein layers together. The number of layers of protein that make up the nanotube walls can be controlled at will by varying the number of alternate protein/glutaraldehyde cycles. After the desired number of layers has been deposited on the pore walls, the alumina template can be dissolved to liberate the protein nanotubes. Glucose oxidase nanotubes (Fig. 8) prepared in this way catalyze glucose oxidation, and hemoglobin nanotubes retain their heme electroactivity. Furthermore, for the glucose oxidase nanotubes, the enzymatic activity increases with the nanotube wall thickness.

DNA has been attached to nano-objects, nanoparticles, and nanotubes, composed of different materials. Nanotubes composed predominantly of DNA, using an alumina-template membrane-synthesis method for preparing the DNA nanotubes, were described [50]. The DNA–nanotubes were liberated by dissolution from the template in conditions such that dehybridization does not occur during membrane dissolution.

A novel electrochemical biosensing platform based on biocompatible, well-ordered, self-assembled diphenylalanine peptide nanotubes was described [51]. The novel peptide nanotubes show remarkable similarity to carbon nanotubes in their morphology and aspect ratio (Fig. 9).

Their assembly as individual entities rather than bundles, however, makes them appealing for use in various nanotechnological applications, and peptide nanotubes can serve as a model for the fabrication of conductive nanowires and other peptide–inorganic composites.

These tubular structures were discovered during the search for the minimal amyloidogenic self-assembled fragment of the β-amyloid polypeptide that relate to Alzheimer's disease. It was found that the diphenylalanine-core recognition of these

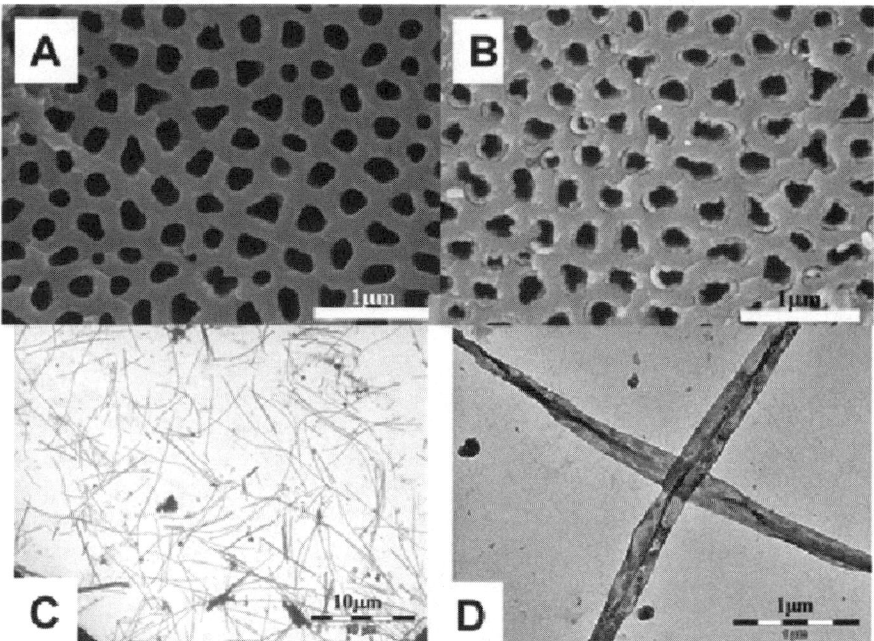

Fig. 8 SEM images of the surface of the alumina template before (**A**) and after (**B**) deposition of six-layer GOD nanotubes. (**C, D**) TEM images of the liberated nanotubes taken at two different levels of magnification. [Reprinted with permission by American Chemical Society ACS from Nano Lett., Vol. 5 (2005), pp. 231–234, S. Hou, J. Wang, C.R. Martin, "Template-synthesized protein nanotubes"]

polypeptides can self-assemble into discrete and well-ordered tubular structures. It was suggested that geometrically restricted aromatic interactions contribute order and directionality and mediate the formation of these well-ordered nanostructures. The diphenylalanine-based peptide nanoassemblies have many attractive properties for various nanotechnological applications, since they are readily self-assembled in soluble nanostructures, are biocompatible, can be easily be modified with biological and chemical elements [52], and show a notable similarity to carbon nanotubes in their morphology and aspect ratio.

The application of discrete and well-ordered self-assembled peptide nanotubes, formed by the diphenylalanine peptide, for electrochemical monitoring was reported [53]. A highly sensitive amperometric enzyme biosensor based on immobilized self-assembled peptide nanotubes attached to a gold electrode surface was built using glucose oxidase (GOx) and ethanol dehydrogenase (ADH) as model enzyme systems to demonstrate the advantages that this novel class of peptide nanotubes provides as an attractive component for future electroanalytical devices.

Fig. 9 Scanning electron microscope images of (**A**) control electrode, (**B**) peptide nanotube-modified electrode, and (**C**) peptide nanotubes after treatment with proteinase K electrode. Scale bar: 100 µm. [Reprinted with permission by American Chemical Society ACS from Nano Lett., Vol. 5 (2005), pp. 183–186, Y. Yemini, M. Reches, J. Rishpon, E. Gazit, "Novel electrochemical biosensing platform using self-assembled peptide nanotubes"]

5.2 *Nanoparticle Magnetic Control of Bioelectrocatalytic Processes*

Electrochemistry of relay-functionalized magnetic nanoparticles reversible "on" and "off" by means of an external magnet has been described [54, 55]. Magnetic particles act as functional components for the separation of biorecognition complexes and for the amplified electrochemical sensing of DNA or antigen/antibody complexes. In addition, electrocatalytic and bioelectrocatalytic processes at electrode surfaces are switched by means of functionalized magnetic particles and in the presence of an external magnet. The system consists of a gold electrode in a solution containing functionalized magnetic particles. The attraction of magnetic nanoparticles to the electrode surface by means of the external magnet activates the electrical contact between redox units, and the redox features of the redox components are switched "on". The electrochemical activity is switched "off" simply by placing the magnet at the top of the electrochemical cell, since the magnetic nanoparticles will be retracted from the surface.

Switching "on" and "off" the electrochemical reaction of the redox-relay groups covalently bound to the magnetic nanoparticles has been applied to the bioelectrocatalytic oxidation of glucose in the presence of glucose oxidase and ferrocene-functionalized magnetic nanoparticles and oxidation of lactate in the presence of lactate dehydrogenase and pyrroloquinoline-*N*-(2-aminoethyl)-β-nicotinamide adenine dinucleotide (PQQ-NAD^+)-functionalized magnetic nanoparticles [54, 55]. The ability to control bioelectrocatalytic transformations by means of an external magnet was also utilized to develop selective dual-biosensing systems, Fig. 10.

The reversible magnetically controlled oxidation of DNA was accomplished in the presence of nucleic-acid-modified magnetic particles. Avidin-modified magnetic particles were functionalized with a biotinylated DNA probe. The hybridized target DNA was separated from the analyzed sample by means of an external magnet. Release of the hybrid DNA under basic conditions, followed by the chronopotentiometric stripping of the released guanine residue enabled the quantitative analysis of the target DNA [56, 57]. The magneto-controlled amplified detection of DNA was demonstrated by using nucleic-acid-modified Au nanoparticles for the electrochemical detection of DNA and semiconductor nanoparticles. The nucleic-acid-functionalized Au nanoparticles associated with the sensing interface undergo electrochemical stripping or chemical dissolution, which enables the electrochemical analysis of the dissolved product. The Au nanoparticle associated with the sensing interface can act as a catalytic site for the deposition of other metals, thus leading to the amplified detection of DNA by the intermediary accumulation of metals that are stripped off, or by generating an enhanced amount of dissolved product that can be electrochemically analyzed [58].

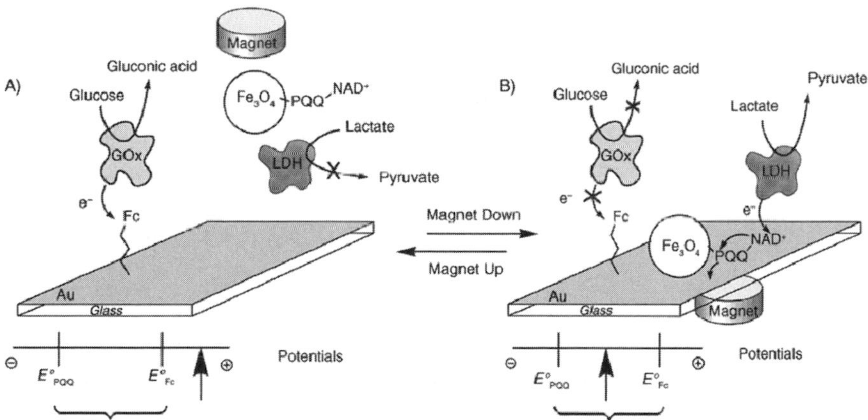

Fig. 10 Magneto-controlled dual biosensing of glucose (**A**) and lactate (**B**) in the presence of GOx, LDH, magnetic particles functionalized with PQQ-NAD^+, and an Au-electrode modified with a monolayer of ferrocene units. [Reprinted with permission by Wiley, from Angew. Chem. Int. Ed., Vol. 42, pp, 4576–4588 (2003), I. Willner, E. Katz, "Magnetic control of electrocatalytic and bioelectrocatalytic processes".]

The use of functional magnetic particles and specifically the external rotation of the magnetic particles provide a novel concept in the development of bioelectrocatalytic systems [54, 55]. The attraction of functionalized magnetic particles to surfaces can be used to concentrate and separate the ingredients involved in the analytical processes. The localization of redox-functionalized magnetic particles on electrodes enables the electronic transduction of the biosensing events. The rotation of the magnetic particles on the electrode surface may enhance biorecognition processes on the particles or amplify the electronic signal that transduces the biosensing events. These features of the magnetic particles may be used separately or sequentially to improve the performance of bioelectrochemical sensor devices. The rotation of such magnetic particles on electrode supports could lead to enhanced electrochemically induced heterogeneous catalysis.

5.3 Detection Limits for Nanoscale Biosensors

The signal-to-noise ratio of developed nanoscale biosensors is often much higher than that of the analogous macroscale sensors, which, in turn, should drive down the lower limit of detection. However, a modeling study [59] suggests that detector size is not the only important factor that needs to be considered when designing a biosensor. Analytical calculations and finite element analysis were used to simulate the behavior of various biosensors as they shrink from the micrometer scale to the nanometer scale. The simulations showed that the rate at which molecules flow past a sensor – the number of molecules that interact with the sensor over a given period of time – is the critical limit to detector sensitivity. This result would suggest that microfluidics device designers can improve the sensitivity of their biosensors by increasing fluid flow past the sensor, but the possible gains decrease as sensor size decreases. Nevertheless, this problem is less severe for disk-shaped sensors than for wire-shaped sensors. The study concluded that the minimum detectable limit for biosensors that rely solely on conventional microfluidics to carry molecules past a detector is unlikely to be lower than femtomolar. Improving on this limit will require new methods for increasing the flux of analytes past a nanoscale detector.

6 Conclusions

Nanobioelectrochemistry opens up exciting new prospects for designing new forms of matter by assembling molecules one by one in bio-inspired structures, materials, and devices. The great increase in surface area and a possible advantage in the intrinsic properties is an immediate consequence of miniaturization of materials, which may be changed for the better when finely divided.

The nano approach makes it possible to think of life as a very complex nanotechnological system constructed as a result of a series of self-assembling processes. Medical benefits can be expected. The importance of early diagnosis, quick analysis through on-chip laboratories, development of targeted and controlled medicine release, and tissue regeneration will lead to great progress. Nanomedicine means carrying out complex repairs at the cellular level inside the human body, in which, due to their size, nanostructures will be able to interact with biomolecules on the surface and penetrate inside the cell.

Feinman, the 1965 Nobel Prize winner, said in 1960, in his talk "*There is plenty of space at the bottom*" that "I am telling you what could be done if the laws *are* what we think; we are not doing it simply because we haven't yet gotten around to it" [60]. Some decades later we are finally doing it!

Acknowledgements Financial support from Fundação para a Ciência e Tecnologia (FCT), POCI (co-financed by the European Community Fund FEDER), and ICEMS (Research Unit 103) are gratefully acknowledged.

References

1. C.M.A. Brett, A.M.C.F. Oliveira Brett, Electrochemistry. Principles, methods and applications, Chapter 17, Bioelectrochemistry, Oxford University Press, 1993, 430.
2. N. Taniguchi, On the basic concept of nano-technology, Proc. Intl. Conf. Prod. Eng. Tokyo, Part II.
3. G. Binnig, H. Rohrer, Scanning Tunneling Microscopy. Helv. Phys. Acta, 1982, 55, 726–735.
4. S. Myhra, A review of enabling technologies based on scanning probe microscopy relevant to bioanalysis. Biosen. Bioelect., 2004, 190, 1345–1354.
5. A.A.G. Requicha, Nanorobots, NEMS, and nanoassembly. Proc. IEEE, 2003, 91(11), 1922–1933.
6. J. Riu, A. Maroto, F.X. Rius, Nanosensors in environmental analysis. Talanta, 2006, 69, 288–301.
7. C.M.A. Brett, The Role of Nanoelectrochemistry in Nanotechnology. Nanotechnol. Percept., 2006, 2, 205.
8. P. Kohli, M. Wirtz, C.R. Martin, Nanotube Membrane Based Biosensors. Electroanalysis, 2004, 16, 9–18.
9. C. Amatore, S. Arbault, I. Bonifas, Y. Bouret, M. Erard, M. Guille, Dynamics of Full Fusion During Vesicular Exocytotic Events: Release of Adrenaline by Chromaffin Cells. ChemPhysChem, 2003, 4, 101–108.
10. C. Amatore, S. Arbault, I. Bonifas, Y. Bouret, M. Erard, A.G. Ewing, L.A. Sombers, Correlation Between Vesicle Quantal Size and Fusion Pore Release in Chromaffin Cell Exocytosis. BioPhys. J., 2005, 88, 4411–4420.
11. A.M. Oliveira Brett, J.A.P. Piedade, A.M. Chiorcea, Voltammetry and AFM imaging of picomoles of adsorbed adriamycin onto glassy carbon and HOPG electrode surfaces. J. Electroanal. Chem., 2002, 538–539, 267–276.
12. F.A. Armstrong, Insights from protein film voltammetry into mechanisms of complex biological electron-transfer reactions. Dalton Trans., 661–671, (2002).
13. F. Baymann, N.L. Barlow, C. Aubert, B. Schoepp-Cothenet, G. Leroy, F.A. Armstrong, Voltammetry of a 'protein on a rope'. FEBS Lett., 2003, 539, 91–94.
14. G.-J. Yang, K. Wang, J.-J. Xu, H.-Y. Chen, Determination of Theophylline in Drugs and Tea on Nanosized Cobalt Phthalocyanine Particles Modified Carbon Paste Electrode. Anal. Lett., 2004, 37, 629–643.
15. S.S. Smith, Nucleoprotein assemblies, in: Encyclopedia of Nanoscience and Technology, H.S. Nalwa (Ed.), 2003, Vol. X, 1–10.
16. N.C. Seeman, Nanotechnology and the double helix. Scientific American, 2004, 290, 35–43.

17. C. Whitehouse, D. Gidalevitz, M. Cahuzac, R.E. Koeppe II, A. Nelson, Interaction of Gramicidin Derivatives with Phospholipid Monolayers. Langmuir, 2004, 20, 9291–9298.
18. B.A. Cornell, G. Krishna, P.D. Osman, R.D. Pace, L. Wieczorek, Tethered-bilayer lipid membranes as a support for membrane-active peptides. Biochem. Soc. Trans., 2001, 29, 613–617.
19. Z. Siwy, L. Troffin, P. Kohli, L.A. Baker, C. Trautmann, C.R. Martin, Protein Biosensors Based on Biofunctionalized Conical Gold Nanotubes. J. Am. Chem. Soc., 2005, 127, 5000–5001.
20. D.B. Allred, M. Sarikaya, F. Baneyx, D.T. Schwartz, Electrochemical Nanofabrication Using Crystalline Protein Masks. Nano Lett., 2005, 5(4), 609–613.
21. M. Handrea, M. Sahre, A. Neubauer, U.B. Sleytr, W. Kautek, Electrochemistry of nano-scale bacterial surface protein layers on gold. Bioelectrochemistry, 2003, 61, 1–8.
22. W. Saenger, Principles of Nucleic Acid Structure, in: Springer Advanced Texts in Chemistry, Ch. R. Cantor (Ed.), Springer-Verlag, New York, 1984.
23. R.F. Service, Biology offers nanotechs a helping hand. Science, 2002, 298, 2322–2323.
24. H. Yan, X. Zhang, Z. Shen, N.C. Seeman, A robust DNA mechanical device controlled by hybridization topology, Nature, 2002, 415, 62–65.
25. I. Willner, Biomaterials for sensors, fuel cells, and circuitry. Science, 2002, 298, 2407–2408.
26. A.M. Oliveira Brett, S.H.P. Serrano, J.A.P. Piedade, Electrochemistry of DNA, Comprehensive Chemical Kinetics, in: Applications of Kinetic Modeling, R.G. Compton, G. Hancock (Eds.), Elsevier, Oxford, UK, 1999, vol. 37, chapter 3, 91–119.
27. A.M. Oliveira Brett, DNA-based biosensors, in: Comprehensive Analytical Chemistry. Biosensors and Modern Specific Analytical Techniques, L. Gorton, (Ed.), 2005, vol. 44, chapter 4, 179–208.
28. I. Willner, Biomaterials for sensors, fuel cells, and circuitry. Science, 2002, 298, 2407–2408.
29. A.M. Oliveira Brett, M. Vivan, I.R. Fernandes, J.A.P. Piedade, Electrochemical detection of in situ adriamycin oxidative damage to DNA. Talanta, 2002, 56, 959–970.
30. E. Palecek, M. Fojta, M. Tomschik, J Wang, Electrochemical biosensors for DNA hybridization and DNA damage. Biosens. Bioelectron., 1998, 13, 621-628.
31. M. Mascini, I. Palchetti, G Marrazza, DNA electrochemical biosensors. Fresenius J. Anal. Chem., 2001, 369, 15–22.
31. D.R. Meldrum, Sequencing genomes and beyond. Science, 2001, 292, 515–517.
32. H.G. Hansma, I. Revenko, K. Kim, D.E. Laney, Atomic force microscopy of long and short double-stranded, single-stranded and triple-stranded nucleic acids, Nucleic Acids Res., 1996, 24713–24720.
33. A.-M. Chiorcea Paquim, T.S. Oretskaya, A.M. Oliveira Brett, Adsorption of synthetic homo- and hetero-oligodeoxynucleotides onto highly oriented pyrolytic graphite. Atomic force microscopy characterization. Biophys. Chem., 2006, 121, 131–141.
34. M. Östblom, B. Liedberg, L.M. Demers, C.A. Mirkin, On the Structure and Desorption Dynamics of DNA Bases Adsorbed on Gold: A Temperature-Programmed Study. J. Phys. Chem. B, 2005, 109, 15150–15160.
35. A. Turberfield, DNA as an engineering material. Phys. World, 2003, 16, 43–36.
36. N.C. Seeman, DNA in a material world. Nature, 2003, 421, 427–431.
37. C. Dekker, M.A. Ratner, Electronic properties of DNA. Phys. World, 2001, 14, 29–33.
38. S.H. Park, H. Yan, J.H. Reif, T.H. LaBean, G. Finkelstein, Electronic nanostrutures templated on self-assembled DNA scaffolds. Nanotechnology, 2004, 15, S525–S527.
39. D. Liu, S.H. Park, J.H. Reif, T.H. LaBean, DNA nanotubes self-assembled from triple-crossover tiles as templates for conductive nanowires. Proc. Nac. Acad. Scien., 2004, 101, 717–722.
40. S.H. Park, P. Yin, Y. Liu, J.H. Reif, T.H. LaBean, H. Yan, Programmable DNA Self-Assemblies for Nanoscale Organization of Ligands and Proteins. Nano Lett., 2005, 5, 729–733.
41. J. Wang, Nanomaterial-based amplified transduction of biomolecular interactions. Small, 2005, 11, 1036–1043.

42. E. Katz, I. Willner, J. Wang, Electroanalytical and Bioelectroanalytical Systems Based on Metal and Semiconductor Nanoparticles. Electroanalysis, 2004, 16, 3–160.
43. E. Katz, I. Willner, Integrated nanoparticle-biomolecule hybrid systems: synthesis, properties, and applications. Angew. Chem. Int. Ed., 2004, 43, 6042–6108.
44. S. Iijima, Helical microtubules of graphitic carbon. Nature, 1991, 354, 56–58.
45. K. Gong, Y. Yan, M. Zhang, S. Xiong, L. Mao, Electrochemistry and electroanalytical applications of carbon nanotubes: a review. Anal. Sci., 2005, 21, 1383–1393.
46. E. Katz, I. Willner, Biomolecule-functionalized carbon nanotubes: applications in nanobioelectronics. Chem. Phys. Chem., 2004, 5, 1084–1104.
47. J. Wang, Carbon-Nanotube Based Electrochemical Biosensors: A Review. Electroanalysis, 2005, 17, 7–14.
48. X. Yu, S.N. Kim, F. Papadimitrakopoulo, J.F. Rusling, Protein immunosensor using single-wall carbon nanotube forests with electrochemical detection of enzyme labels. Mol. Biosyst., 2005, 1, 70–78.
49. S. Hou, J. Wang, C.R. Martin, Template-Synthesized Protein Nanotubes. Nano Lett., 2005, 5, 231–234.
50. S. Hou, J. Wang, C.R. Martin, Template-Synthesized DNA Nanotubes. J. Am. Chem. Soc., 2005, 127, 8586–8587.
51. M. Yemini, M. Reches, J. Rishpon, E. Gazit, Novel Electrochemical Biosensing Platform Using Self-Assembled Peptide Nanotubes. Nano Lett., 2005, 5, 183–186.
52. M. Reches, E. Gazit, Formation of closed-cage nanostructures by self-assembly of aromatic dipeptides. Nano Lett., 2004, 4, 581–585.
53. M. Yemini, M. Reches, E. Gazit, J. Rishpon, Peptide Nanotube-Modified Electrodes for Enzyme-Biosensor Applications. Anal. Chem., 2005, 77, 5155–5159.
54. I. Willner, E. Katz, Magnetic control of electrocatalytic and bioelectrocatalytic processes. Angew. Chem. Int. Ed. 2003, 42, 4576–4588.
55. E. Katz, R. Baron, I. Willner, Magnetoswitchable electrochemistry gated by alkyl-chain-functionalized magnetic nanoparticles: control of diffusion and surface confined electrochemical processes. J. Am. Chem. Soc., 2005, 127, 4060–4070.
56. J. Wang, R. Polsky, D. Xu, Silver-Enhanced Colloidal Gold Electrochemical Stripping Detection of DNA Hybridization. Langmuir, 2001, 17, 5739–5741.
57. J. Wang, Nanoparticles-based electrochemical DNA detection. Anal. Chim. Acta, 2003, 500, 247–257.
58. A.-N. Kawde, J. Wang, Amplified electrical transduction of DNA hybridization based on polymeric beads loaded with multiple gold nanoparticles tags. Electroanalysis, 2004, 16, 101–107.
59. P.E. Sheehan, L.J. Whitman, Detection Limits for Nanoscale Biosensors. Nano Lett., 2005, 5, 803–807.
60. R. Feynman, Nanotechnology. Caltech's Eng. Sci., 1960, XXIII, 22–36.

Self-Organized Oxide Nanotube Layers on Titanium and Other Transition Metals

P. Schmuki

1 Introduction

Self-organized nanostructures of different metals or semiconductors have received considerable attention due to the anticipated high technological potential of these materials. For Al it is long known that anodization in various solutions leads to the formation of ordered porous oxide layers [1–4]. Since the remarkable work of Masuda et al. [1], it is clear that very high degree of order can be achieved with these porous geometries. Many applications of these ordered alumina structures have been shown, for example, using it as a photonic crystal [5], or as a template for the deposition of other materials [6–8]. For a range of other metals such as Ti [9–17], Zr [18–20], Nb [21], W [22, 23], Ta [24, 25], Hf [26], it has recently been found that self-organized porous structures can be formed under optimized electrochemical treatments. Figure 1 shows examples of nanotubular structures produced on Ti and other transition metals. These nanoarchitectured oxide films can have very specific functional properties. For example, nanotubes made of titanium oxide (TiO_2) combine geometrical advantages given by their array structure with the material specific properties of TiO_2. Titanium oxide is used in functional applications such as self-cleaning surfaces [27] and solar cell applications [28], or explored for gas-sensing capabilities [29].

Particular advantages of regular tube arrays as shown in Fig. 1 are the large surface area and the defined geometry. The defined geometry results in a narrow distribution of diffusion path not only for entering the tubular depth (e.g., reactants to be transported to the tube bottom) but also for species to be transported through the tube wall, for example, electrons, holes, and ions. Therefore, the system response of ordered tube arrays in applications such as sensing or photocatalyis is expected to be much more defined than using classical high surface area layers – for example,

P. Schmuki (✉)
University of Erlangen-Nuremberg, Department of Materials Science, LKO, Martensstrasse 7, D-91058 Erlangen, Germany
e-mail: schmuki@ww.uni-erlangen.de

P. Schmuki, S. Virtanen (eds.), *Electrochemistry at the Nanoscale,* Nanostructure Science and Technology, DOI 10.1007/978-0-387-73582-5_12,

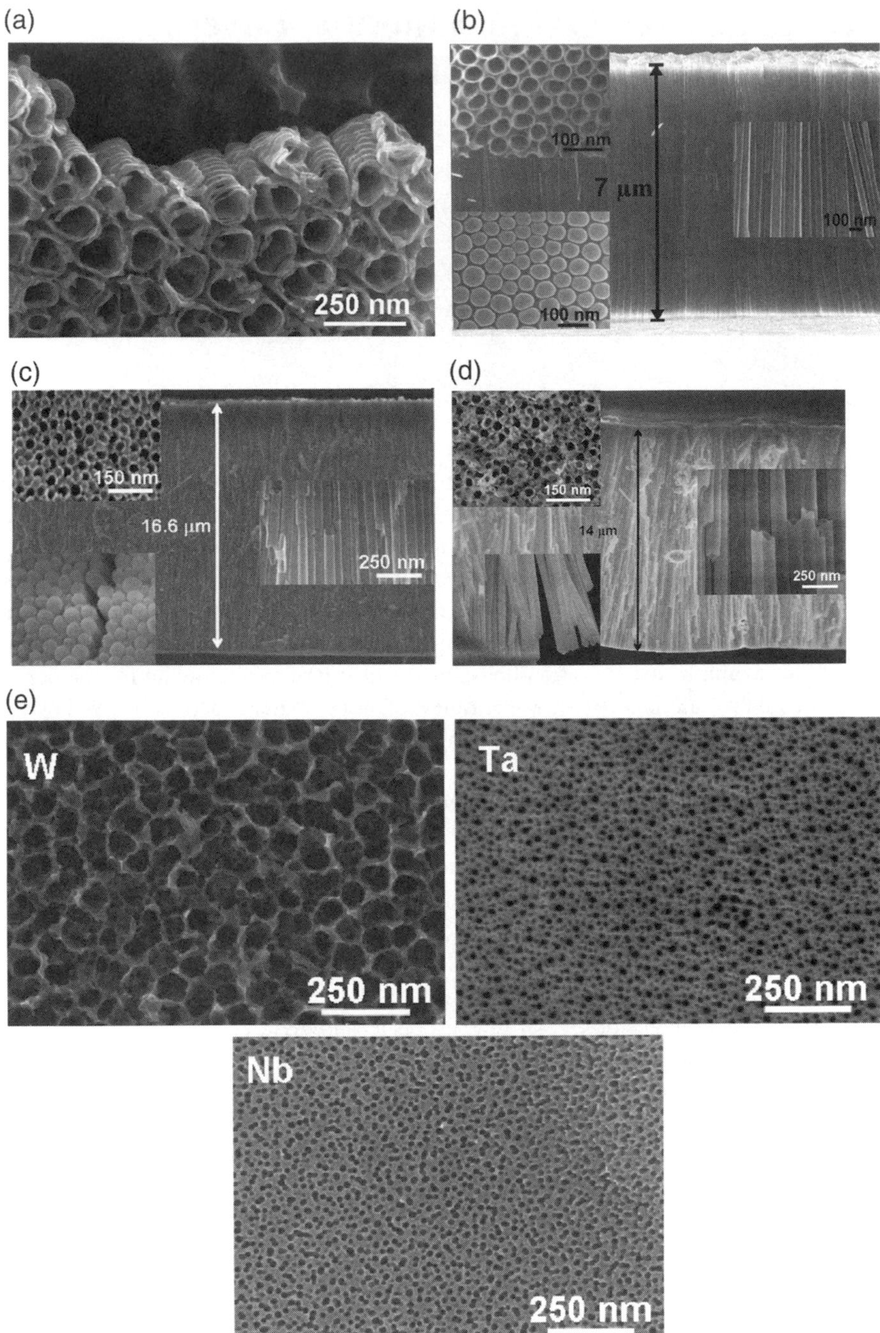

Fig. 1 SEM images showing morphology of various transition metal oxide nanotube layers produced by anodization in fluoride-containing solutions; TiO_2 (**a**, **b**), ZrO_2 (**c**), HfO_2 (**d**), WO_3 + Ta_2O_5 + Nb_2O_5 (**e**); insets show top views (*open tubes*) and bottom views (*closed ends*) and side walls in detail

layers where nanoparticles are compacted or sintered to produce an open porous network.

Ordered nanotubular structures of TiO_2 and other transition metal oxides have been formed by electrochemical anodization of the metal in electrolytes containing small amounts of hydrofluoric acid (HF). In general, the morphology and the structure of porous layers are affected strongly by the electrochemical and the solution parameters. Under optimized conditions self-organized highly ordered nanotubes with a length between some few 10 nm and several 100 μm are formed consisting of arrays with single tube diameter between 10 and 100 nm and a tube wall of some 10 nm thickness. The present chapter discusses factors of the electrochemical pore-formation process and key electrochemical parameters that lead to the formation of these self-organized structures, as well as some properties that can be exploited in various applications.

2 Overview on the Electrochemistry of Valve Metals

For the oxide structures shown in Fig. 1, the base metals (Ti, W, Zr, Hf, Ta, Nb) belong to the class of so-called valve metals. In contrast to other metals, on valve metals it is possible to grow compact oxide layers of considerable thickness (some 100 nm) in aqueous electrolytes. Anodization of valve metals has been widely investigated in a variety of acids at voltages typically up to several hundreds volts. Up to a voltage, where dielectric breakdown of certain spots of an oxide layer occurs, a uniform and compact layer of TiO_2 can be obtained (see e.g., Refs. [30–32] and references therein). For oxide films formed below the breakdown voltage, typically a growth rate of 1–5 nm/V has been reported [33, 34]. A selection of properties of anodic valve metal oxides has been compiled in Table 1. It is apparent that the structural, electronic, and ionic properties cover a wide range. Looking at the electronic properties, that is, from the relatively well-conducting SnO_2 over the semiconductive oxides TiO_2, ZnO, WO_3 to insulating ZrO_2, HfO_2, Al_2O_3. The growth mechanism can be completely dominated by anion inward transport (ZrO_2, HfO_2) to cases with considerable cation outward contribution (Al_2O_3). Also remarkable is that the structure of the grown oxide can be amorphous or crystalline. However, the morphology, structure (and the maximum achievable thickness) of anodic oxide layers formed on valve metals strongly depends on the specific electrochemical parameters such as the applied potential, the time of anodization, or the sweep rate of the potential ramp. For example, the structure of the oxide films on Ti has typically been reported to be amorphous at low voltages (below 20 V [31]), and crystallization to take place at higher voltages. Depending on the anodizing conditions, the crystal structure has been reported to be anatase [33, 34–37], a mixture of anatase and rutile [33, 37, 38], or rutile [33, 37].

At comparably high voltages, if a sufficient electric field is applied to an oxide film, dielectric breakdown (by a Zener or avalanche mechanism) will take place [33, 38]. This is apparent in I–U curves by a rapid current increase, at a distinct breakdown voltage U_{bd} [39–43], or by current or potential fluctuations. Breakdown events can become apparent by sparking or acoustically (crackling noise). For example,

Table 1 Parameters of oxide growth on valve metals

	Oxide				
Metal / Alloy	E_g (eV)	N (cm^{-3})	ϵ_{ox}	type	Ref.
Al	4.5-9	-	7-20	i	69
Al/Cr	-	-	-	n	70
Cr	2.5-3.5	$\approx 10^{20}$	30±20	p(n)?	71-76
Cu	0.6-1.8	?	7-18	p	69, 77
Fe	1.8-2.2	$\approx 10^{20}$	10-35	n	e.g., 78-80
Fe/Cr	1.9-2.1	10^{20}-10^{21}	10-30	n	81
Fe/Cr/Ni (AISI304)	1.9-2.3	10^{20}-10^{21}	10-30	n	82-85
Fe/Cr/Ni/Mo (SMO254, DIN1.4529)	2.3-2.8	$\approx 10^{21}$	10-30	n	84-86
Fe/Ni (xNi<40%)	1.9	$\approx 10^{20}$	10-35	n	81
Ni	2.2-3.7	10^{20}	≈30	p(n)	71,80,87
Sn	3.5-3.7	10^{19}-10^{20}	?	n	77
Ti	3.2-3.8	10^{20}	7-114	n	69
W	2.7-3.1	10^{17}-10^{18}	23-57	n	77
Zn	3.2	$\approx 10^{18}$	8.5	n	77
Zr	4.6-8	-	12-31	i, n	69,77

E_g : band gap energy
N: doping concentration
ϵ_{ox}: dielectric constant
type: conduction type
n: n-type
p: p-type
i: insulator

Wood and Pearson [44] discuss anodic breakdown of oxide films on the valve metals Nb, Ta, Zr, Hf, Al, Ti, W, Mo, and V. It is of interest to note that according to these authors, Ti and W, whose oxide films are semiconductors, did not break down, but turned into conductors leading to oxygen evolution. Generally, breakdown on valve metals results in thickening of films at locations where the breakdown occurred, and thus pinches itself off. Wood and Pearson concluded that avalanche breakdown was being observed at sites where the electrons to initiate the avalanche are supplied by the electrolyte. At high voltages, the breakdown spot may meander over the surface, and for certain applications, extended anodization under these conditions ("spark anodization") is highly useful as it finally leads to a surface covered with an irregular structure of pores in the micrometer-size range – this has found interest to increase the biocompatibility of Ti-based biomedical implants [45, 46].

Under *cathodic* polarization, a range of valve metal oxides show significant cation in-diffusion, accompanying alterations in the electronic structure of the oxide (e.g., incorporation of additional states within the band gap). The most likely mobile species are protons. For example, Dyer and Leach [47] examined oxidized titanium and niobium and found that hydrogen enters the film under cathodic bias in non-negligible amounts. TiO_2, for example, can to a large extent (up to 85%) be converted to TiOOH, as concluded from observations of the change in the index of refraction. Thus cation in-diffusion into passivating films can be substantial under cathodic bias. Hydrogen ingress into Ti oxide layers by cathodic polarization has recently also been studied by *in situ* neutron reflectometry [48].

In general, current passage through oxide layers when the metals are biased cathodically is common on valve metals, for example, zirconium, tantalum, and aluminum, and oxides on silicon can show significant cathodic currents. Schmidt [49, 50] already suggested that these cathodic currents are related to the in-diffusion of protons, since with tantalum and silicon it was found that if an anhydrous electrolyte is used, there is no cathodic current. Vermilyea [51] and others [52] showed that the cathodic currents are laterally inhomogeneous and correspond to flaws in the oxide.

3 Formation of Nanotubular Layers

In the following, the formation of nanotubular (nanoporous) layers focuses on recent findings on their formation in fluoride-containing electrolytes, and will not address the classical case of Al – the interested reader may consider the chapter by Sides and Martin in this book for further references.

3.1 I–U Curves

A convenient way to characterize and compare the electrochemical behavior of valve metals are polarization curves (such as shown in Fig. 2). The electrochemical behavior of most valve metals in typical electrolytes, that is, electrolytes that do not chemically attack the oxide surface, is dominated by spontaneous oxide formation and growth. For example, in H_2SO_4, electrolytes on the valve metals investigated in Fig. 2a, a compact layer forms according to reaction (1) with an increasing thickness the higher the voltage.

$$\mathrm{Me} + 2\mathrm{H_2O} \rightarrow \mathrm{MeO_2} + 4\mathrm{H^+} + 4\mathrm{e^-} \tag{1}$$

However, a clear deviation from the typical valve metal anodization electrochemistry is obtained in fluoride electrolytes [11, 53, 54] as shown in Fig. 2b. This is due to the fact that the oxide formed by reaction (1) is now chemically attacked (dissolved) by the formation of soluble fluoride complexes, for example:

$$\mathrm{MeO_2} + \mathrm{F^-} \rightarrow [\mathrm{MeF_6}]^{2-} \tag{2}$$

If we compare the polarization curves shown in Fig. 2a and b, it becomes clear that the addition of a certain amount of fluorides shows, for the different valve metals, a different effect on the resulting current density. Clearly, on Zr and Hf, the highest steady-state current densities result, thus indicating a rapid dissolution rate of the oxides formed on the metal in HF. For Ti and W, these rates are lower, and for Nb and Ta, these rates are further reduced. This explains that for establishing

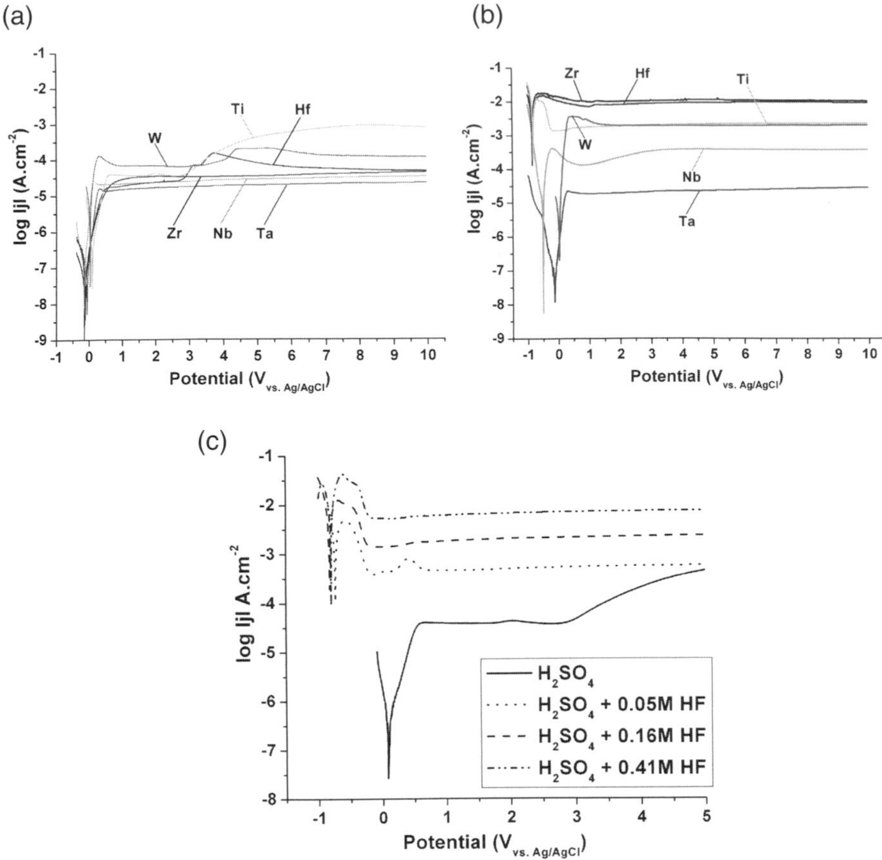

Fig. 2 Polarization curves recorded for valve metals in 1 M H_2SO_4 electrolyte without (**a**) and with 0.5 wt% HF (**b**), and (**c**) for Ti in 1 M H_2SO_4 with different HF concentrations

an electrochemical situation leading to pore formation, the fluoride concentration needs to be optimized for each valve metal.

This is shown for Ti in Fig. 2c. The addition of small amounts of F leads to activation of the surface – the polarization curves show an active/passive transition. The higher the fluoride concentration, the higher is the current density.

Scanning electron microscopy (SEM) observations of the surfaces after acquiring the polarization curves typically lead to the findings that anodization in solutions with a low HF concentration, for Ti typically <0.05 wt% HF, yields a compact oxide layer with a few nonregular pores. On the other hand, anodization in solutions containing >0.5 wt% HF leads to uniform etching of the surface (electropolishing). Therefore, the most promising range to achieve the formation of porous oxide networks is the concentration range between 0.05 and 0.5 wt%.

3.2 I-t Curves and Initiation of Porous Layers

A common observation in porosification experiments is that establishing the final state of the self-organized structure formation requires an induction time. This is also apparent in electrochemical experiments.

Figure 3a shows a schematic current–time transient for anodization of Ti at the "right" voltage in an electrolyte, for example, H_2SO_4 in absence and in presence of the "right" amount of fluoride. In pure background electrolyte (H_2SO_4), the typical exponential current decay is observed due to the growth of a compact oxide layer. For electrolytes containing fluorides, after an initial exponential decay (phase I) the current increases again (phase II) with a time lag that is shorter, the higher the fluoride concentration. Then, the current reaches quasi-steady state (phase III). This steady-state current increases with increasing fluoride concentration.

This type of current–time curves have been previously reported for self-organized pore formation for other materials, as well [55]. Typically, such a current behavior

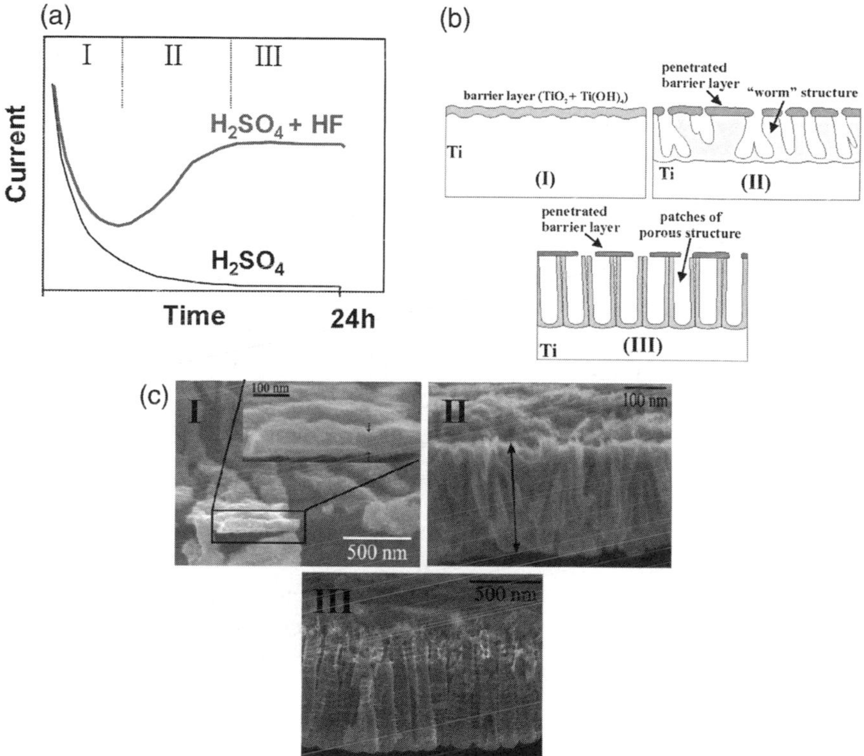

Fig. 3 (a) Schematic current transient for anodization of Ti in H_2SO_4 electrolyte without (**a**) and with HF; (**b**) scheme of pore formation process, (**c**) SEM images of corresponding stages in (**b**). Samples anodized in 1 M $(NH_4)_2SO_4$ + 0.5 wt% NH_4F at 20 V for 0 min (I), 6 min (II), and 20 min (III)

can be ascribed to different stages in the pore-formation process, as schematically illustrated in Fig. 3b. In the first stage, a barrier oxide is formed, leading to a current decay (I). In the next stage, the surface is locally activated and pores start to grow randomly (II). Due to pore growth, the active area increases and the current also increases. After some time, many pores are initiated and a tree-like growth takes place. Therefore, the individual pores start interfering with each other and start competing for the available current. This leads under optimized conditions to a situation where the pores equally share the available current, and self-ordering under steady-state conditions is established (III).

Indeed, if the pore initiation phase is followed in a concrete case by SEM images, the exact sequence described in Fig. 3b can be observed. Figure 3c shows samples removed from the electrolyte at different times – shown here is the evolution of the morphology of the porous titanium oxide in 1 M $(NH_4)_2SO_4$ + 0.5 wt% NH_4F [14]. Immediately after the sweep from the OCP to 20 V, a thin and rough layer (≈50 nm) covers the entire surface of the titanium (I). The results from XPS measurements show that the film at this stage is composed of Ti, O, and some traces of F. The XPS data suggest that the film is composed of an inner layer of TiO_2 and an outer layer of $Ti(OH)_4$. This layer is essentially the fluoride-perforated high-field passive layer – the thickness very well corresponds to a growth factor of 2.5 nm V^{-1}. After some polarization time, the first signs of localized attack become apparent with the FE-SEM; breakdown sites that are randomly distributed over the film surface and round-shaped holes in the substrate can be observed. Then the current increases due to the growth of the pores. At this point, the SEM image reveals that underneath the initial 50-nm-thick top layer some pore structures become visible. This means that the top layer, in fact, must be of a nanoporous nature (beyond the resolution of SEM), permeable to fluorides and electrolyte species. In the cross-section, the mesoporous layer underneath the nanoporous layer shows a thickness of 200–300 nm and a relatively regular but worm-like structure (II). At this time, from the top, patches of the top layer are still visible but they dissolve with time in the F-containing electrolyte. With increasing anodization time, the current drops further, while a steady state of the porous film thickness establishes (III), and a homogeneous self-organized porous structure is obtained.

3.2.1 Current Oscillations

Inspecting in detail the I-t behavior in fluoride-containing electrolytes shows that periodical current fluctuations occur [11, 16]. The oscillations can be very regular and can be maintained for more than 24 h. Typically, the average amplitude and frequency increase with increasing HF concentration. The origin of these oscillations is not understood in detail. However, for other systems such as porous silicon formation, under specific conditions, strong current oscillations have also been reported [56]. A most likely explanation is that these current oscillations can be ascribed to passivation and depassivation reactions on the surface that are competing. Interestingly in the present case, the periodicity of the oscillations can be correlated with some structural features observed in the porous layer. Figure 1a shows an SEM image of a TiO_2 nanotube layer with regularly spaced ripples at the side walls of the tubes. These ripples have a distance of about 50 nm. On the other hand, for the corresponding current oscillations, a frequency of 1/50 s can be determined with a current density of 3 mA/cm^2, corresponding to approximately 1 nm/s dissolution rate. Hence, one period per 50 s corresponds very well to the observed length scale of the ripples in the side walls. Nevertheless, the exact description of the interactions between the current oscillations and structural features remains till date unresolved.

3.3 Steady-State Growth

The features of the tubes shown in Fig. 1a (thickness and side-wall morphology) are typical for nanotube layers grown in acidic electrolytes. In acidic electrolytes, the tubes grow with time only up to a limiting thickness of approximately 500–800 nm. A typical thickness versus time behavior is shown in Fig. 4a.

The fact that the layer thickness (Fig. 4a) and the current density (Fig. 3a) reach a limiting value after a certain polarization time can be explained by a steady-state situation depicted in Fig. 4b. During anodization, permanently growth of oxide takes place at the inner interface, and chemical dissolution of the oxide layer occurs simultaneously. Steady state is established when the pore growth rate at the inner interface is identical to the overall dissolution rate of the oxide film at the outer interface. In this situation, the nanotube oxide layer just permanently "eats" through the titanium substrate without thickening of the oxide layer. As the steady-state current densities are typically considerably high, this occurs even with comparably high velocity. This finding also explains that typical current efficiencies (for oxide formation) in acidic electrolytes range from 3% to 10%, and continuously drop with extended anodization time.

It should be remarked that the picture shown in Fig. 4b is somewhat oversimplified. The chemical dissolution of TiO_2 occurs, of course, over the entire tube length, thus the tubes with extended time become increasingly V shaped in morphology as shown in Fig. 4c. That is, at the top the tubes possess a significantly thinner wall than at their bottom [57, 58].

Several experiments and calculations show that these tubes grow under diffusion control conditions with either the supply of fluoride ions to the tube tip or the transport of the MF_6^{2-} complex away from the tube tip being the rate-determining step.

4 Factors Affecting Tube Morphology

4.1 pH of the Electrolyte

The limiting porous oxide layer thickness can to a large extent be overcome by altering the electrochemical conditions [12–14, 59]. This means, most importantly, that a higher thickness of the porous layer can be obtained if the anodization

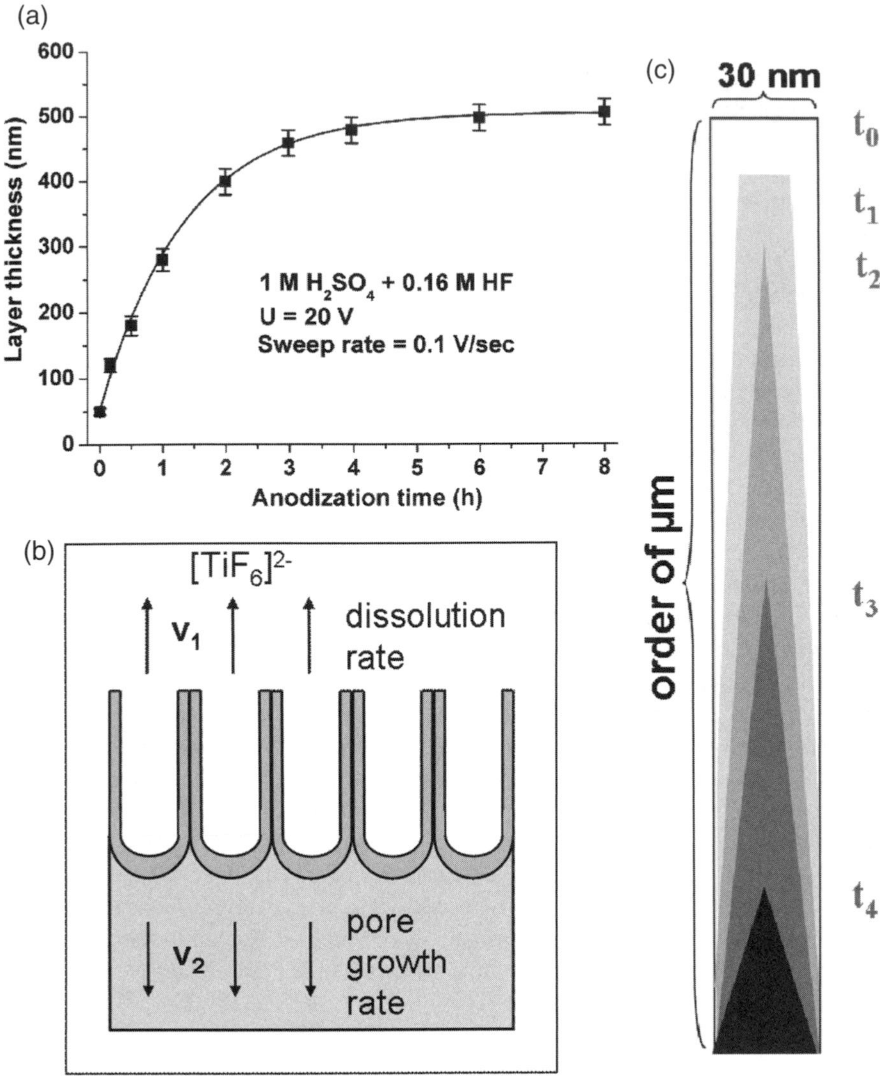

Fig. 4 (**a**) Evolution of tube layer thickness during anodization; (**b**) steady-state competition of TiO_2 dissolution and formation during the pore growth; (**c**) typical V- shape of the tube wall

treatment is carried out in less aggressive (more alkaline) electrolytes. Figure 5a shows the chemical dissolution rate of TiO_2 in fluoride-containing electrolyte as a function of the solution pH. Clearly, the more alkaline the solution, the lower is the dissolution rate, and therefore one could slow down the chemical dissolution of the oxide. However, to maintain pore growth at the pore tip, a certain degree of acidity (dissolution) is needed. The key to longer tubes is to recognize that this acidity can be automatically provided by the water hydrolysis reaction in the

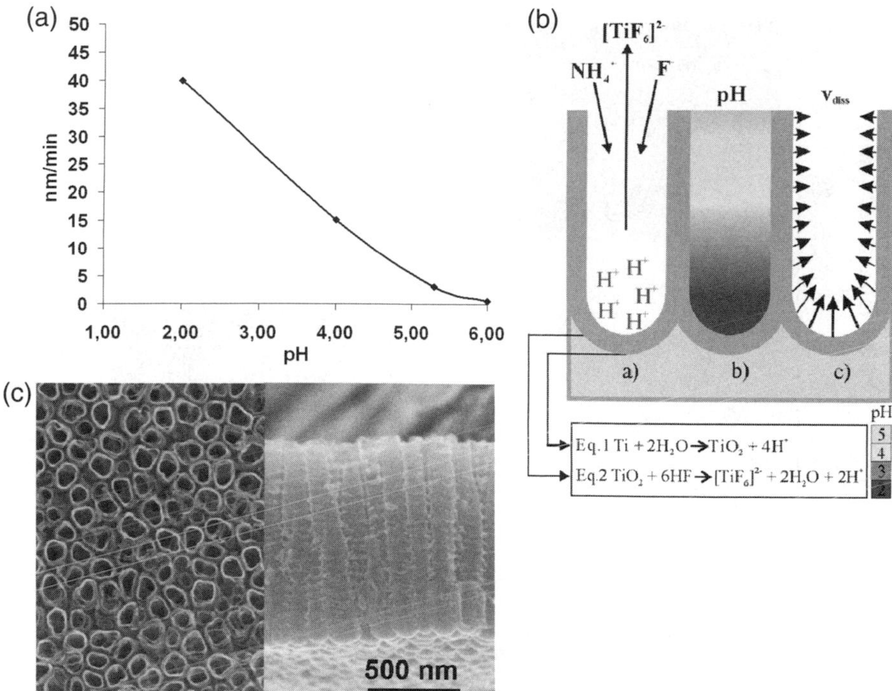

Fig. 5 (**a**) Dissolution rate of TiO_2 measured by XPS depth profiles on flat anodic TiO_2 samples of defined thickness after soaking in F-containing electrolytes with different pH; (**b**) pH profile inside the tubes; (**c**) example of a thick TiO_2 nanotube layer (SEM, top view, and cross-section)

oxide formation as shown in Fig. 5b. If the anodization is carried out in a neutral electrolyte, this acidification accelerates the chemical TiO_2 dissolution exactly at the desired place: at the pore tip (while the rest of the porous structure remains relatively stable due to the more alkaline pH of the background electrolyte) – for more details see Ref. [13]. Figure 5c shows top view and cross-sectional SEM image of an example of a thick TiO_2 nanotube layer produced in a neutral electrolyte.

4.2 Effect of Anodization Voltage

The key factor controlling the tube diameter is the anodization voltage [60]. Experiments carried out in 1 M H_3PO_4 + 0.3 wt% HF – see Fig. 6 – showed that in a potential range between 1 and 25 V, tubes could be grown with diameters from 15 to 120 nm and length from 20 nm to 1 μm, where diameter and length depend linearly on the voltage. Up to potentials of 25 V, self-organized tubular layers are obtained.

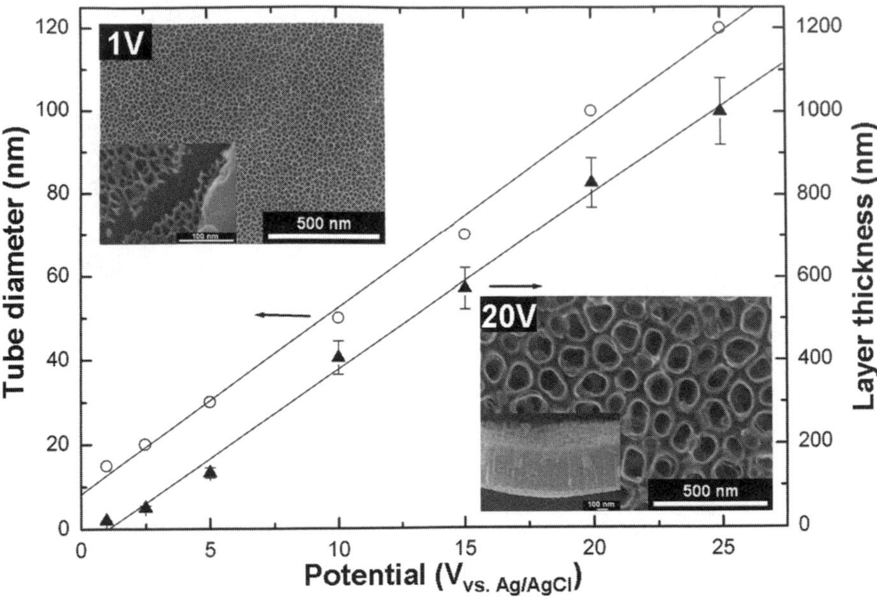

Fig. 6 Tube diameters and length as a function of anodization potential. Electrolyte 1 M H_3PO_4 + 0.3 wt% HF. Insets show SEM images of the nanotube layers formed at 1 and 20 V

At potentials higher than 25 V, the formed layers were no longer self-organized. Particularly remarkable is that self-organized structures are obtained even at potentials as low as 1 V. This may be a specific feature of the H_3PO_4 base electrolyte, as other electrolyte systems used for the formation of TiO_2 nanotubes typically show nonordered etched surfaces at such low voltages. Potentials lower than 1 V cause more or less uniform etching of titanium due to the fact that they reach the active corrosion region of the system.

The morphology features formed at 1 V show rather a web-like structure than a clear tubular morphology. Tubes formed at lower voltages are more connected to each other, whereas the nanotubes formed at 25 V seem to be more isolated from the surrounding ones.

This level of diameter control bears significant potential for applications where the tube diameter needs to be tailored for specific use, such as, for example, when a defined size for embedding of biological species is desired.

4.3 Effect of Viscosity

A series of experiments performed in ethylene glycol and glycerol water mixtures [16, 61] demonstrate clearly a strong influence of the electrolyte viscosity on the tube geometry. The tubes formed in these electrolytes show typically a smooth appearance over their entire length and grow to a significantly higher length – an

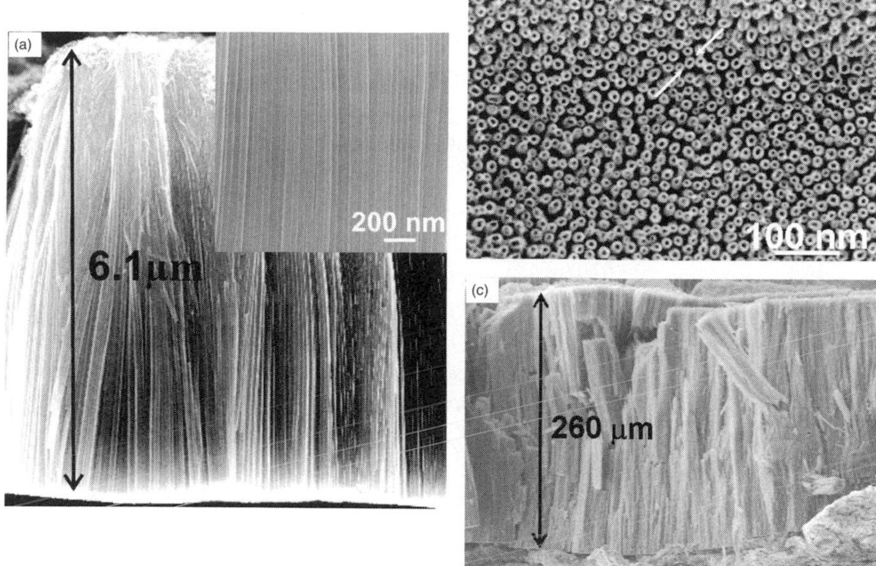

Fig. 7 SEM images of examples of TiO_2 nanotube layers grown in glycerol/NH_4F electrolyte (**a**) acetic acid/NH_4F electrolyte (**b**) and in ethylene glycole/HF electrolyte

example is shown in Fig. 7a. Evidently, the tubes are grouped in very tight bundles and are not connected with one another through any bulk material – this is in contrast to all previous growth attempts in literature, where a connection of the side walls (as in Fig. 1a) was always obtained.

In order to explain the smooth appearance, one may consider that current oscillations occur in aqueous electrolytes of low viscosity [11, 16]. The current transient frequency correlates well with the variations in the wall thickness. This can be explained by the fact that every current transient is accompanied with a pH-burst occurring at the pore tip. In viscous electrolytes, clearly the current–time transient shows much lower current densities than in the purely aqueous electrolyte – this is due to a lower diffusion constant in the electrolyte (considering the Stokes–Einstein relation where the diffusion constant, D, is proportional to $1/\eta$, where η is the solution viscosity). In other words, an increased diffusion constant keeps the acidic range more confined to the pore tip, thus it hampers dissolution of the tube walls and damps oscillations.

4.4 Effect of Water Content

In line with observations of other systems, anodization in (almost) water-free electrolytes can drastically increase the achieved length. Experiments carried out in CH_3COOH systems (example shown in Fig. 7b) showed a remarkable control over the tube diameter [62]. Early work carried out in glycerol showed tubes with length in the range of 7 μm [16] – meanwhile the maximum length reached 260 μm as shown in Fig. 7c [63]. In this case, the tubes were grown in an aged ethylene glycol HF electrolyte with a rest H_2O content of 0.3%. Examination of the structure clearly shows that, at the bottom, a hexagonal cell structure prevails (as in the case of Al), and only the upper part is separated into tubes by chemical dissolution – as discussed in Section 4.1.

4.5 Formation of Multilayers and Free-Standing Membranes

Very recently it has been shown that for a range of valve metal oxides, multilayered structures can be grown [58, 64]. In other words, first, a layer under a first parameter set may lead to a first geometry, and then underneath, a second layer of tubes can be grown with a different parameter set. Interestingly, the underneath layer may be initiated at the bottom of a tube to break through the bottom [64], or in the spaces between the tubes [58]. Such multilayer stacks may have a variety of applications wherever critical tailoring of properties isneeded.

Even more spectacular is the formation of free-standing (open porosity) membranes [65] as a whole range of functionalization may be accessed, including, for example, flow-through photoreactors.

4.6 Different Metal Substrates

Except for Ti, on a range of other valve metals, some self-organized structures have been grown and significantly different morphologies obtained.

Zr, Hf:

In comparison with other valve metals, the most straight-forward formation of nanoporous or nanotubular layers can be achieved on Zr and Hf [18, 19, 26]. On the materials, without a lot of experimental optimization, long and smooth tube morphologies can be easily grown. In contrast to other anodic valve metal nanotubes, the oxide tubes grown on Zr and Hf have a crystalline structure directly after anodization (i.e., without annealing). Zirconium oxides are known to have excellent technological properties, such as chemical and thermal stability, mechanical strength and wear resistance, as well as its good ion-exchange properties. These properties enable zirconium oxide films to be used as industrial catalyst and catalyst supports for its particular acid catalysis. A great effort has been made to prepare porous zirconium oxide with a high surface area to improve the efficiency as a material in catalysis. Therefore, some research efforts have focused on the synthesis of porous zirconium oxide by using templating techniques [66–68] and electrodeposition [67, 69]. Compared with aluminum and titanium, however, only a few early studies report on random patches of porous zirconium oxide structures formed during anodization [70, 71]. The use of fluoride containing acidic electrolytes lead to porous material rather than tubular structures; however, smooth high-aspect-ratio zirconia nanotubes were fabricated by electrochemical anodization of zirconium in neutral buffers [18, 19].

These high-quality zirconia nanotube layers consist of highly regular arrays of straight nanotubes with a diameter of 50 nm and a length of 17 μm, with pore walls that are completely smooth and straight. The grown nanotubes were found by XRD and TEM-SAED to have a cubic crystalline structure without any further heat treatment.

Similar findings were obtained for HfO_2 in H_2SO_4 electrolytes containing NaF [26]. Under a range of experimental conditions, highly ordered tube-like structures can be formed. The pore diameter increases with increasing potential from ∼15 to 90 nm. The porous oxide layers can be grown to a thickness of several-tens micrometer in a voltage range from 10 to 60 V.

Ta, Nb:

As already discussed in the section on I–U curves, it is more difficult to grow ordered nanotubular layers on Ta and Nb. Typically, only layers of some 100-nm thickness are obtained. Porous oxide structures on Nb bear high application potential as gas sensors [72], catalysts [73], optical [74], and electrochromic [75] devices. First investigations [21] reported porous Nb_2O_5 layers that consist of ordered-pore arrays with single-pore diameters ranging from 20 to 30 nm. The layers were formed in H_2SO_4/HF electrolytes, and it was found that the pore morphology and the layer thickness strongly depend on the HF concentration and the time of anodization. Well-ordered uniform porous layers of up to 500 nm in thickness were formed using 1–2 wt% HF. Later work by Choi et al. [76] has shown the possibility of growing porous Nb_2O_5 layers also in H_3PO_4/HF mixtures. Furthermore, they have increased their thicknesses by repeated porosification, that is, Nb_2O_5 layers were grown and annealed, and then a second anodization was used to form a second layer [77]. In recent work, additionally, Karlinsey has demonstrated that by anodization of Nb in HF electrolytes (0.25–2.5 wt% HF) Nb_2O_5 microcones can be grown on Nb surfaces [78].

Ta_2O_5 has received considerable attention as a protective coating material for chemical equipment, as part of optical devices, and as suitable material for storage capacitors in very large-scale integrated circuits [79–84].

Similar to Nb, on Ta the formation of porous Ta_2O_5 on Ta was investigated in H_2SO_4 electrolytes containing different concentrations of HF (0.1–5 wt%) [24, 25]. Porous (nanotubular) tantalum oxide that consists of pore arrays with single-pore diameters of about 20 nm and pore spacing of about 15 nm was obtained up to thicknesses of several 100 nm for HF concentrations of around 2% and several hours of anodization.

W:

Experimentally, the formation of porous (nanotubular) WO_3 is most challenging, but it is also most highly rewarding as tungsten is one of the most important valve metals, and its oxide has electrical and optical properties that are exploited for a variety of applications such as photolysis [85], electrochromic devices [86–88], and gas sensors [89–92]. Several preparation techniques have been reported for the fabrication of tungsten oxides, for example, thermal evaporation [90, 91], chemical vapor deposition [92], and sol–gel coating [86–88]. Though anodization of tungsten in electrolytes is one of the methods to fabricate tungsten oxide and has been intensively studied, the formed films typically show a compact structure [93, 94]. In fluoride electrolytes (e.g., 0.2 wt% NaF) at elevated potentials such as 40 V and 60 V, well-structured porous arrays can be grown with 59–70 nm tube diameter and a thickness of several 100 nm [22]. As-formed layers show an amorphous structure, but the layers can be altered to a crystalline monoclinic structure by thermal annealing at 500°C. The annealed porous WO_3 layers show a very high specific photocurrent-conversion efficiency, reaching 20–30% in the UV range [95].

Alloys:

Attempts have been made to grow self-organized anodic tube layers on technologically relevant substrates such as Ti6Al4V and Ti6Al7Nb [9, 96, 97]. However, these approaches suffered typically from two problems: (1) the selective dissolution of less stable elements and (2) the different reaction rates on different phases of an alloy. In designing an "ideal" alloy for nanotube formation, one desires a single-phase microstructure and a composition that only contains elements that show similar electrochemical oxidation rates. Therefore, an alloy such as the recently developed TiNb, TiZr, or biomedical alloys such as Ti29Nb13Ta4.6Zr [98, 99] have promising composition as they can be tuned to be a single phase (β-phase), and they consist only of elements on which nanotubes have successfully been grown in fluoride-containing media.

For TiNb, surprising synergistic effects on the growth morphologies of the oxide nanotubes were found [100]. It was shown that the range of achievable diameters and lengths of TiO_2-based nanotubes can be significantly expanded, if a binary Ti–Nb alloy, rather than pure Ti, is used as a substrate for oxide nanotube growth in an aqueous electrolyte. The length of the resulting mixed oxide nanotubes can be adjusted from 0.5 to 8 μm, and the diameter from 30 to 120 nm. The morphology of the tubes differs significantly from that of the nanostructures grown under the same conditions on pure Ti or Nb substrates: only considerably shorter tubes grow on Ti, whereas irregular porous structures grow on Nb.

The tubes consist of an amorphous mixed TiO_2–Nb_2O_5 oxide structure. After annealing at 450°C, anatase-type TiO_2 appears in XRD. After annealing at 650°C, an additional peak assigned to rutile appears, but the major TiO_2 phase remains anatase with a crystallite size of 25 nm. This result shows that the anatase–rutile transition takes place at a higher temperature in the polycrystalline mixed-oxide nanotubes than in pure TiO_2.

Weak reflections assigned to pseudo-hexagonal TT–Nb_2O_5 can be detected after annealing at 650°C. Apart from the retardation of the anatase–rutile transition by Nb_2O_5, it is also remarkable that the mixed-oxide nanotubes are stable to annealing at 650°C. For pure TiO_2 nanotubes treated at this temperature, a substantial collapse of the tubular morphology has been reported [101]. Very recently. high-efficient intercalation properties have been reported for this alloy [102].

For anodic nanotubes formed on TiZr alloys [57, 64, 103], the morphological character of the oxide nanotubes is between those of titanium oxide and zirconium oxide nanotubes. However, the nanotubes have a straight and smooth morphology with a diameter ranging from 15 to 470 nm and a length up to 21 μm depending on the terminal anodization potential (i.e., they show a largely expanded structural flexibility compared with nanotubes formed on the individual elements). In contrast to nanotubes formed on TiNb, the tubes grown on TiZr consist of a zirconium titanate oxide – as formed, with an amorphous structure that can be crystallized after adequate heat treatment [103].

Such nanotube structures are interesting in view of several applications, as zirconium titanate (ZT) is used as microwave-resonant components and frequency-stable oscillators [104–106], optical devices [107, 108], refractory ceramics [109], and template for lead zirconium titanate (PZT).

The alloy Ti29Nb13Ta4.6Zr has been developed by Niinomi et al. for biomedical applications in order to reduce the elastic modulus of titanium alloys to the level of living bone [110]. Self-organized oxide nanotubes grown on this alloy [98, 99], except for a high degree of structural flexibility, can show a very spectacular feature, that is, multiscale self-organization (oxide nanotube arrays with two discrete sizes and geometries).

Additionally, one may note that recently it was demonstrated that also for the Al case, neutral F-containing solutions can be used to achieve fast and highly ordered oxide structures [111].

5 Structure and Chemistry

5.1 Crystallographic Structure

As-formed TiO_2 tubes typically have an amorphous structure. However, different TiO_2 applications require specific crystallographic structures for an optimized performance. For example, the anatase form of TiO_2 shows the highest solar energy conversion efficiency [112] and has also the highest activity for catalysis [113]. Several studies show that the tubes can be converted to anatase or a mixture of anatase and rutile at temperatures higher than approximately 280°C in air [101, 114, 115].

Figure 8a shows a comparison of XRD patterns of nanotube samples after formation (amorphous), after annealing at 450°C (anatase), and after annealing at 550°C (mixture of anatase and rutile).

Figure 8b shows high-resolution transmission electron microscope (HRTEM) images of the bottom of the nanotubes before and after annealing and selected area diffraction patterns (SAED) of the tube bottoms taken from the corresponding images (as insets).

Figure 8b confirms the XRD measurements that upon thermal annealing the samples, a transformation from amorphous to crystalline structure occurs. From XRD measurements, one can obtain that the major anatase is the (101) plane. Other

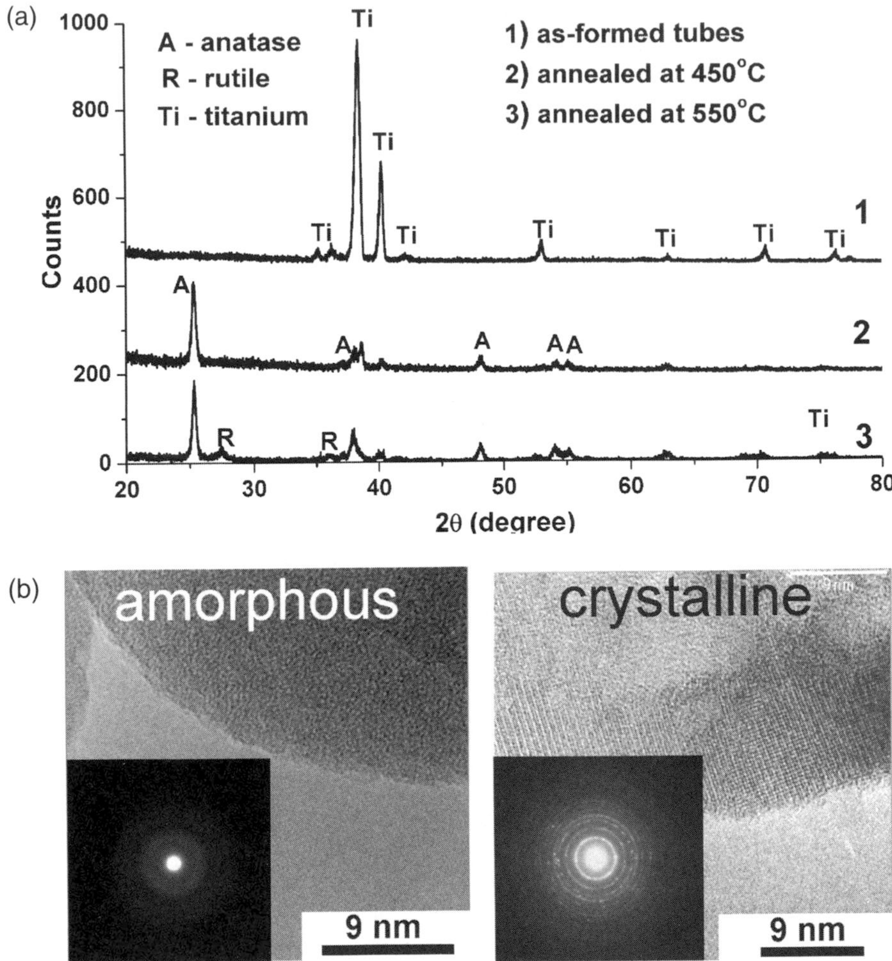

Fig. 8 (**a**) XRD of as-formed amorphous TiO_2 nanotube layer and crystalline layers annealed at 450°C and 550°C; (**b**) HRTEM images of amorphous(as-formed) and crystalline (annealed at 450°C) tube bottoms, insets shows diffraction patterns (SAED) taken at the same locations. Annealing performed for 3 h with heating and cooling rate of 30°C min^{-1}

planes, such as (200) and (105) are only present in a minor amount. For the nanotube layers, typically, crystal growth starts at the tube bottom via interface nucleation, due to the larger space available for crystal growth, than in the side wall [101]. For certain tubes, it was found that essentially over the entire length, only one plane – (101) – is present along the walls. This points to the possibility of growing single crystalline tubes [115, 116].

Annealing under oxygen-free conditions (e.g., in argon atmosphere) leads to a blackening of the tube layers due to a significant reduction of the Ti(IV)-species in the oxide to Ti(III). Such structures typically have a very limited mechanical stability [114].

Annealing in air at temperatures above 450°C typically leads to increasing rutile content. At temperatures around 650°C and higher, the tube layers start losing their morphological integrity – that is, they start collapsing [101, 114].

Introducing alloying elements such as Nb [117] or C [118] into the TiO_2 increases significantly the temperature of the anatase-to-rutile conversion and also shifts the temperature of structural collapse to a higher value.

5.2 Chemical Composition

In XPS and EDX investigations, as-formed tubes show a composition of TiO_2 with minor contents of hydroxides on the surface of the tube walls [14]. Different background ions in the electrolyte are integrated into the tube structure at different concentration levels. While ClO_4^- ions are hardly incorporated, SO_4^{2-} and particularly PO_4^{3-} are incorporated into the entire tube to significant levels (some few at.%). It is very clear that significant amounts of F^- (≈1–5 at-%) are entering the TiO_2 structure. This is in line with some earlier work on the anodization of Ti in fluoride containing electrolytes [119]. An example of a XPS sputter depth profile is shown for nanotube TiO_2 layer in Fig. 9.

Annealing leads to almost an complete loss of the fluorides at around 300°C [116] and clearly the amount of surface hydroxides is reduced [120, 121].

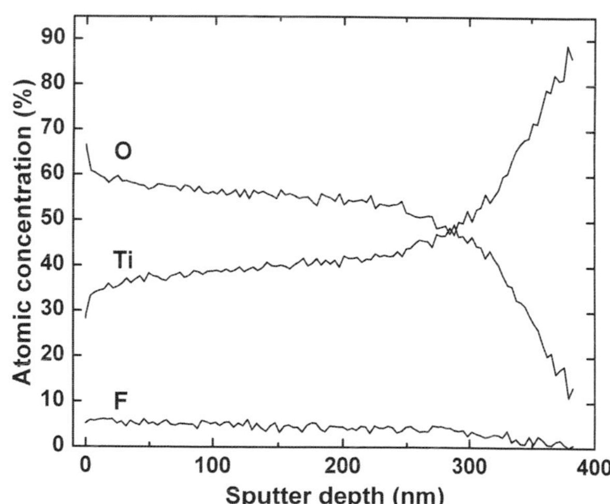

Fig. 9 XPS depth profile of nanotubular TiO_2 layer formed in 1 M H_2SO_4 + 0.16 M HF at 20 V. Thickness is approximately 350 nm, fluoride uptake is visible

6 Properties of the Tubes

6.1 Photoresponse of the Tubes

Typical photocurrent characteristics for the as-formed and annealed samples are provided in Fig. 10 [114, 122]. The comparison shows IPCE plots for the annealed nanotubular TiO_2 and the annealed compact TiO_2. The inset figure gives the corresponding $(IPCE\ h\nu)^{1/2}$ versus $h\nu$ plots for these samples. Clearly, the band gap results also for the annealed samples as 3.15 ± 0.05 eV which is in line with typical value reported for anatase [123, 124]. It is evident that the IPCE is drastically increased for the tubes – that is, for the annealed tubes a value of almost 50% at short wavelength (compared to 10% for the annealed compact layer and 3–4% for the unannealed tubular or compact samples). In fact, for the as-formed tubes, most of the photocurrent is generated at the bottom of the tubes, and the tube wall contribution is negligible. This is in line with the expectation that the amorphous structure provides a high number of defects that lead to a high carrier recombination rate. The drastic increase of the photocurrent after annealing indicates that by conversion to anatase the tube walls are activated and contribute to the photocurrent. The voltage dependence of the photocurrents recorded for annealed nanotubular and compact TiO_2 layers showed that an enhanced photocurrent is obtained essentially over the entire potential range. However, at potentials (E) close to the flat-band situation (the optical flat-band potential E_{fb} is approximately at $-0.2\ V_{Ag/AgCl}$ in neutral solutions), the absolute photocurrent response for the nanotubes can be up to 20 times higher

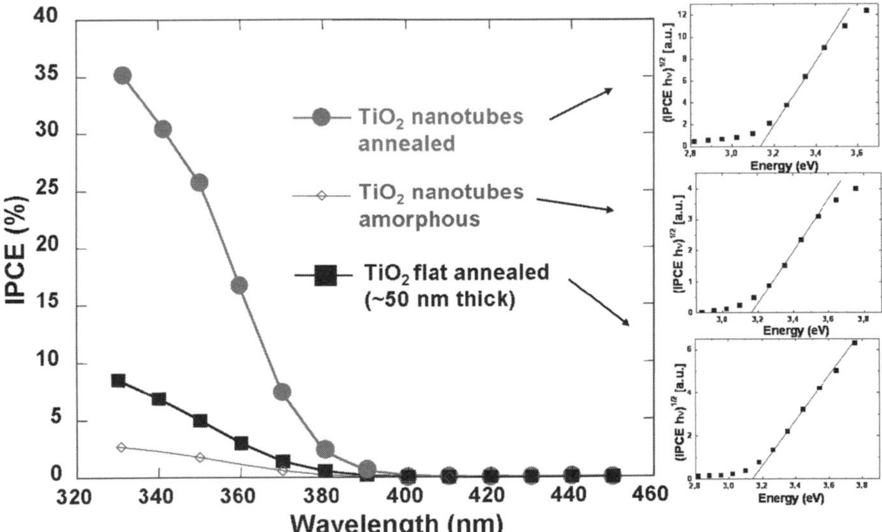

Fig. 10 Photocurrent spectra recorded for as-formed and annealed (as in Fig. 8) TiO_2 nanotube layers and for comparison also for compact anodic TiO_2 layer. Insets show evaluation of the band-gap energy (E_g) of the respective layers

than for the compact oxide. At higher anodic potentials ($E > 1$ V), a saturation of the photoresponse is observed. The potential dependence is very well in line with expectations of $i_{ph} \alpha E^{0.5}$.

Further details of optimizing the tube geometry (length, diameter, and tube wall thickness), structure (anatase and anatase/rutile) need to be carried out. Presently, highest conversion efficiencies are achieved with tubes annealed to anatase and a length of approximately 300 nm^{-1} μm. This is due to a competition between most-efficient light-absorption length and increased recombination with increased length.

6.2 Doping, Dye Sensitization

In general, in order to enhance the efficiency of TiO_2 in the visible range, considerable efforts have been made involving dye sensitization [28, 112] of the electrode with suitable species, which showed that certain organic dyes (mainly Ru-complexes) fixed on TiO_2 can inject electrons into the conduction band of TiO_2 upon light excitation (as the LUMO overlays with the conduction band edge of TiO_2). As the HOMO/LUMO distance of these dyes typically is only 1–2 eV, the reaction can be triggered by visible light. The dye can be attached on TiO_2 surfaces by –COOH coupling – the oxidized dye can be reduced by a suitable redox species to again form the active (reduced) state on the surface. Also with the nanotube layers, successful dye sensitization has been demonstrated using a commercial Ru-dye as shown in Fig. 11a [125].

A completely different approach is the so-called doping of TiO_2 (Fig. 11b). Asahi et al. [126] reported that doping TiO_2 with nitrogen by sputtering in a nitrogen-containing gas mixture improves the photoelectrochemical reactivity of TiO_2 films toward organic molecules under visible light illumination. Other doping species,

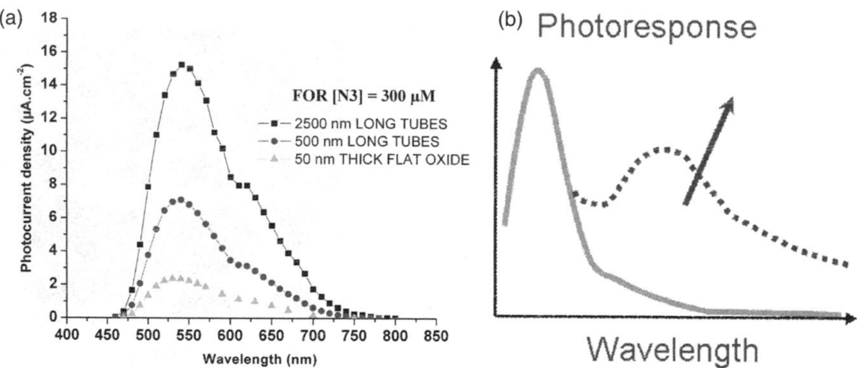

Fig. 11 (**a**) Photocurrent spectra of dye-sensitized samples (N3 dye) measured in acetonitrile + KI + I_2 electrolyte; (**b**) schematic picture of doping the TiO_2 toward visible light photoresponse

such as a number of transition metals [127, 128], or nonmetals such as phosphor [129], fluorine [130], carbon, [131], sulfur [132, 133], and boron, [134] have been introduced into TiO_2 compact layers or powders using various techniques. Ion implantation is the most straightforward approach for doping, but, up to now, efforts carried out on TiO_2 (by transition metal implantation) were hampered by the accompanying structural damage [128].

Considerable nitrogen doping of the tube layers was achieved by ion implantation [135, 136] and by thermal treatment in NH_3 [137, 138], wet chemical approaches [139] show only limited success. Other elements such as Cr [140] or C [141, 142] have also led to structures with a considerable visible light response.

6.3 Insertion Properties for Li and H^+ and Strong Electrochromic Effects

Another specific property of TiO_2 is its ability to serve as a host for hydrogen ion or lithium ion insertion [143–146]. These insertion reactions take place under cathodic polarization and are accompanied by reduction of Ti^{4+} to Ti^{3+} and a change of the color of the material [147].

The kinetics and magnitude of ion insertion and the electrochromic reaction (contrast) strongly depend on the ion diffusion length and therefore on geometry of the electrode surface. Due to the specific geometry of the TiO_2 nanotubes, a very high contrast can be obtained using vertically oriented nanotubes [148].

For example, Fig. 12a shows the cyclic voltammograms in a 0.1 M $HClO_4$ electrolyte for a compact anodic TiO_2 layer and a TiO_2 nanotube layer. The cathodic

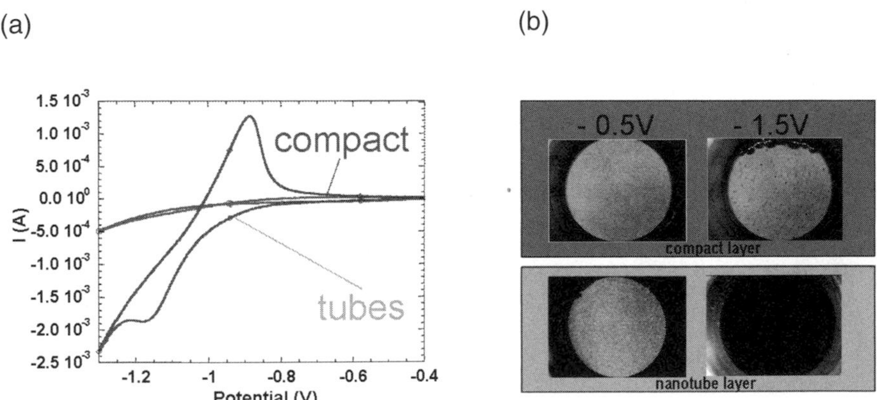

Fig. 12 (**a**) Cyclic voltammograms recorded for nanotube and flat TiO_2 layers in $LiClO_4$ and (**b**) corresponding optical images of the samples showing strong electrochromic switching effect observed at different potentials

peak can be ascribed to Ti^{4+} reduction combined with H^+ intercalation (Eq. (1)) while the anodic peak can be ascribed to the counter-reaction (i.e., the H^+ release reaction) [149]:

$$Ti^{4+}O_2 + H^+ + e^- \leftrightarrow Ti^{3+}O(OH)$$

The reaction occurs relatively sluggish (and hardly at all) on the compact oxide layer and on the as-formed nanotubular TiO_2 structure, while it is more pronounced for the annealed (anatase) structure.

While scanning from anodic to cathodic values, a very clear modification of the color can be observed for the nanotube layer by the naked eye – in these cases, the color changed from a very light gray to a solid black. On the compact oxide surface, however, hardly any color change could be observed by eye. This is shown in Fig. 12b. Similar findings were observed for Li insertion reactions [150].

6.4 Photocatalysis

After Fujishima and Honda reported, for the first time, on light-induced water splitting on TiO_2 surfaces, the material has been intensively investigated for applications in heterogeneous catalysis [27]. Since then, TiO_2 has shown to be an excellent photocatalyst [113, 151, 152] with a long-term stability, low-cost preparation, and a strong-enough oxidizing power to be useful for the decomposition of unwanted organic compounds [153–155].

The principle of the photocatalytic decomposition is that photons from a light source, which have sufficient energy, that is, higher than the band-gap energy – E_g – of the TiO_2, excite electrons from the valence band to the conduction band of the semiconductor and charge carrier pairs (consisting of hole h^+ and electron e^-) are formed [156]. These charge carriers either recombine inside the particle, or migrate to its surface, where they can react with adsorbed molecules. In aqueous solutions, positively charged valence band holes from TiO_2 typically form $\cdot OH$ radicals, while electrons in the conduction band mainly reduce dissolved molecular oxygen to superoxide $\cdot O_2^-$ radical anions. Organic molecules present in the solution may react with these oxidizing agents, inducing their oxidative degradation to inorganic compounds, including carbon dioxide and water [157]. In order to achieve a maximum decomposition efficiency, except for adequate band-edge positions, rapid charge separation, and high-quantum yield, a large area of the catalyst is desired.

Recently it has been demonstrated that the annealed TiO_2 nanotubes show considerably higher decomposition efficiency than a compacted Degussa P25 layer (20–30-nm diameter nanopowder composed of anatase and rutile) under comparable conditions [158], and even more recently, a flow-through membrane has been fabricated that is highly interesting for a large number of photocatalytic applications [65].

6.5 Highly Adjustable Wetting Properties

The ability to decompose organics on the surface combined with a nanotubular geometry can be used to tailor surface-wetting properties of TiO_2 from super hydrophilic to superhydrophobic [159, 160].

Titanium is a widely used biocompatible material; therefore, alterations of surface topography and wetting behavior are of great importance for its biomedical application. Several approaches using UV light [161, 162] or organic monolayers [163] have been reported to control the surface wettability of TiO_2 structures and nanoparticles.

The water-contact angle for freshly prepared flat TiO_2 samples is around $49 \pm 2°$, while the nanotube structures showed a superhydrophilic behavior, that is, complete spreading of water on the entire surface and into the tubes. After the samples had been stored in the dark for 3 weeks, the water-contact angle for compact film increased up to $92 \pm 2°$, whereas it remains unchanged for the tubular layers at $0 \pm 1°$ – that is, the nanoporous layer still shows complete wettability (Fig. 13a). The change in wetting behavior of the compact TiO_2 with time has been ascribed to a change in surface termination [162] – that is, by ageing the hydrophilic OH groups convert into less hydrophilic oxide.

In order to alter the surface properties, octadecylsilane or octadecylphosphonic acid molecules can be attached to the TiO_2 surface [159, 160]. The originally completely hydrophilic surface becomes superhydrophobic, with a water-contact angle of about $165 \pm 2°$ for the silane-SAMs and $167 \pm 2°$ for phosphonic acid-SAMs (Fig. 13b). For comparison, the compact layers showed $107 \pm 2°$ by organic modification.

Using UV illumination, the wetting behavior can be altered. After about 12 min for the silane-SAM and 5 min for the phosphonic acid-SAM (Fig. 13c), the porous surface changed from super hydrophobic (contact angle $\sim 165°$) to complete hydrophilic behavior (contact angle $\sim 0°$). Figure 13d shows the contact-angle change under UV illumination for different exposure times. This alteration can be stopped at any time leading to a very well-adjusted and stable surface with a well-defined wettability. XPS studies revealed, in line with literature [164], that during UV light treatment, organic monolayers start to decompose by chain scission at the functional end of the C chain leaving -Si-O^-, -P-O^- groups behind (see Fig. 13e).

6.6 Biomedical Applications

Titanium and its alloys such as Ti-6Al-4 V and Ti-6Al-7Nb are widely used in biomedical applications for orthopedic or dental implants due to their good mechanical properties and biochemical compatibility [165]. Therefore, studies on the interaction of the nanotube material in regard to a biorelevant environment are of a very high significance. Mainly, two directions have so far been explored with TiO_2 nanotube coated substrates:

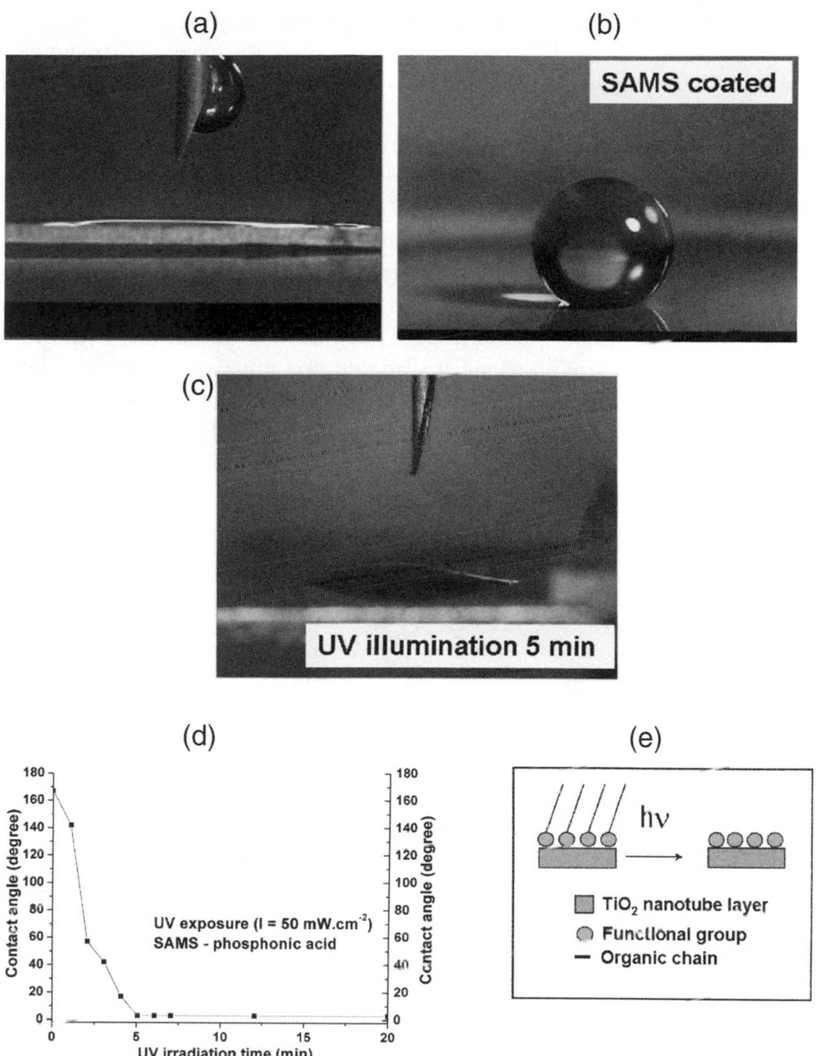

Fig. 13 Optical images of (**a**) superhydrophilic TiO_2 nanotube surface, (**b**) superhydrophobic surface coated with organic monolayer-phosphonic acid, (**c**) the same surface as in (**b**) after 5 min of exposure to UV light, (**d**) evaluation of the contact angle for sample in (**b**) upon exposure of UV light, (**e**) mechanism of photocatalytic decomposition of the organic monolayer on TiO_2

(a) hydroxyapatite growth
(b) cell interactions

(a) Hydroxyapatite growth

Apatite formation is considered to be essential for the bone-binding ability of biomaterials. In order to improve bioactivity of titanium and to enhance biocompatibility, surface treatments such as hydroxyapatite coating or chemical treatments have been exploited [166–173]. In particular, the chemical treatments in NaOH have been extensively examined [170, 171]. An electrochemical method to increase biocompatibility of titanium surfaces is spark anodization [45, 46, 172, 173], which typically leads to a formation of a rough random porous TiO_2 layers.

Recently, hydroxyapatite formation on TiO_2 nanotube layers with different tube lengths was investigated [174]. The nanotube layers could strongly enhance apatite formation compared with compact TiO_2 layers, as shown in Fig. 14. Surprisingly, the apatite coverage of the nanotube layers was dependent on the nanotube length – this was attributed to a different surface roughness of the different-length nanotubes influencing nucleation of hydroxyapatite precipitation. Annealing the nanotube layers (from amorphous structure) to anatase, or anatase and rutile, further enhanced apatite formation. The induction time for apatite formation on TiO_2 nanotube layers becomes comparable to the best other known treatments of Ti surfaces. It should be noted that also on alloys such as Ti6Al4V, Ti6Al7Nb, Ti29Nb13Ta4.6Zr (Niinomi alloy), formation of an ordered and robust oxide nanotube layer is possible [9, 96–99].

(b) Cell response to nanotube layers

Recently, studies were carried out using rat mesenchymal stem cells and primary human osteoblast-like cells that were seeded on TiO_2 nanotube structures,

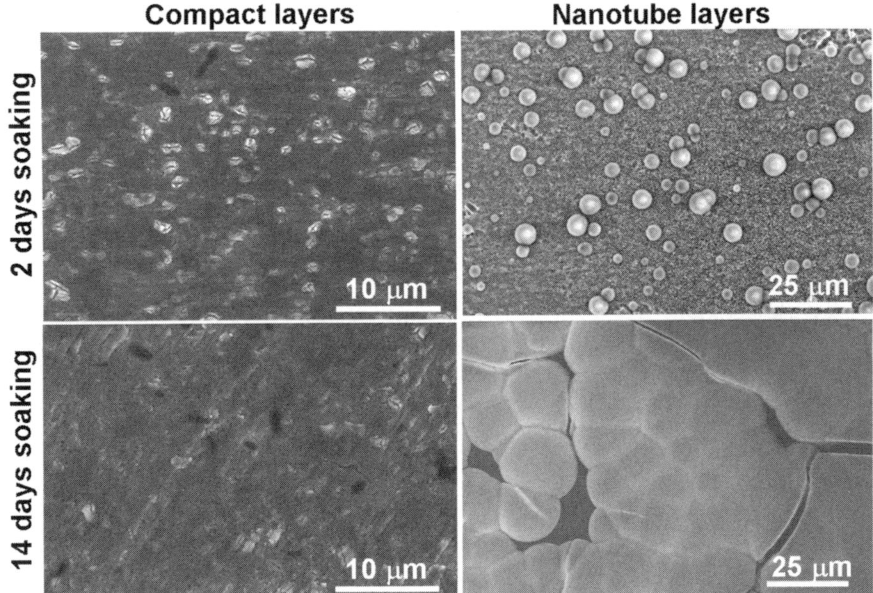

Fig. 14 SEM images of nanotube and anodic TiO_2 layers after 2 and 14 days of soaking in SBF solution

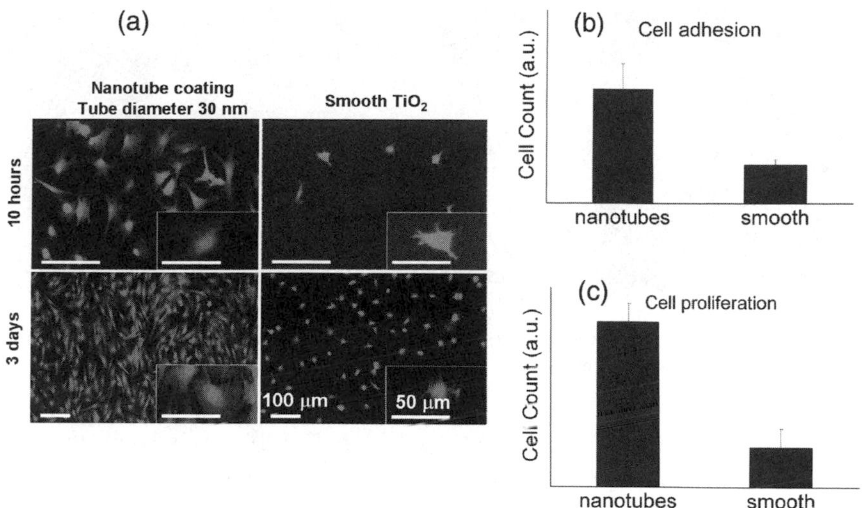

Fig. 15 (**a**) Cell behavior on nanotube TiO_2 layer (tube diameter ≈30 nm) compared with flat (smooth) TiO_2 layer. Tests done at 37°C with mesenchymal stem cell (GFP-labeled) using a cell density of 5000 cells cm^{-2} and 5% CO_2; (**b**) cell adhesion after 24 h; (**c**) cell proliferation after 3 days

with inner diameters ranging from 15 to 100 nm [175]. Cell adhesion, proliferation, and migration were significantly affected by the nanotube size. Clearly, geometries with a spacing of approximately 15 nm were most stimulating for cell growth and differentiation, whereas diameters of ~100 nm lead to a drastically increased cell apoptosis (see Fig. 15). This drastic effect of the nanoscale microenvironment on cell fate was ascribed to specific interactions between a specific nanotube size with the focal adhesion (FA) complex.

6.7 Other Aspects

Other applications of the tubes target H-sensing. For instance, Grimes et al. has shown manifold increase in electrical conductivity of the TiO_2 nanotube layer upon exposure to H_2 environments. For example, response in order of several magnitudes has been determined for 1000 ppm H_2 containing nitrogen atmospheres [176, 177].

For many applications, thin nanotubular layer of TiO_2 films on a foreign substrate is desired such as on Si-wafer or on conductive glass (ITO). Several groups reported successful fabrication of oxide nanotube layers from sputter-deposited thin titanium films [178–180]. The key is to alter the dissolution rate as much as possible, that is, to achieve high current efficiency (e.g., by lowering the electrolyte temperature).

The applications of TiO_2 nanotubes can significantly be expanded if secondary material can successfully be deposited into the tubes. Recently, such an approach has been reported that leads to selective electrodeposition of Cu into the tubes

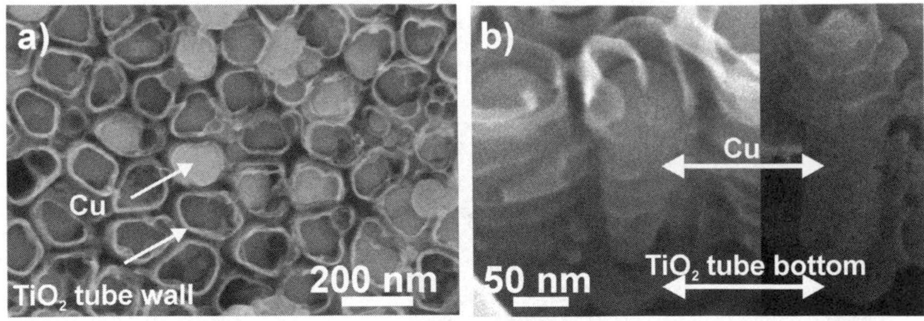

Fig. 16 SEM images showing top view (**a**) and cross-section (**b**) of nanotube layer with electrodeposited Cu filling the entire tube space

[181] leading to completely filled tubular layers (Fig. 16). This is an important step toward magnetic nanotube materials, solid-junction solar cells, or biomedical release systems.

References

1. H. Masuda, K. Fukuda, Science 268 (1995) 1466.
2. H. Masuda, K. Yada, A. Osaka, Jap. J. Appl. Phys. 37 (1998) L1340.
3. H. Masuda, K. Fukuda, Appl. Phys. Lett. 71 (1997) 2770.
4. O. Jessensky, F. Muller, U. Gösele, Appl. Phys. Lett. 72 (1998) 1173.
5. H. Masuda, K. Yada, A. Osaka, Jap. J. Appl. Phys. 38 (1999) L1403.
6. D. Al Mawiawi, N. Coombs, M. Moskovits, J. App. Phys. 70 (1991) 4421.
7. C. R. Martin, Science 266 (1994) 1961.
8. M. Steinhart, J. H. Wendorff, A. Greiner, R. B. Wehrspohn, K. Nielsch, J. Schilling, J. Choi, U. Gösele, Science 296 (2002) 1997.
9. V. Zwilling, E. Darque-Ceretti, A. Boutry-Forveille, Electrochim. Acta 45(1999) 921.
10. D. Gong, C. A. Grimes, O. K. Varghese, W. Hu, R. S. Singh, Z. Chen, E. C. Dickey, J. Mater. Res. 16 (2001) 3331.
11. R. Beranek, H. Hildebrand, P. Schmuki, Electrochem. Solid-State Lett. 6 (2003) B12.
12. J. M. Macak, K. Sirotna, P. Schmuki, Electrochim. Acta 50 (2005) 3679.
13. J. M. Macak, H. Tsuchiya, P. Schmuki, Angew. Chem. 44 (2005) 2100.
14. L. V. Taveira, J. M. Macak, H. Tsuchiya, L. F. P. Dick, P. Schmuki, J. Electrochem. Soc. 152 (2005) B405.
15. A. Ghicov, H. Tsuchiya, J. M. Macak, P. Schmuki, Electrochem. Commun. 7 (2005) 505.
16. J. M. Macak, H. Tsuchiya, L. Taveira, S. Aldabergerova, P. Schmuki, Angew. Chem. Int. Ed. 44 (2005) 7463.
17. P. Schmuki, H. Tsuchiya, L. V. Taveira, L. F. P. Dick, J. M. Macak, in Proc. of the Symposium "Pits and Pores: Formation, Properties and Significance for Advanced Materials", P. Schmuki, D. J. Lockwood, H. S. Isaacs, Y. Ogata, M. Seo, Eds., The Electrochemical Society, PV of ECS Hawaii meeting 2004, PV 2004-19, 70.
18. H. Tsuchiya, P. Schmuki, Electrochem. Commun. 6 (2004) 1131.
19. H. Tsuchiya, J. M. Macak, I. Sieber, P. Schmuki, Small 1 (2005) 722.
20. W. J. Lee, W. H. Smyrl, Electrochem. Solid-State Lett. 8 (2005) B7.
21. I. Sieber, H. Hildebrand, A. Friedrich, P. Schmuki, Electrochem. Commun. 7 (2005) 97.
22. H. Tsuchiya, J. M. Macak, I. Sieber, L. Taveira, A. Ghicov, K. Sirotna, P. Schmuki, Electrochem. Commun. 7 (2005) 295.

23. N. Mukherjee, M. Paulose, O. K. Varghese, G. K. Mor, C. A. Grimes, J. Mater. Res. 18 (2003) 2296.
24. I. Sieber, B. Kannan, P. Schmuki, Electrochem. Solid-State Lett. 8 (2005) J10.
25. I. Sieber, P. Schmuki, J. Electrochem Soc.152 (2005) C639.
26. H. Tsuchiya, P. Schmuki, Electrochem. Commun. 7 (2005) 49.
27. A. Fujishima, K. Honda, Nature 238 (1972) 37
28. B. O'Regan, M. Grätzel, Nature 353 (1991) 737.
29. G. Sberveglieri Ed., Gas Sensors, Kluwer Academic Publishing, Dordrecht, 1992.
30. C. K. Dyer, J. S. L. Leach, J. Electrochem. Soc. 125 (1978) 1032.
31. J. W. Schultze, M. M. Lohrengel, D. Ross, Electrochim. Acta 28, (1983) 973.
32. P. Schmuki, J. Solid State Electrochem. 6 (2002) 145.
33. J. C. Marchenoir, J. P. Loup, J. Masson, Thin Solid Films 66 (1980) 357.
34. L. Arsov, M. Froehlicher, M. Froment, A. Hugot-le-Goff, Journal de Chimie Physique 3 (1975) 275.
35. G. Blondeau, M. Froehlicher, M. Froment, A. Hugot-le-Goff, Journal de Microscopie et de spectroscopie electronique 2 (1977) 27.
36. J. Yahalom, J. Zahavi, Electrochim. Acta 15 (1970) 1429.
37. J.-C. Marchenoir, J. Gautron, J. P. Loup, Metaux, corrosion-industrie (1977) 83.
38. J.-L. Delplancke, R. Winand, Electrochim. Acta 33 (1988) 1551.
39. S. R. Morrison, Electrochemistry at semiconductor and oxidized metal electrodes, Plenum Press, New York (1980).
40. S. M. Sze, VLSI Technology, McGraw-Hill, New York (1983).
41. S. Ikonopisov: Electrochim. Acta 22 (1977) 1077.
42. S. Ikonopisov, A Girginov, M. Machkova, Electrochimica Acta 24 (1979) 451.
43. I. Montero, J. M. Albella, J. M. Martinez-Duart, J. Electrochem. Soc. 132 (1985) 814.
44. G. C. Wood, C. Pearson, Corros. Sci. 7 (1967) 119.
45. Y. T. Sul, C. B. Johansson, S. Petronis, A. Krozer, Y. Jeong, A. Wennerberg, T. Albrektsson, Biomaterials 23 (2002) 491.
46. Y. Mueller, S. Virtanen, in Pits and Pores II, P. Schmuki, D. J. Lockwood, Y. H. Ogata, H. S. Isaacs, Eds., The Electrochemical Society Proccedings Series, Pennington, NJ (2000), PV 2000-25, p. 294.
47. C. K. Dyer, J. S. L. Leach, J. Electrochem. Soc. 125 (1978) 23.
48. Z. Tun, J. J. Noel, D. W. Shoesmith, J. Electrochem. Soc. 146 (1999) 988.
49. P. F. Schmidt, J. Appl. Phys. 28 (1957) 278.
50. P. F. Schmidt, J. Electrochem. Soc. 115 (1968) 167.
51. D. A. Vermilyea, J. Appl. Phys. 27 (1956) 963.
52. A. Ward, A. Damjanovic, E. Gray, M. O'Jea, J. Electrochem. Soc. 123 (1976) 1599.
53. H. Böhni, PhD thesis 4020, ETH Zürich, 1967.
54. A. Singh, J. M. Macak, H. Hildebrand, P. Schmuki, in preparation.
55. V. P. Parkhutik, V. I. Shershulsky, J. Phys. D. 25 (1992) 1258.
56. V. Parkutik, Pits and Pores II, Proc. Vol. ECS 2000-25, (2000) 168.
57. K. Yasuda, P. Schmuki, Electrochimica Acta 52 (2007) 4053.
58. K. Yasuda, P. Schmuki, Electrochem. Commun. 9 (2007) 615.
59. X. Feng, J. M. Macak, P. Schmuki, Chem. Mater. 19 (2007) 1534.
60. S. Bauer, S. Kleber, P. Schmuki, Electrochem. Commun. 8 (2006) 1321.
61. J. M. Macak, P. Schmuki, Electrochimica Acta 52 (2006) 1258.
62. H. Tsuchiya, J. M. Macak, L. Taveira, E. Balaur, A. Ghicov, K. Sirotna, P. Schmuki, Electrochem. Commun. 7 (2005) 576.
63. S. Albu, A. Ghicov, J. M. Macak, P. Schmuki, Phys. Stat. Sol. (RRL) 1 (2007) R65
64. J. M. Macak, S. Albu, D. H. Kim, I. Paramasivam, S. Aldabergerova, P. Schmuki, Electrochem. Solid-State Lett., 10(2007)K28.
65. S. Albu, A. Ghicov, J. M. Macak, R. Hahn, P. Schmuki, Nano Lett., 7(2007)1286.
66. U. Ciesla, M. Froba, G. Stucky, F. Schüth, Chem. Mater. 11 (1999) 227.

67. B. T. Holland, C. F. Blanford, T. Do, A. Stein, Chem. Mater. 11 (1999) 795.
68. H.-R. Chen, J.-L. Shi, J. Yu, L.-Z. Wang, D.-S. Yan, Micropor. Mesopor. Mater. 39 (2000) 171.
69. P. Stefanov, D. Stoychev, M. Stoycheva, J. Ikonomov, Ts. Marinova, Surf. Interface Anal. 30 (2000) 628.
70. R. A. Ploc, M. A. Miller, J. Nucl. Mat. 64 (1977) 71.
71. B. Cox, J. Electrochem. Soc. 117 (1970) 654.
72. M. Ohtaki, J. Peng, K. Eguchi, H. Arai, Sens. Actuat. B13–14 (1993) 495.
73. G. H. Hutching, S. H. Taylor, Catal. Today 49 (1999) 105.
74. R. I. Aagard, Appl. Phys. Lett. 27 (1975) 605.
75. B. Ohtani, K. Iwai, S. Nishimoto, T. Inui, J. Electrochem. Soc.141 (1994) 2439.
76. J. Choi, J. H. Lim, S. Ch. Lee, J. H. Chang, K. J. Kim, M. A. Cho, Electrochimica Acta 51 (2006) 5502.
77. J. Choi, J. H. Lim, J. Lee, K. J. Kim, Nanotechnology 18 (2007) 055603.
78. R. Karlinsey, Electrochem. Commun. 7 (2005) 1190.
79. Y. Masuda, S. Wakamatsu, K. Koumoto, J European Ceram Soc. 24 (2004) 301.
80. K. Kamada, M. Mukai, Y. Matsumoto, Electrochimica Acta 49 (2004) 321.
81. A. Mozalev, M. Sakairi, I. Saeki, H. Takahashi, Electrochimica Acta 48 (2003) 3155.
82. A. I. Vorobyova, E. A. Outkina, Thin Solid Films 324 (1998) 1.
83. W. S. Kim, J. H. Kim, J. H. Kim, K. H. Hur, J. Y. Lee, Mater. Chem. 79 (2003) 204.
84. C. Wang, L. Fang, G. Zhang, D. M. Zhuang, M. S. Wu, Thin Solid Films 458 (2003) 246.
85. M. A. Butler, R. D. Nasby, R. K. Quinn, Solid State Commun. 19 (1976) 1011.
86. A. E. Aliev, H. W. Shin, Solid State Ionics 154–155 (2002) 425.
87. S. Badilescu, P. V. Ashrit, Solid State Ionics 158 (2003) 187.
88. E. Ozkan, S.-H. Lee, P. Liu, C. E. Tracy, F. Z. Tepehan, J. R. Pitts, S. K. Deb, Solid State Ionics 149 (2002) 139.
89. Y.-K. Chung, M.-H. Kim, W.-S. Um, H.-S. Lee, J.-K. Song, S.-C. Choi, K.-M. Yi, M.-J. Lee, K.-W. Chung, Sensor. Actuator. B 60 (1999) 49.
90. C. Cantalini, M. Pelino, H.-T. Sun, M. Faccio, S. Santucci, L. Lozzi, M. Passacantando, Sensor. Actuator. B 35–36 (1996) 112.
91. H.-T. Sun, C. Cantalini, L. Lozzi, M. Passacantando, S. Santucci, M. Pelino, Thin Solid Films 287 (1996) 258.
92. M. Tong, G. Dai, Y. Wu, X. He, D. Gao, J. Mater. Sci. 36 (2001) 2535.
93. B. Reidhman, A. J. Bard, J. Electrochem. Soc. 126 (1979) 583.
94. M.-G. Verge, C.-O. A. Olsson, D. Landolt, Corros. Sci. 46 (2004).
95. S. Berger, H. Tsuchiya. A. Ghicov, P. Schmuki, Appl. Phys. Lett. 88 (2006) 203119.
96. V. Zwilling, E. Darque-Ceretti, A. Boutry-Forveille, D. David, M. Y. Perrin, M. Aucouturier, Surf. Interface Anal. 27 (1999) 629.
97. J. M. Macak, H. Tsuchiya, L. Taveira, A. Ghicov, P. Schmuki, J. Biomed. Mat. Res. 75A (2005) 928.
98. H. Tsuchiya, J. M. Macak, A. Ghicov, P. Schmuki, Small 2 (2006) 888
99. H. Tsuchiya, J. M. Macak, A. Ghicov, Y. Ch. Tang, S. Fujimoto, M. Niinomi, T. Noda, P. Schmuki, Electrochimica Acta 52 (2006) 94.
100. A. Ghicov, S. Aldabergerova, P. Schmuki, Angew. Chem. Int. Ed. 44 (2005) 7463.
101. O. K. Varghese, D. Gong, M. Paulose, C. A. Grimes, E. C .Dickey, J. Mater. Res. 18 (2003) 156.
102. A. Ghicov, M. Yamamoto, P. Schmuki, Angew. Chem. Int. Ed. 47(2008) 7934.
103. K. Yasuda, P. Schmuki, Adv. Mater., 19 (2007) 1757.
104. K. Wakino, K. Minai, H. Tamura, J. Am. Ceram. Soc. 36 (1984) 278.
105. C. L. Wang, H. Y. Lee, F. Azough, R. Freer, J. Mater. Sci. 32 (1997) 1693.
106. A. Bianco, G. Gusmano, R. Freer, P, Smith, J. Eur. Ceram. Soc. 19 (1999) 959.
107. D.-A. Chang, P. Lin, T. Tseng, J. Appl. Phys. 77 (1995) 4445.
108. X. Z. Fu, L. A. Clark, Q. Yang, M. A. Anderson, Environ. Sci. Technol. 30 (1996) 647.

109. F. H. Simpson, Mater. Eng. 52 (1960) 16.
110. D. Kuroda, M. Niinomi, M. Morinaga, Y. Kato, T. Yashiro, Mater. Sci. Eng. 1998, A243, 244.
111. H. Tsuchiya, S. Berger, J. M. Macak, A. G. Munoz, P. Schmuki, Electrochem. Commun. 9 (2007) 545.
112. M. Grätzel, Nature 414 (2001) 338.
113. A. L. Linsebigler, G. Lu, J. T. Yates, Chem. Rev. 95 (1995) 735.
114. A. Ghicov, H. Tsuchiya, J. M. Macak, P. Schmuki, Phys. Stat. Sol. A 203 (2006) R28.
115. J. Zhao, X. Wang, T. Sun, L. Li, Nanotechnology 16 (2005) 2450.
116. J. M. Macak, S. Aldabergerova, A. Ghicov, P. Schmuki, Phys. Stat. Sol. A 203 (2006) R67.
117. J. Arbiol. J. Cerda, G. Dezanneau, A. Cicera, F. Peiro, A. Cornet, J. R. Moraste, J. Appl. Phys. 92 (2002) 853.
118. B. Tryba, A. W. Moravski, M. Inagaki, Appl. Catal. B: Environ. 52 (2003) 203.
119. K. Shimizu, K. Kobayashi, J. Surf. Finish. Soc. Jpn. 46 (1995) 402.
120. J. Kunze, A. Ghicov, H. Hildebrandt, J. M. Macak, L. Taveira, P. Schmuki Z. Phys. Chem. 129 (2006) 1561.
121. J. Lausmaa, J. Electron Spectrosc. Relat. Phenom. 81 (1996) 343.
122. R. Beranek, H. Tsuchiya, T. Sugishima, J. M. Macak, L. Taveira, S. Fujimoto, H. Kisch, P. Schmuki, Appl. Phys. Lett. 87 (2005) 243114.
123. S. Tanemura, L. Miao, P. Jin, K. Kaneko, A. Terai, N. Nabatova-Gabain, Appl. Surf. Sci. 212–213 (2003) 654.
124. D. Mardare, G. I. Rusu, Mater. Lett. 56 (2002) 210.
125. J. M. Macak, H. Tsuchiya, A. Ghicov, P. Schmuki, Electrochem. Commun. 7 (2005) 1138.
126. R. Asahi, T. Morikawa, T. Ohwaki, A. Aoki, Y. Taga, Science 293 (2001) 269.
127. K. Wilke, H. D. Breuer, J. Photochem. Photobiol., A 127 (1999) 107.
128. L. Lin, W. Lin, Y. Zhu, B. Zhao, Y. Xie, Chem. Lett. 34 (2005) 284.
129. M. Anpo, Catal. SurV. Jpn. 1 (1997) 169.
130. T. Yamaki, T. Umebayashi, T. Sumita, S. Yamamoto, M. Maekawa, A. Kawasuso, H. Itoh, Nucl. Instrum. Methods Phys. Res. Sec. B 206 (2003) 254.
131. S. Sakthivel, H. Kisch, Angew. Chem. Int. Ed. 42 (2003) 4908.
132. T. Ohno, T. Mitsui, M. Matsumura, Chem. Lett. 32 (2003) 364.
133. T. Umebayashi, T. Yamaki, S. Tanaka, K. Asai, Chem. Lett. 32 (2003) 330.
134. W. Zhao, W. Ma, C. Chen, J. Zhao, Z. Shuai, J. Am. Chem. Soc. 126 (2004) 4782.
135. A. Ghicov, J. M. Macak, H. Tsuchiya, J. Kunze, V. Heaublein, L. Frey, P. Schmuki, Nano Lett. 6 (2006) 1080.
136. A. Ghicov, J. M. Macak, H. Tsuchiya, J. Kunze, V. Haeublein, S. Kleber, P. Schmuki, Chem. Phys. Lett. 419 (2005) 426.
137. R. P. Vitiello, J. M. Macak, A. Ghicov, H. Tsuchiya, L. F. P. Dick, P. Schmuki, Electrochem. Commun. 8 (2006) 544.
138. J. M. Macak, A. Ghicov, R. Hahn, H. Tsuchiya, P. Schmuki, J. Mater. Res. 21 (2006) 2824.
139. K. Shankar, K. Ch. Tep, G. K. Mor, C. A. Grimes, J. Phys. D: Appl. Phys. 39 (2006) 2361.
140. A. Ghicov, B. Schmidt, J. Kunze, P. Schmuki, Chem. Phys. Lett. 433 (2007) 323.
141. J. H. Park, S. Kim, A. J. Bard, Nanoletters 6 (2006) 24.
142. R. Hahn, A. Ghicov, J. Salonen, P. Schmuki, Nanotechnology 18 (2007) 105604.
143. I. I. Philips, P. Poole, L. L. Shreir, Corros. Sci. 12 (1972) 855.
144. Z. A. Foroulis, J. Electrochem. Soc. 128 (1981) 219.
145. E. Brauer, R. Gruner, F. Rauch, Ber. Bunsen. Phys. Chem. 87 (1983) 341.
146. D. J. Blackwood, L. M. Peter, H. E. Bishop, P. R. Chalker, D. E. Williams, Electrochim. Acta 34 (1989) 1401.
147. H. Tokudome, M. Miyauchi, Angew. Chem., Int. Ed. 44 (2005) 1974.
148. A. Ghicov, H. Tsuchiya, R. Hahn, J. M. Macak, A. G. Munoz, P. Schmuki, Electrochem. Commun. 8 (2006) 528.
149. N. Sakai, A. Fujishima, T. Watanabe, K. Hashimoto, J. Electrochem. Soc. 148 (2001) E395.

150. R. Hahn, A. Ghicov, H. Tsuchiya, J. M. Macak, A. G. Munoz, P. Schmuki, Phys. Stat. Sol. A 204 (2007) 1281.
151. M. R. Hofmann, S. T. Martin, W. Choi, D. W. Bahnemann, Chem. Rev.95 (1995) 69.
152. A. Mills, S. Le Hunte, J. Photochem. Photobiol. A 108 (1997) 1.
153. F. Kiriakidou, D. I. Kondarides, X. E. Verykios, Catalysis Today 54 (1999) 119.
154. F. Zhang, J. Zhao, T. Shen, H. Hidaka, E. Pelizzetti, N. Serpone, Appl. Catal. B 15 (1998) 147.
155. J. Krýsa, M. Keppert, G. Waldner, J. Jirkovský, Electrochim. Acta 50 (2005) 5255.
156. N. Serpone, E. Pelizzetti, Photocatalysis – Fundamentals and Applications, Wiley, New York, 1989.
157. D. F. Ollis, H. Al-Ekabi, Photocatalytic Purification and Treatment of Water and Air, Elsevier, Amsterdam, (1993).
158. J. M. Macak, M. Zlamal, J. Krysa, P. Schmuki, Small 3 (2007) 303.
159. E. Balaur, J. M. Macak, L. Taviera, H. Tsuchiya, P. Schmuki, Electrochem. Commun. 7 (2005) 1066.
160. E. Balaur, J. M. Macak, H. Tsuchiya, P. Schmuki, J. Mater. Chem. 15 (2005) 4488.
161. R. Wang, K. Hashimoto, A. Fujishima, M. Chikuni, E. Kojima, A. Kitamura, M. Shimuhigoshi, T. Watanabe, Nature 388 (1997) 431.
162. R. Wang, K. Hashimoto, A. Fujishima, M. Chikuni, E. Kojima, A. Kitamura, M. Shimohigoshi, T. Watanabe, Adv. Mater. 10 (1998) 135.
163. J. P. Lee, H. K. Kim, C. R. Park, G. Park, H. T. Kwak, S. M. Koo, M. M. Sung, J. Phys. Chem. B 107 (2003) 8997.
164. J. P. Lee, H. K. Kim, C. R. Park, G. Park, H. T. Kwak, S. M. Koo, M. M. Sung, J. Phys. Chem. B 107 (2003) 8997.
165. D. M. Brunette, P. Tengvall, M. Textor, P. Thomsen, Titanium in Medicine. Springer, Berlin (2001).
166. X. Liu, P. K. Chu, C. Ding, Mater. Sci. Eng. R 47 (2004) 49.
167. C. M. Roome, C. D. Adam, Biomaterials 16 (1995) 69.
168. J. Weng, Q. Liu, J. G. C. Wolke, X. Zhang,K. de Grood, Biomaterials 18 (1997) 1027.
169. P. Li, C. Ohtsuki, T. Kokubo, K. Nakanishi, N. Soga, K. de Groot, J Biomed. Mater. Res. 28 (1994) 7.
170. P. Li, I. Kangasniemi, K. de Groot, T. Kokubo, J. Am. Ceram. Soc. 5 (1994) 1307.
171. L. Jonasova, F. A. Müller, A. Helebrant, J. Strnad, P. Greil, Biomaterials 23 (2002) 3095.
172. B. Yang, M. Uchida, H. M. Kim, X. Zhang, T. Kokubo, Biomaterials 25 (2004) 1003.
173. W. H. Song, Y. K. Jun, Y. Han, S. H. Hong, Biomaterials 25 (2004) 3341.
174. H. Tsuchiya, J. M. Macak, L. Muller, J. Kunze, F. Muller, S. P. Greil, S. Virtanen, P. Schmuki, J. Biomed. Mat. Res. 2006, 77A, 534.
175. J. H. Park, S. Bauer, K. von der Mark, P. Schmuki, Nano Lett. 7 (2007) 1686.
176. O. K. Varghese, D. Gong, K. G. Ong, C. A. Grimes, Sensors and Actuators B93, (2003) 338.
177. O. K. Varghese, D. Gong, M. Paulose, K. G. Ong, E. C. Dickey, C. A. Grimes, Adv. Mater. 15 (2003) 624.
178. J. M. Macak, H. Tsuchia, S. Berger, S. Bauer, S. Fujimoto, P. Schmuki Chem. Phys. Lett. 428 (2006) 421.
179. X. Yu, Y. Li, W. Ge, Q. Yang, N. Zhu, K. K. Zadeh, Nanotechnology 17 (2006) 808.
180. Y. D. Premchand, T. Djenizian, F. Vacandio, P. Knauth, Electrochem. Commun. 8 (2005) 1840.
181. J. M. Macak, B. G. Gong, M. Hueppe, P. Schmuki, Adv. Mater. 19 (2007) 3027.

Index